Texts in Philosophy
Volume 5

Causality and Probability in the Sciences

Volume 1
Knowledge and Belief
Jaakko Hintikka

Volume 2
Probability and Inference: Essays in Honour of Henry E. Kyburg
Bill Harper and Greg Wheeler, eds

Volume 3
Monsters and Philosophy
Charles T. Wolfe, ed.

Volume 4
Computing, Philosophy and Cognition
Lorenzo Magnani and Riccardo Dossena, eds

Volume 5
Causality and Probability in the Sciences
Federica Russo and Jon Williamson, eds

Texts in Philosophy Series Editors
Vincent F. Hendricks vincent@ruc.dk
John Symons jsymons@utep.edu

Causality and Probability in the Sciences

edited by

Federica Russo
and
Jon Williamson

© Individual author and College Publications 2007 All rights reserved.

ISBN 1-904987-35-4
Published by College Publications
Scientific Directors: Dov Gabbay, Vincent F. Hendricks and John Symons
Managing Director: Jane Spurr
Department of Computer Science
King's College London
Strand, London WC2R 2LS, UK

http://www.collegepublications.co.uk

Original cover design by Richard Fraser
Cover produced by orchid creative   www.orchidcreative.co.uk
Printed by Lightning Source, Milton Keynes, UK

All rights reserved. No part of this publication may be reproduced, stored in a retrieval system or transmitted, in any form, or by any means, electronic, mechanical, photocopying, recording or otherwise, without prior permission, in writing, from the publisher.

CONTENTS

Preface ix
FEDERICA RUSSO AND JON WILLIAMSON
Introduction 1

PART I   CAUSALITY AND PROBABILITY IN ARTIFICIAL INTELLIGENCE 15

SAM MAES, STIJN MEGANCK, PHILIPPE LERAY
An integral approach to causal inference with latent variables 17

ALEX A. FREITAS, KEN MCGARRY, ELON CORREA
Integrating Bayesian networks and Simpson's paradox in data mining 43

PART II   CAUSALITY AND PROBABILITY IN THE PHYSICAL SCIENCES 63

MAURICIO SUÁREZ
Causal inference in quantum mechanics: a reassessment 65

PART III   CAUSALITY AND PROBABILITY IN THE SOCIAL SCIENCES 107

ALESSIO MONETA
Mediating between causes and probabilities: the use of graphical models in econometrics 109

STEPHEN LEROY
Causality in economics 131

DAMIEN FENNELL
Causality, mechanisms and modularity: structural models in econometrics 161

JULIAN REISS
Time series, nonsense correlations and the principle of the common cause 179

ERIK WEBER
Conceptual tools for causal analysis in the social sciences 197

PART IV CAUSALITY AND PROBABILITY IN THE BIOMEDICAL SCIENCES 215

FEDERICA RUSSO AND JON WILLIAMSON
Interpreting probability in causal models for cancer 217

BERT LEURIDAN
Galton's blinding glasses. Modern statistics hiding causal structure in early theories of inheritance 243

VANESSA DIDELEZ AND NUALA A. SHEEHAN
Mendelian randomisation: why epidemiology needs a formal language for causality 263

PART V CAUSALITY AND PROBABILITY 293

MARIANNE BELIS
The causal roots of probability 295

ANDREA L'EPISCOPO
Causality and the axiomatic probability calculus 319

FRANÇOISE LONGY
Two probabilities of dysfunction and two kinds of chance 335

PART VI CAUSAL PLURALISM 361

MONIKA M. DULLSTEIN
Causal dualism: which position? Which arguments? 363

AMIT PUNDIK
Can one deny both causation by omission and causal pluralism? The case of legal causation 379

PART VII GENERAL FRAMEWORKS FOR CAUSAL ANALYSIS 413

FRIEDEL WEINERT
A conditional view of causality 415

AVIEZER TUCKER
The inference of common cause naturalized 439

MARGHERITA BENZI
Contexts for causal models 467

ISABELLE DROUET
Causal inference. How can Bayesian networks contribute? 487

A. PHILIP DAWID
Counterfactuals, hypotheticals and potential responses: a philosophical examination of statistical causality 503

INDEX 533

Preface

This volume is a product of the project *Causality and the interpretation of probability in the social and health sciences* funded by the British Academy and the UCLouvain Special Funds for Research.

Many papers in this volume were presented at the associated conference 'Causality and probability in the sciences', held at the University of Kent at Canterbury, 14-16 June 2006. We would like to thank all the conference participants, and also all the referees for providing very helpful comments and criticisms of earlier drafts of the papers. We are grateful to the British Academy, the Mind Association, the British Society for the Philosophy of Science and the Kent Institute for Advanced Studies in the Humanities, who supported the conference financially.

Thanks, too, to Jane Spurr of College Publications for much needed advice and support. The manuscript was prepared in LaTeX by the authors and Federica Russo. Copyright of each paper rests with its authors.

Introduction

FEDERICA RUSSO AND JON WILLIAMSON

Causality and probability in the sciences

Towards the end of the nineteenth century Karl Pearson noted that a probabilistic dependence between two variables does not necessarily imply that the two variables are causally connected (Pearson, 1897). This led Pearson, in the third (1911) edition of his book Pearson (1892), to argue that talk of cause and effect should be eradicated from science in favour of talk of probabilistic dependence (in the form of contingency tables). Around the same time, Bertrand Russell threw his intellectual weight behind this purge of causality (Russell, 1913). In the same vein, Ernst Mach (1905) argued that causality, understood as a way to explain phenomena, should be replaced by the concept of *relation*, which is a way to merely *describe* phenomena. These attacks had a profound influence on much of twentieth century science. Although scientists continued to reason causally—e.g., to find causes of phenomena, to devise experiments to measure interventions, to inform policy decisions—explicit mention of causality met with disapproval.

Then came the 1980s. As explained below, causal methods developed in Artificial Intelligence (AI) in the 1980s helped to rehabilitate the concept of cause. While causality was no less controversial from a philosophical perspective, new formalisms for handling causality and probability together helped mathematise the notion of cause. Fig. 1 and Fig. 2 show the resulting transformation. A search of the *Web of Science* databases for papers whose titles include a word beginning with 'caus-' (e.g., 'causality', 'causation', 'causal') revealed a stark increase in the numbers of such papers after about 1990. This was so for the *Science Citation Index Expanded* (SCIE) database, which deals mainly with physical, biological and computational sciences, and for the *Social Sciences Citation Index* (SSCI) database. The growth in papers on causality in the *Arts and Humanities Citation Index* (AHCI), which covers philosophy, is rather more gradual. (Of course, the volume of all academic papers increased markedly in this period. In an effort to compensate for this general growth, for each of the three databases Fig. 2 portrays the yearly number of papers involving causal terms divided by the yearly number of papers with an author whose name begins with the letter 'J', a rather arbitrary indicator of the general volume of papers in the database in question.)

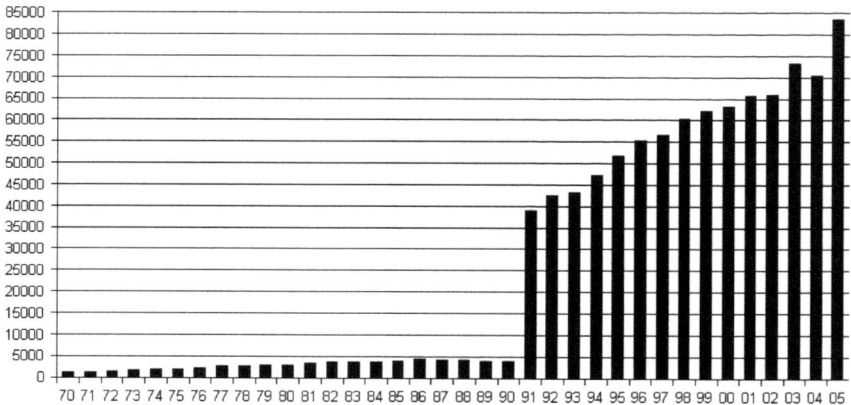

Figure 1. Total yearly numbers of papers involving causal terms.

Given this rehabilitation of causal talk it is all the more important to further our understanding of the notion of cause. As a step in this direction, the papers in this volume seek to shed light on the relationship between causality and probability. Methodologically, the work presented here is *science-driven*: the papers seek to learn lessons about causality and probability motivated by actual scientific practice.

This volume

Causality and probability in artificial intelligence

Researchers in AI have been responsible for many of the key developments in causal modelling since the early 1980s. Broadly speaking this line of research is motivated by the following question: given a dataset containing a series of past observations, what is the most appropriate causal model of the domain in question? This question led to the development of *causal nets*, also called *causally interpreted Bayesian nets* (Pearl, 1988; Williamson, 2005), which represent qualitative causal connections by means of a directed acyclic graph (DAG) and which represent a joint probability distribution by means of the probability distribution of each variable conditional on its direct causes together with an independence assumption, the *Causal Markov Condition*, which says that each variable is probabilistically independent of its non-

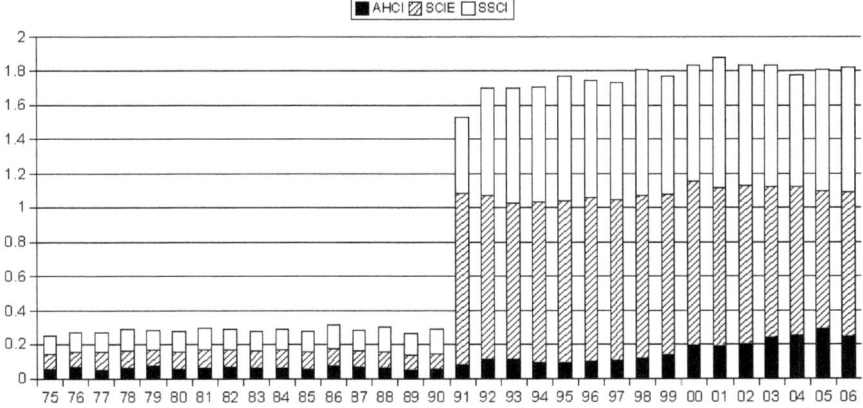

Figure 2. Papers involving causal terms, according to citation database and adjusted by overall volume.

effects conditional on its direct causes. Closely related are *structural equation models*, which can be thought of as causal nets with functional equations instead of conditional probability distributions (Spirtes et al., 1993; Pearl, 2000). Causal nets and structural equation models are often called *graphical causal models*.

In 'An integral approach to causal inference with latent variables', Sam Maes, Stijn Meganck and Philippe Leray contribute to the development of the causal net framework. As originally devised, causal net learning algorithms produce the causal net on the variables of the given dataset that best fits the dataset. Now it may be that a probabilistic dependence amongst two variables in the dataset is induced, not by causal relationships amongst the variables in the dataset, but because the two variables are effects of a common cause that is not measured in the dataset. Such an unmeasured common cause is called a *latent variable*. The question thus arises as to which causal net, involving latent variables as well as variables in the dataset, best fits the dataset. Causal nets have been extended in various ways in order to try to answer this question, and this chapter develops the *semi-Markovian causal model* approach. It also extends the causal net framework in another direction, by allowing experimental as well as observational data to help determine the appropriate causal model.

Alex Freitas, Ken McGarry and Elon Correa forge an interesting connection between causal nets and knowledge discovery in 'Integrating Bayesian networks and Simpson's paradox in data mining'. In AI, knowledge discovery has two types of application. It may be that an artificial agent needs to extract generalities from data that will be useful to it in its activities: an automated system for drug synthesis may need to determine laws about the structure of molecules from experimental and observation data, for example, in order to synthesise molecules of the appropriate shape. By and large, although useful to the machine in question, these laws are of less interest to human chemists (though as Gillies (1996) points out, this is not always the case). On the other hand, artificial agents often work less autonomously, liaising with humans: for instance a financial system may need to sift through past data to offer general investment advice to human investors. In this second type of application it is important that the knowledge gleaned from the data be of interest to the humans. This chapter exploits Simpson's paradox, a conundrum that often crops up in the literature on probabilistic causality, to help isolate interesting pieces of knowledge. The authors also argue that Simpson's paradox can be exploited to help learn causal nets from data.

Causality and probability in the physical sciences

Bertrand Russell (1913) maintained that the physical sciences do not appeal to the concept of cause, but instead deal with functional equations. These days, causal claims are not considered to be a world apart from functional equations; a structural equation model, for example, is a type of causal model. The question then arises as to how the functional equations of physics might be given a causal interpretation. Thus the ideal gas law, $PV = nRT$ (where P is pressure, V is volume, n is the number of moles of gas, R is the universal gas constant, and T is temperature), is typically taken to encapsulate a number of causal relationships, with each of P, V and T causally dependent on the other two.

The functional equations of quantum mechanics are often thought to pose a special problem for causality, with several philosophers arguing that the Einstein-Podolsky-Rosen thought experiment can not be interpreted causally. Mauricio Suárez, in 'Causal inference in quantum mechanics: a reassessment', takes these philosophers to task. Suárez puts forward five different causal models that could be taken to underlie this experiment. The problem isn't so much that quantum mechanics is incompatible with causality, but rather that the most plausible model (model II in this chapter) forces a reassessment of the relationship between causality and probability. Under this model, the Causal Markov Condition, often taken to be an invariant or even defining feature of causality, fails. If so, the Causal Markov Condition may have to be relegated to the status of a *default* rule (Williamson, 2005).

Causality and probability in the social sciences

Karl Popper (1934) put forward a hypothetico-deductive account of how one should discover causal relationships: first hypothesise a causal law, then deduce its consequences, rejecting the law if these predictions are not borne out. In economics, the Cowles Commission, founded by Alfred Cowles in 1932 with the aim of promoting a mathematical approach to economic theory, suggested that economic theory be used to provide the hypothesis and statistics be used to determine the parameters of a causal model and to derive predictions.

In contrast, the approaches to causal modelling developed in AI open up the possibility of an inductivist approach to discovery: from a dataset of past observations, directly induce causal laws (Spirtes et al., 1993). In 'Mediating between causes and probabilities: the use of graphical models in econometrics', Alessio Moneta argues that the graphical models of AI are better seen as involving aspects of both hypothetico-deductivism and inductivism. Moneta takes seriously the idea that the assumptions behind causal models—such as the Causal Markov Condition—need not hold invariably and should be treated as default working assumptions. In which case, the models yielded by inductive methods can at best be viewed as tentative hypotheses, in need of further testing.

Stephen LeRoy compares the approach of the Cowles Commission with more recent developments in 'Causality in economics'. LeRoy argues that graphical causal models are a point of departure from the more traditional approach in economics due to Herbert Simon, who began his economic career in the Cowles Commission. This is because the functional equations of the traditional approach do not admit a straightforward causal interpretation by treating a variable on the left-hand side of the equation as the effect and those variables on the right-hand side as its direct causes, while structural equation models as defined by Pearl (2000) do admit such an interpretation. LeRoy favours the traditional Cowles approach.

Damien Fennell, in 'Causality, mechanisms and modularity: structural models in econometrics', takes this comparison a step further. Fennell argues that the graphical model approach differs from Simon's analysis in another respect: advocates of the graphical model approach tend to assume *modularity*, i.e., that one can intervene to fix the value of any variable without changing the values of its causes in the model and without changing the nature of the causal relationships in the model (a so-called *divine intervention* or *perfect intervention*), while Simon makes no such assumption. Modularity does not always hold—Fennell argues that this gives another reason to prefer the traditional Cowles approach over the graphical causal model approach.

In 'Time series, nonsense correlations and the principle of the common cause', Julian Reiss discusses a further way in which the Causal Markov Condition can fail. The Causal Markov Condition implies the *Principle of the Common Cause*, which says that if two variables are probabilistically dependent then either one causes the other or they are effects of a common cause. But it is well known that time series give rise to correlations that admit no such causal explanation. Yule (1926), for instance, cited a correlation between the proportion of Church of England marriages and the mortality rate, in the years 1866-1911; both are decreasing but for different reasons, not because one causes the other or because they are effects of a common cause. Advocates of graphical causal models often try to hold out against such counterexamples to the Causal Markov Condition, either by maintaining that these counterexamples dissolve under closer scrutiny, or by claiming that they do not make a practical difference on the use of graphical models. Reiss argues against both these moves and concludes that the Principle of the Common Cause is a fallible assumption.

The previous papers focus on methodological aspects in the fields of economics and econometrics. In 'Conceptual tools for causal analysis in the social sciences', Erik Weber attempts to provide social scientists with useful tools for causal analysis. Weber distinguishes two different tasks of a philosophical investigation into causality. On the one hand, a conceptual analysis develops a definition of causality to adequately represent our everyday causal talk. On the other, we can develop a set of concepts that are supposed to help scientists. Weber confines his discussion to the second task. As a conceptual pluralist, Weber puts forward three concepts for social scientists. The first is the causal relation at the population level, as defined by standard average effect theories; the second is the causal interaction at the individual level, and the third is the specification of spontaneous preservation that takes place after causal interactions. The most original part of the paper consists in the modification of Salmon's concept of causal interaction and the definition of spontaneous preservation. In this way, the concept of causal interaction, originally thought for physics, is now also well suited to the social sciences.

Causality and probability in the biomedical sciences

When trying to assess frameworks for causal modelling, two key questions arise. First, what notion of cause is employed in the model? Causality can be interpreted in a variety of ways—e.g., mechanistic, probabilistic, counterfactual, agency, epistemic—and the choice of interpretation can have a bearing on whether the modelling assumptions hold. Second, what notion of probability is employed in the model? Probability also admits of a variety of interpretations—e.g., frequency, propensity, chance, classical, logical and several Bayesian interpretations—and modelling assumptions such as the Causal Markov Condition depend on the chosen notion of probability as well as that of causality.

In 'Interpreting probability in causal models for cancer', Federica Russo and Jon Williamson argue that cancer epidemiology is distinctive in that it is concerned with both generic and single-case probabilities, since it deals with both causal laws and with particular patient diagnoses and prognoses. Consequently, we claim, it requires a twin interpretation of probability: generic probabilistic claims should be given a frequency interpretation while single-case claims should be interpreted using degrees of belief. In particular, objective Bayesianism turns out to be the most appropriate interpretation in the single-case. If we are right, then this has an important consequence for modelling: the Causal Markov Condition can be proved to hold by default under an objective Bayesian interpretation (Williamson, 2005), so graphical causal modelling becomes a plausible methodology in cancer epidemiology.

The choice of statistical framework can have crucial repercussions. A good framework can lead to perspicuous models with well-articulated claims and assumptions, while a poor framework can obfuscate the problem in hand and even lead to situations in which no model in the framework adequately captures what is going on. Bert Leuridan, in 'Galton's blinding glasses: modern statistics hiding causal structure in early theories of inheritance', argues that the statistical context in which Francis Galton worked prevented him from finding the causal story behind inheritance. This is because Galton's choice of model was constrained by having to account for occurrences of the normal distribution and of regression towards the mean. In contrast Gregor Mendel, who made little use of statistical theory, hit upon essentially the right causal picture. The lesson, Leuridan maintains, is that we should be very careful in our use of contemporary frameworks such as graphical causal models: we should test their assumptions and be aware of their possible blinding influence.

Vanessa Didelez and Nuala A. Sheehan advocate the graphical causal modelling framework in 'Mendelian randomisation: why epidemiology needs a formal language for causality'. They show that when it is not possible, either in principle or for practical reasons, to perfectly intervene on a variable, one can instead use observational data and instrumental variables to decide causal claims, and that this procedure can be nicely represented via the graphical causal model approach. Didelez and Sheehan argue that by casting a problem in this formal language one can clarify the key causal questions that one is trying to answer, and isolate the conditions that make such causal inferences possible. Thus graphical causal models can be illuminating rather than blinding.

Causality and probability

In philosophy, causality has been a central theme since the ancient Greeks and has never lost its appeal. However, two events in the sciences have radically changed, and indeed invigorated, the debate: the advent of probability theory and the discovery of indeterministic phenomena. These two events gave a new flavour to old questions and raised completely new ones. For instance, is causality essentially indeterministic or deterministic? If probabilities are at the heart of causal processes, what is the relation between causes and probabilities? If probabilities characterise physical and social processes, how are probabilities to be interpreted?

As a consequence of these changes, a notion of cause in terms of necessary and sufficient conditions has gradually been replaced by a probabilistic notion. Thus Hans Reichenbach (1956) put forward the Principle of the Common Cause and Patrick Suppes (1970) provided a probabilistic account of causality that was very general in scope. The development of graphical causal models in AI can be viewed as a continuation of this tradition.

In the 'The causal roots of probability', Marianne Belis investigates the relation between causality and probability, particularly focusing on the single case. Pursuing the idea that singular causes have ontological priority over Humean regularities, Belis argues that two concepts elucidate the sought-after relation: 'propensity' and 'capacity', borrowed from Popper (1959) and Cartwright (1989) respectively. Belis defends the propensity notion of probability in spite of the well-known difficulty of their measurement—a difficulty that often sees propensities labelled as 'metaphysical rather than scientific'. Instead, Belis argues that there is a way to measure propensities—that is through the algebraic sum of all the strengths exerted upon them. Moreover, propensities 'reveal the ontological roots of probability', i.e., the inner *causal* character of probability in the single case.

Andrea L'Episcopo, in 'Causality and the axiomatic probability calculus', narrows down on Phil Dowe's conserved quantity (CQ) theory and probabilistic theories of causality. The main claim of the paper is that any theory of causality should be evaluated in its proper domain of application. For this reason, argues L'Episcopo, many counterexamples and criticisms to the major accounts of causality are misdirected. This claim hinges upon two distinctions. The first is between an intuitive versus a physical notion of causality, and the second is between an empirical versus a conceptual analysis of causality. For instance, Dowe's CQ theory is an empirical analysis of the physical notion of causality, and therefore, L'Episcopo argues, causation by omission or prevention do not constitute genuine problems for it.

In 'Two probabilities of dysfunction and two kinds of chance' Françoise Longy draws our attention to the interesting case of the probability of dysfunctions of artifacts. The probability of dysfunction of artifacts may be given two different interpretations. Consider, for instance, the probability that a light bulb will burn out within five minutes after its first use. The first reading concerns the *physical* probability that this particular object (i.e., the *physical* object) with this particular physical structure will burn out within five minutes—let us call this probability DYSF-PHYS probability; the second concerns the probability that the object, as an artifact produced in such and such factory, will burn out within five minutes—let us call this probability DYSF-ART probability. According to Longy, the question is whether or not these are two different sorts of objective probability of dysfunction, interpretable as chances rooted in some particular feature of the world. Indeed they are. In particular, the DYSF-ART probability is rooted in a number of conditions that determine the object as an artifact but not as the specific physical make-up of the object. Longy supports this claim by relying on the concept of function as is developed in the biological sciences.

Causal pluralism

In recent decades many different views of causality have been proposed. Among the most influential are Suppes' probabilistic theory (Suppes, 1970), Lewis' counterfactual approach (Lewis, 1973), the Salmon-Dowe process theory (Salmon, 1998; Dowe, 2000), and agency and manipulability approaches (Menzies and Price, 1993; Woodward, 2003). These approaches provides us with a variety of accounts of what causality is. This raises the problem of whether causality is genuinely a plurality of different concepts or whether it is a single concept.

This abundance of accounts has made pluralism a fashionable stance. Pluralists argue that different concepts of cause fit different contexts and that, therefore, there is no real incompatibility between, say, the probabilistic and process approach, exactly because they employ different concepts of cause in different domains.

In 'Causal dualism: which position? Which argument?', Monika Dullstein focuses on Hall's recent argument for causal dualism. Hall argues that 'cause' has two different meanings: the first relates to the concept of production (causes are physically linked to and produce their effects), and the second to difference-making (causes are responsible for differences, either probabilistic or counterfactual, in the occurrence of their effects). The reason to defend this dualist position lies in the fact that, on the one hand, no production account can deal with negative causation, and, on the other, no difference-making account can deal with overdetermination. Dullstein points out that such a pluralist stance can be simply read as a way of thinking of causation, or, more interestingly, as a specific tenet about the metaphysics of causation. Dullstein then raises the question of whether Hall's argument for causal dualism can provide any reason to believe in *metaphysical* dualism. Her answer is that Hall's argument fails in this respect.

Amit Pundik, in 'Can one deny both causation by omission and causal pluralism? The case of legal causation' draws the philosophers' attention to the intricate case of causation in the law. This ambitious paper aims at establishing a number of points. In the first place, Pundik shows that in the legal context omissions are often regarded as genuine cases of causation. Secondly, that causation by omission and causal pluralism cannot be denied coherently. In other words, if we opt for causal pluralism and accept that law has its specific concept of causation, then this concept must include causation by omission. Thirdly, he tries to convince those who might dismiss legal causation as a genuine philosophical problem that, instead, this is an extremely relevant and interesting matter.

General frameworks for causal analysis

A philosophical account of causality can have different purposes. For instance, the metaphysics of causality investigates what causality in fact is; epistemology is interested in how we come to know about causal relations; and methodology explores new methods for causal reasoning and inference. Phil Dowe (2000) introduced the distinction alluded to above between a conceptual and an empirical analysis of causality. Conceptual analysis is concerned with our talk about causality, i.e., about the meaning of cause in ordinary language. Empirical analysis is instead concerned with the meaning of cause in science.

Friedel Weinert, in 'A conditional view of causality', puts forward a framework for causal analysis that fits both the natural and the social sciences. His conditional model is based on Mackie's INUS model, involving conditions that are Insufficient but Necessary components of a set of Unnecessary but Sufficient conditions. In Weinert's account, causal relations are facts about conditional dependencies between antecedent conditions and consequent conditions. To show the wide applicability of such a conditional view, Weinert analyses the Franck-Hertz experiment in quantum mechanics and Max Weber's adequate causation in the social sciences.

Aviezer Tucker, in 'The inference of common cause naturalized', draws our attention once more to the Principle of Common Cause, one of the most debated principles in the philosophy of causality. In this paper, Tucker shows that Reichenbach failed to deduce the principle from the second law of thermodynamics, and points out that much of the literature on the topic failed to distinguish between the inference that *some* common cause existed (without specifying the particular properties of such a common cause) and the inference that a *concrete* common cause existed (with a unique set of properties). He then proposes a naturalised theory of common causes, relying on examples from various disciplines—for instance, biology, linguistics and philology. The inference of the common cause involves three consecutive stages of comparisons: first, a comparison of likelihoods given some common cause, whose properties are unknown, and given separate causes; second, if the common cause makes the evidence more likely, five types of common cause hypotheses compete; and third, if it is possible to prove which of the five types is the most probable, scientists attempt to infer the actual properties of common causes.

In 'Contexts for causal models' Margherita Benzi distinguishes between two possible approaches to causal modelling. Whilst the first relies on the idea of an underlying omnicomprehensive causal network, the second relativises the construction of the causal model to the context of inquiry. Although the context-sensitivity of causal models generally attracts a consensus among philosophers and practising scientists, there is little agreement as to what this context exactly is. Benzi proposes a taxonomy of contexts that is based on the idea that different levels of analysis reflect an increasing specificity of the factors to be taken into account. At the lowest level, we find generic background knowledge; this is refined in the context of inquiry, where factors are selected according to the needs of the specific study at hand; the causal model further refines the context of inquiry and picks out the factors to include in the model; the last level involves those factors that are more directly relevant for the assessment of the causal relation. Benzi argues that approaches that relativise to the context of inquiry ought to be preferred even if the price to pay is to give up the hope of the reduction of causes to probabilities.

As mentioned above, causally interpreted Bayesian nets have shed new light on causal inference in the last two decades. The major problem with Bayesian nets is that they will deliver successful results *if* some basic assumptions—e.g., the Causal Markov and Faithfulness conditions—are satisfied. In 'Causal inference. How can Bayesian nets can contribute?' Isabelle Drouet investigates the extent to which Bayesian nets algorithms satisfying those assumptions can be integrated into the traditional path analytic methodology. Drouet proposes a mixed methodology for causal inference, in which Bayesian nets algorithms are run only after good reasons that basic assumptions are satisfied have been provided.

Philip Dawid, in 'Counterfactuals, hypothetical and potential responses: a philosophical examination of statistical causality', runs a detailed and careful analysis of the frameworks and tools developed for causal inference by statisticians. In particular, he focuses on the potential response model and on decision theory. The potential response model (PR) is perhaps the dominant methodology. However, Dawid gives reasons to prefer the decision theoretic approach (DT). The reasons to prefer DT lie in the distinction between hypothetical and counterfactual queries. Hypothetical and counterfactuals, in turn, relate to the different tasks of inferring effects of causes or causes of effects, respectively. Dawid's main criticism of PR, which includes structural equations, is that, although it can handle the inference of effects of causes, it faces serious troubles in inferring causes of effects, especially when only observational data is available. The superiority of DT is claimed on both philosophical and pragmatic grounds.

The anti-causal correlation-mongers have been deposed. Causality is no relic of a bygone age; the renewed interest in causality from the 1980s onward proves the importance of causal reasoning both for cognitive and action-oriented purposes. On the one hand, knowledge of causes is essential to the intellectual enterprise of understanding and explaining the world. On the other, the action-oriented goal—e.g., to guide and inform policies, prescribe treatments, etc.—needs a solid causal grounding, not the quicksands of mere correlation.

Federica Russo
Philosophy, University of Kent, UK.
f.russo@kent.ac.uk

Jon Williamson
Philosophy, University of Kent, UK.
j.williamson@kent.ac.uk

BIBLIOGRAPHY

Cartwright, N. (1989). *Nature's capacities and their measurement*. Clarendon Press.
Dowe, P. (2000). *Physical causation*. Cambridge University Press.
Gillies, D. (1996). *Artificial intelligence and scientific method*. Oxford University Press, Oxford.
Lewis, D. K. (1973). Causation. In *Philosophical papers*, volume 2, pages 159–213. Oxford University Press (1986), Oxford.
Mach, E. (1905). *Knowledge and error*. Reidel Publishing Company.
Menzies, P. and Price, H. (1993). Causation as a secondary quality. *British Journal for the Philosophy of Science*, 44:187–203.
Pearl, J. (1988). *Probabilistic reasoning in intelligent systems: networks of plausible inference*. Morgan Kaufmann, San Mateo CA.

Pearl, J. (2000). *Causality: models, reasoning, and inference.* Cambridge University Press, Cambridge.
Pearson, K. (1892). *The grammar of science.* Black, London, third (1911) edition.
Pearson, K. (1897). Mathematical contributions to the theory of evolution.—on a form of spurious correlation which may arise when indices are used in the measurement of organs. *Proceedings of the Royal Society of London,* 60:489–498.
Popper, K. (1959). The propensity interpretation of probability. *British Journal for Philosophy of Science,* 10:25–42.
Popper, K. R. (1934). *The Logic of Scientific Discovery.* Routledge (1999), London. With new appendices of 1959.
Reichenbach, H. (1956). *The direction of time.* University of California Press.
Russell, B. (1913). On the notion of cause. *Proceedings of the Aristotelian Society,* 13:1–26.
Salmon, W. C. (1998). *Causality and explanation.* Oxford University Press, Oxford.
Spirtes, P., Glymour, C., and Scheines, R. (1993). *Causation, Prediction, and Search.* MIT Press, Cambridge MA, second (2000) edition.
Suppes, P. (1970). *A probabilistic theory of causality.* North Holland Publishing Company, Amsterdam.
Williamson, J. (2005). *Bayesian nets and causality: philosophical and computational foundations.* Oxford University Press, Oxford.
Woodward, J. (2003). *Making things happen: a theory of causal explanation.* Oxford University Press, Oxford.
Yule, G. U. (1926). Why do we sometimes get nonsense-correlations between time series? A study in sampling and the nature of time-series. *Journal of the Royal Statistical Society,* 89(1):1–63.

PART I

CAUSALITY AND PROBABILITY IN ARTIFICIAL INTELLIGENCE

An integral approach to causal inference with latent variables

SAM MAES, STIJN MEGANCK, PHILIPPE LERAY

ABSTRACT. This article discusses graphical models that can handle latent variables without explicitly modeling them quantitatively. In artificial intelligence, there exist several paradigms for such problem domains. Two of them are semi-Markovian causal models and maximal ancestral graphs. The application of these techniques to a given problem domain consists of several steps, typically: structure learning from observational and experimental data, parameter learning, probabilistic inference, and finally, quantitative causal inference. A problem is that each of the existing approaches only focuses on one or on a few steps involved in the process of modeling a problem including latent variables. The goal of this article is to investigate the integral process from observational and experimental data into different types of efficient inference.

1 Introduction

This article discusses graphical models that can handle latent variables without explicitly modeling them quantitatively. In the *uncertainty in artificial intelligence* area there exist several paradigms for such problem domains. Two of them are *semi-Markovian causal models* and *maximal ancestral graphs*. Applying these techniques to a problem domain consists of several steps, typically: structure learning from observational and experimental data, parameter learning, probabilistic inference, and, quantitative causal inference.

The main problem is that each of the existing approaches only focuses on one or a few of all the steps involved in the process of modeling a problem including latent variables. The goal of this article is to investigate the integral process from observational and experimental data unto different types of efficient inference.

Semi-Markovian causal models (SMCMs) are an approach developed by Pearl (2000); Tian and Pearl (2002a). They are specifically suited to performing quantitative causal inference in the presence of latent variables.

However, at this time no efficient parametrisation of such models is provided and there are no techniques for performing efficient probabilistic inference. Furthermore there are no techniques to learn these models from data issued from observations, experiments or both.

Maximal ancestral graphs (MAGs) are an approach developed by Richardson and Spirtes (2002). They are specifically suited for structure learning in the presence of latent variables from observational data. However, the techniques only learn up to Markov equivalence and provide no clues on which additional experiments to perform in order to obtain the fully oriented causal graph. See Eberhardt et al. (2005); Meganck et al. (2006) for that type of result for Bayesian networks without latent variables. Furthermore, as of yet no parametrisation for discrete variables is provided for MAGs and no techniques for probabilistic inference have been developed. There is some work on algorithms for causal inference, but it is restricted to causal inference of quantities that are the same for an entire Markov equivalence class of MAGs (Spirtes et al., 2000; Zhang, 2006).

We have chosen to use SMCMs as a representation in our work, because they are the only formalism that allows us to perform causal inference while fully taking into account the influence of latent variables. However, we will combine existing techniques to learn MAGs with newly developed methods to provide an integral approach that uses both observational data and experiments in order to learn fully oriented semi-Markovian causal models.

Furthermore, we have developed an alternative representation of the probability distribution represented by a SMCM, together with a parametrisation for this representation, where the parameters can be learned from data with classical techniques. Finally, we discuss how probabilistic and quantitative causal inference can be performed in these models with the help of the alternative representation and its associated parametrisation.[1]

The next section introduces the necessary notation and definitions. It also discusses the semantical and other differences between SMCMs and MAGs. In section 3, we discuss structure learning for SMCMs. Then we introduce a new representation for SMCMs that can easily be parametrised. We also show how both probabilistic and causal inference can be performed with the help of this new representation.

---

[1] By the term parametrisation we understand the definition of a complete set of parameters that describes the joint probability distribution which can be efficiently used in computer implementations of probabilistic inference, causal inference and learning algorithms.

2 Notation and definitions

We start this section by introducing notation and defining concepts necessary in the rest of this article. We will also clarify the differences and similarities between the semantics of SMCMs and MAGs.

2.1 Notation

In this work uppercase letters are used to represent variables or sets of variables, i.e., $V = \{V_1, \ldots, V_n\}$, while corresponding lowercase letters are used to represent their instantiations, i.e., v_1, v_2 and v is an instantiation of all v_i. $P(V_i)$ is used to denote the probability distribution over all possible values of variable V_i, while $P(V_i = v_i)$ is used to denote the probability of the instantiation of variable V_i to value v_i. Usually, $P(v_i)$ is used as an abbreviation of $P(V_i = v_i)$.

The operators $Pa(V_i), Anc(V_i), Ne(V_i)$ denote the observable parents, ancestors and neighbors respectively of variable V_i in a graph and $Pa(v_i)$ represents the values of the parents of V_i. Likewise, the operator $LPa(V_i)$ represents the latent parents of variable V_i. If $V_i \leftrightarrow V_j$ appears in a graph then we say that they are spouses, i.e., $V_i \in Sp(V_j)$ and vice versa.

When two variables V_i, V_j are independent we denote it by $(V_i \perp\!\!\!\perp V_j)$, when they are dependent by $(V_i \not\perp\!\!\!\perp V_j)$.

2.2 Modeling latent variables

First of all, consider the model in Figure 1(a), it is a problem with observable variables $V_1, \ldots, V_6$ and latent variables L_1, L_2 and it is represented by a directed acyclic graph (DAG). As this DAG represents the actual problem henceforth we will refer to it as the **underlying DAG**.

One way to represent such a problem is by using this DAG representation and modeling the latent variables explicitly. Quantities for the observable variables can then be obtained from the data in the usual way. Quantities involving latent variables however will have to be estimated. This involves estimating the cardinality of the latent variables and this process can be difficult and lengthy. One of the techniques to learn models in such a way is the structural EM algorithm (Friedman, 1997).

Another method to take into account latent variables in a model is by representing them implicitly. With that approach, no values have to be estimated for the latent variables, instead their influence is absorbed in the distributions of the observable variables. In this methodology we only keep track of the position of the latent variable in the graph if it would be modeled, without estimating values for it. Both the modeling techniques that we will use in this article belong to that approach, they will be described in the next two sections.

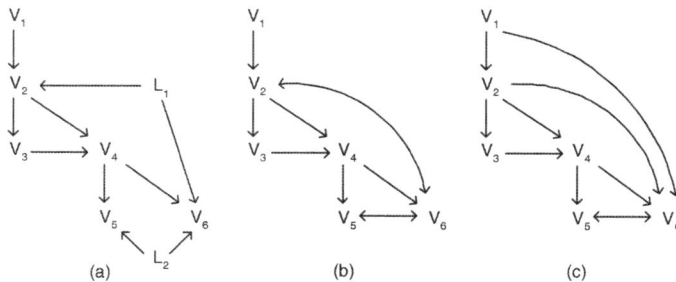

Figure 1. (a) A problem domain represented by a causal DAG model with observable and latent variables. (b) A semi-Markovian causal model representation of (a). (c) A maximal ancestral graph representation of (a).

2.3 Semi-Markovian causal models

The central graphical modeling representation that we use are the semi-Markovian causal models. They were first used by Pearl (2000), and Tian and Pearl (2002a) have developed causal inference algorithms for them.

Definitions

DEFINITION 1. A **Semi-Markovian causal model** (SMCM) is an acyclic causal graph G with both directed and bi-directed edges. The nodes in the graph represent observable variables $V = \{V_1, \ldots, V_n\}$ and the bi-directed edges implicitly represent latent variables $L = \{L_1, \ldots, L_{n'}\}$.

See Figure 1(b) for an example SMCM representing the underlying DAG in (a).

The fact that a bi-directed edge represents a latent variable, implies that the only latent variables that can be modeled by a SMCM can not have any parents, i.e., root nodes have exactly two children that are both observed. This seems very restrictive, however it has been shown that models with arbitrary latent variables can be converted into SMCMs, while preserving the same independence relations between the observable variables (Tian and Pearl, 2002b).

Semantics

In a SMCM each directed edge represents an immediate autonomous causal relation between the corresponding variables. Our operational definition of causality is as follows: a relation from variable C to variable E is causal in a certain context, when a manipulation in the form of a randomised controlled experiment on variable C induces a change in the probability distribution of variable E, in that specific context (Neapolitan, 2003).

In a SMCM a bi-directed edge between two variables represents a latent variable that is a common cause of these two variables.

The semantics of both directed and bi-directed edges imply that SMCMs are not maximal, meaning that not all dependencies between variables are represented by an edge between the corresponding variables. This is because in a SMCM an edge either represents an immediate causal relation or a latent common cause, and therefore dependencies due to a so called *inducing path*, will not be represented by an edge.

DEFINITION 2. An **inducing path** is a path in a graph such that each observable non-endpoint node is a collider, and an ancestor of at least one of the endpoints.

Inducing paths have the property that their endpoints can not be separated by conditioning on any subset of the observable variables. For instance, in Figure 1(a), the path $V_1 \rightarrow V_2 \leftarrow L_1 \rightarrow V_6$ is inducing.

Parametrisation

SMCMs cannot be parametrised in the same way as classical Bayesian networks (i.e., $\forall V_i : P(V_i|Pa(V_i))$), since variables that are connected via a bi-directed edge have a latent variable as a parent. E.g. in Figure 1(b), associating $P(V_5|V_4)$ with variable V_4 only would lead to erroneous results, as the dependence on variable V_6 via the latent variable L_2 in the underlying DAG is ignored. As mentioned before, using $P(V_5|V_4, L_2)$ as a parametrisation and estimating the cardinality and the values for latent variable L_2 would be a possible solution. However we choose not to do this as we want to leave the latent variables implicit for reasons of efficiency.

In Tian and Pearl (2002a), a factorisation of the joint probability distribution over the observable variables of an SMCM was introduced. We will derive a representation for the probability distribution represented by a SMCM based on that result.

Learning

In the literature no algorithm for learning the structure of an SMCM exists, in this article we develop techniques to perform that task, given some simplifying assumptions, and with the help of experiments.

Inference

Since as of yet no efficient parametrisation for SMCMs is provided in the literature, no algorithm for performing probabilistic inference exists. We will show how existing probabilistic inference algorithms for Bayesian networks can be used together with our parametrisation to perform that task.

SMCMs are specifically suited for another type of inference, i.e., causal inference.

DEFINITION 3. **Causal inference** is the process of calculating the effect of manipulating some variables X on the probability distribution of some other variables Y, this is denoted as $P(Y = y|do(X = x))$.

An example causal inference query in the SMCM of Figure 1(a) is $P(V_6 = v_6|do(V_2 = v_2))$.

Causal inference queries are calculated via the Manipulation Theorem (Spirtes et al., 2000), which specifies how to change a joint probability distribution (JPD) over observable variables in order to obtain the post-manipulation JPD. Informally, it says that when a variable X is manipulated to a fixed value x, the parents of variables X have to be removed by dividing the JPD by $P(X|Pa(X))$, and by instantiating the remaining occurrences of X to the value x.

Tian and Pearl have introduced theoretical causal inference algorithms to perform causal inference in SMCMs (Pearl, 2000; Tian and Pearl, 2002a). However, these algorithms assume the availability of a subset of all the conditional distributions that can be obtained from the JPD over the observable variables. We will show that with our representation these conditional distributions can be obtained in an efficient way in order to apply this algorithm.

2.4 Maximal ancestral graphs

Maximal ancestral graphs are another approach to modeling with latent variables developed by Richardson and Spirtes (2002). The main research focus in that area lies on learning the structure of these models and on representing exactly all the independences between the observable variables of the underlying DAG.

Definitions

Ancestral graphs (AGs) are graphs that are complete under marginalisation and conditioning. We will only discuss AGs without conditioning as is commonly done in recent work.

DEFINITION 4. An **ancestral graph** without conditioning is a graph with no directed cycle containing directed $\rightarrow$ and bi-directed $\leftrightarrow$ edges, such that there is no bi-directed edge between two variables that are connected by a directed path.

DEFINITION 5. An ancestral graph is said to be a **maximal ancestral graph** if, for every pair of non-adjacent nodes V_i, V_j there exists a set Z such that V_i and V_j are d-separated given Z.

A non-maximal AG can be transformed into a unique MAG by adding some bi-directed edges (indicating confounding) to the model. See Figure 1(c) for an example MAG representing the same model as the underlying DAG in (a).

Semantics

In this setting a directed edge represents an ancestral relation in the underlying DAG with latent variables. I.e., an edge from variable A to B represents that in the underlying causal DAG with latent variables, there is a directed path between A and B.

Bi-directed edges represent a latent common cause between the variables. However, if there is a latent common cause between two variables A and B, and there is also a directed path between A and B in the underlying DAG, then in the MAG the ancestral relation takes precedence and a directed edge will be found between the variables. $V_2 \rightarrow V_6$ in Figure 1(c) is an example of such an edge.

Furthermore, as MAGs are maximal, there will also be edges between variables that have no immediate connection in the underlying DAG, but that are connected via an inducing path. The edge $V_1 \rightarrow V_6$ in Figure 1(c) is an example of such an edge.

These semantics of edges make some causal inferences in MAGs impossible. As we have discussed before the Manipulation Theorem states that in order to calculate the causal effect of a variable A on another variable B, the immediate parents (i.e., the old causes) of A have to be removed from the model. However, as opposed to SMCMs, in MAGs an edge does not necessarily represent an immediate causal relationship, but rather an ancestral relationship and hence in general the modeler does not know which are the real immediate causes of a manipulated variable.

An additional problem for finding the original causes of a variable in MAGs is that when there is an ancestral relation and a latent common cause between variables, the ancestral relation takes precedence and the confounding is absorbed in the ancestral relation.

Learning

There is a lot of recent research on learning the structure of MAGs from observational data. The Fast Causal Inference (FCI) algorithm (Spirtes et al., 2000), is a constraint based learning algorithm. Together with the rules discussed in Zhang and Spirtes (2005a), the result is a representation of the Markov equivalence class of MAGs. This representation is referred to as a *complete partial ancestral graph* (CPAG) and in Zhang and Spirtes (2005a) it is defined as follows:

DEFINITION 6. Let $[G]$ be the Markov equivalence class for an arbitrary

MAG G. The **complete partial ancestral graph** (CPAG) for $[G]$, P_G, is a graph with possibly the following edges $\rightarrow, \leftrightarrow, o\!\!-\!\!o, o\!\!\rightarrow$, such that

1. P_G has the same adjacencies as G (and hence any member of $[G]$) does;

2. A mark of arrowhead ($>$) is in P_G if and only if it is invariant in $[G]$; and

3. A mark of tail ($-$) is in P_G if and only if it is invariant in $[G]$.

4. A mark of (o) is in P_G if not all members in $[G]$ have the same mark.

Parametrisation and inference
At this time no parametrisation for MAGs with discrete variables exists that represents all the properties of a joint probability distribution, (Richardson and Spirtes, 2002), neither are there algorithms for probabilistic inference.

As mentioned above, due to the semantics of the edges in MAGs, not all causal inferences can be performed. However, there is an algorithm due to Spirtes et al. (2000) and refined by Zhang (2006), for performing causal inference in some restricted cases. More specifically, they consider a causal effect to be identifiable if it can be calculated from all the MAGs in the Markov equivalence class that is represented by the CPAG and that quantity is equal for all those MAGs. This severely restricts the causal inferences that can be made, especially if more than conditional independence relations are taken into account during the learning process, as is the case when experiments can be performed. In the context of this causal inference algorithm, Spirtes et al. (2000) also discuss how to derive a DAG that is a minimal I-map of the probability distribution represented by a MAG.

In this article we introduce a similar procedure, but for a single SMCM instead of for an entire equivalence class of MAGs. In that way a larger class of causal inferences can be calculated, as the quantities do not have to be equal in all the models of the equivalence class.

2.5 Assumptions

As is customary in the graphical modeling research area, the SMCMs we take into account in this article are subject to some simplifying assumptions:

1. *Stability*, i.e., the independencies in the CBN with observed and latent variables that generates the data are structural and not due to several influences exactly cancelling each other out (Pearl, 2000).

2. Only a *single immediate connection* per two variables in the underlying DAG. I.e., we do not take into account problems where two

variables that are connected by an immediate causal edge are also confounded by a latent variable causing both variables. Constraint based learning techniques such as IC* (Pearl, 2000) and FCI (Spirtes et al., 2000) also do not explicitly recognise multiple edges between variables. However, Tian and Pearl (2002a) presents an algorithm for performing causal inference where such relations between variables are taken into account.

3. *No selection bias.* Mimicking recent work, we do not take into account latent variables that are conditioned upon, as can be the consequence of selection effects.

4. *Discrete variables.* All the variables in our models are discrete.

3 Structure learning

Just as for learning a graphical model in general, learning a SMCM consists of two parts: structure learning and parameter learning. Both can be done using data, expert knowledge and/or experiments. In this section we discuss structure learning.

3.1 Without latent variables

Learning the structure of Bayesian networks without latent variables has been studied by a number of researchers: Pearl (2000); Spirtes et al. (2000). The results of these algorithms is a representative of the Markov equivalence class.

In order to perform probabilistic or causal inference, we need a fully oriented structure. For probabilistic inference this can be any representative of the Markov equivalence class, but for causal inference we need the correct causal graph that models the underlying system. In order to obtain this, additional experiments have to be performed.

In previous work (Meganck et al., 2006), we studied learning the completely oriented structure for causal Bayesian networks without latent variables. We proposed a solution to minimise the total cost of the experiments needed by using elements from decision theory. The techniques used could be extended to the results of this article.

Furthermore, recently a lot of attention has been given to developing methods for score-based learning in the space of Markov equivalent models. An example of such a method for BNs is *greedy equivalence search* (Chickering, 2002).

3.2 With latent variables

In order to learn graphical models with latent variables from observational data the Fast Causal Inference (FCI) algorithm (Spirtes et al., 2000) has been constructed. Recently this result has been extended with the complete tail augmentation rules introduced in Zhang and Spirtes (2005a). The results of this algorithm is a CPAG, representing the Markov equivalence class of MAGs consistent with the data.

Recent work in the area consists of characterising the equivalence class of CPAGs and finding single-edge operators to create equivalent MAGs (Ali and Richardson, 2002; Zhang and Spirtes, 2005a,b). One of the goals of these advances is to create methods that search in the space of Markov equivalent models instead of the space of all models, mimicking results in the case without latent variables (Chickering, 2002).

As mentioned before for MAGs, in a CPAG the directed edges have to be interpreted as representing ancestral relations instead of immediate causal relations. More precisely, this means that there is a directed edge from V_i to V_j if V_i is an ancestor of V_j in the underlying DAG and there is no subset of observable variables D such that $(V_i \perp\!\!\!\perp V_j | D)$. This does not necessarily mean that V_i has an immediate causal influence on V_j, it may also be a result of an inducing path between V_i and V_j. For instance in Figure 1(c), the link between V_1 and V_6 is present due to the inducing path V_1, V_2, L_1, V_6 shown in Figure 1(a).

Inducing paths may also introduce $o\rightarrow$ or $o\!-\!o$ between two variables indicating either a directed or bi-directed edge, although there is no immediate influence in the form of an immediate causal influence or latent common cause between the two variables. An example of such a link is $V_3 o\!-\!o V_4$ in Figure 2.

A consequence of these properties of MAGs and CPAGs is that they are not very suited for general causal inference, since the immediate causal parents of each observable variable are not available as is necessary according to the manipulation theorem. As we want to learn models that can perform causal inference, we will discuss how to transform a CPAG into a SMCM in the next sections. Before we start, we have to mention that we assume that the CPAG is correctly learned from data with the FCI algorithm and the extended tail augmentation rules, i.e., each result that is found is not due to a sampling error or insufficient sample size.

3.3 Transforming the CPAG

Our goal is to transform a given CPAG in order to obtain a SMCM that corresponds to the underlying DAG. Remember that in general there are four types of edges in a CPAG: $\leftrightarrow$, $\rightarrow$, $o\rightarrow$, $o\!-\!o$, in which o means either a tail mark $-$ or a directed mark $>$. So one of the tasks to obtain a valid SMCM is to disambiguate those edges with at least one o as an endpoint. A second task will be to identify and remove the edges that are created due to an inducing path.

In the next section we will first discuss exactly which information we obtain from performing an experiment. Then, we will discuss the two possibilities $o \rightarrow$ and $o-o$. Finally, we will discuss how we can find edges that are created due to inducing paths and how to remove these to obtain the correct SMCM.

Performing experiments

The experiments discussed here play the role of the manipulations discussed in Section 2.3 that define a causal relation. An experiment on a variable V_i, i.e., a randomised controlled experiment, removes the influence of other variables in the system on V_i. The experiment forces a distribution on V_i, and thereby changes the joint distribution of all variables in the system that depend directly or indirectly on V_i but does not change the conditional distribution of other variables given values of V_i. After the randomisation, the associations of the remaining variables with V_i provide information about which variables V_i influences (Neapolitan, 2003). To perform the actual experiment we have to cut all influence of other variables on V_i. Graphically this corresponds to removing all incoming arrows into V_i from the underlying DAG.

We then measure the influence of the manipulation on variables of interest by obtaining samples from their post-experimental distributions.

More precisely, to analyse the results of an experiment on a variable V_{exp}, we compare for each variable of interest V_j the original observational sample data D_{obs} with the post-experimental sample data D_{exp}. The experiment consists of manipulating the variable V_{exp} to each of its values v_{exp} a sufficient number of times in order to obtain sample data sets that are large enough to analyse in a statistically sound way. The result of an experiment will be a data set of samples for the variables of interest for each of the values v_{exp} of variable V_{exp}, we denote such a data set by $D_{exp,i}$.

In order to see whether an experiment on V_{exp} made an influence on another variable V_j, we compare each post-experimental data set $D_{exp,i}$ with the original observational data set D_{obs} (with a test like χ^2). Only if at least one of the data sets is statistically significantly different, can we conclude that variable V_{exp} causally influences variable V_j.

However, this influence does not necessarily have to be immediate between the variables V_{exp} and V_j, but can be mediated by other variables, such as in the underlying DAG: $V_{exp} \rightarrow V_{med} \rightarrow V_j$.

In order differentiate between a direct influence and a potentially mediated influence via V_{med}, we will no longer compare the complete data sets $D_{exp,i}$ and D_{obs}. Instead, we will divide both data sets in subsets based on the values of V_{med}, or in other words condition on variable V_{med}. Then we compare each of the smaller data sets $D_{exp,i}|v_{med}$ and $D_{obs}|v_{med}$ with each other and this is for all values of V_{med}. By conditioning on a potentially mediating variable, we block the causal influence that might go through that variable and we obtain the immediate relation between V_{exp} and V_j.

Note that it might seem that if the mediating variable is a collider, this approach will fail, because conditioning on a collider on a path between two variables creates a dependence between those two variables. However, this approach will still be valid and this is best understood with an example: imagine the underlying DAG is of the form $V_{exp} \cdots \to V_{med} \leftarrow \ldots V_j$. In this case, when we compare each $D_{exp,i}$ and D_{obs} conditional on V_{med}, we will find no significant difference between both data sets, and this for all the values of V_{med}. This is because the dependence that is created between V_{exp} and V_j by conditioning on the collider V_{med} is present in both the original underlying DAG and in the post-experimental DAG, and thus this is also reflected in the data sets $D_{exp,i}$ and D_{obs}.

In order not to overload that what follows with unnecessary complicated notation we will denote performing an experiment at variable V_i or a set of variables W by $exp(V_i)$ or $exp(W)$ respectively, and if we have to condition on some other set of variables D while performing the experiment, we denote it as $exp(V_i)|D$ and $exp(W)|D$.

In general if a variable V_i is experimented on and another variable V_j is affected by this experiment, i.e., has another distribution after the experiment than before, we say that V_j *varies with* $exp(V_i)$, denoted by $exp(V_i) \rightsquigarrow V_j$. If there is no variation in V_j we note $exp(V_i) \not\rightsquigarrow V_j$.

Before going to the actual solutions we have to introduce p.d. paths:

DEFINITION 7. A **Potentially directed path** (p.d. path) in a CPAG is a path made only of edges of types $o\to$ and $\to$, with all arrowheads in the same direction. A p.d. path from V_i to V_j is denoted as $V_i \dashrightarrow V_j$.

Solving $o\to$

An overview of the different rules for solving $o\to$ is given in Table 1.

For any edge $V_i o\to V_j$, there is no need to perform an experiment at V_j because we know that there can be no immediate influence of V_j on V_i, so we will only perform an experiment on V_i.

If $exp(V_i) \not\rightsquigarrow V_j$, then there is no influence of V_i on V_j so we know that there can be no directed edge between V_i and V_j and thus the only remaining possibility is $V_i \leftrightarrow V_j$ (Type 1(a)).

If $exp(V_i) \rightsquigarrow V_j$, then we know for sure that there is an influence of V_i on V_j, we now need to discover whether this influence is immediate or via some intermediate variables. Therefore we differentiate as to whether there is a potentially directed (p.d.) path between V_i and V_j of length ≥ 2, or

$Ao{\rightarrow}B$	Type 1(a)	Type 1(b)	Type 1(c)
Exper. result	$exp(A) \not\leadsto B$	$exp(A) \leadsto B$ $\nexists$p.d. path $A \dashrightarrow B$ $(length \geq 2)$	$exp(A) \leadsto B$ $\exists$p.d. path $A \dashrightarrow B$ $(length \geq 2)$
Orient. result	$A \leftrightarrow B$	$A \rightarrow B$	Block all p.d. paths by conditioning on blocking set D: $exp(A)\|D \leadsto B$: $A \rightarrow B$ $exp(A)\|D \not\leadsto B$: $A \leftrightarrow B$

Table 1. An overview of how to complete edges of type $o{\rightarrow}$.

$Ao{-}oB$	Type 2(a)	Type 2(b)	Type 2(c)
Exper. result	$exp(A) \not\leadsto B$	$exp(A) \leadsto B$ $\nexists$p.d. path $A \dashrightarrow B$ $(length \geq 2)$	$exp(A) \leadsto B$ $\exists$p.d. path $A \dashrightarrow B$ $(length \geq 2)$
Orient. result	$A \leftarrow oB$ ($\Rightarrow$Type 1)	$A \rightarrow B$	Block all p.d. paths by conditioning on blocking set D: $exp(A)\|D \leadsto B$: $A \rightarrow B$ $exp(A)\|D \not\leadsto B$: $A \leftarrow oB$ ($\Rightarrow$Type 1)

Table 2. An overview of how to complete edges of type $o{-}o$.

not. If no such path exists, then the influence has to be immediate and the edge is found $V_i \rightarrow V_j$ (Type 1(b)).

If at least one p.d. path $V_i \dashrightarrow V_j$ exists, we need to block the influence of those paths on V_j while performing the experiment, so we try to find a blocking set D for all these paths. If $exp(V_i)|D \leadsto V_j$, then the influence has to be immediate, because all paths of length ≥ 2 are blocked, so $V_i \rightarrow V_j$. On the other hand if $exp(V_i)|D \not\leadsto V_j$, there is no immediate influence and the edge is $V_i \leftrightarrow V_j$ (Type 1(c)).

A blocking set D consists of one variable for each p.d. path. This variable can be chosen arbitrarily as we have explained before that conditioning on a collider does not invalidate our experimental approach.

Solving o–o

An overview of the different rules for solving $o{-}o$ is given in Table 2.

For any edge $V_i o{-}o V_j$, we have no information at all, so we might need to perform experiments on both variables.

If $exp(V_i) \not\rightsquigarrow V_j$, then there is no influence of V_i on V_j so we know that there can be no directed edge between V_i and V_j and thus the edge is of the following form: $V_i \leftarrow\!oV_j$, which then becomes a problem of Type 1.

If $exp(V_i) \rightsquigarrow V_j$, then we know for sure that there is an influence of V_i on V_j, and like with Type 1(b) we make a difference whether there is a potentially directed path between V_i and V_j of length ≥ 2, or not. If no such path exists, then the influence has to be immediate and the edge becomes $V_i \rightarrow V_j$.

If at least one p.d. path $V_i \dashrightarrow V_j$ exists, we need to block the influence of those paths on V_j while performing the experiment, so we find a blocking set D like with Type 1(c). If $exp(V_i)|D \rightsquigarrow V_j$, then the influence has to be immediate, because all paths of length ≥ 2 are blocked, so $V_i \rightarrow V_j$. On the other hand if $exp(V_i)|D \not\rightsquigarrow V_j$, there is no immediate influence and the edge is of the following form: $V_i \leftarrow\!oV_j$, which again becomes a problem of Type 1.

Removing inducing path edges

In the previous phase only o-parts of edges of a CPAG have been oriented. The graph that is obtained in this way can contain both directed and bi-directed edges, each of which can be of two types. For the directed edges:

- an immediate causal edge that is also present in the underlying DAG,
- an edge that is due to an inducing path in the underlying DAG.

For the bi-directed edges:

- an edge that represents a latent variable in the underlying DAG,
- an edge that is due to an inducing path in the underlying DAG.

When representing the same underlying DAG, a SMCM and the graph obtained after orienting all unknown endpoints of the CPAG have the same connections except for edges due to inducing paths in the underlying DAG, these edges are only represented in the experimentally oriented graph.

When an inducing path between two variables V_i and V_j creates an edge between these two variables during learning because the two are dependent conditional on any subset of observable variables, we will call such an edge an *i-false* edge.

For instance in Figure 1(a), the path V_1, V_2, L_1, V_6 is an inducing path, which causes the FCI algorithm to find an i-false edge between V_1 and V_6, see Figure 1(c). Another example is given in Figure 2 where the SMCM is given in (a) and the result of FCI in (b). The edge between V_3 and V_4

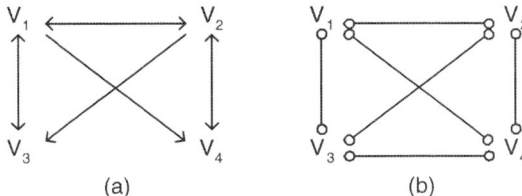

Figure 2. (a) A SMCM. (b) Result of FCI, with an i-false edge $V_3 o\text{--}o V_4$.

in (b) is a consequence of the inducing path via the observable variables V_3, V_1, V_2, V_4.

In order to be able to apply a causal inference algorithm we need to remove all i-false edges from the learned structure. The substructures that can indicate this type of edge can be identified by looking at any two variables that (i) are connected by an edge and (ii) have an inducing path between them.

To check whether the immediate connection needs to be present we have to block all inducing paths by performing one or more experiments on an inducing path blocking set (i-blocking set) D^{ip} and block all other open paths by conditioning on a blocking set D. If V_i and V_j are dependent, i.e., $(V_i \not\!\perp V_j)$, under these circumstances then the edge is correct and otherwise it can be removed.

In the example of Figure 1(c), we can block the inducing path by performing an experiment on V_2, and hence can check that V_1 and V_6 do not covary with each other in these circumstances, so the edge can be removed.

An i-blocking set consists of a collider on each of the inducing paths connecting the two variables of interest. Here a blocking set D is a set of variables that blocks each of the other open paths between the two variables of interest.

In Table 3 an overview of the actions to resolve i-false edges is given.

3.4 Example

We will demonstrate a number of steps to discover the completely oriented SMCM (Figure 1(b)) based on the result of the FCI algorithm applied on observational data generated from the underlying DAG in Figure 1(a). The result of the FCI algorithm can be seen in Figure 3(a). We will first resolve problems of Type 1 and 2, and then remove i-false edges. The result of each step is explained in Table 4 and indicated in Figure 3.

After resolving all problems of Type 1 and 2 we end up with the structure shown in Figure 3(f), this representation is no longer consistent with the

Given	A MAG with a pair of connected variables V_i, V_j, and a set of inducing paths $V_i, \ldots, V_j$
Action	Block all inducing paths $V_i, \ldots, V_j$ by performing experiments on i-blocking set D^{ip}. Block all other open paths between V_i and V_j by conditioning on blocking set D. When performing all $exp(D^{ip})\|D$: if $(V_i \not\!\perp\!\!\!\perp V_j)$: - confounding is real - else remove edge between V_i, V_j

Table 3. Removing i-false edges.

Exper.	Edge before	Experiment result	Edge after	Type
$exp(V_5)$	$V_5 o\!\!-\!\!o V_4$	$exp(V_5) \not\leadsto V_4$	$V_5 \leftarrow\!\!o\, V_4$	Type 2(a)
	$V_5 o\!\!\rightarrow V_6$	$exp(V_5) \not\leadsto V_6$	$V_5 \leftrightarrow V_6$	Type 1(a)
$exp(V_4)$	$V_4 o\!\!-\!\!o V_2$	$exp(V_4) \not\leadsto V_2$	$V_4 \leftarrow\!\!o\, V_2$	Type 2(a)
	$V_4 o\!\!-\!\!o V_3$	$exp(V_4) \not\leadsto V_3$	$V_4 \leftarrow\!\!o\, V_3$	Type 2(a)
	$V_4 o\!\!\rightarrow V_5$	$exp(V_4) \leadsto V_5$	$V_4 \rightarrow V_5$	Type 1(b)
	$V_4 o\!\!\rightarrow V_6$	$exp(V_4) \leadsto V_6$	$V_4 \rightarrow V_6$	Type 1(b)
$exp(V_3)$	$V_3 o\!\!-\!\!o V_2$	$exp(V_3) \not\leadsto V_2$	$V_3 \leftarrow\!\!o\, V_2$	Type 2(a)
	$V_3 o\!\!\rightarrow V_4$	$exp(V_3) \leadsto V_4$	$V_3 \rightarrow V_4$	Type 1(b)
$exp(V_2)$	$V_2 o\!\!-\!\!o V_1$	$exp(V_2) \not\leadsto V_1$	$V_2 \leftarrow\!\!o\, V_1$	Type 2(a)
	$V_2 o\!\!\rightarrow V_3$	$exp(V_2) \leadsto V_3$	$V_2 \rightarrow V_3$	Type 1(b)
$exp(V_2)\|V_3$	$V_2 o\!\!\rightarrow V_4$	$exp(V_2)\|V_3 \leadsto V_4$	$V_2 \rightarrow V_4$	Type 1(c)

Table 4. Example steps in disambiguating edges by performing experiments.

MAG representation since there are bi-directed edges between two variables on a directed path, i.e., V_2, V_6. However, this structure is not necessarily a SMCM yet, as there is a potentially i-false edge $V_1 \leftrightarrow V_6$ in the structure with inducing path V_1, V_2, V_6, so we need to perform an experiment on V_2, blocking all other paths between V_1 and V_6 (this is also done by $exp(V_2)$ in this case). Given that the original structure is as in Figure 1(a), performing $exp(V_2)$ shows that V_1 and V_6 are independent, i.e., $exp(V_2) : (V_1 \perp\!\!\!\perp V_6)$. Thus the bi-directed edge between V_1 and V_6 is removed, giving us the SMCM of Figure 1(b).

4 Parametrisation of SMCMs

As mentioned before, in his work on causal inference, Tian provides an algorithm for performing causal inference given knowledge of the structure of

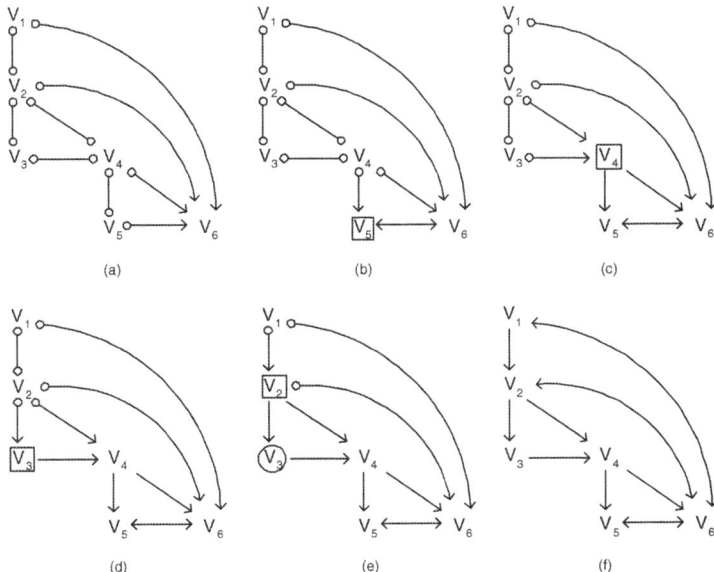

Figure 3. (a) The result of FCI on data of the underlying DAG of Figure 1(a). (b) Result of an experiment at V_5. (c) Result after experiment at V_4. (d) Result after experiment at V_3. (e) Result after experiment at V_2 while conditioning on V_3. (f) Result of resolving all problems of Type 1 and 2.

an SMCM and the joint probability distribution (JPD) over the observable variables. However, a parametrisation to efficiently store the JPD over the observables is not provided.

We start this section by discussing the factorisation for SMCMs introduced in Tian and Pearl (2002a). From that result we derive an additional representation for SMCMs and a parametrisation of that representation that facilitates probabilistic and causal inference. We will also discuss how these parameters can be learned from data.

4.1 Factorising with Latent Variables

Consider an underlying DAG with observable variables $V = \{V_1, \ldots, V_n\}$ and latent variables $L = \{L_1, \ldots, L_{n'}\}$. Then the joint probability distribution can be written as the following mixture of products:

$$P(v) = \sum_{\{l_k | L_k \in L\}} \prod_{V_i \in V} P(v_i | Pa(v_i), LPa(v_i)) \prod_{L_j \in L} P(l_j). \qquad (1)$$

Remember that in a SMCM the latent variables are implicitly represented by bi-directed edges, then consider the following definition.

DEFINITION 8. In a SMCM, the set of observable variables can be partitioned into disjoint groups by assigning two variables to the same group iff they are connected by a bi-directed path. We call such a group a **c-component** (from "confounded component") (Tian and Pearl, 2002a).

E.g. in Figure 1(b) variables V_2, V_5, V_6 belong to the same c-component. Then it can be readily seen that c-components and their associated latent variables form respective partitions of the observable and latent variables. Let $Q[S_i]$ denote the contribution of a c-component with observable variables $S_i \subset V$ to the mixture of products in equation 1. Then we can rewrite the JPD as follows: $P(v) = \prod_{i \in \{1,\ldots,k\}} Q[S_i]$.

Finally, Tian and Pearl (2002a) proved that each $Q[S]$ could be calculated as follows. Let $V_{o_1} < \ldots < V_{o_n}$ be a topological order over V, and let $V^{(i)} = V_{o_1}, \ldots, V_{o_i}$, $i = 1, \ldots, n$ and $V^{(0)} = \emptyset$.

$$Q[S] = \prod_{V_i \in S} P(v_i | (T_i \cup Pa(T_i)) \setminus \{V_i\}) \qquad (2)$$

where T_i is the c-component of the SMCM G reduced to variables $V^{(i)}$, that contains V_i. The SMCM G reduced to a set of variables $V' \subset V$ is the graph obtained by removing all variables $V \setminus V'$ from the graph and the edges that are connected to them.

In the rest of this section we will develop a method for deriving a DAG from a SMCM. We will show that the classical factorisation $\prod P(v_i | Pa(v_i))$ associated with this DAG, is the same as the one that is associated with the SMCM as above.

4.2 Parametrised representation

Here we first introduce an additional representation for SMCMs, then we show how it can be parametrised and finally, we discuss how this new representation could be optimised.

PR-representation

Consider $V_{o_1} < \ldots < V_{o_n}$ to be a topological order O over the observable variables V, and let $V^{(i)} = V_{o_1}, \ldots, V_{o_i}$, $i = 1, \ldots, n$ and $V^{(0)} = \emptyset$. Then

	Given a SMCM G and a topological order O, the PR-representation has these properties:
1.	The nodes are V, the observable variables of the SMCM.
2.	The directed edges that are present in the SMCM are also present in the PR-representation.
3.	The bi-directed edges in the SMCM are replaced by a number of directed edges in the following way: Add an edge from node V_i to node V_j iff: a) $V_i \in (T_j \cup Pa(T_j))$, where T_j is the c-component of G reduced to variables $V^{(j)}$ that contains V_j, b) except if there was already an edge between nodes V_i and V_j.

Table 5. Obtaining the parametrised representation from a SMCM.

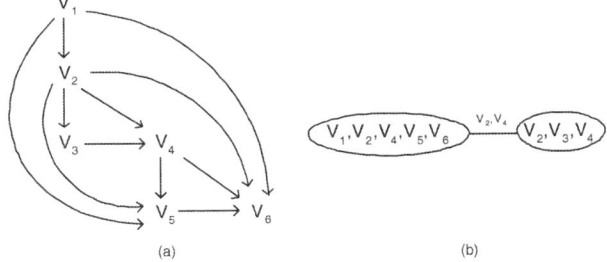

Figure 4. (a) The PR-representation applied to the SMCM of Figure 1(b). (b) Junction tree representation of the DAG in (a).

Table 5 shows how the parametrised (PR-) representation can be obtained from the original SMCM structure.

What happens is that each variable becomes a child of the variables it would condition on in the calculation of the contribution of its c-component as in Equation (2).

In Figure 4(a), the PR-representation of the SMCM in Figure 1(a) can be seen. The topological order that was used here is $V_1 < V_2 < V_3 < V_4 < V_5 < V_6$ and the directed edges that have been added are $V_1 \to V_5$, $V_2 \to V_5$, $V_1 \to V_6$, $V_2 \to V_6$, and, $V_5 \to V_6$.

The resulting DAG is an I-map Pearl (1988), over the observable variables of the independence model represented by the SMCM. This means that all the independencies that can be derived from the new graph must also be present in the JPD over the observable variables. This property can be more formally stated as the following theorem.

THEOREM 9. *The PR-representation PR derived from a SMCM S is an **I-map** of that SMCM.*

Proof. Proving that PR is an I-map of S amounts to proving that all independences represented in PR (A) imply an independence in S (B), or $A \Rightarrow B$. We will prove that assuming both A and $\neg B$ leads to a contradiction.

Assumption $\neg B$: consider that two observable variables X and Y are dependent in the SMCM S conditional on some (possible empty) set of observable variables Z: $X \not\perp_S Y | Z$.

Assumpion A: consider that X and Y are independent in PR conditional on Z: $X \perp_{PR} Y | Z$.

Then based on $X \not\perp_S Y | Z$ we can discriminate two general cases:

1. $\exists$ a path C in S connecting variables X and Y that contains no colliders and no elements of Z.

2. $\exists$ a path C in S connecting variables X and Y that contains at least one collider Z_i that is an element of Z. For the collider there are three possibilities:

    (a) $X \ldots C_i \to Z_i \leftarrow C_j \ldots Y$
    (b) $X \ldots C_i \leftrightarrow Z_i \leftarrow C_j \ldots Y$
    (c) $X \ldots C_i \leftrightarrow Z_i \leftrightarrow C_j \ldots Y$

Now we will show that each case implies $\neg A$:

1. Transforming S into PR only adds edges and transforms double-headed edges into single headed edges, hence the path C is still present in S and it still contains no collider. This implies that $X \perp_{PR} Y | Z$ is false.

2. (a) The path C is still present in S together with the collider in Z_i, as it has single headed incoming edges. This implies that $X \perp_{PR} Y | Z$ is false.

    (b) The path C is still present in S. However, the double-headed edge is transformed into a single headed edge. Depending on the topological order there are two possibilities:

    - $C_i \to Z_i \leftarrow C_j$: in this case the collider is still present in PR, this implies that $X \not\perp_{PR} Y | Z$

- $C_i \leftarrow Z_i \leftarrow C_j$: in this case the collider is no longer present, but in PR there is the new edge $C_i \leftarrow C_j$ and hence $X \not\perp_{PR} Y|Z$

(c) The path C is still present in S. However, both double-headed edges are transformed into single headed edges. Depending on the topological order there are several possibilities. For the sake of brievity we will only treat a single order here, for the others it can easily be checked that the same holds.

If the order is $C_i < Z_i < C_j$, the graph becomes $C_i \rightarrow Z_i \rightarrow C_j$, but there are also edges from C_i and Z_i to C_j and its parents $Pa(C_j)$. Thus the collider is no longer present, but the extra edges ensure that $X \not\perp_{PR} Y|Z$.

This implies that $X \perp\!\!\!\perp_{PR} Y|Z$ is false and therefore we can conclude that PR is always an I-map of S under our assumptions. ∎

Parametrisation

For this DAG we can use the same parametrisation as for classical BNs, i.e., learning $P(v_i|Pa(v_i))$ for each variable, where $Pa(v_i)$ denotes the parents in the new DAG. In this way the JPD over the observable variables factorises as in a classical BN, i.e., $P(v) = \prod P(v_i|Pa(v_i))$. This follows immediately from the definition of a c-component and from Equation (2).

Optimising the Parametrisation

We remark that the number of edges added during the creation of the PR-representation depends on the topological order of the SMCM.

As this order is not unique, giving precedence to variables with a lesser amount of parents, will cause less edges to be added to the DAG. This is because added edges go from parents of c-component members to c-component members that are topological descendants.

By choosing an optimal topological order, we can conserve more conditional independence relations of the SMCM and thus make the graph more sparse, leading to a more efficient parametrisation.

Note that the choice of the topological order does not influence the correctness of the representation, Theorem 9 shows that it will always be an I-map.

Learning parameters

As the PR-representation of SMCMs is a DAG as in the classical Bayesian network formalism, the parameters that have to be learned are $P(v_i|Pa(v_i))$. Therefore, techniques such as ML and MAP estimation (Heckerman, 1995) can be applied to perform this task.

4.3 Probabilistic inference

Two of the most famous existing probabilistic inference algorithms for models without latent variables are the $\lambda - \pi$ algorithm (Pearl, 1988) for tree-structured BNs, and the *junction tree* algorithm (Lauritzen and Spiegelhalter, 1988) for arbitrary BNs.

These techniques cannot immediately be applied to SMCMs for two reasons. First of all until now no efficient parametrisation for this type of models was available, and secondly, it is not clear how to handle the bi-directed edges that are present in SMCMs.

We have solved this problem by first transforming the SMCM to its PR-representation which allows us to apply the junction tree (JT) inference algorithm. This is a consequence of the fact that, as previously mentioned, the PR-representation is an I-map over the observable variables. And as the JT algorithm only uses independencies in the DAG, applying it to an I-map of the problem gives correct results. See Figure 4(b) for the junction tree obtained from the parametrised representation in Figure 4(a).

Note that any other classical probabilistic inference technique that only uses conditional independencies between variables could also be applied to the PR-representation.

4.4 Causal inference

In Tian and Pearl (2002a), an algorithm for performing causal inference was developed, however as mentioned before they have not provided an efficient parametrisation.

In Spirtes et al. (2000); Zhang (2006), a procedure is discussed that can identify a limited amount of causal inference queries. More precisely only those whose result is equal for all the members of a Markov equivalence class represented by a CPAG.

In Richardson and Spirtes (2003), causal inference in AGs is shown in an example, but a detailed approach is not provided and the problem of what to do when some of the parents of a variable are latent is not solved.

By definition in the PR-representation, the parents of each variable are exactly those variables that have to be conditioned on in order to obtain the factor of that variable in the calculation of the c-component, see Table 5 and Tian and Pearl (2002a). Thus, the PR-representation provides all the necessary quantitative information, while the original structure of the SMCM provides the necessary structural information, for Tian's algorithm to be applied.

5 Conclusions and perspectives

In this article we have proposed a number of solutions to problems that arise when using SMCMs in practice.

More precisely we showed that there is a big gap between the models that can be learned from data alone and the models that are used in theory. We showed that it is important to retrieve the fully oriented structure of a SMCM, and discussed how to obtain this from a given CPAG by performing experiments.

As our learning approach relies on randomized controlled experiments, in general it is not scalable to problems with a large number of variables, due to the associated large number of experiments. Furthermore, it cannot be applied in application areas where such experiments are not feasible due to practical or ethical reasons.

For future work we would also like to relax the assumptions made in this article. First of all we want to study the implications of allowing two types of edges between two variables, i.e., confounding as well as an immediate causal relationship. Another direction for possible future work would be to study the effect of allowing multiple joint experiments in cases other than when removing inducing path edges.

Furthermore, we believe that applying the orientation and tail augmentation rules of Zhang and Spirtes (2005a) after each experiment might help to reduce the number of experiments needed to fully orient the structure. In this way we could extend our previous results (Meganck et al., 2006) on minimising the total number of experiments in causal models without latent variables, to SMCMs. This allows to compare practical results with the theoretical bounds developed in Eberhardt et al. (2005).

SMCMs have not been parametrised in any other way than by the entire joint probability distribution. We showed that using an alternative representation, we can parametrise SMCMs in order to perform probabilistic as well as causal inference. Furthermore this new representation allows to learn the parameters using classical methods.

We have informally pointed out that the choice of a topological order when creating the PR-representation, influences the size and thus the efficiency of the PR-representation. We would like to investigate this property in a more formal manner. Finally, we have started implementing the techniques introduced in this article into the structure learning package (SLP)[2] of the Bayesian networks toolbox (BNT)[3] for MATLAB.

---

[2] http://banquiseasi.insa-rouen.fr/projects/bnt-slp/
[3] http://bnt.sourceforge.net/

Acknowledgements

This work was partially funded by a IWT-scholarship. This work was partially supported by the IST Programme of the European Community, under the PASCAL network of Excellence, IST-2002-506778. This publication only reflects the authors' views. We thank the reviewers for useful comments and suggestions.

Sam Maes
INSA Rouen - LITIS, St-Etienne-du-Rouvray, France.
sammaes@vub.ac.be

Stijn Meganck
Vrije Universiteit Brussel, Belgium.
smeganck@vub.ac.be

Philippe Leray
INSA Rouen - LITIS, St-Etienne-du-Rouvray, France.
philippe.leray@insa-rouen.fr

BIBLIOGRAPHY

Ali, A. and Richardson, T. (2002). Markov equivalence classes for maximal ancestral graphs. In *Proc. of the 18th Conference on Uncertainty in Artificial Intelligence (UAI)*, pages 1–9.

Chickering, D. (2002). Learning equivalence classes of Bayesian-network structures. *Journal of Machine Learning Research*, 2:445–498.

Eberhardt, F., Glymour, C., and Scheines, R. (2005). On the number of experiments sufficient and in the worst case necessary to identify all causal relations among n variables. In *Proc. of the 21st Conference on Uncertainty in Artificial Intelligence (UAI)*, pages 178–183.

Friedman, N. (1997). Learning belief networks in the presence of missing values and hidden variables. In *Proc. of the 14th International Conference on Machine Learning*, pages 125–133.

Heckerman, D. (1995). A tutorial on learning with Bayesian networks. Technical report, Microsoft Research.

Lauritzen, S. L. and Spiegelhalter, D. J. (1988). Local computations with probabilities on graphical structures and their application to expert systems. *Journal of the Royal Statistical Society, series B*, 50:157–244.

Meganck, S., Leray, P., and Manderick, B. (2006). Learning causal Bayesian networks from observations and experiments: A decision theoretic approach. In *Modeling Decisions in Artificial Intelligence, LNCS*, pages 58–69.

Neapolitan, R. (2003). *Learning Bayesian Networks*. Prentice Hall.

Pearl, J. (1988). *Probabilistic Reasoning in Intelligent Systems*. Morgan Kaufmann.

Pearl, J. (2000). *Causality: Models, Reasoning and Inference*. MIT Press.

Richardson, T. and Spirtes, P. (2002). Ancestral graph Markov models. Technical Report 375, Dept. of Statistics, University of Washington.

Richardson, T. and Spirtes, P. (2003). *Causal inference via ancestral graph models*, chapter 3. Oxford Statistical Science Series: Highly Structured Stochastic Systems. Oxford University Press.

Spirtes, P., Glymour, C., and Scheines, R. (2000). *Causation, Prediction and Search*. MIT Press.

Tian, J. and Pearl, J. (2002a). On the identification of causal effects. Technical Report (R-290-L), UCLA C.S. Lab.

Tian, J. and Pearl, J. (2002b). On the testable implications of causal models with hidden variables. In *Proc. of the 18th Conference on Uncertainty in Artificial Intelligence (UAI)*, pages 519–527.

Zhang, J. (2006). *Causal Inference and Reasoning in Causally Insufficient Systems*. PhD thesis, Carnegie Mellon University.

Zhang, J. and Spirtes, P. (2005a). A characterization of Markov equivalence classes for ancestral graphical models. Technical Report 168, Dept. of Philosophy, Carnegie-Mellon University.

Zhang, J. and Spirtes, P. (2005b). A transformational characterization of Markov equivalence for directed acyclic graphs with latent variables. In *Proc. of the 21st Conference on Uncertainty in Artificial Intelligence (UAI)*, pages 667–674.

Integrating Bayesian networks and Simpson's paradox in data mining

ALEX A. FREITAS, KEN MCGARRY, ELON CORREA

ABSTRACT. This paper proposes to integrate two very different kinds of methods for data mining, namely the construction of Bayesian networks from data and the detection of occurrences of Simpson's paradox. The former aims at discovering potentially causal knowledge in the data, whilst the latter aims at detecting surprising patterns in the data. By integrating these two kinds of methods we can hopefully discover patterns which are more likely to be useful to the user, a challenging data mining goal which is under-explored in the literature. The proposed integration method involves two approaches. The first approach uses the detection of occurrences of Simpson's paradox as a preprocessing for a more effective construction of Bayesian networks; whilst the second approach uses the construction of a Bayesian network from data as a preprocessing for the detection of occurrences of Simpson's paradox.

1 Introduction

Data mining consists of the (semi-)automatic extraction of interesting patterns from real-world data-sets. Data mining is usually considered the core step in a broader process called knowledge discovery, which includes several steps related to preprocessing of the data to be mined, the data mining step, and other steps related to the post-processing of the discovered patterns. A well-known and informative definition of knowledge discovery is as follows (Fayyad et al., 1996):

> Knowledge Discovery in Databases is the non-trivial process of identifying valid, novel, potentially useful, and ultimately understandable patterns in data.

Although this definition is often quoted in the literature, in general it has not been taken very seriously by the data mining and knowledge discovery research community. This claim is supported by the fact that the vast majority of the data mining literature focuses on discovering patterns that are valid—or accurate—ignoring the other aforementioned pattern-quality criteria. Unfortunately, the focus on the maximization of predictive accuracy often hinders the discovery of surprising, novel patterns, which is the kind of pattern that tends to be more interesting and more useful to the user (Silberchatz and Tuzhilin, 1996).

One possible explanation for this focus on accuracy in the literature seems to be that discovering novel, surprising patterns is in general a lot harder than discovering accurate patterns. A couple of examples can illustrate this point, as follows.

Brin et al. (1997) found, in a Census data set, several rules which were very accurate but were also useless, because they represented obvious patterns in the data, such as "five-year olds don't work", "unemployed residents don't earn income from work" and "men don't give birth". Tsumoto (2000) found 29,050 rules, out of which only 220 (less than 1% of them) were considered interesting or unexpected by the user.

These two works are examples of the fact that high accuracy is not a sufficient condition for the interestingness (novelty or surprisingness) or usefulness of a pattern. In addition, although high accuracy is clearly a very desirable property of a discovered pattern, high accuracy is not always a necessary condition for the usefulness or interestingness of a pattern. For instance, Wong and Leung (2000) found rules with just 40-60% confidence that were considered, by senior medical doctors, novel and more accurate than the knowledge of some junior doctors.

This paper focuses on Bayesian networks, an increasingly popular data mining technique. In terms of the aforementioned pattern-quality criteria, methods for constructing Bayesian networks from data tend to discover patterns that satisfy the criteria of good accuracy (due to the solid mathematical basis of probability theory) and good comprehensibility (due to the graphical representation of Bayesian networks). However, methods for constructing Bayesian networks are not designed to discover surprising patterns. Hence, it is quite possible that a certain Bayesian network constructed from data be accurate and comprehensible to the user, yet not very interesting, because it is only representing well-known correlations in the data, without representing any novel, surprising pattern to the user. The goal of this paper is to discuss how to remedy this situation, by integrating methods for constructing Bayesian network from data with a method for discovering surprising patterns from data, based on the detection of Simpson's paradox.

The remainder of this paper is organized as follows. Section 2 presents an overview of methods for constructing Bayesian networks from data. Section 3 presents an overview of methods for the discovery of interesting patterns and Simpson's paradox. Section 4 describes the proposed method for integrating the two aforementioned kinds of methods, and Section 5 presents the conclusions.

2 An Overview of Methods for Constructing Bayesian Networks from Data

A Bayesian network is essentially a directed acyclic graph (DAG) where each node represents a random variable (an *attribute* in the context of data mining) and an edge pointing from node X_i to node X_j means that the value of variable X_j is directly dependent on the value of the variable X_i. Assuming discrete variables (which is the focus of this paper), the strength of the dependence between two variables X_i and X_j connected by an edge is quantified by the conditional probability of the child variable X_j given the parent variable X_i.

A very simple example of a Bayesian network is illustrated in Figure 1, showing hypothetical relationships between four variables: the amount of unhealthy food eaten by a person, whether or not certain genetic factors are present in a person, the level of bad cholesterol of a person and the probability of a person having a heart attack. The hypothetical network in Figure 1 basically indicates that the probability of heart attack is directly dependent only on the level of the cholesterol of a person; whilst the latter variable is directly dependent on the amount of unhealthy food and genetic factors associated with that person. The conditional probabilities associated with the strengths of dependence between the variables are not shown, for the sake of simplicity.

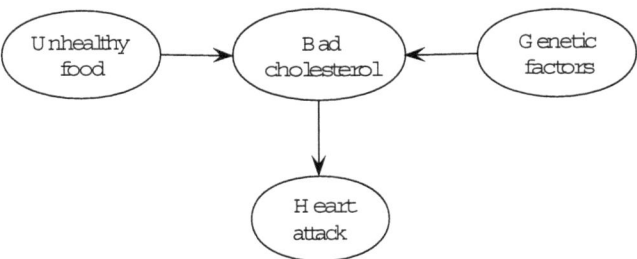

Figure 1. A very simple, hypothetical example of a Bayesian network involving 4 variables.

A Bayesian network summarizes the (in)dependencies in the data set being mined in the sense that the joint probability distribution of the set of attributes in the data set being mined—denoted by $p(X)$—is factorized into the following expression:

$$p(x) = \prod_{i=1}^{M} p(X_i|Par(X_i)), \qquad (1)$$

where $Par(X_i)$ is the set of variables that are parent nodes of the X_i node in the DAG representing the Bayesian network and M is the number of variables (attributes) in the data set being mined.

There are two major kinds of methods for constructing a Bayesian network from data, namely methods based on conditional independence tests and methods based on a search guided by a scoring function (Blanco, 2005; Korb and Nicholson, 2004). The former is based on the assumption that the results of a statistical independence test matches the true independence relationships in the data. A popular method following this approach is the PC algorithm (Spirtes et al., 1993). This algorithm uses the concept of the order of a conditional independence, which means the number of variables on which the independence between two variables is conditioned (Shipley, 2000). Hence, a zero-order conditional independence means an independence between two variables without conditioning in any other variable, a first-order conditional independence means an independence between two variables conditioning on just one other variable, and so on. This algorithm starts with the complete undirected graph. Then it reduces the graph by removing edges with zero-order conditional independence. Next it reduces the graph again by removing first-order conditional independencies, and so on. The main problem with this kind of method is that it is very computationally expensive and it does not scale up well to data sets with a large number of attributes, because, for each pair of variables candidate to have an independence relationship, it may have to test the conditional independence for all possible order sizes.

As a result, in the last few years methods based on a different approach, viz. search guided by a scoring function, have significantly grown in popularity. Methods following this approach can be classified according to different criteria, such as the kind of search engine that they use, the kind of scoring function that they use, etc. In our brief review of these methods we focus on the search engine only, which is more relevant for an understanding of the discussion in Section 4. A more comprehensive discussion about methods based on a search guided by a scoring function can be found e.g., in Blanco (2005), Korb and Nicholson (2004).

Concerning their search engine, methods following this approach can be broadly classified into sequential or population-based methods. Sequential methods work by considering a single candidate solution (a candidate DAG) at a time. They iteratively modify the current candidate solution, trying to improve it as assessed by a given scoring function, until a stopping criterion is satisfied.

The most popular sequential method for constructing Bayesian networks seems to be the B algorithm, which is essentially a hill-climbing, greedy method. This algorithm starts with an empty DAG and at each iteration it adds, to the current candidate solution, the edge that maximizes the value of the scoring function.

Another popular method is the K2 algorithm, which is also essentially a hill-climbing, greedy method. Unlike the B algorithm, K2 requires that the variables be ordered and it requires, as a user-defined parameter, the maximum number of parents of each variable in the DAG to be constructed. Hence, K2 performs a more restricted search than the B algorithm.

Population-based methods work by considering a population of candidate solutions at a time. They iteratively use information from the current population of candidate solutions to create new candidate solutions, again guided by a scoring function, until a stopping criterion is satisfied. In general population-based methods perform a global search in the search space, reducing the risk of getting stuck in local optima in the search space—which often happens with greedy, hill-climbing methods such as the B and K2 algorithms.

In addition, population-based methods normally are stochastic (non-deterministic) methods, whereas sequential methods can be either stochastic or deterministic. (Both the B algorithm and K2 are deterministic methods.)

A major kind of population-based method is evolutionary algorithms (EAs), and there are several methods designed for estimating the joint probability distribution $p(X)$ from data in EAs (Larranaga and Lozano, 2002). These methods vary from the simplest ones—in which $p(X)$ is simply factorized as the product of independent univariate marginal distributions—to the most complex ones—which can construct an arbitrarily complex Bayesian network.

Finally, we mention in passing that another important kind of method for constructing Bayesian networks from data consists of using Markov Chain Monte Carlo sampling (Husmeier, 2003; Korb and Nicholson, 2004). However, this kind of method is not discussed here because it is not relevant to the proposed method for integrating Bayesian network construction and Simpson's paradox detection, to be discussed in Section 4.

3 On the Discovery of Interesting (Novel or Surprising) Patterns and Simpson's Paradox

There are two basic approaches to discover novel or surprising (unexpected) patterns in the context of data mining, namely the user-driven (or "subjective") approach and the data-driven (or "objective") approach (Silberchatz and Tuzhilin, 1996; Freitas, 2006). In essence, the user-driven approach is based on using the domain knowledge, beliefs or preferences of the user; whilst the data-driven approach is based on statistical properties of the patterns. Hence, the data-driven approach is more generic, independent of the application domain. This makes it easier to use this approach, avoiding difficult issues associated with the manual acquisition of the user's background knowledge and its transformation into a computational form suitable for a data mining algorithm. On the other hand, the user-driven approach tends to be more effective at discovering truly novel or surprising knowledge to the user, since it explicitly takes into account the user's background knowledge.

To illustrate these approaches, let us mention one simple example of each of them. The two following examples will be based on the knowledge representation of IF-THEN rules, i.e., rules of the form:

> IF (a-set-of-conditions-on-some-attributes-is-true)
> THEN (predict-a-certain-value-for-another-attribute)

Although this knowledge representation is quite different from Bayesian networks (the focus of this paper), IF-THEN rules are used in the following examples because the vast majority of work on the discovery of interesting patterns have focused on this kind of representation.

An example of user-driven method for discovering interesting patterns is the use of user-defined general impressions (Liu et al., 1997; Romao et al., 2004). In this case the user specifies general impressions in the form of IF-THEN rules, such as "IF (job_contract_length = long_term) AND (salary = high) THEN (credit = good)". Note that this is a general impression because its conditions are not precisely defined. By contrast, the data mining algorithm is supposed to produce rules with well-defined conditions, such as "job_contract_length > 4 years" or "salary > £50K". Once such rules are produced by the data mining algorithm, the system can match the rules with the general impressions, in order to find surprising rules. In particular, if a rule and a general impression have similar antecedents ("IF part") but different consequents ("THEN part"), the rule can be considered surprising, in the sense of contradicting a user's belief (general impression). For instance, the rule "IF (job_contract_length > 4 years) AND (salary >

£50k) AND (Mortgage = yes) THEN (credit = bad)" would be considered surprising with respect to the aforementioned general impression.

One kind of data-driven method consists of using a data-driven measure of rule interestingness, which assigns a numerical degree of interestingness to a rule based on some kind of statistical property of the rule. A classic example of this approach is the data-driven rule interestingness measure proposed by (Piatetsky-Shapiro, 1991), defined as Interest = $|A \cap C|$ - $(|A| \times |C|)$ / N, where $|A \cap C|$ is the number of data instances (database records) satisfying both the rule antecedent A and the rule consequent C, $|A|$ and $|C|$ are the number of data instances satisfying the rule antecedent A and rule consequent C respectively, and N is the total number of data instances in the data set being mined. Hence, Interest is a measure of the deviation from statistical independence between A and C. Note that it measures the symmetric correlation between A and C, and not an asymmetric implication, i.e., Interest has the same value for the two "opposite" rules: IF A THEN C, IF C THEN A. There are more than 50 data-driven measures of rule quality that have been called rule "interestingness" measures in the literature. A review of these measures is out of the scope of this paper—the interested reader is refereed to Hilderman and Hamilton (2001), Tan et al. (2002)—but it is important to point out that recent results have questioned the effectiveness of such data-driven rule interestingness measures (Ohsaki et al., 2004; Carvalho et al., 2005). These recent results support the intuitive argument that it is difficult to use a purely data-driven approach for discovering patterns that are truly novel or surprising to the user.

There is, however, another kind of data-driven approach for discovering surprising patterns which is not based on statistical properties of rules, but rather based on the idea of detecting occurrences of Simpson's paradox. This is the approach followed in this paper, and although it is mainly a data-driven approach—since occurrences of the paradox are extracted from the data without the need for background knowledge specified by the user— it is explicitly designed for discovering surprising patterns to users, based on the fact that instances of Simpson's paradox tend to be very surprising to users in general—almost by definition, due to the nature of the "paradox". Hence, this approach tries to combine the best of both worlds (data-driven and user-driven measures of interestingness) (McGarry, 2005).

An occurrence of Simpson's paradox can be described as follows (Pearl, 2000). Let the event C be the apparent "cause" of an event E, the "effect". Simpson's paradox occurs if the event C increases the probability of the event E in a given population Pop and, at the same time, decreases the probability of event E in every sub-population of Pop. Let F and $\neg F$ denote two opposite values of a confounding variable, representing complementary properties describing two sub-populations of Pop. Then, mathematically Simpson's paradox occurs if the following 3 inequalities hold for a given data set:

$$P(E|C) > P(E|\neg C), \qquad (2)$$

$$P(E|C, F) < P(E|\neg C, F), \qquad (3)$$

$$P(E|C, \neg F) < P(E|\neg C, \neg F), \qquad (4)$$

where $P(X|Y)$ denotes the conditional probability of X given Y.

To illustrate these concepts, consider the hypothetical example involving Tables 1 and 2 (Pearl, 2000). Table 1 shows the number of patients who recovered (E) or not ($\neg E$), given that they received a drug (C) or no drug ($\neg C$), as well as the corresponding recovery rate. Table 2 shows the data considering the sub-populations of males and females separately.

Table 1. Recovery rates for the entire population.

Combined		Recovery (E)	($\neg E$)	Total	Recovery rate
Drug	(C)	20	20	40	50%
	($\neg C$)	16	24	40	40%
	Total	36	44	80	

Table 2. Recovery rates for the sub-populations of males and females separately.

Males		Recovery (E)	($\neg E$)	Total	Recovery rate
Drug	(C)	18	12	30	60%
	($\neg C$)	7	3	10	70%
	Total	25	15	40	

Females		Recovery (E)	($\neg E$)	Total	Recovery rate
Drug	(C)	2	8	10	20%
	($\neg C$)	9	21	30	30%
	Total	11	29	40	

Observing Table 1 only we would conclude that receiving the drug *improves* the recovery rate. However, when we observe the data partitioned by sub-population, as shown in Table 2, we observe a reversal of the effect of receiving the drug, because *in each sub-population*—i.e., in both the male and the female sub-populations—receiving the drug *reduces* the recovery rate.

This kind of reversal of the effect of the apparent "cause" seems paradoxical (under a causal interpretation) and tends to be very surprising to users. However, there is actually an explanation for the paradox. In the examples of Tables 1 and 2, the drug seems beneficial overall (in the entire population) due to a combination of two factors, namely males have higher recovery rates than females—both in the case of people who receive the drug and in the case of people who do not receive the drug—and more males receive the drug.

It is important to note that, strictly speaking, Simpson's paradox is not really a paradox in the context of probability theory, because it does not involve a real contradiction, just an apparent contradiction that is eliminated by providing an explanation for the occurrence of the paradox, as just discussed in the case of the paradox occurrence shown in Tables 1 and 2.

A second important and related point is that Simpson's paradox is well-known by statisticians in general. Despite these two points, it should be noted that the apparent contradiction associated with Simpson's paradox looks very puzzling (actually, it *looks like* a real contradiction) to most users, and, as pointed out by (Fabris and Freitas, 2006), in practice Simpson's paradox is usually very surprising to *users*, who typically have no formal statistical training. Hence, the discovery of Simpson's paradox instances is a valid approach for discovering surprising patterns from data, since the goal of data mining is to discover patterns that are surprising to users, rather than to statisticians or data analysts.

A number of occurrences of Simpson's paradox in real-world data are reported in (Fabris and Freitas, 1999, 2006; Kohavi, 2005). The two works by Fabris & Freitas also describe algorithms that systematically search for occurrences of Simpson's paradox in data.

An analysis of the computational complexity of an algorithm for detecting occurrences of Simpson's paradox in data was presented in (Fabris and Freitas, 2006). In summary, that analysis showed that the computational time taken by the algorithm: (a) grows linearly with respect to the number of data instances (records) in the data; (b) in the best (worst) case, it has a linear (cubic) growth with respect to the number of categorical (or discrete) attributes in the data—the algorithm ignores continuous (real-valued) attributes; (c) grows linearly with respect to the number of values per categorical attribute.

4 Integrating Bayesian networks and Simpson's paradox

So far the construction of Bayesian networks from data and the detection of Simpson's paradox in data have been considered as independent tasks in the data mining literature. Each of these two kinds of methods ignores the other, and they have been proposed to solve very different data mining problems.

This paper proposes to integrate these two kinds of methods. The basic idea of this integration is to combine the advantages of both kinds of methods, as follows. On one hand, Bayesian networks provide a graphical, easy-to-interpret representation of the structure of important relationships in the data, and their causal interpretation is potentially more useful for intelligent decision making than other knowledge representations that do not even attempt to represent causal knowledge. On the other hand, algorithms for detecting Simpson's paradox potentially discover very surprising patterns to the user. Hence, a synergistic combination of these two kinds of methods improves our chances of discovering patterns that are both potentially useful and surprising to the user, satisfying the two hardest-to-satisfy criteria mentioned in Fayyad et al.'s definition of knowledge discovery—quoted in the beginning of the Introduction.

In this paper we assume "causal sufficiency", i.e., all relevant variables involved in the underlying causal process being modeled are present in the data, so that there are no latent variables.

The remainder of this section is organized as follows. Section 4.1 discusses some limitations of Bayesian network construction algorithms, which served as the foundational ideas for the design of the proposed method for integrating Bayesian network construction algorithms and algorithms for detecting occurrences of Simpson's paradox. Section 4.2 discusses the proposed method itself.

4.1 The Framework for the Proposed Method

First of all, in general, Bayesian networks are Independence-maps (I-maps) of the true probability distribution (Pearl, 1988; Korb and Nicholson, 2004). This means that—if the Bayesian network was correctly constructed (see below)—every independence between variables represented in the network corresponds to an actual independence in the true probability distribution, but the converse is not true, i.e., dependencies between variables represented in the network are not guaranteed to correspond to actual dependencies in the true probability distribution.

Another limitation of conventional Bayesian network learning algorithms is that such algorithms learn only up to the Markov equivalence class of

Bayesian networks (Neapolitan, 2003). All DAGs that are members of that class are an I-map of the data, but only one of them is the truly causal DAG.

In addition to the just-mentioned "theoretical limitations" of Bayesian networks, there are two other "practical limitations" of Bayesian network construction methods. First, the problem of learning the optimal topology of a Bayesian network is NP-hard (Chickering et al., 1994; Blanco, 2005), and the size of the search space grows exponentially with the number of variables in the data set being mined. More precisely, the number of DAGs that can be generated from a set of M variables—denoted $NumDAG(M)$—is given by the following recursive equation (Correa, 2005):

$$NumDAG(M) = \sum_{i=1}^{M}(-1)^{i-1}C_{M,i}2^{i(M-1)}NumDAG(M-1), \qquad (5)$$

where the recursion stopping criterion is given by $NumDAG(0) = 1$ and $C_{M,i}$ is the number of combinations of i elements that can be taken from M elements.

Hence, when mining data sets with a large number of attributes which are commonplace in the area of data mining—we usually have to accept that it will not be possible to construct the optimal network within a certain reasonably-constrained amount of time. This justifies the use of heuristic methods for constructing a good network (rather than the ideal, optimal network) within the time constraints determined by the application domain or the user, and explains the popularity of the heuristic methods reviewed in Section 2, despite the fact that they usually discover suboptimal solutions.

The second practical problem is that, even if there was a computational method that could construct the optimal Bayesian network within an acceptable amount of time in the target application domain, in practice the actual discovery of the optimal Bayesian network for the target application domain would still depend on to what extent the following assumption is satisfied: the probability distribution of the observed data (*which is just a sample of the underlying population*) is the same as the probability distribution of the population (Shipley, 2000). In practice, the data usually has some sampling variation and/or it is noisy, making it even more difficult for a computational method to discover the optimal Bayesian network. In addition, even if there is no sample variation or noise in the observed data, there can be cases where not all independencies in the data are mirrored in the structure of the Bayesian network because, in the underlying "causal process" that produced the data, two causal paths exactly cancel each other out, thus making the learning of the Bayesian network with classical techniques impossible.

4.2 The Proposed Method

A major type of spurious dependence in a Bayesian network is the apparent dependence between events C and E (apparent cause and effect) when in reality this dependence is due to a confounding event F (Pearl, 2000). This kind of spurious dependence can potentially be discovered by detecting occurrences of Simpson's paradox, as follows. (This assumes that the confounding event is observed in the data, of course—recall that we are assuming there are no latent variables.)

An occurrence of Simpson's paradox involving a triple of events C, E, F—whose meanings were explained in Section 3—is evidence that we should not always believe that C is a cause of E. This is the basic idea of the method proposed here. To implement this idea we propose two approaches for integrating an algorithm for detecting Simpson's paradox (using algorithms described in Fabris and Freitas (2006)) and algorithms for constructing Bayesian networks (reviewed in Section 2). It should be noted that both approaches involve a "loosely-coupled" integration between the two aforementioned kinds of algorithm, in the sense that in each of these approaches first one of the algorithms is run as usual, and the results of that run are passed to the second algorithm, which is somewhat modified to take advantage of those results. The development of a more "tightly-coupled" integration is left for future research.

The first approach, here called "paradox detection before Bayesian network construction", consists of running a Simpson's paradox detection algorithm as a kind of "preprocessing step" for the Bayesian network construction algorithm, producing a list of occurrences of the paradox found in the data. This list of paradox occurrences can then be used to modify the Bayesian network construction algorithms' procedures for generating candidate networks. More precisely, consider a potential dependence of the form $C \rightarrow E$ (where C and E are thought to be cause and effect events). If C and E are associated in an occurrence of Simpson's paradox where F is a confounding event, this indicates that the effect E can be caused by the confounding event F, rather than by the apparent cause C, which suggests that the dependence $C \rightarrow E$ might be an apparent one. So, the Bayesian network construction algorithms could be modified to include, in the network being constructed, not only the edge $C \rightarrow E$, but also the edge $F \rightarrow E$. This would avoid the constructed network having just that former edge, and not the latter, which would miss the (potentially causal) effect of F on E.

The second approach for integrating an algorithm for detecting Simpson's paradox and algorithms for constructing Bayesian networks is here called "paradox detection after Bayesian network construction". This approach consists of using the result of a previously constructed Bayesian network to prune the search space for the Simpson's paradox detection algorithm. More precisely, the Simpson's paradox detection algorithm will focus its search on the pairs of variables for which there is a direct dependence represented by an edge from the potential cause C to the effect E in the Bayesian network. For each such pair of variables, the paradox detection method will try to find an occurrence of the paradox involving those two variables—by trying to find a third variable that acts as a confounding variable F between those two variables. If an occurrence of the paradox is found involving the two variables and a third confounding variable, intuitively this occurrence of the paradox is likely to be particularly surprising to the user. The detection of this occurrence of Simpson's paradox would be particularly interesting if there is no edge in the network pointing from the confounding F to the effect E, because in this case the paradox detection method would have detected a pattern not present in the Bayesian network.

At this point it should be noted that the proposed integration method (in both approaches) has a natural limitation. In particular, it is possible that the data to be mined does not contain any occurrence of Simpson's paradox. If this is the case, then the proposed method will have a limited usefulness. However, even in this case the application of the method can be considered to some extent useful, because, if no occurrence of Simpson's paradox was detected in the data, we would have an increased degree of confidence that the dependencies represented in the network are true (rather than spurious) dependencies, since the candidate dependencies represented in the network would have passed an additional test—i.e., no confounding variable related to the dependence was detected. This additional test complements (and not replaces) conventional methods for evaluating Bayesian networks.

4.3 Preliminary Computational Results

In this section we report preliminary computational results for the proposed method—in the approach of "paradox detection after Bayesian network construction"—in the Congressional Voting data set. This is a well-known public domain data set often used in machine learning research, available from the UCI Machine Learning Repository.[1] Each record (example) contains the votes of a United States Congressperson with respect to 16 key questions—each vote is represented by a binary attribute. In addition, each record is assigned to one out of two classes: democrat or republican. This data set is typically used for evaluating a classification algorithm, where the goal is to predict the Class (party affiliation) of a Congressperson based on

---

[1] University of California at Irvine, UCI Machine Learning Repository, World Wide Web address: http://www.ics.uci.edu/~mlearn/MLRepository.html

her/his votes with respect to the 16 questions (attributes). In the context of this paper, however, we are interested in constructing a Bayesian network from this data set, detecting occurrences of Simpson's paradox in the data set, and then integrate the results of these two kinds of data mining techniques.

To construct this Bayesian network (BN) we used a search procedure and a scoring metric. The scoring metric evaluates the goodness-of-fit of a candidate BN structure to the data. The search procedure generates alternative structures and selects the best one based on the scoring metric. We use a greedy search algorithm to generate BN structures. As a rule, the greedy search algorithm starts with an empty network. At each step, it then adds the edge, considering all possible pairs of nodes, that most increases the scoring metric of the current network structure. The search terminates when none of the possible edge additions improve the score of the BN. To reduce the search space of networks, only candidate networks in which each node has at most k inward edges (parents) are considered—k is a parameter determined by the user. For the experiments reported in this paper $k = 5$.

The scoring metric assigns a score to each candidate BN. Its purpose is to measure how well that BN describes the given data set. In this work we use the $K2$ scoring metric (Cooper and Herskovits, 1991; Heckerman, 1995) because its requirements exactly match our assumptions about the data set: (1) that the process that generated the database can be accurately modeled as a Bayesian network; (2) that given a Bayesian network model data instances (records) occur independently; and (3) that there are no latent variables.

The Voting data set has numerous missing values. To cope with this problem we used the following approach. When computing a given probability referring to a set of attributes X, a data instance (record) was ignored, i.e., it was not counted for probability-computation purposes, if the data set instance had a missing value for any of the attributes in the set X. This approach to cope with missing values was also used in (Fabris and Freitas, 1999), where, in the computation of probabilities associated with an occurrence of Simpson's paradox involving variables C, E and F, a data instance was ignored if it had a missing value for any of those three variables. (The results reported in Fabris and Freitas (1999) will be used later in this paper when analyzing the results of the constructed Bayesian network.)

Once the Bayesian network for the Voting data set has been constructed, we can ask two related questions: (a) Is there any occurrence of Simpson's paradox in this data set? (b) If the answer to (a) is "yes", is any of the paradox occurrences referring to a certain relationship between a triple of variables (C, E and F in the notation of Section 3) which is not a relationship observed in the constructed Bayesian network? In order to answer these questions we can use some computational results about the detection of Simpson's paradox reported in Fabris and Freitas (1999). In particular, that work reports 4 occurrences of Simpson's paradox in the Voting data set, so that the answer to the above question (a) is clearly "yes". The answer to question (b) is more elaborate, as follows. In all the 4 reported occurrences of the paradox, the "effect" (E) variable is the "Class" attribute. Hence, we looked at the constructed Bayesian network to identify which attributes are the parents of the Class attribute in that network. The parent attributes are "*El-Salvador-aid*", "*Anti-satellite-test-ban*", "*Aid-to-Nicaraguan-Contras*", "*MX-missile*", "*Immigration*". Out of these attributes, just one, Anti-satellite-test-ban, occurs as variable C (potential cause) in a paradox reported in (Fabris and Freitas, 1999). Hence, our analysis here focuses on this occurrence of the paradox, as shown in Tables 3 and 4.

Looking at Table 3, with data combined for the entire population, it seems that Congress Members voting "yes" to Anti-satellite-test-ban are much more likely to be democrats than Congress Members voting "no" to the same question. However, looking at Table 4, with data partitioned into two sub-populations based on the kind of vote ("yes" or "no") to the Physician-fee-freeze question, there is a reversal of the relationship shown in Table 3. In Table 4 Congress Members voting "yes" to Anti-satellite-test-ban are *less* likely to be democrats than Congress Members voting "no" to the same question, in both sub-populations—i.e., for both values "yes" and "no" of the attribute Physician-fee-freeze.

Note that in the paradox occurrence reported in Tables 3 and 4 the potential cause variable (C) is "Anti-satellite-test-ban", the effect variable (E) is Class (party affiliation which can be Democrat or Republican), and the confounding variable (F) is "Physician-fee-freeze". Now, let us focus on two alternative causal models of the relationships between these 3 variables, as shown in Figure 2.

In the causal model of Figure 2(a) Physician-fee-freeze affects both Anti-satellite-test-ban and Class. This suggests that Table 4 (the sub-population-specific table) better represents the causal process underlying the data. By contrast, in the causal model of Figure 2(b) it is Anti-satellite-test-ban that affects both Physician-fee-freeze and Class, which suggests that Table 3 (the entire-population table) better represents the causal process underlying the data. This is because in Figure 2(b) Physician-fee-freeze is in the middle of the causal path from Anti-satellite-test-ban to Class, so we should not condition on Physician-fee-freeze when determining the effect of Anti-satellite-test-ban on Class. For an analogous and more detailed discussion of these points in the context of the artificial data presented in Tables 3 and 4, (see Pearl, 2000, Section 6.1)[2].

Table 3. Democrat rates for the entire population.

		Democrat			
	Anti-sat-ban	(E)	(¬E)	Total	Democrat rate
Vote	Yes (C)	197	39	236	83.5%
	No (¬C)	55	122	177	31.1%
	Total	252	161	413	

Table 4. Democrat rates for sub-populations based on different values of Physician-fee-freeze.

Phys-fee-free = yes (F)					
		Democrat			
	Anti-sat-ban	(E)	(¬E)	Total	Democrat rate
Vote	Yes (C)	2	37	39	5.1%
	No (¬C)	12	122	134	9.0%
	Total	14	159	173	
Phys-fee-free = no (¬F)					
		Democrat			
	Anti-sat-ban	(E)	(¬E)	Total	Democrat rate
Vote	Yes (C)	195	2	197	99.0%
	No (¬C)	43	0	43	100%
	Total	238	2	240	

Note that neither the causal model in Figure 2(a) nor the causal model in Figure 2(b) are represented in the Bayesian network constructed from the data. Actually, the subset of the network containing the variables Anti-satellite-test-ban, Class and Physician-fee-freeze, as well as the edge connecting these variables in the network, is shown in figure 3. The network in that figure suggests that Class affects Physician-fee-freeze, whilst Figures 2(a) and 2(b) suggest that it is actually the Physician-fee-freeze that affects the Class.

---

[2] Actually, it should be noted that the causal models in Figure 2(a) and Figure 2(b) are analogous to the causal models in Figure 6.2(a) and 6.2(b) in (Pearl, 2000). By "analogous" we mean that, once the variables Anti-satellite-test-ban, Class and Physician-fee-freeze of our Figure 2 are mapped into the variables C, E, F of Pearl's Figure 6.2, the network of our Figure 2(a) has the same structure (the same directed edges) as the network of Pearl's Figure 6.2(a), and the network of our Figure 2(b) has the same structure as the network of Pearl's Figure 6.2(b).

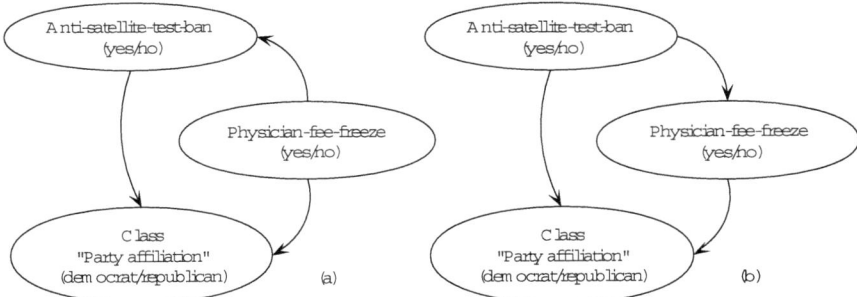

Figure 2. Two alternative causal models for the data in Tables 3 and 4.

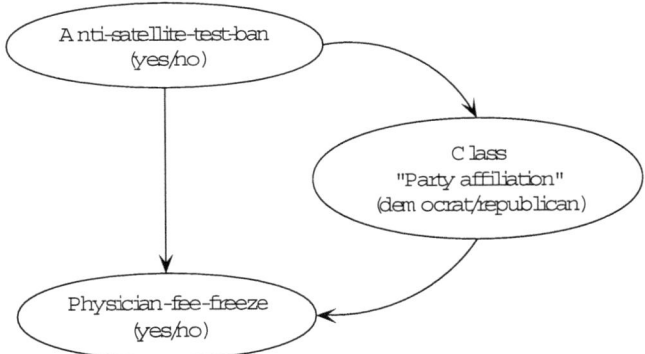

Figure 3. Relationship among the variables *Physician-fee-freeze*, *Anti-satellite-test-ban* and *Class* (Party affiliation) as observed in the Bayesian network constructed for the Congressional Voting data set.

It is beyond the scope of this paper to decide which causal model - Figure 2(a), 2(b) or 3—better represents the causal process underlying the data. Such a decision is left to an expert in the meaning of the variables and American politics, who would make the role of the user in data mining. In particular, if it is feasible, it would be advisable to perform randomized controlled experiments to study more precisely the causal dependence between variables. Note, however, that such controlled experiments are not always feasible (Shipley, 2000). It should be recalled that data mining is mainly used as a decision *support* technology, rather than as a decision *making* technology—the actual decision is made by human beings using their expert background knowledge about the application domain.

The main point of the computational results reported in this section was just to show a kind of "proof of existence" of a situation where the detection of Simpson's paradox can discover a (potentially causal) pattern in the data that is not represented in a Bayesian network constructed from the data. Such proof of existence constitutes, of course, a very preliminary result. Much more extensive experiments will be necessary to evaluate to what extent the proposed method for integrating Simpson's paradox detection and Bayesian network construction is really useful in practice, when mining real-world data sets where a user expert in the data and the application domain will analyze the discovered patterns.

5 Conclusions

This paper proposed a method for integrating two very different kinds of algorithms, namely algorithms for constructing Bayesian networks from data and algorithms for detecting occurrences of Simpson's paradox in data. The basic idea of this integration is to combine the advantages (from a data mining perspective) of both kinds of algorithms, as follows. First, the causal interpretation of Bayesian networks is potentially more useful for intelligent decision making than other knowledge representations used in data mining—which typically do not even attempt to represent causal knowledge. Second, algorithms for detecting Simpson's paradox potentially discover very surprising patterns to the user, almost by definition—due to the nature of the "paradox".

Hence, intuitively the proposed method improves our chances of (but of course does not guarantee) discovering patterns that are both potentially useful and surprising to the user, satisfying two very demanding criteria to evaluate the quality of the patterns discovered by a data mining algorithm— an area of research clearly under-explored in the data mining literature. Since only a preliminary computational result was reported in this paper, much more extensive computational experiments and analyses of the results by users, in several real-world data sets, are required in the future.

Alex A. Freitas
Computing Laboratory, University of Kent, UK.
A.A.Freitas@kent.ac.uk

Ken McGarry
School of Computing & Technology, University of Sunderland, UK.
ken.mcgarry@sunderland.ac.uk

Elon Correa
Computing Laboratory, University of Kent, UK.

E.S.Correa@kent.ac.uk

BIBLIOGRAPHY

Blanco, R. (2005). *Learning Bayesian networks from data with factorization and classification purposes. Applications in Biomedicine.* PhD thesis, Dept. of Computer Science and Artificial Intelligence, University of the Basque Country, Spain.

Brin, S., Motwani, R., Ullman, J., and Tsur, S. (1997). Dynamic itemset counting and implication rules for market basket data. In *Proc. 3rd Int. Conf. on Knowledge Discovery and Data Mining (KDD-97)*. AAAI Press.

Carvalho, D., Freitas, A., and Ebecken, N. (2005). Evaluating the correlation between objective rule interestingness measures and real human interest. In *Proc. 9th European Conf. on Principles and Practice of Knowledge Discovery in Databases (PKDD-2005)*, Lecture Notes in Artificial Intelligence 3721, pages 453–461. Springer.

Chickering, D., Geiger, D., and Heckerman, D. (1994). Learning bayesian networks is NP-hard. Technical Report MSR-TR-94-17, Microsoft Research Technical Report.

Cooper, G. F. and Herskovits, E. (1991). A Bayesian method for induction of probabilistic networks from data. Technical Report SMI-91-01, University of Pittsburgh, Pittsburgh, PA, USA.

Correa, E. S. (2005). *Model complexity and convergence pressure in estimation of distribution algorithms.* PhD thesis, Faculty of Engineering and Physical Sciences, School of Computer Science, University of Manchester, Manchester, United Kingdom.

Fabris, C. and Freitas, A. (1999). Discovering surprising patterns by detecting instances of Simpson's paradox. In *Research and Development in Intelligent Systems XVI*, pages 148–160. Springer.

Fabris, C. and Freitas, A. (2006). Discovering surprising instances of Simpson's paradox in hierarchical multi-dimensional data. In *Int. J. on Data Warehousing & Mining, 2(1)*, pages 26–48.

Fayyad, U., Piatetsky-Shapiro, G., and Smyth, P. (1996). From data mining to knowledge discovery: an overview. In *Advances in Knowledge Discovery and Data Mining, 1-34*. AAAI Press.

Freitas, A. (2006). Are we really discovering "interesting" knowledge from data? *Expert Update Magazine.* Specialist Group on Artificial Intelligence - British Computer Society, in press.

Heckerman, D. (1995). A tutorial on learning with Bayesian networks. Technical Report MSR-TR-94-09, Microsoft Research, Redmond, WA, USA.

Hilderman, R. and Hamilton, H. (2001). *Knowledge Discovery and Measures of Interest.* Kluwer.

Husmeier, D. (2003). Sensitivity and specificity of inferring genetic regulatory interactions from microarray experiments with dynamic Bayesian networks. *Bioinformatics, 19(17)*, pages 2271–2282.

Kohavi, R. (2005). Focusing the mining beacon: lessons and challenges from the world of e-commerce. Invited talk at PKDD-2005. www.kohavi.com, Visited on Jan. 2006.

Korb, K. and Nicholson, A. (2004). *Bayesian Artificial Intelligence.* Chapman.

Larranaga, P. and Lozano, J. (2002). *Estimation of Distribution Algorithms: a new tool for evolutionary computation.* Kluwer.

Liu, B., Hsu, W., and Chen, S. (1997). Using general impressions to analyze discovered classification rules. In *Proc. 3rd Int. Conf. on Knowledge Discovery and Data Mining (KDD-97)*, pages 31–36. AAAI Press.

McGarry, K. (2005). A survey of interestingness measures for knowledge discovery. *Knowledge Engineering Review*, 20(1):39–61.

Neapolitan, R. E. (2003). *Learning Bayesian networks.* Prentice Hall, first edition.

Ohsaki, M., Kitaguchi, S., Okamoto, K., Yokoi, H., and Yamaguchi, T. (2004). Evaluation of rule interestingness measures with a clinical dataset on hepatitis. In *Proc. 8th European Conf. on Principles and Practice of Knowledge Discovery in Databases (PKDD-2004)*, pages 362–373. Springer.

Pearl, J. (1988). *Probabilistic Reasoning in Intelligent Systems*. Morgan Kaufmann.

Pearl, J. (2000). *Causality: models, reasoning and inference*. Cambridge University Press.

Piatetsky-Shapiro, G. (1991). Discovery, analysis and presentation of strong rules. pages 229–248. AAAI/MIT Press.

Romao, W., Freitas, A., and Gimenes, I. (2004). Discovering interesting knowledge from a science & technology database with a genetic algorithm. In *Applied Soft Computing 4*, pages 121–137.

Shipley, B. (2000). *Cause and Correlation in Biology: a user's guide to path analysis, structural equations and causal inference*. Cambridge University Press.

Silberchatz, S. and Tuzhilin, A. (1996). What makes patterns interesting in knowledge discovery systems. In *IEEE Trans. Knowledge and Data Engineering, 8(6)*.

Spirtes, P., Glymour, C., and Scheines, R. (1993). *Causation, Prediction and Search*. Springer-Verlag.

Tan, P.-N., Kumar, V., and Srivastava, J. (2002). Selecting the right interestingness measure for association patterns. In *Proc. ACM SIGKDD Int. Conf. on Knowledge Discovery and Data Mining (KDD-2002)*, pages 32–41. ACM Press.

Tsumoto, S. (2000). Clinical knowledge discovery in hospital information systems: two case studies. In *Proc. 4th European Conf. on Principles and Practice of Knowledge Discovery in Databases (PKDD-2000)*, Lecture Notes in Artificial Intelligence 1910, pages 652–656. Springer.

Wong, M. and Leung, K. (2000). *Data mining using grammar based genetic programming and applications*. Kluwer.

PART II

CAUSALITY AND PROBABILITY IN THE
PHYSICAL SCIENCES

Causal inference in quantum mechanics: a reassessment

MAURICIO SUÁREZ

ABSTRACT. There has been an intense discussion, albeit largely an implicit one, concerning the inference of causal hypotheses from statistical correlations in quantum mechanics ever since John Bell's first statement of his notorious theorem in 1966. As is well known, its focus has mainly been the so-called Einstein-Podolsky-Rosen (EPR) thought experiment, and the ensuing observed correlations in real EPR-like experiments. But although implicitly the discussion goes as far back as Bell's work, it is only in the last two decades that it has become recognizably and explicitly a debate about causal inference in the quantum realm. The bulk of this paper is devoted to a review of three influential arguments in the philosophical literature that aim to show that causal models for the EPR correlations are impossible, due to Bas Van Fraassen, Daniel Hausman and Huw Price. I contend that all these arguments are inconclusive since they contain premises or presuppositions that are false, unwarranted, or at least controversial. Five different causal models are outlined that seem perfectly viable for the EPR correlations. These models are then employed to illustrate various difficulties with the premises and presuppositions underlying Van Fraassen's, Hausman's and Price's arguments. In all cases it is argued that the difficulties cut deep against these authors' own theories of causation and causal inference. My conclusions are that causal models for the EPR correlations certainly remain viable, that philosophical work is still required to assess their relative virtues, and that in any case the mere theoretical conceivability and empirical possibility of these models sheds doubts over Van Fraassen's, Hausman's and (important elements in) Price's theories of causation and causal inference.

1 The EPR correlations

In 1935 Einstein wrote and published jointly with two collaborators a notorious paper describing a thought experiment with correlated entangled pairs of particles.[1] The stated aim of the paper is to demonstrate that the quantum theory is incomplete since it does not describe fully all the "elements" of quantum reality, yet the paper is nowadays celebrated as the source of the burgeoning literature on what is known as "quantum non-locality". On the one hand, as Arthur Fine[2] and other have shown, the real conclusion of the EPR argument is rather a dilemma between locality (in EPR's own characteristic definition this entails that a disturbance of the state of the nearby particle can exert no change of any of the properties of the distant particle) and completeness. On the other hand Bell's theorem is taken by many to demonstrate that any empirically adequate completion of the quantum theory is committed to the existence of EPR-like correlations between the measurement events of certain properties of distant particles that have interacted in the past. So the conclusion that actually does follow from the EPR argument is the existence of distant correlations. Indeed these correlations have been positively tested experimentally on numerous occasions.

David Bohm's version of the EPR thought experiment is most often discussed, and it is this version that provides the model for most of the real experiments that have actually been carried out.[3] In this Einstein-Podolsky-Rosen-Bohm (EPR-Bohm) experiment two particles ("1" and "2") move in opposite directions, after either interacting in the past, or having been created simultaneously in some past decay event "E". As a result of their interacting history, quantum mechanics describes their composite state as an entangled singlet state. The initial angular momentum is zero, so their values of spin must be correlated throughout. Any particle's spin can be measured by means of a Stern-Gerlach apparatus. This is a magnetometer that impresses a force upon the particle proportional to its spin value, thereby correlating perfectly the particle's position with its spin value at the time the particle interacts with the magnetometer. A Stern-Gerlach apparatus can be rotated along 360 degrees, in order to measure the particle's spin value along any direction.

In a Minkowski space-time diagram, both particles describe symmetric paths along the time axis (see figure 1). The Stern-Gerlach apparati that measure these particles' spin at each wing of the experiment are at rest in the laboratory frame so their world lines are represented by vertical lines "A_1" and "A_2" in that frame. Each time the experiment is repeated, laboratory technicians are at freedom to select a particular orientation of the

---

[1] Einstein et al. (1935). The exposition in this section borrows from a previous paper of mine, Suárez (2004).
[2] (Fine, 1987)
[3] (Bohm, 1951, chapter 22).

measurement apparatus which will result in a measurement of spin along the corresponding direction. Such setting events are denoted by a and b. The two arrows pointing towards such events represent the fact that each of those setting events is controllable by experimental means. (In the language of the causal inference literature: they are exogenous variables, and are moreover controllable by agents.) Each particle's spin is measured by means of a measurement interaction between the particle and the associated measuring device on the corresponding wing. The outcomes that are produced are denoted by "s_1" and "s_2" respectively, and are known as the "outcome-events".

An important feature of the EPR-Bohm experiments is that these outcome-events are spacelike connected, i.e., they lie outside each other's lightcone. Thus a signal from one event to the other must travel at speed greater than the speed of light, during a finite part of its trajectory at least. The implications of this fact regarding the special theory of relativity are both deep and complex, and have been the object of an intense debate.[4] Although this debate is not directly relevant to much of what I will say here, it is nonetheless important for the overall assessment of the prospects of causal inference in quantum mechanics. But the importance lies in the possibility of direct-cause models for the EPR correlations, as will be seen later, and I will center the bulk of my discussion upon the other type of causal models available, namely common cause models. It seems legitimate not to enter the debate in full here (other than by mentioning some options in the interpretation of relativity) since doing so can only strengthen the position defended in this paper, namely that there is a very large range of different causal models available for the EPR correlations.

Let us now return to the quantum mechanical description of the EPR-Bohm experiment. According to quantum mechanics, for the type of particles involved, there are only two possible values of spin in any direction of measurement (θ): positive spin ($\uparrow_\theta$) and negative spin ($\downarrow_\theta$). Quantum mechanics describes the spin states of the composite system of both particles at either the time of emission or measurement by means of what is known as the singlet state (here "1" and "2" refer to each particle):

$$\Psi = \frac{1}{\sqrt{2}}(|\uparrow_{1\theta}\rangle|\downarrow_{2\theta}\rangle - |\downarrow_{1\theta}\rangle|\uparrow_{2\theta}\rangle).$$

The theory offers two kinds of probabilistic predictions. First, it offers predictions about the outcomes of measurements performed on each particle. To calculate these, we must first apply what is known as the axiom of

---

[4]See e.g., Maudlin (1995).

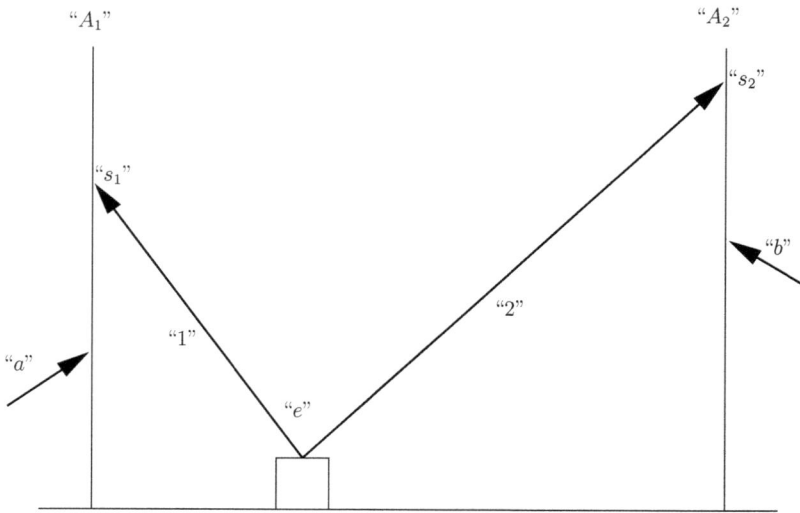

Figure 1. EPR experiment

reduction, which allows us to derive the state of each particle, individually taken:[5]

$$W_1 = \frac{1}{2}|\uparrow_{1\theta}\rangle\langle\uparrow_{1\theta}| + \frac{1}{2}|\downarrow_{1\theta}\rangle\langle\downarrow_{1\theta}|,$$
$$W_2 = \frac{1}{2}|\uparrow_{2\theta}\rangle\langle\uparrow_{2\theta}| + \frac{1}{2}|\downarrow_{2\theta}\rangle\langle\downarrow_{2\theta}|.$$

We can apply the quantum statistical algorithm to W_1 and W_2 in order to calculate the probabilities of outcomes of measurements performed on each particle in any direction θ:

$$prob(|\uparrow_{1\theta}\rangle) = \langle\Psi|\uparrow_{1\theta}\rangle\langle\uparrow_{1\theta}|\Psi\rangle = \left(\frac{1}{\sqrt{2}}\right)^2 = \frac{1}{2}.$$
$$prob(|\downarrow_{1\theta}\rangle) = prob(|\uparrow_{2\theta}\rangle) = prob(|\downarrow_{2\theta}\rangle) = prob(|\uparrow_{1\theta}\rangle) = \frac{1}{2}.$$

It is important to emphasise that W_1 and W_2 are the states of each of the particles individually taken, *consistently* with the fact that the state of the composite is the singlet state. In other words the axiom of reduction allows

---
[5] See e.g., Hughes (1989, pp. 149-150).

us to uniquely derive W_1 and W_2 from Ψ. There is however no similar axiom in quantum mechanics that would allow us to derive Ψ uniquely from W_1 and W_2. And this is at the heart of the notorious fact that the singlet state of the composite system is underdetermined by the states of its component subsystems, which often gives rise to the claims of quantum holism and quantum non-separability.[6] So our calculations of predictions for measurement results on each particle on the basis of W_1 and W_2 are perfectly consistent with the singlet state of the composite, and in fact required by it. But the predictions do not require the singlet state since they are consistent with all kinds of states of the composite that result from a mere phase difference from the singlet state (i.e., the minus sign in the singlet state could for all we care here be replaced by a plus sign).

In other words, the description offered by the singlet state Ψ of the composite system contains the greatest possible amount of information about both systems. By contrast, if we only consider the states of the systems individually taken, W_1 and W_2, we can see that we have lost relevant information. Erwin Schrödinger was perhaps the first to note that "a portion of knowledge of the composite system" is found "squandered on conditional statements that operate between the subsystems" in the form of correlations between the measurement events that we can perform on each system.[7] Indeed by applying the quantum statistical algorithm to the entangled pair of particles in the singlet state Ψ we find the following conditional probabilities of outcomes of measurements on either particle, conditional on any particular outcome of any measurement made on the other particle:

$$prob(\uparrow_{2\theta'} / \downarrow_{1\theta}) = \frac{prob(\uparrow_{2\theta'} \wedge \downarrow_{1\theta})}{prob(\downarrow_{1\theta})} = \frac{1}{2}\sin^2\frac{1}{2}\theta\theta'.$$

In the specific case $\theta = \theta'$ we obtain the following conditional probabilities, which we can immediately see imply a case of anticorrelation:

$$prob(\uparrow_{2\theta} / \downarrow_{1\theta}) = 1 = prob(\downarrow_{2\theta} / \uparrow_{1\theta}),$$
$$prob(\downarrow_{2\theta} / \downarrow_{1\theta}) = 0 = prob(\uparrow_{2\theta} / \uparrow_{1\theta}).$$

---

[6] I do not pursue these claims further in this paper. But a referee helpfully pointed out that both the arguments I describe in section 3, and the models I present in response in section 4 presuppose the assumption of separability, roughly: that in EPR-like situations it is perfectly legitimate to postulate the existence of two distinct physical systems at the wings of the experiment, however entangled their states. It might be harder to articulate some causal explanations for EPR correlations under a contrary assumption of non-separability, but then it would also be harder, I think, to articulate any arguments against such types of explanations. Moreover there is a sense in which any non-separable model of EPR is by definition causal: it merely postulates correlations between properties of one and the same entity.

[7] (Schrödinger, 1933, p. 161).

This means that if we measure both particles' spin along the same direction, the singlet state predicts an anti-correlation between the spin values. If we measure the first particle's spin in the θ direction, and we find the outcome corresponding to "positive" spin ($\uparrow_\theta$), we can predict that the outcome of a later measurement of the second particle's spin in the same direction will be "negative" ($\downarrow_\theta$) with certainty.

The kind of necessity expressed by these conditional statements, according to quantum mechanics, is merely nomological, since quantum mechanics does not describe any physical process capable of transmitting the information required from one system to the other. This of course is not to say that such a mechanism does not really exist. It is consistent to affirm both horns of the EPR argument's dilemma: i.e., that there are correlations between distant particles and that the theory is incomplete. What it means is that a causal explanation of these correlations would have to introduce some type of mechanism, or additional physical hypothesis, to explain these conditional statements. In a causal model the "additional portion of knowledge" would not be "squandered in conditional statements". For instance, in a model where the causes operate directly between the wings of the experiment, the "extra" portion of knowledge could be transmitted directly from one subsystem to the other by means of "mark-transmitters".[8] In a common cause model by contrast the causal influence might well follow the very same particle trajectories. In any case, causal structure will need to be postulated that is not described by quantum theory, but will hopefully be consistent with it.

2 Reichenbach's principle of the common cause

The EPR-Bohm experiment yields a typical case of statistical correlation, which is both predicted by a theory and experimentally verified; it is somewhat surprising in retrospect that it took so long for it to be seen as fertile ground for the application of techniques of causal inference from statistical data. Reichenbach's principle of the common cause provides one of the earliest and most influential techniques, together with a complex theory of probabilistic causation and causal structure. In deriving his notorious theorem Bell essentially employed Reichenbach's techniques, unbeknownst to him. Bell did not identify his statistical conditions as techniques of causal inference, but instead took them to be conditions on physical "locality". It was the philosopher Bas Van Fraassen instead who, in a couple of influential and important papers in the 1980's, first explicitly analysed Bell's theorem

---

[8] "Mark-transmitter" is the term employed in both Hans Reichenbach's (Reichenbach, 1956, p. 198) and Wesley Salmon's (Salmon, 1984, pp. 148-150) theories of causality. Nothing I say in this paper however hinges on such accounts.

in terms of Reichenbachian conditions of causal inference. Van Fraassen's argument has been immensely influential in drawing philosophers of physics to sceptical conclusions regarding causation in the quantum realm. In this section I intend to describe Reichenbach's conditions for causal inference, and Van Fraassen's analysis of Bell's theorem in terms of these conditions; in the next section I will analyse Van Fraassen's full argument against causal models for the EPR correlations, together with related arguments to the same effect by Daniel Hausman and Huw Price. My conclusions will be critical, and I will argue that this scepticism is premature.

At its heart Reichenbach's theory of causal inference is extremely simple. Its central principle (the "principle of the common cause") asserts that "if an improbable coincidence has occurred, there must exist a common cause".[9] Thus for Reichenbach the search for causes underlying correlation phenomena is a methodological maxim. Two comments however, regarding the principle, are in order. First, by "improbable" Reichenbach does not mean a coincidence between two token events with a low prior probability. Rather what he has in mind is a statistical correlation between two event-types A and B that is robust both theoretically and experimentally, i.e., (a) that it is predicted by some established theory and (b) that it has been verified empirically, or at least not refuted by experiment. But it must also be the case that the correlation between the event types can not be explained as a mere direct causal relation between those types. The methodological maxim to unearth common cause structure is not applicable for event types that have already been explained by means of a direct causal connection: those are not *"improbable"*.

Hence the first condition for causal inference, according to Reichenbach's theory, is correlation between two event types A and B which are not directly causally related:

$$prob(A \wedge B) \neq prob(A)prob(B) \qquad \text{(Correlation)}$$

The second condition is the existence of a open fork, i.e., a third variable C representing an event type in the past of A and B that makes the correlation between A and B vanish:[10]

$$prob(A \wedge B|C) = prob(A|C)prob(B|C) \qquad \text{(Open Fork)}$$

---

[9](Reichenbach, 1956, pp. 157 ff).

[10]The qualification "in the past of A and B" is anachronistic, and very much my own. Reichenbach thought that open forks could be used to define the direction of time, so to say of an open fork that it is oriented towards the future (or, as I say above, that the screener off must lie in the common past of A and B) would just amount, in Reichenbach's theory, to the trivial truism that an open fork is oriented as an open fork.

This condition can be expressed in a mathematically equivalent way as a screening off condition. If C, A, B form an open fork as described above, then it follows that C screens off A from B and vice versa:[11]

$$prob(A|B \wedge C) = prob(A|C)$$
$$prob(B|A \wedge C) = prob(B|C) \qquad \text{(Screening off)}$$

Reichenbach was very aware that we must tread carefully at this point, since screening off is not a sufficient condition on common causes but only a necessary one, and only under a very strong presumption of completeness. Let me explain. To see that screening off is not sufficient for common causes it is enough to observe that for any correlation, there will always exist some variable D that is not common cause, but satisfies the screening off condition, i.e some variable D such that:

$$prob(A \wedge B|D) = prob(A|D)prob(B|D).$$

For example a common effect D lying in the future of A and B will screen off A from B and viceversa (figure 2). These cases can be dispensed with by means of the "in the common past of A and B" qualification, which rules out a screener off in the future of either A or B. However, any common effect D lying to the future of C but to the past of A and B will also satisfy screening off (figure 3).

Hence screening off is not necessary for a common cause, unless we insist on a very strong assumption of completeness: i.e., unless we assure ourselves that C is the only causally relevant variable for A and B; but to know this would be to know precisely what Reichenbach's inferential techniques were meant to allow us to learn in the first place—i.e., that C is the common cause of A and B.

Thus causal inference for Reichenbach was to proceed negatively: by discovering violations of the screening off condition. In other words screening off was to be taken to be merely necessary for a common cause: not all screeners off are common causes but all common causes screen off. Hence, roughly, if conditioning upon some variable C does not render A and B statistically independent then we can at least be sure that C is not the common cause. Yet, even this simple statement does not turn out to be generally true. To see why consider the following structure where two common causes C and D acting independently underlie the correlation between A and B (figure 4).

---

[11] And, conversely, if C screens off A from B, *and C lies in the past of A and B*, then C, A, B form an open fork.

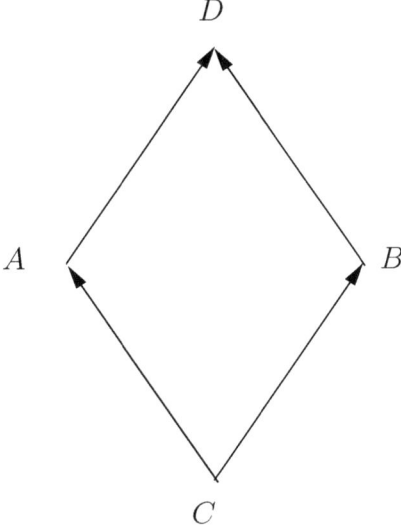

Figure 2. Common Effect

In this structure neither C nor D will screen off on their own. It is instead the conjunction of C and D that makes the correlation vanish:

$$prob(A \wedge B | C \wedge D) = prob(A|C \wedge D) prob(B|C \wedge D).$$

Hence a violation of this condition, in a structure with these four variables only, allows us to safely infer that either C, or D, or both fail to be common causes. A violation of the corresponding screening off conditions for C and D disjointly would allow us to infer nothing safely at all about C and D other than the very minimal conditional fact that *if C (D) is a common cause of A and B, then C (D) certainly is not the only cause.* And that again presupposes precisely some of the causal knowledge that Reichenbach's methods were supposed to allow us to discover.

Things actually get worse. The only piece of causal knowledge that we can possibly discover on the basis of statistical analysis by means of Reichenbach's screening off condition turns out to be conditional once again on a strong assumption of completeness. We concluded above that if the conjunction of C and D fails to screen off A from B, then we can be sure that $C \wedge D$ is not the common cause, *in a structure with four variables only.* But of course the same reasoning that led us to consider four instead of

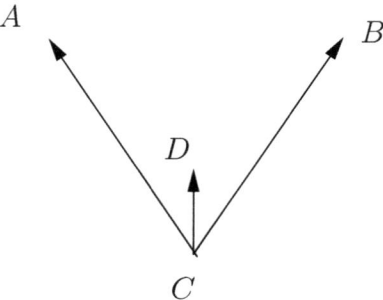

Figure 3. Future Effect

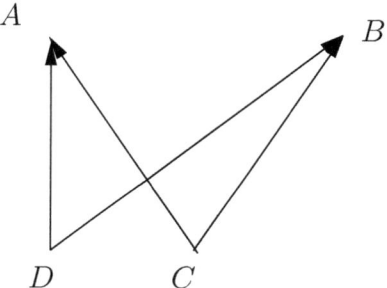

Figure 4. Two Common Causes

three variables might well lead us now to consider five. Consider the following structure (figure 5), with three putative common causes C, D and E of the correlation between A and B. In this structure the conjunction of C and D will not necessarily screen off A from B. We can only expect the conjunction of D, C and E to do so. So the only violation of screening off that would be informative about the actual causal structure would be:

$$prob(A \wedge B | C \wedge D \wedge E) \neq prob(A | C \wedge D \wedge E) prob(B | C \wedge D \wedge E).$$

But in turn this violation of the screening off condition will be informative about the actual causal structure, only *in a structure with only five variables* A, B, C, D, E. Once again this implies a strong completeness condition is in place for causal structure. So it requires us to know a fair amount about the causal structure before we can apply Reichenbach's methods in order

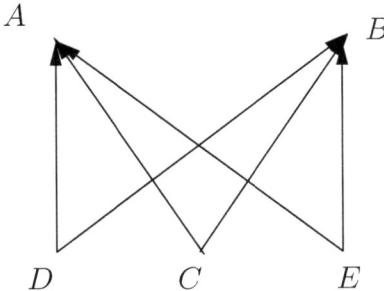

Figure 5. Three Common Causes

to discover any facts about the causal structure. And so on. By means of these simple examples, we can already easily appreciate why one of the often repeated main lessons of the recent literature on causal structure has been the insight that there is no causal discovery without background causal knowledge: *no causes in, no causes out*. It is not possible to learn anything regarding causal structure from knowledge of statistical correlations alone. Additional causal knowledge is essential for informative inference from statistical correlations to causal hypotheses.

This is an insight that will turn out to be important for the rest of this paper. Let me quickly recapitulate its implications towards Reichenbach's Principle of the Common Cause. Two different commitments are often conflated under this rubric. On the one hand there is a commitment to the general maxim: "if an improbable coincidence has occurred, there must exist a common cause". By itself this says nothing about whether common causes necessarily screen off. Reichenbach's second commitment is that a common cause structure will satisfy screening off. This commitment conversely seems to be independent of the first, since there could be unexplained correlations even if all common causes that do in fact exist screen off.

Let us refer to these independent commitments as Reichenbach's *principle of the common cause*, and Reichenbach's *criterion for common causes*. The *principle* is then the assertion that every well established correlation must have causal explanation. This is a metaphysical statement regarding the nature of correlations, lacking any methodological implications in the absence of a more concrete algorithm for causal discovery. Reichenbach's *criterion* on the other hand establishes that common causes necessarily screen off. The criterion is supposed to provide the principle with methodological bite, and to allow causal discovery to proceed on the basis of statistical analysis. But as we have seen the form of the criterion that seems sound already builds in causal knowledge from the start since it can only establish that *complete* common causes screen off. This is the main insight to be learnt from the preceding discussion in this section. However, we should be clear that this insight only compromises Reichenbach's *criterion*. In and by itself the insight says nothing at all about Reichenbach's metaphysical *principle*, which might well be true in spite of our failure to find any grounds for causal discovery that would allow us to test it. In other words, Reichenbach's principle of the common cause does not stand or fall with Reichenbach's screening off criterion for common causes. The principle might be true even if the criterion turns out to be flawed.

3 The arguments against causal models

In this section I review what I consider to be the three outstanding arguments against causal accounts of the EPR correlations, due to Bas Van Fraassen, Daniel Hausman and Huw Price. In the subsequent sections I shall attempt to rebut these arguments.

3.1 Van Fraassen's Reichenbachian argument

The most influential argument against causal models for the EPR correlations is due to Bas Van Fraassen. In a set of two overlapping papers in the 1980's Van Fraassen argued that the EPR correlations can not receive a causal explanation in either the direct-cause or common cause varieties.[12] It would be difficult to overestimate the argument's influence, even if it is not always explicitly acknowledged. I believe that this argument is historically the main source of many philosophers' scepticism towards causal accounts of the EPR correlations. Van Fraassen's papers have also deeply influenced the way philosophers of physics have come to analyse and understand the nature of quantum non-locality and its possible conflict with relativistic causation.[13]

Van Fraassen begins by establishing an analysis of the main statistical condition at the heart of Bell's theorem (the notorious "factorizability" condition) in terms of three distinct and independent conditions called "causality", "hidden locality" and "hidden autonomy". Factorizability is a necessary condition for deriving Bell's inequalities, which almost everyone agrees

---

[12] (van Fraassen, 1982, 1989).

[13] An instance is Jon Jarrett's influential distinction between parameter and outcome independence Jarrett (1984), which tracks Van Fraassen's "causality" and "hidden locality" conditions, and which has been widely adopted among philosophers of physics (for an acute dissenting criticism of Jarrett's conditions see (Maudlin, 1995, chapter 4).

have been refuted by experiment.[14] Van Fraassen takes this to imply that it is an empirical fact that "factorizability" is false. It thus follows from his analysis that at least one of the three conditions that factorizability can de decomposed into must necessarily be empirically false. His argument purports to put the blame entirely on "causality", which Van Fraassen then takes both to imply that no causal model is viable for the EPR correlations, and that Reichenbach's principle of the common cause is false as a matter of fact: not all well established correlations admit of a causal model.

Let us now look into this argument in greater detail. Van Fraassen first aims to establish that the EPR correlations constitute an example of "improbable coincidence" in Reichenbach's sense. So he aims to show that the measurement outcome event on each wing of the experiment can not be directly causing the outcome event on the other wing. Let us suppose that in the laboratory rest frame the measurement on particle "1" is carried out before the measurement on particle "2"; in other words measurement outcome s_1 occurs before measurement outcome s_2. A direct cause model would be one in which the measurement outcome event in one wing, s_1, is a direct partial cause of the measurement outcome event in the other wing, s_2 (figure 6). This is just the kind of model that Van Fraassen aims first to rule out.

In this spacetime representation of the EPR experiment A_1 and A_2 represent the worldlines of the measurement devices; "1" and "2" represent those of the particles; and the line comprised between s_1 and s_2 is the worldline of a direct causal process between the wings of the experiment. s_1, s_2, and e denote event-types, where e is the particles' emission event, and s_1 and s_2 are the outcome-events that result from measurements on particle "1" by device A_1, and on particle "2" by device A_2, respectively. s_2 is in addition the reception event by particle "2" of the causal influence emitted by particle "1".

Van Fraassen rules out direct cause models by appealing to relativity theory, since this theory implies that events s_1 and s_2, which lie outside each others' light cone, are not absolutely oriented in time. This entails, according to Van Fraassen, that any direct causal model for the EPR correlations in

---

[14]Not everyone agrees that Bell's inequalities have been experimentally refuted. However, contrary to what some uninformed physicists seem to believe, philosophers have not been at all prominent among those disputing the experimental results—on the contrary philosophers of physics on the whole have shown at least as great, if not greater, a readiness to accept the standard understanding of the experimental results as any physicists. For some dissenting views among physicists see for instance Marshall, Santos and Selleri (1983), Foadí and Selleri (2000).

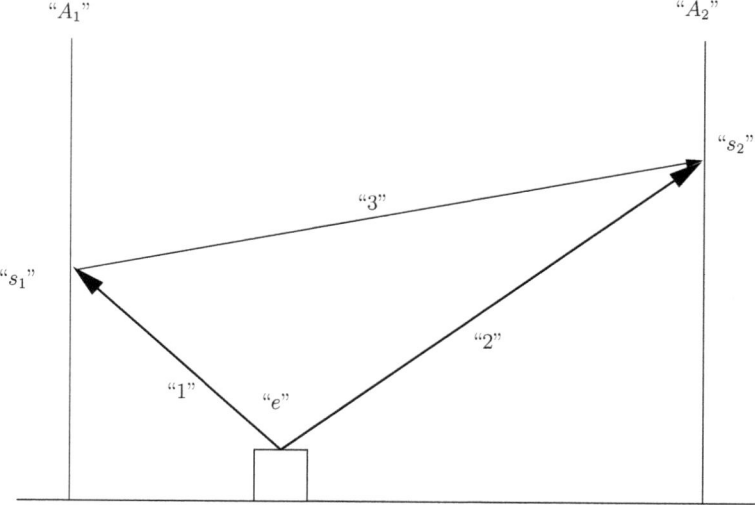

Figure 6. Direct Cause

these circumstances could only provide an "explanation by coordination":[15]

> "By *coordination* I mean a correspondence effected by signals (in a wide sense): some energy or matter travelling from one location to another, and acting as a partial producing factor for the corresponding event. The situation need not be deterministic—there can be indeterministic signalling if the signal is not certain to arrive and/or not certain to have the required effect. But the word "travel" must be taken seriously. Hence this explanation cannot work for corresponding events with spacelike separation. To speak of instantaneous travel from X to Y is a mixed or incoherent metaphor, for the entity in question is implied to be simultaneously at X and at Y—in which case there is no need for travel, as it is at its destination already."

In other words special relativity entails that there exists some frame of reference, equally valid for the description of the physical facts, where the emission of the causal influence is simultaneous with its reception (figure 7).

---

[15] (van Fraassen, 1982, in 1989, p. 112).

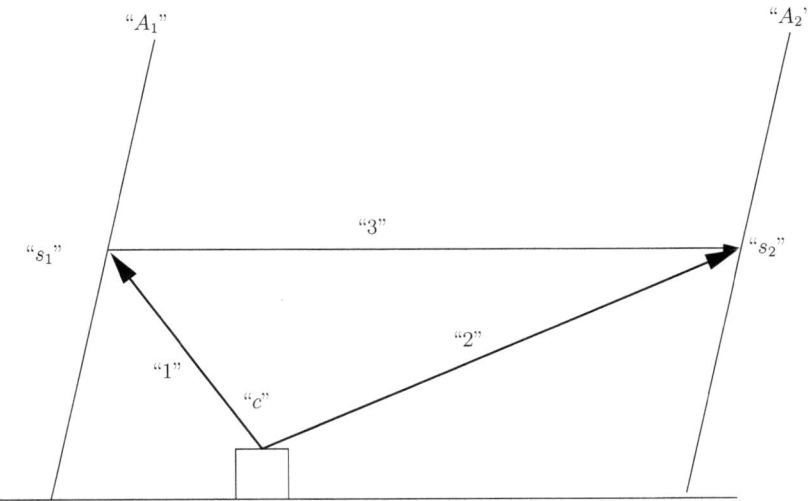

Figure 7. Instantaneous Direct Cause

In this frame of reference the material process that transmits the causal influence must travel at an infinite speed, which raises Van Fraassen's question regarding the inappropriate use of the word "travel" in this context. This part of Van Fraassen's argument is controversial (see section 4) but let us accept it here for the sake of argument. Let us then suppose for the sake of argument that a direct-cause model for the EPR correlations is impossible. This would indeed mean that the EPR correlations are precisely a case of Reichenbach's "improbable coincidence" and thus, if Reichenbach's *principle* is to hold, require a common cause explanation.

As has already been noted a common cause model for the EPR correlations is precluded, according to Van Fraassen, by the experimental violation of Bell's inequalities. His analysis starts from a consideration of the so-called "factorizability" condition that lies at the heart of the Bell inequalities:

$$prob(s_1 \wedge s_2 | a \wedge b \wedge \Psi) = prob(s_1 | a \wedge \Psi) prob(s_2 | b \wedge \Psi) \qquad \text{(FACT)}$$

Factorizability has sometimes been identified with a condition of physical locality, but this is nowadays considered a very contentious identification, so I will stick to the more neutral terminology in this paper.[16] Van

---
[16]Some philosophers (as well as physicists) have contested that the experimental refu-

Fraassen shows that this condition is the conjunction of three distinct statistical conditions, which he calls "causality", "hidden locality" and "hidden autonomy":

$$prob(s_1|s_2 \wedge a \wedge b \wedge \Psi) = prob(s_1|a \wedge b \wedge \Psi)$$
$$prob(s_2|s_1 \wedge a \wedge b \wedge \Psi) = prob(s_2|a \wedge b \wedge \Psi) \qquad \text{(Causality)}$$

This condition is a straightforward application of Reichenbach's screening off criterion for common causes, which Van Fraassen adopts. It states that the conjoint event $(a \wedge b \wedge \Psi)$ makes event s_2 statistically irrelevant to the probability of s_1, and viceversa. It has already been noted that according to Reichenbach's criterion, screening off is a necessary condition on a common cause. Thus Reichenbach's criterion implies the following conditional: if (Causality) is false in the EPR experiment then the conjunction $(a \wedge b \wedge \Psi)$ can not be a common cause of the correlations.

The second condition, "hidden locality", is also an application of the screening off condition. It states, for each wing of the experiment, that the conjunction of the creation event at the source and the corresponding setting event at one wing together screen off the outcome event at that wing from the setting event at the distant wing:

$$prob(s_1|a \wedge b \wedge \Psi) = prob(s_1|a \wedge \Psi)$$
$$prob(s_2|a \wedge b \wedge \Psi) = prob(s_2|b \wedge \Psi) \qquad \text{(Hidden Locality)}$$

The conjunction of (Causality) and (Hidden Locality) is sufficient for factorizability. The proof is trivial and begins with the observation that it is generally the case that:

$$prob(s_1 \wedge s_2|a \wedge b \wedge \Psi) = prob(s_1|s_2 \wedge a \wedge b \wedge \Psi)prob(s_2|a \wedge b \wedge \Psi).$$

By (Causality) it follows that:

$$prob(s_1 \wedge s_2|a \wedge b \wedge \Psi) = prob(s_1|a \wedge b \wedge \Psi)prob(s_2|a \wedge b \wedge \Psi)$$

And by (Hidden Locality) we obtain factorizability (FACT):

$$prob(s_1 \wedge s_2|a \wedge b \wedge \Psi) = prob(s_1|a \wedge \Psi)prob(s_2|b \wedge \Psi)$$

tation of Bell's inequalities entails the existence of physical "non-locality". But on the whole philosophers accept that (FACT) has been violated by experiment. They just question that (FACT) can be identified in any straightforward way with a condition of physical locality (see most prominently, Fine (1982)).

It then seems surprising that Van Fraassen invokes a third condition, namely "hidden autonomy", which guarantees that the state at the source is statistically independent of the apparatus setting-events:

$$prob(\Psi | a \wedge b) = prob(\Psi) \qquad \text{(Hidden Autonomy)}$$

This condition establishes that the probability of the particles to be in a particular state Ψ at the time of their emission is independent of the selection of the setting-events in either wing. Note first that were a violation of this condition to entail a causal influence it would necessarily entail an influence backwards in time in the rest frame of the laboratory since in that frame the source event is prior to any of the setting events. (Hidden Autonomy) can be appealed to in this analysis for a couple of reasons. The first is simply that it is necessary for (FACT), just like the others. That is (Causality) and (Hidden Locality) are each entailed by (FACT) but so is (Hidden Autonomy). On the other hand it is clear that a violation of (Hidden autonomy) must in turn entail that either (Causality) or (Hidden Locality) is false since otherwise, given what we just proved above, (FACT) would hold—even if *ex hypothesis* (Hidden Autonomy) is false. The most natural culprit is (Hidden Locality): if the state statistically depends on the settings, then it seems that the outcomes—which in turn depend on the state—must statistically depend on the settings. This is the second and most important reason to include (Hidden Autonomy) explicitly among the conditions: There is an interesting case of failure of (Hidden Locality), and consequently of (FACT), that turns on a violation of (Hidden Autonomy). And it is precisely this interesting case that will become relevant later on in assessing another argument against causality in quantum physics, namely the argument due to Huw Price.

Thus we have established, following Van Fraassen's analysis, that the violation of the Bell inequalities requires that at least one among (Causality), (Hidden Locality) and (Hidden Autonomy) be false. Van Fraassen then goes to argue that the EPR correlations themselves show (Causality) to be false. This is because the state of the particles at the time of their emission does not screen off the outcome-events from each other. For, let us suppose that $a = b = \theta$, without loss of generality. Then (Causality) reduces to:

$$prob(s_1 | s_2 \wedge \Psi) = prob(s_1 | \Psi)$$
$$prob(s_2 | s_1 \wedge \Psi) = prob(s_2 | \Psi)$$

And this condition is certainly false, since according to quantum mechanics:

$$prob(s_1|s_2 \wedge \Psi) = 1 \neq prob(s_1|\Psi) = \frac{1}{2}$$
$$prob(s_2|s_1 \wedge \Psi) = 1 \neq prob(s_2|\Psi) = \frac{1}{2}$$

Van Fraassen thus concludes that a common cause model for EPR of the sort envisioned by Reichenbach is not viable: "The conclusion is surely inevitable: there are well attested phenomena which cannot be embedded in any common-cause model." (van Fraassen, 1982, in 1989, p. 108).

3.2 Hausman's Independence Argument

A different argument against causal accounts of the EPR correlations has been provided by Daniel Hausman, who aims to reproduce Van Fraassen's negative conclusion by applying his own distinct theory of causation. At the heart of Hausman's theory there are a couple of anti-Humean principles. The first one, partly inspired by Reichenbach, exerts the connection between causes and probabilities dependencies, and Hausman refers to it as the Necessary Connection Principle (N-Connection or NC Principle):[17]

N-Connection Principle (NC): Events a and b are n-connected if and only if they are distinct and (1) a causes b or b causes a or (2) a and b are effects of a common cause.

The relationship between (NC) and Reichenbach's *principle* of the common cause is, according to Hausman, akin to the relation between tokens and types of causally related events. The (NC) principle applies to token events, and Hausman defends the view that this relation manifests itself as the kind of probabilistic dependence (correlation) among types that prompts Reichenbach's principle.[18] Another important point of clarification regarding this definition has to do with the notion of "distinctness". For Hausman, two events are *distinct* if they are neither logically related nor do they have any part in common. Thus Hausman's definition unpacks some of the commitments underlying Reichenbach's notion of "improbable coincidence" among event types. Finally it is important to note that to have wide application, the (NC) principle must be relativised to some causal "field", otherwise the principle would hold trivially for all events in relation to the big bang.[19]

---

[17](Hausman, 1999, p. 81). For Hausman's own theory of causation see also Hausman (1998) where the (NC) principle appears as "The Connection Principle".
[18](Hausman, 1998, p. 59).
[19](Hausman, 1999, p. 81; 1998, p. 40 and p. 60).

The (NC) principle is in essence just an anti-Humean assertion of causation as an independent relation between token events which might give rise to and explain, but must not be confused with, probabilistic association among the corresponding types. The most distinct among Hausman's tenets, which identifies his theory, is another principle. The *independence* condition asserts that every effect must have a distinct and individual cause, unrelated to all its other causes (hence often represented by an exogenous variable), which in principle at least allows us to fix its value independently. I will in this paper call such distinct and individual causes a "handle" or "leverer", since at least in principle they allow us to control the presence of token effects and the corresponding probability of their associated types.

Independence Condition (I): If a causes b or a and b are n-connected only as effects of some common cause, then b has a cause that is distinct from a and not n-connected to a.

Hausman claims that (I) is a conceptual truth about causation: "a boundary condition on the possibility of causal attributions" (1999, p. 83); "a necessary condition for the possibility of causal attributions and causal explanations" (1998, p. 64); and "when all the same things have an to a and b, causal concepts are inapplicable" (1999, p. 83). The intuition is that causal concepts are inapplicable to cases that do not satisfy (I). Prominent among these are putative cases of singular proximate causes. Hence Hausman is ruling out the standard understanding of, for instance, radioactive decay—according to which the nuclear structure of the radioactive element is itself the sole and proximate cause of its decay with a certain probability. He states in response to this putative counterexample to (I) that "the phenomena that we identify as causes and effects are not at all like this" (1998, p. 69). The view that I will defend on the contrary is that the EPR correlations precisely show that we are quite prepared to entertain a causal relation that ascribes to an effect a sole and proximate cause.

Together with an assumption of transitivity of causation (i.e., if c is caused by b and b is caused by a then c is caused by a), (NC) and (I) jointly entail the following "theory of causation":

Independence Theory of Causation (C): a causes b if and only if (i) a is n-connected to b, (ii) everything n-connected to a and distinct from b is n-connected to b, and (iii) something n-connected to b is causally independent of a.

The notion of causal independence that Hausman appeals to here is a straightforward application of the other terms already defined in his theory:

"I shall say that events are causally independent if and only if they are both distinct and not causally connected [i.e., n-connected]" (1998, section 4.4). Thus substituting our prior definitions we obtain the following paraphrase of the (C)'s main implication: We can only meaningfully claim that some token event a is the cause of some other event b if 1) a and b are distinct events 2) causally connected to each other by either 3i) directly causing each other or 3ii) as effects of some common cause, and 4) such that all causes and effects of a are also causes or effects of b, while 5) b has at least one cause or effect that is neither cause nor effect of a. It is worth noting that each commitment in (C) essentially responds to either (NC), (I) or transitivity. Thus (NC) is essentially responsible for 1), 2) and 3); transitivity applied to (NC) yields 4); while condition 5) is essentially the result of applying the independence condition (I).

We may now return to Hausman's analysis of the EPR correlations. An application of (C) to EPR yields the conclusion that the measurement wing on one wing can not be said to be the cause of the measurement event in the other wing (see figures 1 and 6).[20] The reason is that condition (5) is apparently violated: in other words the independence condition (I) fails. To check this claim in an EPR background, we must first translate condition (I) into that setting, as follows:

<u>Independence Condition for EPR</u> (I for EPR): If s_1 causes s_2 or s_1 and s_2 are n-connected only as effects of some common cause, then s_2 has a cause that is distinct from s_1 and not n-connected to s_1.

A straightforward application of this condition to the simplest commoncause model—with causal influences travelling along the particles' worldlines, and the setting events rendered irrelevant by setting them both in the same direction θ, as in figure 1—illustrates nicely the failure of (I) that Hausman alludes to. In that scenario there is no event that is a cause of s_2 that is not also n-connected to s_1. In other words there is no independent "handle" on s_2 that would allow us to control its value independently of the causal relation between s_1 and s_2. The failure of (I for EPR) entails,

---

[20] Hausman's most detailed analysis appeals instead to the so-called GHZ experiment, which requires a smaller range of setting-events in order to generate the contradiction with a Bell-like inequality. But as far as I can see this is an unnecessary complication and detour since the reason why Hausman thinks that no causal model will apply to GHZ is exactly the same that leads him to the same scepticism in the EPR case, namely the failure of the independence (I) condition as described above (Hausman, 1999, p. 86). And my proposal in section 5 of the applicability of (I) to some possible causal models of EPR applies *mutatis mutandis* to causal models for GHZ.

according to Hausman, that causal concepts are inapplicable in this scenario: our putative common cause model for EPR is not a genuinely *causal* model.[21]

3.3 Huw Price's Asymmetry Argument

Yet another argument against causal models for the EPR correlations is due to Huw Price.[22] Price's views are close to Hausman's in the following regard:[23] both seem to think that even if models for the EPR correlations may look causal, they nonetheless fail to be *genuinely* causal—since they do not employ fully articulate causal concepts. But they differ as to what they consider the key to a full articulation of the concept of causality. And they correspondingly differ in their analysis of what is lacking in putative causal models for the EPR correlations. For Hausman, as was noted in the previous section, the key is the failure of the Independence (I) condition, while for Price the key is the absence of a time-asymmetry in the relation of causal dependence.

It might seem odd to include Price in the list of critics of causal models, since he is well known for defending an explanation of the EPR correlation along the lines of the *zigzag* model of Costa de Beauregard, employing the notion of backwards in time influences; and this is often understood to be a causal model. In my view Price's theses on causal asymmetry make it explicit that his conception of causation does not really fit a *causal* understanding of such backwards-in-time explanations of the EPR correlations. And his most recent defense of causal perspectivalism makes it plain that for him the notion of causation is unsuitable for microphysics altogether.[24] Hence the conclusion to be extracted is that the kind of explanation that Price is advocating for EPR correlations is not, on his own account, genuinely causal.

The type of model advocated by Price for the EPR correlations can be represented in a spacetime diagram of the sort that we have been employing as follows (8):

---

[21] Hausman then goes on to devote a fair amount of work to provide a revised version of the (NC) principle that accommodates the existence of nomological but non-causal correlations of the sort that he finds in EPR (Hausman, 1999, pp. 88ff.); but this is irrelevant to our purposes here, since the failure of (I) still impugns the application of his theory of causation (C) to the EPR correlations.

[22] (Price, 1996, chapters 7-9).

[23] Or rather "Hausman's views are close to Price's", since Price's work in this area precedes Hausman's.

[24] (Price, 2005).

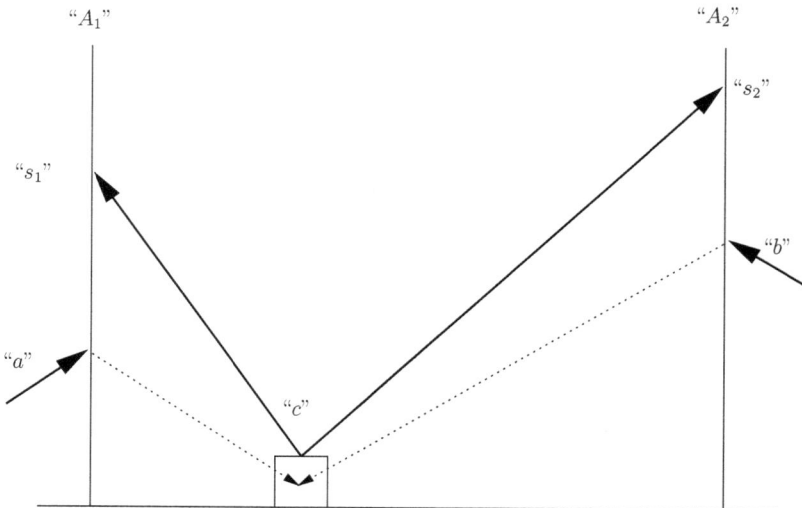

Figure 8. Zigzag Model

In this model the setting events influence the creation event at the source. (The continuous arrows represent causal influences; the discontinuous arrows by contrast represent "influences" that need not be considered causal). They partly determine the state of the particle pair at the source. Hence the state of the particles by the time they reach their corresponding measurement devices is not the singlet state but the corresponding mixture over the possible values of spin in the direction selected by the settings. This explains causally the manifestation of both values of spin at the measurement outcome-events, and yields the appropriate statistical correlation over a large number of similarly prepared particles. And the model has the great advantage of being local, in the sense specified by both contiguity and relativity theory. All influences are transmitted by continuous worldlines in spacetime, and between events lying within the relativistic light cones (the arrows back from the setting events can be made to point as far back in the past of the creation event as we like—and thus can be pushed back into the shared part of the past lightcones of s_1 and s_2).

Yet the model has some apparent counterintuitive features from the point of view of our ordinary experience of the macroscopic world. Since in a typical EPR experiment the setting events a and b take place after the creation event c in the rest frame of the laboratory, it follows that were this influence causal it would constitute an instance of backwards-in-time causation. But Price does not characterise it this way, preferring instead to refer to a backwards-in-time *influence*; and he defends this model as an instance of a time-symmetric explanation of physical phenomena that violates the *microscopic innocence* (μ-*innocence*) principle.

μ-*innocence* establishes that the states of systems in microphysics record all their past interactions but none of their future ones.[25] So that the states of systems that have interacted in the past might well be entangled, but not so the states of systems that will interact in the future. In other words systems are innocent of any interactions that lie in their future, which can have no effect upon their present state. The corresponding principle applied to macroscopic phenomena (the macroscopic innocence principle) is of course intuitive, given the arrow of entropy defined by the second law of thermodynamics—so intuitive in fact that Price thinks that it is rarely made explicit. But it is not appropriate for microphysics, according to Price, since it is in open conflict with the time-symmetry of the equations of fundamental physics: "There really is a conflict in the intuitive picture of the world with which contemporary physics operates. [...] Our intuitive commmitement to μ-*innocence* is incompatible with T-symmetry" (Price, 1996, p. 123). Given the time-reversal invariance of the fundamental laws of microphysics, Price rejects μ-*innocence*, and finds support for his rejection in the backwards-in-time influences model in figure 7.

But it is important to note that for Price μ-*innocence* is not a principle of causation, and the arrow of time that it defines does not thereby provide the required asymmetry of causation. It is rather the other way round: the reason we find μ-*innocence* intuitive in general is grounded upon our causal perspective upon the (macroscopic) world. Price adopts an agency based theory of causation, roughly: a causes b if and only if some agent can bring about b by producing a. Since, according to Price, agents are macroscopic creatures acting in the macroscopic world it follows from the time-oriented character of macroscopic phenomena that causes precede their effects, by definition.

So it is in a way misleading, according to Price, to refer to the zigzag model as a "backwards causation" model. For Price the idea of backwards causation only makes sense as a projection from our ordinary forward-oriented causal perspective of the macroworld. Price defends a model of this sort for the EPR correlations as the only type of model that preserves locality and a nearly fully classical understanding of the quantum world; but

---

[25](Price, 1996, pp. 120ff).

far from defending this as a *causal* model for the EPR correlations, he uses it precisely to show the limits of the causal perspective. The model shows that the causal concepts that are properly employed to describe our experience of the macroscopic world, do not ultimately reflect any real properties of the physical world itself. These concepts are not properly applicable to the fundamental description of the processes underlying quantum correlation phenomena. Thus Price ends up embracing the same type of scepticism regarding causal accounts of the EPR correlations, and of quantum phenomena in general, as we have seen defended by Hausman and Van Fraassen.

4 Five causal models for the EPR correlations

In this section I describe five different causal models, and I argue that they have not been refuted by any experimental results or theoretical considerations. (In the next section I will argue that several of these models are in no way compromised by the philosophical arguments previously reviewed). The first model is a more general version of the direct-cause model of figure 6.

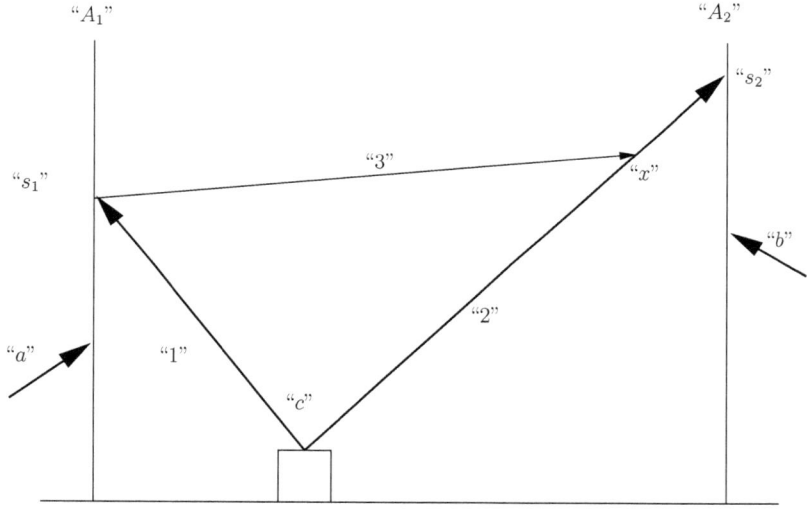

Figure 9. Causal Model I (Direct Cause)

In this model, unlike the model represented in figure 6, the causal influence does not "travel" directly from event s_1 to event s_2, but it is assumed instead for the sake of generality that s_1 is the cause of some change x in the state of particle "2" on its way towards the distant wing. This event x is in turn a partial cause of the subsequent spin measurement outcome event on that particle, s_2. Causal Model I is represented in figure 9. I have argued elsewhere that this model is not actually ruled out by relativity theory, contrary to what some philosophers, including Van Fraassen, have sometimes thought in the past. On the contrary there are at least three different readings of this model that allow it to successfully overcome any potential conflict with the theory of special relativity: an account of causation that does not require the transmission of energy or mass, such as a counterfactual account; an account in terms of hypothetical physical entities with superluminal velocities (tachyons); or an account that fixes a privileged frame of reference and abandons Lorentz invariance.[26] I will not rehearse these accounts here but will just reiterate that they remain experimentally and theoretically viable; so however controversial, Causal Model I remains a live option.

Hence it is not clear that, on Reichenbach's Principle of the Common Cause, the EPR correlations are a case of improbable coincidence that requires explanation by means of common causes; for as we saw in section 2 correlations between directly causally implicated events are not "improbable coincidences" on Reichenbach's definition. But let us go along with Van Fraassen in supposing so. We must then notice that Van Fraassen's argument requires Reichenbach's principle but also what in section 2 I referred to as Reichenbach's criterion. Van Fraassen assumes that a Reichenbachian analysis of causation requires that correlations be explained by causal models and also that common causes necessarily screen off. But in the recent literature on causal inference the criterion has turned into an enormously controversial assumption, which is tightly related to the controversy regarding the causal Markov condition and its applicability to indeterministic phenomena.[27] I will not rehearse the debate here, but only mention that a straightforward non-screening off common cause model remains viable for the EPR correlations (figure 10).

In a common cause model of this sort the condition (Causality) need not generally hold:

$$prob(s_1|s_2 \wedge a \wedge b \wedge c) \neq prob(s_1|a \wedge b \wedge c)$$
$$prob(s_2|s_1 \wedge a \wedge b \wedge c) \neq prob(s_2|a \wedge b \wedge c)$$

Nor does it need to hold in the particular case where the settings a and

---

[26] For the details see Suárez (2004). These issues are admirably treated in (Maudlin, 1995, chapter 5).
[27] Hausman and Woodward (1999), Cartwright (2000), Hofer-Szabó et al. (1999).

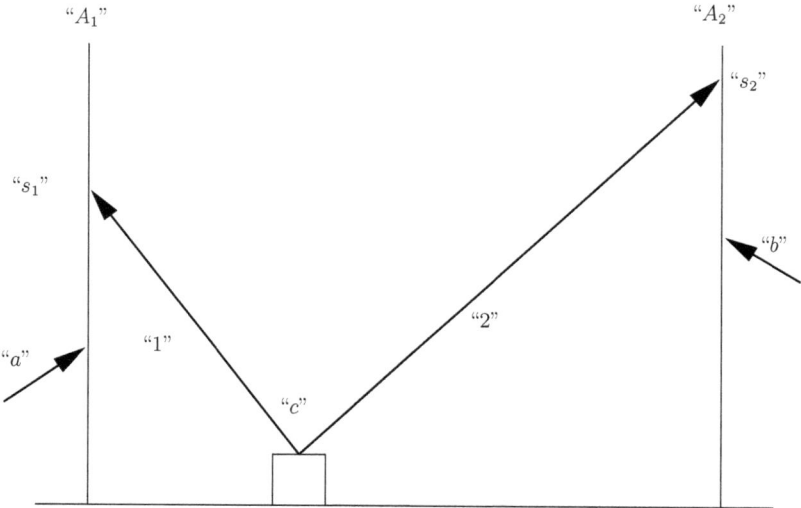

Figure 10. Causal Model II (Common Cause).

b are fixed to the same value and thus rendered irrelevant:

$$prob(s_1|s_2 \wedge c) \neq prob(s_1|c)$$
$$prob(s_2|s_1 \wedge c) \neq prob(s_2|c)$$

Other common cause models are possible.[28] For instance, a model is possible where the common cause is not c, the emission event at the source, but some prior event, d. Indeed the common cause structure underlying the EPR correlations could be quite complex (figure 11).

In this causal structure: d is the partial common cause of c, but might also be a partial cause of a, b, s_1, s_2; c is the partial common cause of s_1, s_2; a is the partial cause of s_1; b is the partial cause of s_2. The continuous lines represent causal influence, while the discontinuous lines represent *possible* causal influence. Thus the figure captures not just one, but a whole family of causal models. Individually taken, these "common causes" can not be expected to screen off their effects, on pain of a violation of factorizability. That is:

$$prob(s_1 \wedge s_2|c) \neq prob(s_1|c)prob(s_2|c)$$
$$prob(s_1 \wedge s_2|d) \neq prob(s_1|d)prob(s_2|d)$$

---

[28] Suárez (2004).

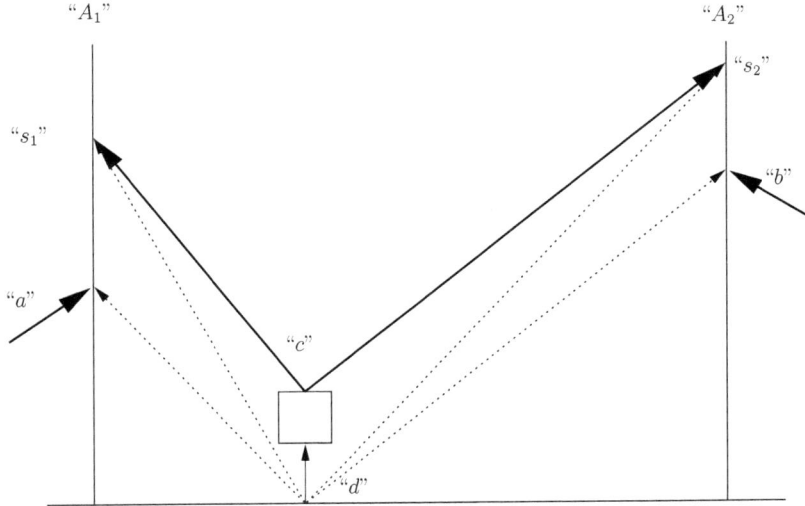

Figure 11. Causal Model III (Complex Common Cause).

However, the conjunction of both c and d might (although it need not) satisfy factorizability:

$$prob(s_1 \wedge s_2 | a \wedge b \wedge c \wedge d) = prob(s_1 | a \wedge c \wedge d) prob(s_2 | b \wedge c \wedge d).$$

In this case the failure of factorizability with respect to the initial state that yields the experimental violation of Bell's inequalities might well be just the result of focusing our attention on a small part of the complex whole common cause structure.

Yet another possibility is a model where the common cause is not a discrete event but a whole part of a spacelike hypersurface with a value of the time parameter prior to the measurement events in both wings—we might for instance consider the hypersurface that includes the last setting event in the laboratory rest frame[29] (figure 12).

Other, more nuanced possibilities emerge once we realise that there is no real threat from relativity theory. For instance, a model becomes possible for the EPR correlations where the failure of factorizability is due to a failure of (Hidden Locality) rather than (Causality) as in (figure 13).

---

[29] Some of these space-time options are described in Butterfield (1989).

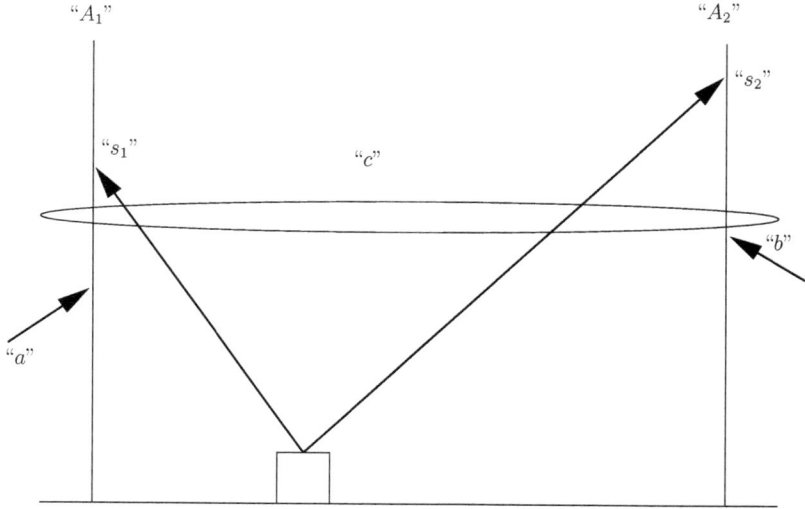

Figure 12. Causal Model IV (Hypersurface Common Cause).

In this model a setting-event in one wing is a partial cause of the outcome-event in the opposite wing, so (Hidden Locality) is false:

$$prob(s_1|a \wedge b \wedge \Psi) \neq prob(s_1|a \wedge \Psi)$$
$$prob(s_2|a \wedge b \wedge \Psi) \neq prob(s_2|b \wedge \Psi)$$

Yet the model is compatible with Reichenbach's criterion for common causes, since (Causality) might well be satisfied, while keeping with the empirical predictions of quantum mechanics:

$$prob(s_1|s_2 \wedge a \wedge b \wedge c) = prob(s_1|a \wedge b \wedge c)$$
$$prob(s_2|s_1 \wedge a \wedge b \wedge c) = prob(s_2|a \wedge b \wedge c)$$

Finally, we must recall the type of model advocated by Huw Price, presented in section 3.3. In this model (regardless of whether we call the backwards influences causal or not) the (Hidden Autonomy) condition is violated, since the initial state of the particles at the source is statistically dependent upon the setting events, i.e.,

$$prob(\Psi|a \wedge b) \neq prob(\Psi)$$

In a thoroughly causal reading of this model (in contrast to Price's own non-causal reading), the setting-event in any of the wings is a partial cause

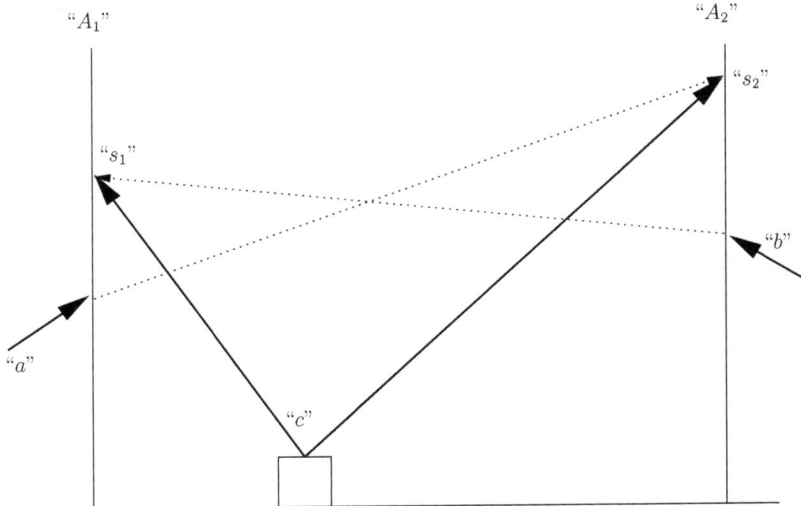

Figure 13. Causal Model V (Hidden Non-Locality).

of the state of the particles as they are emitted at the source—an earlier event in the laboratory frame (14).

The differences between this model and the one presented in figure 8 are twofold. First, this model makes it the case that the influences from the settings to the creation event are causal in nature (therefore represented as continuous lines); so we are indeed assuming a case of genuine backwards causation. Second, we are assuming for the sake of generality that these events' influence upon c, the creation event at the source, is indirect and goes via a previous cause of this state d. This both provides the model with full generality and represents the fact, announced in section 3, that the model need in no way conflict with any intuitions from relativity theory, under any interpretation of the special theory, since all transmission of causal influence can be infraluminal, including the causal influence of the setting events upon the creation event at the source.

5 Replies to the arguments

I am now in a position to state and defend the main claim of this paper: The arguments so far advanced against causal models for the EPR correlations all include unwarranted assumptions and premises, and their scepticism about quantum causation is premature. But before showing this

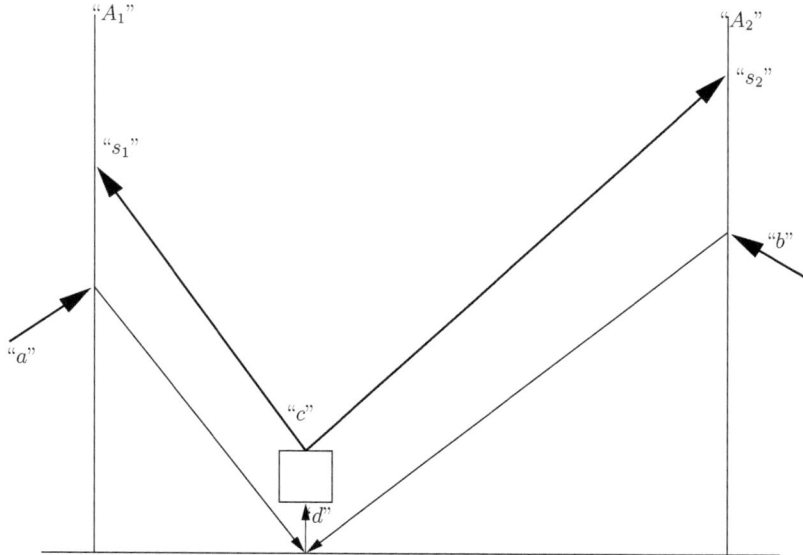

Figure 14. Causal Model VI (Hidden Non-Autonomy).

in detail, let me advance a disclaimer: I am not defending the view that causal concepts are necessary or compulsory in the description of empirical phenomena. I am not even defending the view that *there is* a causal model underlying the quantum correlations; on the issue of causation in quantum mechanics (and elsewhere in physics) the position adopted in this paper will be a thoroughly pragmatist one instead.[30]

The starting point is that Price's pragmatism does not go far enough in relation to causation: for the pragmatist the causal perspective ought to be a pragmatic option, which we might choose to adopt for particular purposes

---

[30] In other words I oppose Nancy Cartwright's causal fundamentalism just as much as I oppose Van Fraassen's causal scepticism. Instead I adopt, here and elsewhere, the particularism and pragmatism of Arthur Fine's NOA. The present paper is yet another NOA-inspired exercise in my career: It presupposes that whether or not causality can be applied to the EPR correlations depends greatly on the details of the theory of causation employed, and on those of the particular model of the EPR correlations adopted. And there is no point attempting to answer the question in the abstract, independently of such details. For Fine's criticisms of Cartwright's causal fundamentalism see Fine (1991). My own criticism of Cartwright's causal fundamentalism is Suárez (2002)—a paper that is regrettably still in press four years after it was written! I also share Price's sceptical comments on Cartwright's causal fundamentalism (Price, 2005, section 10).

of explanation, prediction or coherence. So unlike Price, I will not take the causal perspective to be compulsory in macrophysics, but neither will I see it as incompatible with microphysics. Causal models can be made available for virtually any correlation phenomena. Provided that the completeness assumption discussed in section 2 is rejected (and provided enough imagination!) any causal structure can be suitably expanded to provide some causal model for any correlation phenomena, including the EPR correlations. We are not compelled to accept any of these models (and we might be forced to reject some of them on empirical grounds) but we are at freedom to adopt any of those not ruled out by the physics if we wish for pragmatic purposes.

The models discussed in the previous section are all permitted by the physics, and we are therefore at freedom to adopt any of them on pragmatic grounds. These models are very helpful in displaying the unwarranted presuppositions and premises in the arguments, reviewed in section 3, against causal accounts for the EPR correlations. I will discuss each set of arguments against each of these models in turn.

5.1 Reply to Van Fraassen's Reichenbachian arguments

Causal model I shows that Van Fraassen's relativistic argument against direct-cause models is too harsh. The issues are threefold. First, it is unclear whether the concept of "travelling" is at all appropriate for this type of model - there are accounts of causation that do not require any material object to travel from cause to effect. Second, the existence of superluminal entities (tachyons), albeit controversial, has not been ruled out yet by the physics. Third, it is always possible that the EPR experiment fixes a privileged frame of reference (other interpretations of quantum mechanics do too, notably Bohm's), in which case there is no issue of "infinite speed" travelling and the concept "travel" may apply fully. Since causal model I remains viable under any of these options, Van Fraassen's brief argument against superluminal causation models is inconclusive.

Causal model II shows that Van Fraassen's assumption of Reichenbach's criterion, in addition to Reichenbach's principle, is problematic too. Van Fraassen conflates principle and criterion, but twenty years on these two commitments seem clearly distinct. For there is now a large body of literature on the topic of causal inference that finds the criterion deeply controversial while agreeing fully with the principle.[31] Moreover model II is not

---

[31] And since the criterion is controversial, the strongest defence today of the principle will abandon the criterion altogether. Thus the strongest defence of the principle of common cause takes it that common causes do not generally screen off. For a defence of this point of view see Cartwright (1988, 1989). Although I am adopting the principle for the purposes of this paper, I am aware of possible limitations—such as Sober's well known

ruled out by the physics. First the causal influences in the model are all infraluminal, so no relativistic issues arise. Second, since the common cause does not screen off, (Causality) will not hold in this causal model, and the failure of factorizability that gives rise to the violation of Bell's inequalities can be accommodated in a straightforward and simple way.

Causal model III shows that the failure of (FACT) with respect to the creation event at the source (or more generally, the quantum state of the particle pair) can in principle be accommodated within a larger common cause structure. Moreover this larger structure might even satisfy Reichenbach's criterion, not just his principle. Again the physics does not yet rule this out, although no credible candidates for the deeper common cause structure have been suggested to date.

Finally, models V and VI show that Van Fraassen is too rashly blaming the failure of factorizability upon (Causality) alone. (Hidden Locality) could be the culprit instead, and this would be so in a causal model in which the setting events in one wing are causes, either directly of the measurement outcome events in the distant wing or indirectly via the common cause. The latter case is a particularly interesting case of failure of (Hidden Locality) since it also entails the failure of (Hidden Autonomy) and requires backwards causation. Model V by contrast does not entail backwards causation in the laboratory rest frame, but the relativistic considerations must once again be addressed. But neither case is precluded by the physics itself.

Hence Van Fraassen's argument contains three sets of presuppositions that require far more discussion that has been provided in the philosophy of quantum mechanics literature so far. First, it is in no way automatic that relativity rules out any of these causal models. Second, it seems at least necessary to distinguish between Reichenbach's principle and Reichenbach's criterion and to discuss their implications for each of these models. Third, it seems premature to put the blame for the failure of factorizability solely upon the (Causality) condition. I am suggesting here that there are definite avenues for defending that the models are not ruled out by relativity, that they might conflict with Reichenbach's criterion but not with his principle, and that it might well be the case that (Hidden Non-Locality) and not (Non-Causality) is to be blamed for the failure of factorizability in some of these causal models.

---

argument against the application of the principle to arbitrary monotonically increasing time series. Sober's argument is not relevant to the present discussion because the EPR correlations are not correlations between time-series, and his rejection of the principle in general is consistent with my adoption of it for the purposes of this paper.

5.2 Reply to Hausman's independence argument

Let us now turn to Daniel Hausman's argument, reviewed in section 3.2. Hausman claims that the independence condition fails in general in quantum mechanics, and hence the quantum mechanical correlations are an exception to his theory of causation (C): they are established law-like correlations that can not be given a causal explanation. I believe the argument is mistaken on two grounds. First, I will argue that there are causal models for the EPR correlations that obey Hausman's independence (I) condition.[32] This might look like good news for Hausman, since it points out that some of these models are amenable to his theory; if so quantum mechanics is not the exception to his theory that he has taken it to be. But secondly, I will argue that at least some of the causal models for the EPR correlations are not amenable to this type of treatment. In these models the independence condition fails. However, the models are clearly causal, or at least they have been taken as such by virtually anyone in the literature. Hence, regrettably for Hausman's theory, causal concepts have been applied to situations that Hausman's theory rules out as non-causal. My conclusion will therefore be that Hausman's theory does not capture all conceptual truths about causation. If this is correct then, contrary to what he claims, the independence condition (I) can not a conceptual truth about causation.

I mantain that model II satisfies Hausman's theory of causality, including the independence condition (I). The contrast is clearest with model IV which having a similar structure, nonetheless fails to satisfy Hausman's independence condition (I). In model II the creation event at the source "c" is the common cause of both measurement outcome events s_1 and s_2. We already noticed in section 3.2 that in this scenario the independence condition becomes:

<u>Independence Condition for EPR</u> (I for EPR) : If s_1 causes s_2 or s_1 and s_2 are n-connected only as effects of some common cause, then s_2 has a cause that is distinct from s_1 and not n-connected to s_1.

This condition is satisfied because b is a cause of s_2, distinct from s_1, and not n-connected to s_1. The condition can be conversely applied to s_1 in relation to s_2, with a as the independent cause of s_1. It is important to notice how the setting events a, b are genuine independent causes, that can be used to affect the values of s_1 and s_2. We can choose the setting events as late as we like. Suppose that event s_1 takes place first, then we can choose at a later setting event b the orientation of the magnetometer that we desire,

---

[32] I conjecture that these models can be extended in a straightforward manner to cover the GHZ correlations.

therefore fixing the decomposition of the singlet state for that particle, and the corresponding probabilities. Since the measurement event outcomes are just one of two (plus or minus) we effectively alter the probability of each possible outcome. On any probabilistic theory of causation c is a cause of e if c's occurrence changes the value of the probability of e all other things being equal. Thus each of the causally related events s_1 and s_2 have their own "leverers" or "handles" that can help to fix the probability for their values independently, namely the setting events a, b in their respective wings.

There is a crucial difference between model II and the specific scenario that I described in section 3.2 as an illustration of Hausman's theory. In that scenario all the settings are fixed in advance, so there is no "handle" *ex hypothesis*. But in the scenario described by model II there is a well defined handle on the effect's wing for any setting and outcome event on the distant wing. True, an agent can make use of this "handle" to bring about differences in the probabilities as desired only if he or she has full information regarding the setting and outcome on the distant wing. However, the notion of "handle" presupposed by Hausman is ontological, and unaffected by the lack of knowledge of a particular situated agent, so this fact is irrelevant to the causal nature of model II.

The independence condition fails, by contrast, in model IV, even if this model has a very similar common cause structure. The main difference is that in model IV the common cause is a complete hypersurface of space-time that contains b and (all the effects of) a. So it is impossible in this model to control the value of a and b independently of the common cause c itself. Independence fails here because there is no cause of s_2 that is not n-connected to s_1. So, according to Hausman's theory model IV is not a genuine causal model.

Then there are versions of model I that satisfy Hausman's conditions too, and are thus fully causal under his own theory. For instance, take the version of model I discussed by Van Fraassen (figure 6), where the event s_1 is a direct partial cause of the event s_2. In this model the causal influence takes some time to reach s_2 since this event occurs some time after its cause s_1. Now any setting event b on the mesuring device in the wing of the experiment of particle "2" that takes place after event s_1 will also be an effective leverer or handle for s_2, just as previously discussed. In the absence of any setting events, however (or if the settings have been preordained to be the same well before the experiment is run) the model will too fail to satisfy the independence condition. In the more general kind of model I described in figure 9, the causal relation between the intermediate event x and its partial effect s_2 on that wing will satisfy independence, but not so the causal relation between s_1 and x.

Similarly some versions of model III comply with independence, while others do not. A version of model III with continuous arrows between the deeper common cause d and the measurement outcome events s_1 and s_2, and between d and the setting events a and b will not satisfy independence, as in this model every cause of s_1 is n-connected to every cause of s_2 and viceversa. However, a version of model III in which there are no arrows between d and a and b satisfies independence since there is some cause of s_1, namely a, which is not n-connected to s_2, and conversely there is some cause of s_2, namely b, which is also not n-connected to s_1. The details are crucial here to assess whether or not independence obtains.

On the other hand causal models V and VI unambiguously fail Hausman's test. For instance in model V there is no cause of s_2 that is not n-connected to s_1 *including* the setting event b which in this model is ex-hypothesis a partial direct cause of s_1. Or take model VI in its full causal reading: no cause of s_1 (s_2) fails in this model to be n-connected to s_2 (s_1), even if it is via a backwards in time causal influence upon the common cause c. Regardless of how we interpret the model as long as the influences are taken to be causal, the independence condition will fail, and Hausman's independence theory of causation (C) will be violated.

But notice what a bizarre consequence this seems to be. We have concluded that models IV, V, VI and some admissible versions of models I and III all fail the independence test. Hausman's analysis would unambiguously classify them as failed causal models, not because they conflict with the physics or the experiments, but simply because they fail his independence test. And this is supposed to be as a result of some prior conceptual truths about causation, such as the independence condition. I find it then extremely implausible that these models would have been the subject of any debate regarding their possible incompatibility with special theory of relativity. For notice that the only reason why the threat of incompatibility has bothered physicists and philosophers alike is the causal reading of the special relativity theory, and particularly the light cone structure. If models I, III, IV, V and VI are not properly causal why would anyone have cared about their incompatibility with special relativity in the first place? Yet we intensely care about the incompatibility issue. What this shows, I believe, is that these models are eminently causal, and universally thought to be so among both philosophers and physicists. There seems to be nothing in our concept of causation to prevent us thinking that these models are causal (we might disagree on whether they are *plausible* models, but we do not disagree over the fact that they are causal in nature). Put in another way: if it fell naturally out of our concept of causation that these models for the EPR correlations are not causal, why would any causal sceptic like Van Fraassen (who presumably shares our concept) spend so much time and effort trying to rule them out as impossible or implausible?

To sum up, it seems to me that Hausman was wrong to rule out all models of the EPR correlations as non-causal, since at least three of these models satisfy his conditions. Instead of using his theory to rule out all causal models for the EPR correlations, the independence condition can be more helpfully applied to distinguish between these models. But then the fact that another four of these models are ruled out as non-causal by Hausman's theory is, I believe, just as much of an indictment of his theory—for it shows that our concept of causation, which we seem perfectly prepared to apply to microscopic physics, is neither exhausted by nor in agreement with Hausman's independence condition. These models seem to refute Hausman's claim that independence (the (I) condition) is a conceptual truth about causality.

5.3 Reply to Price's asymmetry argument

For the purposes of analysis it is helpful to explicitly list the premises in Huw Price's argument, reviewed in section 3.3. In my view, Price assumes that

i) causation is agent based notion, roughly: a causes b if and only if an agent could bring about b by producing a.

ii) this is explicitly a modal notion: what counts is what agents would be able to do in the appropriate circumstances, and not what a particular agent is capable of doing.

iii) Since causation is agent-based, it is limited to the macroscopic world that agents operate in.

iv) In particular the agent-based theory is not directly applicable to microphysics.

v) The only sense of causation that would be applicable in microphysics is a projection from our macroscopic based concepts of causation.

vi) Our macroscopic concept of causation is time-oriented, because agency in the macroscopic world always is: we act in order to bring about effects in the future, not in the past.

vii) In other words, the macroscopic concept of causation is tied up to a notion of macroscopic innocence, according to which systems that interact keep a record of the interaction in their future history but not in their past history.

viii) But the equivalent notion of innocence in microphysics (μ-*innocence*) is false, since it conflicts with the fact that the fundamental laws of microphysics are invariant under time reversal, or T-symmetric.

ix) Hence properly speaking there is no causation in microphysics.

x) It is possible to build models of the EPR correlations that abandon μ-*innocence*, such as model VI in section 4.

xi) This model is not properly speaking causal, since it lacks the temporal asymmetry of our ordinary agent-based concept of causation.

xii) But the model uniquely solves the problem of explaining the EPR correlations, without any appeal to non-locality.

xiii) This in turn strengthens the claim (ix) that there is no genuine causation in microphysics.

Let us first consider Price's claim regarding model VI. Price assumes that this is not properly speaking a causal model. The arrows pointing back from the setting events a and b towards the common cause d are not to be thought of as causal influences (and should not be represented as continuous world lines). But it is very hard to see why, from the point of view of an agent-based theory of causation, these influences should not be counted as perfectly causal. After all if an agent was interested in bringing about a particular outcome of the spin measurement on particle "2", the best they could possibly do would be to orient the magnetometers in a particular direction. True, they would have to have some knowledge about the orientation of the distant magnetometer, in order to be able to change the probabilities of the possible outcomes in their own wing. And if they knew both distant setting and distant outcome, then they would actually be able to bring about a plus or a minus outcome, with probability one, as the measurement outcome event in their own wing, simply by orienting the magnetometer appropriately. It is hard to see why this would fail to qualify as a proper intervention to bring about an effect. The fact that the causal influence must first go backwards in time to change the initial state of the particle pair before affecting the final outcome on the distant wing seems, from the strict point of view of an agency theory, neither here nor there. At least in this scenario the fact that some of the causing is backwards seems irrelevant to the possibility of the kind intervention demanded by an agency theory.

What I think this reflection shows is that Price conflates two different causal commitments (or axioms) into one. There is the commitment strictly embraced by an agency theory, which is expressed in our premises i) and ii) above. Then there is the commitment deriving from the time-symmetry of causation in the macroscopic domain, namely that causes always precede effects. In the macroscopic world, given the second law of thermodynamics, these commitments coincide; and the agency theory then builds in an asymmetry in its concept of causation. This is in effect the implication of premises iii) to vii) in Price's argument. But the two commitments come apart in the microscopic domain, where the second law no longer applies. Price extracts the conclusion that agency theories are not applicable there, and therefore causation overall fails (premise ix). But one could extract precisely the opposite implication, namely that agency theories no longer carry a commitment to temporal asymmetry in that domain. In other words, the microscopic domain forces us to distinguish between our two commitments regarding the nature of causation, and then choose which of them we will take as primary. If we take the temporal asymmetry commitment as primary then model VI is not a causal model, by definition, since the influences from the settings to the source are back-in-time. But if we take the agent-based commitment to be primary then model VI is a *causal* model, since agents set on bringing about particular outcome-values in the distant wing can affect the probabilities for those outcomes by means of their settings (as long as the settings in the other wing are known to them in advance).

It is unclear to me which of these two commitments is primary for Price. What is clear is that he has run them together in the microscopic domain in a way that I believe to be both mistaken and unhelpful. If instead we distinguish these commitments we can have a far more detailed and nuanced analysis of the status of agent based causality in the EPR experiment in particular, and in the domain of microphysics in general.

We can now approach all the other causal models in the light of this distinction. We will see that in some of these models we are in effect forced to make a similar choice between agency-based account and a temporal asymmetry account. Models I, V and VI, and some versions of model III are causal in the agency-based sense but not in the temporal asymmetric sense if we apply relativity theory and consider time order in *all* frames of reference. Model IV is causal in the time-asymmetry sense but not in the agent-based sense (since clearly no agent can control a whole hypersurface of spacetime).

Model II is a curiously problematic model for Price's argument. This model is causal in both the agency-based sense and the temporal asymmetric sense. The cause c precedes the effects s_1 and s_2, and the state can be altered so as to bring about different statistics (by e.g., replacing the singlet with a different kind of entangled state). Indeed in this model the conflation between the agency-based commitment and the time-asymmetric commitment that we saw was operating in the macroscopic domain is fully restored even though it is a model of phenomena in the microscopic domain. A perfectly legitimate causal model of microphysical correlations that obeys both of Price's causal axioms! Model II can hardly be said to be less explanatory than Price's own model, among other things because it does away with relativistic non-locality altogether, like Price's model. So it is difficult to see why anyone who is commited to both axioms regarding causality would opt for model VI at the expense of model II. The problem for Price is that he is one of those people apparently commited to both axioms, which suggests that on the grounds of his own preferred theory of causation he should abandon the model that he has been defending for the EPR correlations for the last twenty years, and adopt instead the simpler model II.[33]

In situations where the settings are predetermined, however, model II is not causal in the agency-based sense, since there is no handle for agents to operate to bring about effects independently. So the model becomes causal only in the temporal asymmetric sense. If Price preferred to choose model VI instead of model II he would be showing that his real commitment lies with the agency-based axiom, not the temporal asymmetric axiom. In any case, the choice between all these different models, in their different interpretations, opens up a route to test the real commitments underlying Price's theory of causation. My guess is that he would try to stick to both for as long as possible but if forced to choose might opt for temporal asymmetry. In either case, once the choice has been made between these two commitments, plenty of possibilities are still available to explain causally the EPR correlations: the causal perspective has been shown to be just as good for microphysics (and microphysicsts!) as it is for ordinary cognition and practice.

6 Conclusions and prospects

I hope to have shown that all arguments so far to the effect that causation is ruled out in microphysics on account of the EPR correlations are unwarranted and premature. Much more philosophical work still needs to be done

---

[33]But of course the price to pay would be Price's argument for causal perspectivalism which requires us to accept that causation is not part of the ultimate furniture of the universe as described by fundamental physics. A thorough pragmatist should have no fear to pay that price, I contend, since a causal description of phenomena is a pragmatic option, available to physicists as well as anyone else.

in order to establish the ways in which the different theories of causation can be applied to explain quantum correlation phenomena. It seems also clear that the question "are the EPR correlations causal?" in general has no informative answer. To answer this question we have to engage with the details both of the different theories of causation and the different possible models for the EPR correlations. Different combinations of causal theories and empirical models will yield different answers to this question.

In his brilliant 1998 book on quantum probability and non-locality, Michael Dickson showed how typical metaphysical questions regarding the notion of locality in quantum mechanics (such as "is the nature of the quantum world non-local?") have no general or universal answers. These questions can only be informatively answered by considering how the different notions of locality fare with respect to each and every interpretation of quantum mechanics. And we should expect the answers to be different and even contrary in different cases.[34] The kind of non-locality built into Bohm's theory is not the same as is built into Ghirardi-Rimini-Weber theories, or the modal interpretation. The same lesson I think applies to the nature of causation in quantum mechanics. We are unlikely to learn much from attempts to completely rule causation out, or completely rule it in. We will probably learn much more if we proceed in an unprejudiced and cautious way to a detailed and piecemeal study of all the different concrete possibilities instead.

Acknowledgments

I want to thank the audience at my talk to the Causality and Probability in the Sciences conference at the University of Kent, as well as two anonymous referees for helpful questions and comments. Thanks also to Iñaki San Pedro for his feedback, and his help with the production of the paper. Research towards this paper has been funded by project HUM2005-07187-C03-01 of the Spanish Ministry of Education and Science.

Mauricio Suárez
Universidad Complutense de Madrid, Spain.
msuarez@filos.ucm.es

BIBLIOGRAPHY

Bohm, D. (1951). *Quantum Theory*. Prentice Hall.
Bohm, D. (1952). A suggested interpretation of the quantum theory in terms of hidden variables, i and ii. *Physical Review*, 85:369–396.

---

[34]Dickson (1998). In my review of this book (Suárez, 2000) I embraced the position, which I referred to as a deflationism, and argued that Dickson actually did not go far enough in its defence!

Bovens, L. and Hartmann, S., editors (2006). *Nancy Cartwright's Philosophy of Science*. Routledge.

Butterfield, J. (1989). A space-time approach to the Bell inequality. In Cushing, J. and McMullin, E., editors, *Philosophical Consequences of Quantum Theory*, pages 114–144. University of Notre Dame Press.

Cartwright, N. (1988). How to tell a common cause: Generalizations of the conjunctive fork criterion. In Fetzer, J. H., editor, *Probability and Causality*, pages 181–188. Reidel Pub. Co.

Cartwright, N. (1989). *Nature's Capacities and their Measurement*. Oxford University Press.

Cartwright, N. (2000). Against modularity, the causal Markov condition, and any link between the two: Comments on Hausman and Woodward. *The British Journal for the Philosophy of Science*, 53:411–453.

Cushing, J. and McMullin, E., editors (1989). *The Philosophical Consequences of Quantum Theory*. Notre Dame University Press.

de Beauregard, O. C. (1977). Time symmetry and the Einstein paradox. *Il Nuovo Cimento*, 42B:41–64.

Dickson, M. (1998). *Quantum Non-Locality and Probability*. Cambridge University Press.

Dummett, M. (1954). Can an effect precede its cause? *Proceedings of the Aristotelian Society, Supp. Volume*, 38:27–44.

Einstein, A., Podolsky, B., and Rosen, N. (1935). Can quantum-mechanical description of physical reality be considered complete? *Physical Review*, 47:777–780. Reprinted in Wheeler and Zurek (1985), pp. 138-141.

Fine, A. (1982). Hidden variables, joint probability and the Bell inequalities. *Physics Review Letters*, 48:291–295.

Fine, A. (1987). *The Shaky Game: Einstein Realism and the Quantum Theory*. Chicago University Press.

Fine, A. (1989). Do correlations need to be explained. In Cushing, J. and McMullin, E., editors, *Philosophical Consequences of Quantum Theory*, pages 175–194. University of Notre Dame Press.

Fine, A. (1991). Piecemeal realism. *Philosophical Studies*, 61:79–96.

Hausman, D. (1998). *Causal Asymmetries*. Cambridge University Press.

Hausman, D. (1999). Lessons from quantum mechanics. *Synthese*, 121:79–92.

Hausman, D. M. and Woodward, J. (1999). Independence, invariance and the causal Markov condition. *The British Journal for the Philosophy of Science*, 50:521–583.

Hofer-Szabó, G., Rédei, M., and Szabó, L. (1999). On Reichenbach's common cause principle and Reichenbach's notion of common cause. *The British Journal for the Philosophy of Science*, 50:377–99.

Hughes, R. I. G. (1989). *The Structure and Interpretation of Quantum Mechanics*. Harvard University Press.

Jarrett, J. (1984). On the physical significance of the locality conditions in the Bell arguments. *Nous*, 18:569–589.

Maudlin, T. (1995). *Quantum Non-Locality and Relativity*. Oxford Blackwells.

Price, H. (1996). *Time's Arrow and Archimedes' Point*. Oxford University Press.

Price, H. (2005). Causal perspectivalism. In Price, H. and Corry, R., editors, *Causation, Physics and the Constitution of Reality: Russell's Republic Revisited*. Oxford University Press.

Price, H. and Corry, R., editors (2005). *Causation, Physics and the Constitution of Reality: Russell's Republic Revisited*. Oxford University Press.

Reichenbach, H. (1956). *The Direction of Time*. University of California Press.

Salmon, W. (1984). *Scientific Explanation and the Causal Structure of the World*. Princeton University Press.

Schrödinger, E. (1933). The present situation in quantum mechanics: A translation of Schrödinger's 'cat paradox' paper. *Proceedings of the American Philosophical Society*, 124:323–38. Reprinted in Wheeler and Zurek (1985).

Suárez, M. (2000). The many faces of non-locality: Dickson on the quantum correlations. *British Journal for the Philosophy of Science*, 51:882–892.

Suárez, M. (2002). Experimental realism defended: How inference to the most likely cause might be sound. Delivered at the Konstanz conference in honour of Nancy Cartwright, December 2002, and forthcoming in Bovens and Hartmann (eds).

Suárez, M. (2004). Causal processes and propensities in quantum mechanics. *Theoria*, 19:271–300.

van Fraassen, B. C. (1982). The charybdis of realism: Epistemological implications of Bell's inequality. *Synthese*, pages 25–38. Reprinted with corrections in J. Cushing and E. McMullin, eds. (1989).

Wheeler, J. and Zurek, W., editors (1985). *Quantum Theory and Measurement*. Princeton University Press.

PART III

CAUSALITY AND PROBABILITY IN THE SOCIAL SCIENCES

Mediating between causes and probabilities: the use of graphical models in econometrics

ALESSIO MONETA

ABSTRACT. The development of macro-econometrics has been persistently fraught with a tension between "deductivist" and "inductivist" approaches to causal inference. The former conceives causes as something that economic theory must provide and that statistical methods must measure. The latter opens the possibility of inferring causes from statistical properties of the data alone. I argue that these conceptions can be interpreted as two opposite responses to the problem of under-determination of theoretical causal relations by statistical properties (the problem of identification). Econometrics offers a clear example as to how the general problem of causal inference can be solved only by delicately mediating between background knowledge and the statistical properties of the data. I show how graphical causal models, appropriately interpreted, can serve this purpose.

1 Introduction

Econometrics represents a privileged locus for studying both the problem of causal inference from observational data and for verifying to what extent philosophical theories about causation apply to special sciences. Indeed, many well-studied problems in the philosophy of science—such as that the problem of underdetermination of causal theoretical models by data, of identifying causal relationships that are invariant under intervention, of differentiating between causation and correlation—have all been rigorously addressed by econometricians. The aim of this paper is to show that the debate about causal inference in econometrics contains some useful lessons for the philosophy of science.[1]

---

[1] The focus of this paper is on macro-econometrics, which originally coincided with econometrics itself, but which should now be distinguished by the parallel discipline of micro-econometrics, that is econometrics applied to microeconomic data. This is also a discipline in which the problem of causal inference is crucial, but the nature of data is quite different between the two sub-disciplines. While micro-econometrics deals with cross-

The development of methods for causal inference in macro-econometrics has been fraught with a tension between what I call a "deductivist" approach and an "inductivist" approach. The first conceives of causes as something that economic theory must provide and that statistical methods must measure. The second considers economic theory a not very reliable source of causal knowledge and opens the possibility of inferring causes from statistical properties of the data "without pretending to have too much a priori theory" (Sargent and Sims, 1977). The first conception was advocated by some exponents of the Cowles Commission during 1950s and is fashionable among the calibration approach to econometrics. The second conception was formalised by Granger's (1969) test of causality and by Sims (1980) vector autoregressive models, methods which are still very popular in econometrics.

These conceptions can be interpreted as two opposite solutions to the same problem of underdetermination of theoretical causal relations. In econometrics this is called the problem of identification. This first approach risks the commitment to an apriorist strategy, while the second approach is impeded by the well-known difficulties of the probabilistic theories of causality. I argue that econometrics offers a clear example of how only a delicate mediation between background knowledge and the statistical properties of data can solve the general problem of causal inference. The methods for this careful handling are much dependent upon the discipline considered.

With respect to macro-econometrics, graphical models, that is the methods for causal inference developed by Pearl (2000) and Spirtes et al. (2000), can be very useful in mediating between probabilistic and causal knowledge. Indeed, graphical models permit us to take into account the maximum amount of probabilistic information (partial correlations of all possible orders), which can be used to exclude false causal relations. Partial correlations, however, are never sufficient to isolate the unique true causal relations, except in very exceptional circumstances. Indeed, background knowledge always has to be incorporated and this approach permits the use of background causal knowledge in a very efficient way.

In the next section I consider the tension associated with the problem of causal inference in macro-econometrics; in the third section I discuss how the use of graphical models can mediate such tension; in the fourth section I present an empirical example that shows how graphical models can perform that task; the fifth section concludes.

---

section or panel data, macro-econometrics prevalently deals with time series variables in which experiments (or quasi-experiments) are not feasible.

2   Causal inference in macro-econometrics

The *Econometric Society* was founded in 1933 with the aim of unifying two approaches to economic problems that divided (and perhaps still divide) economists into those devoted to develop formalised theory without measurement and those devoted to develop measurement without theory. In fact, although Frisch (1933, p. 2) advocated a "mutual penetration of quantitative economic theory and statistical observation," it is possible to identify a similar tension inside econometrics itself. This is the tension between econometrics as an instrument of empirical application of theory and econometrics as an instrument of discovery of theoretical economic relationships. It is also reflected on the debate on causal inference in macro-econometrics.

The basic ingredients of any econometric study are data and models. The role of an econometric model, which is usually an algebraic model, is to abstract particular features of the world by means of a system of equations (Intriligator, 1983). An actual process or phenomenon is represented by the model for the sake of forecasting, explanation (understanding), and intervention. For each of these purposes econometricians have implicitly considered and sometimes made explicit a notion of causation. For example, if macroeconomists use the model to advise policymakers, they are looking for causal relations invariant under intervention. If they are just seeking macroeconomic forecasts, perhaps a weaker notion of causation is sufficient. There are, in other words, different ontological conceptions of causation involved in macro-econometrics, but I am not facing this issue here (the reader is referred to Moneta 2005). The focus here is on the different epistemological strategies for causal inference.

The typical macroeconometric model consists of a system of equations involving a number of *endogenous* variables (whose values depend upon the values of the other variables in the model), *exogenous* variables (whose values are determined outside the system but which influence it by affecting the values of the endogenous variables), and random shocks (which account for the omission of relevant variables, specification and measurement errors, etc.). The idea is to use the data to estimate (or *fit*) the model. The typical linear macroeconometric model takes the following form:[2]

(1) $A_0 Y_t + A_1 Y_{t-1} + \ldots + A_m Y_{t-m} + B_0 X_t + B_1 X_{t-1} + \ldots + B_n X_{t-n} = \epsilon_t,$

where Y_t is a $(l \times 1)$ vector of endogenous variables, X_t is a $(k \times 1)$ vector of exogenous variables, and ϵ_t is a vector of stochastic disturbances. The matrices A_i's are each $(l \times l)$; the B_j's are $(l \times k)$. The vector ϵ_t is a white

---

[2] See, for example, Intriligator (1983, pp. 187-195). A more complicated model would be one with shocks entering in the equation with lags, or a non-linear model. But this would not change the substance of the present discussion.

noise, which means that is serially uncorrelated with a mean of zero and variance-covariance matrix Σ_ϵ. Moreover, by definition of exogeneity, X_t is uncorrelated with ϵ_s for every t and s.

The model (1) is a system of l equations (equal to the number of endogenous variables) in which the relationships are interpreted as causal and invariant under intervention for the sake of policy evaluation. It can easily be normalised so that one is able to write each equation which specifies one endogenous variable as a function of other endogenous variables, exogenous variables, and a stochastic disturbance term, with a unique such endogenous variable for each equation:

$$y_{1t} = f_1(y_{2t}, \ldots, y_{lt}, y_{1(t-1)}, \ldots, y_{l(t-m)}, x_{1t}, \ldots, x_{l(t-m)}, \epsilon_{1t})$$
$$y_{2t} = f_2(y_{1t}, y_{3t}, \ldots, y_{lt}, y_{1(t-1)}, \ldots, y_{l(t-m)}, x_{1t}, \ldots, x_{l(t-m)}, \epsilon_{2t})$$
$$\ldots$$
$$y_{lt} = f_l(y_{1t}, \ldots, y_{(l-1)t}, y_{1(t-1)}, \ldots, y_{l(t-m)}, x_{1t}, \ldots, x_{l(t-m)}, \epsilon_{lt})$$

This normalised system of equations, however, cannot be estimated via ordinary least squares regression, because the ϵ's are in general correlated with some of the endogenous variables entering in the equation, since some right hand side variables can be caused by some left hand side variables.

In general, the structural model (1) can be solved for Y_t in terms of lagged Y's, X's, and current ϵ's. Multiplying (1) by A_0^{-1} and solving for Y_t yields

(2) $Y_t = -A_0^{-1}A_1 Y_{t-1} - \ldots - A_0^{-1}A_m Y_{t-m} - A_0^{-1}B_0 X_t - \ldots - A_0^{-1}B_n X_{t-n} + A_0^{-1}\epsilon_t.$

Introducing the matrices $P_i = -A_0^{-1}A_i$ and $Q_i = -A_0^{-1}B_i$, and the vector of disturbances $u_t = A_0^{-1}\epsilon_t$, equation (2) can be re-written

(3) $Y_t = P_1 Y_{t-1} + \ldots + P_m Y_{t-m} + Q_0 X_t + \ldots + Q_n X_t + u_t.$

Equation (3) is called the *reduced form* and can be estimated consistently using least squares regression, because it is clear that the left hand side variables cannot cause anyone of the right side variables, which are either exogenous or lagged variables (and it is assumed that the future cannot cause the past).

In general, it is not possible to deduce the estimates of the structural parameters A's and B's from the estimates of P's and Q's, because there are infinitely many matrices like A's and B's which are compatible with a single set of P's and Q's. This is what econometricians call the *problem of identification*. It corresponds to what philosophers of science call "the problem of under-determination of theory by data": any theory that makes reference to unobservable features of the world will always encounter rival theories incompatible with the original theory but equally compatible with the currently available data. This problem is particularly relevant in econometrics for two reasons. First, theoretical relations in economics are always approximate and "the error in approximation constitutes an auxiliary hypothesis of typically unknown dimension" (Sawyer et al., 1997, p. 21). Second, and crucially connected with the topic of this chapter, econometricians try to confirm causal relations using statistical properties (like correlations). This raises the problem of differentiating between an asymmetric relation like causation and a symmetric relation like correlation. Econometricians have reflected on this problem for a long time and indeed "[a]n important contribution of econometric thought was the formalization of the notion developed in philosophy that many different causal interpretations may be consistent with the same data" (Heckman, 2000, p. 47).[3]

2.1 Deductivist Approaches

Haavelmo (1944) presents some algebraic conditions that a system of equations like (1) must satisfy to be identifiable. These conditions refer to the number of endogenous variables relative to the number of exogenous variables ("order condition for identification") and to the rank of the reduced form matrix P's and Q's ("rank condition for identification"). I am not going to go into details of these conditions: it is important here to highlight the fact that structural parameters (coefficients of equation 1) are identified by the imposition of several types of a priori restrictions on the A's and B's. In the so-called Cowles Commission approach to econometrics, of which Haavelmo was one of the founders, these restrictions consist in a priori setting of many of the elements of A's and B's (in equation 1) to zero, and in a priori classifying of variables as exogenous and endogenous (considering the fact that a relatively high number of exogenous variables aids identification). It is important to notice that these restrictions correspond to causal auxiliary hypotheses. Indeed, setting a priori some structural coefficients to zero in equation (1) corresponds to a priori assuming that a particular variable is not causally influencing a particular endogenous variable. Moreover, a priori assuming that a particular variable is exogenous corresponds to a priori assuming that that variable is not causally influenced by any other variable in the system.

The solution pursued by the Haavelmo-Cowles program was that these causal restrictions had to be derived from economic theory. The theory in consideration was Keynes's macroeconomics, but filtered by the neoclassi-

---

[3]However, the econometric literature on structural analysis and the problem of identification was in part anticipated by the work of the geneticist Sewall Wright on path analysis in the 1920s.

cal synthesis, which introduced the Walrasian notion of general equilibrium. However, the object of the Haavelmo-Cowles program was more general: it was not explicitly specified which theory one had to use in order to get restrictions. The crucial issue was that restrictions had to be derived from economic theory. Once the model was identified, it could be estimated using sound statistical methods and tested against the empirical evidence. But both the problem of confirming theoretical causal models and of choosing between competing models are not central issues in the Cowles Commission methodology. Statistical techniques such as regression analysis are mainly designed to estimate the importance of each causal factor that is dictated by economic theory, and only to a lesser degree to perform empirical validation. However, as Hoover (1994 and 2006) points out, the Cowles Commission methodology is subject to alternative interpretations. Koopmans in his debate with Vining about the possibility (denied by Koopmans) of "measurement without theory" demonstrated what Hoover (1994) calls a strong apriorist view. This corresponds to considering theory prior to data and to denying the possibility of interpreting data without theoretical presuppositions. According to this view, econometric models have to be built imposing restrictions derived from a well-articulated theory accepted a priori. Thus, the object of econometrics would be one of measurement of causal relationships and not of validation or discovery of causal hypotheses. This view would correspond to a very strong interpretation of the under-determination thesis denying the possibility of any induction from correlations to causation. But the problem with Koopmans' position is, as argued by Hoover (2006, p. 74), that it "places the empiricist in a vicious circle: how do we obtain empirically justified theory if empirical observation can only take place on the supposition of a true background theory?"

The position of Haavelmo, however, was quite different from Koopmans's one. Although also Haavelmo maintained that empirical investigations were to be founded on a priori theoretical restrictions, he favored statistical testing of causal hypotheses. Thus he endorsed a view of econometrics, called by Hoover (1994) weak apriorism, which recognizes the need for an interplay of theoretical models with empirical results. This permits one to partially avoid the danger, implicit in the strong apriorist view, of being committed to a set of a priori causal assumptions without having the possibility of empirically confirm them.

Lucas's (1976) article, "Econometric Policy Evaluation: A Critique," is a crucial step in the development of causal inference in econometrics. In fact, the Lucas critique was an attack more directed to the economic theory commonly used to derive the a priory restrictions necessary for the identification of the model, than the general Haavelmo-Cowles methodology for causal inference. However, the research program pursued by Lucas with his critique, which shaped the basis of the "New Classical Macroeconomics," yielded new and alternative econometric methodologies for causal inference. The point raised by the Lucas critique was, in few words, the following: large-scale econometric models based on the Cowles Commission methodology and using restrictions derived by Keynesian macroeconomic theory (filtered by the neoclassical synthesis) could not be used for policy evaluation. This is because the estimated coefficients of such models were unlikely to remain invariant to the policy interventions that are object of evaluation. In other words, the causal relationships identified by the Haavelmo-Cowles methodology were not invariant under intervention, according to Lucas. They were not invariant, or stable, because the standard macroeconometric models inspired by the Haavelmo-Cowles approach did not take into account the fact that people have forward-looking behaviour (rational expectations) which prompts them to change behaviour as soon as the intervention takes place, in order to take advantage of the new policy regime associated with the intervention.

Thus, the object of Lucas's attack was not the general deductivist approach in which causal relations are identified in the Cowles Commission methodology, but the lack of foundation of the a priori theoretical restrictions used to identify the models. Moving from this criticism, Lucas focused on micro-founded theoretical assumptions that were able, in his view, to dictate structural (causal) relations invariant to changes in policy. The first assumption was the rational expectations hypothesis mentioned above: individual agents have forward-looking and perfectly rational behaviour, which permits them to take the maximum advantage of the available information, without making any systematic error. The second principle was that, in line with the Walrasian tradition, markets continuously clear, so that all observed output are the results of a continuous state of (short and long-run) equilibrium. Moreover, theoretical models do not need to formalize the behaviour of every agent, but, thanks to the homogeneity of individual rationality, just the behaviour of typically one representative agent, which stands in for the behaviour of all agents. In other words, the problem of aggregation of the causal relations among microeconomic agents into causal relations among macroeconomic aggregates is simply bypassed (Moneta, 2005).

A first response to the Lucas critique was completely consistent with the Haavelmo-Cowles methodology. The idea was to supplement economic theory with the rational expectation hypothesis, from which it could be possible to derive cross-equations restrictions on the matrices A's and B's in equation (1), in order to identify the structural model (Hansen and Sargent, 1980).

Thus causal relations are inferred, once again, in a general methodological approach in which theory is prior to data. Although testing of a theoretical causal hypothesis is still pursued, the theoretical assumptions used to restrict the estimatable equations are not questioned. Therefore, this approach shares with all the forms of apriorism the problem of obstructing an empirically disciplined knowledge of causal relations.

Even more apriorist and deductivist approaches to causal inference, however, have been developed in the wake of the Lucas critique. I am referring to the calibration approach that has been developed as the method of empirical assessment of equilibrium real business cycle models (see Kydland and Prescott, 1982). But its roots are in the method proposed by Lucas (1980): "[o]ne of the functions of theoretical economics is to provide fully articulated, artificial economic systems that can serve as laboratories in which policies that would be prohibitively expensive to experiment with in actual economies can be tested out at much lower cost. ... Any model that is well articulated to give clear answers to the questions we put to it will necessarily be artificial, abstract, patently 'unreal' " (Lucas, 1980, p. 271).

A theoretical model, which can be thought as representing a set of causal relations invariant under interventions, need not fit the data according criteria dictated by statistical theory, according to the calibration approach. Indeed, it would be easily rejected, since it is built upon very idealised assumptions that do not take into account all the contingencies, which are not related with the deep structure, whose knowledge is essential to answer a limited set of policy questions. Then, such disturbing factors, unaccounted in the model, but present in the reality, would deform parameter estimates. Thus, the model has to be calibrated, instead. A model is calibrated when its parameters are not estimated in the context of their own model, but are picked in micro-econometric unrelated empirical investigations, or are chosen to guarantee that the simulated model matches some particular and unrelated features of the historical data, drawn from considerations of national accounting, etc. Once calibrated, the model is validated via simulation. The model is validated if it matches moments of the data or reproduces some stylised facts obtained by independent empirical analysis of the data. In fact, this approach seems to appeal to the sound principle that a theory is better supported when validated on information not used in the formulation (Hoover, 1995a). But the acceptance of this principle is not clear, at least in two respects. First, the collection of stylised facts through statistical analysis of data is only partially an independent exercise. Indeed the so-called stylised facts express more or less implicitly causal relations (saying, for example, that a monetary shock is neutral, which means that it does not have any causal impact on income, in the long-run), which also need some a priori assumptions to be identified. Second, equilibrium business cycle models, for which the calibration approach has been developed, are based on the simplification of the representative agent. Thus, when the models are calibrated using parameters derived from microeconomic investigations, it is tacitly assumed that aggregation does not fundamentally alter the structure of the aggregate model. Such assumption is hardly defensible, as Forni and Lippi (1997) and Kirman (1992), among others, have shown.

Thus, the main characteristic of the calibration methodology is a strong commitment to economic theory (with the typical new classical features: general equilibrium, rational expectations, perfect aggregation), taken for granted a priori. This is a form of apriorism even stronger than Koopman's one, because it rules out likelihood-based statistical estimates of model parameters, which are standard in any version of the Cowles Commission methodology. This raises the question as to how to judge between competing calibrated models. And, more important: is there any possibility of growth of knowledge at all, if the hard core of the new classical theory, thanks to the protection of the calibration methodology, is immune from revision? (Hoover, 1995a).

To conclude, all the approaches just presented share several features in common with the hypotetico-deductive method for causal confirmation. Theoretical statements about causal relations together with the data (which can be thought as initial conditions) imply the event to be explained. Haavelmo's methodology is quite well in tune with Popper's falsificationism: a theoretical causal statement is hypothesised, from which consequences are deduced, and if these consequences do not fit the data, the theoretical causal statement is re-formulated. In fact, the Haavelmo's apparent falsificationism is beset with the problem of under-determination of theory by data, but Haavelmo, as mentioned above, recognizes the fundamental importance of the empirical testing of causal hypotheses. The other approaches, Koopmans's one and calibration in particular, represent a deductivist approach without falsificationism. The possibility of rejecting theoretical causal statements are reduced to the minimum.[4]

In general, all these approaches share the difficulties of the hypothetico-deductive approaches to causal discovery (Williamson, 2005), which amount to failing to explain how causal relationships are to be hypothesised (to what extent is economic theory a reliable source of causal hypotheses?), and to failing to explain how predictions can be reliably deduced from the causal

---

[4]The strong apriorist approach corresponds very closely to a scientific research program, as defined by Lakatos, in which a large set of assumptions, constituting the *hard core*, is never confronted with the data.

statements in spite of the under-determination (identification) problem.

2.2 Inductivist Approaches

Sims' (1980) article, "Macroeconomics and Reality," pursued the criticism of traditional macroeconometric models in another direction, with respect to Lucas (1976). Sims claimed that econometricians inspired by the Cowles Commission methodology "imposed large numbers of restrictions that were incredible in the sense that they did not arise from sound economic theory or institutional or factual knowledge, but simply from the need of the econometrician to have enough restrictions to secure identification" (Hoover, 1995b, p. 6). But the reaction is different to the rational-expectations econometrics approach. While Hansen and Sargent (1980), as mentioned in the last section, continued to pursue identification of structural models, by using restrictions grounded in individual decision-making, Sims argued that economic relations are in principle not identifiable. "Sims proposed that macroeconometrics give up the impossible task of seeking identification of structural models and instead ask only what could be learned from macroeconomic data without imposing restrictions" (Hoover, 1995b, p. 6). The approach proposed by Sims deals with unrestricted reduced form equations, namely vector autoregressive models (VARs). Each variable is considered as endogenous and it is regressed on lagged values of itself and of all the other variables. This corresponds to the reduced form considered in equation (3), devoid of the exogenous variables X's:

(4) $\quad Y_t = P_1 Y_{t-1} + \ldots + P_m Y_{t-m} + u_t$

Once the model (4) is estimated, it is possible to study the dynamic causal effect of a single shock on each variable of Y_t. However, it is not possible to isolate the effect of a single shock u_{jt}, since u_{jt} is in general correlated with the other components of u_t. Sims (1980) proposed to orthogonalize the residuals u_t by multiplying both sides of equation (4) by a particular matrix Γ, obtained by the Choleski factorization of the covariance matrix of the residuals u_t. Indeed this is one of the most simple ways to transform equation (4) into another equation in which the shocks are orthogonal, like the following:

(5) $\quad A_0 Y_t = A_1 Y_{t-1} + \ldots + A_m Y_{t-m} + \epsilon_t$

But there are many ways of obtaining equation (5), and the one with A_0 is just a particular case. In other words, the problem of identification reappears. Indeed, the transformation of equation (4) in another equation in which residuals are orthogonal—residuals orthogonalization, in short—is equivalent to imposing a contemporaneous causal structure on the variables (Stock and Watson, 2001). The method of orthogonalization proposed by Sims (1980), mentioned above, corresponds to impose on the system a strictly recursive causal structure among the contemporaneous variables.[5]

Sims' method is atheoretical and inductivist: the idea is to impose the most common and most simple causal structure in order to obtain identification, and to learn causal relationships directly from data. The causal relationships that are objects of interest in this method are the relationships between exogenous shocks and the components of Y_t at any lead and lag, and not the relationships among the components of Y_t as in the Haavelmo-Cowles framework. But Sims (1980) solution to the VAR identification problem is highly arbitrary because he picks up a very special causal structure (the recursive causal ordering) among a very big number ($l!$) of possible causal structures.

The so-called structural VAR literature recognizes this arbitrariness and focuses its efforts on the imposition of restrictions on the contemporaneous causal structure, derived, entirely consistently with the Cowles Commission methodology, from economic theory or institutional knowledge. However, it is not clear to what extent the restrictions suggested by economic theory are reliable. Thus, the structural VAR approach recovers some issues of the deductivist methodology, included some of its problems. In the next section I will show how such problems can be faced using graphical causal models. In general, the VAR approach is atheoretical, in the sense of letting the data speak as much as possible, and so at odds with an apriorist methodology. But the problems of the other approach are replicated here in a new form: since measurement without theory (because of the underdetermination problem mentioned above) is very a difficult task, strong a priori assumptions turn out to be hidden behind implicit (but often arbitrary) assumptions.

In this general framework Granger's conception of causality has flourished. Granger (1969, 1980) defined causal relationships in the following way: a time series variable x_t causes *prima facie* another time series variable y_t if the probability of y_t conditioned on its own past values and the past values of x_t (besides the set Ω of the relevant information) does not equal the probability of y_t conditional on its own past history alone (and Ω). More formally, x_t Granger-causes y_t if and only if:

(6) $P(y_t|y_{t-1}, y_{t-2}, \ldots, x_{t-1}, x_{t-2}, \ldots, \Omega) \neq P(y_t|y_{t-1}, y_{t-2}, \ldots, \Omega)$.

The intuition behind this definition is that x_t renders y_t more likely, or, in a more epistemological sense, x_t contains some special information which

---

[5]With strictly recursive causal structure, I mean a causal chain among the components of Y_t, according to which the only causal connections are: y_{1t} causes y_{2t}, y_{2t} causes y_{3t}, ..., y_{l-1} causes y_l.

helps predict y_t. Indeed, another way of reading (6) is that x_t Granger-causes y_t if the knowledge of the past and present values of x_t contributes to forecasting y_t. Based on this idea and definition, Granger was able to devise very simple tests of this conception of causality. Indeed, the "incremental predictability" of a variable is easily measured as a reduction of the variance of the prediction error. In the VAR framework it is straightforward to test the absence of Granger causality. In order to test the Granger *non*-causality from y_{it} to y_{jt}, it is sufficient to test that the (j, i) entries of the matrices $P_1, \ldots, P_m$ in equation (4) are significantly close to zero.

In fact, Granger causality was devised before the formulation of the VAR approach. Moreover, Hansen and Sargent (1980) claimed that Granger causality played a "natural role" in rational expectations models. Nevertheless, the methodological approach behind Granger causality is extremely inductivist and is well in tune with the VAR framework.

The closeness of Granger causality with probabilistic theories of causality developed in the philosophy of science is evident. In particular, Spohn (1984) highlights the closeness with Suppes (1970) account. Indeed, Granger causality shares with any other probabilistic account of causality all its difficulties, well studied in the philosophy of science. First, merely probabilistic accounts are not able to identify causation as an asymmetric relation. This is because if A renders B more likely ($P(B|A) > P(B)$), the probability calculus implies that also B renders A more likely ($P(A|B) > P(A)$). Granger (like Suppes and Hume) solves this difficulty imposing the condition that causes must temporally precede the effect. But this is not sufficient to solve the second difficulty: mere probabilistic accounts are not able to distinguish between statistical association and direct causation. The typical example is that the barometer helps predict the weather, but is not causing it. This problem can be solved assuming a common cause (e.g. pressure) which is causing both the barometer index and the weather, but how does one know that all possible common causes are included in the set Ω? Thus, unless one can appeal to some background knowledge of the causal structure, the dependence on the set Ω of all relevant information makes the concept of Granger causality non operational.

In sum, the VAR approach and Granger causality share all the difficulties of the inductivist approaches to causal learning: either they are not able to identify a causal structure under-determined by the statistical properties of the data (that is, there may be other causal structures observationally equivalent), or they are able to do that with implicit background assumptions, which are typically not validated.

3 Graphical models

The deductivist and inductivist approaches can be thought as two opposite responses to the problem of under-determination or problem of identification. The risk of the first approach is the commitment to an apriorist strategy, while the second approach is impeded by the typical difficulties of the probabilistic theories of causality. I suggest that econometrics offers a clear example as to how the general problem of causal inference can be solved only by delicately mediating between background knowledge and statistical properties of the data. I want to argue that graphical causal models can be helpful for this purpose.[6]

Graphical causal models developed by Pearl (2000) and Spirtes et al. (2000) are a suitable tool for the task of mediating between causes and probabilities. These techniques (I refer in particular to Spirtes et al. (2000)) have been shown to be very useful to infer partial information about causal structures from observational data. A graphical causal model consists of a graph[7] whose vertices are random variables with a joint probability distribution subject to some restrictions. The graph is given a causal interpretation (a directed link from A to B means that A causes B) and in many cases it is assumed that the graph is a directed acyclic graph (DAG), excluding feedbacks and loops. In the DAG case, the restriction on the probability distribution is the *causal Markov condition*, which limits the pairing of DAGs and probabilities: each variable is independent of its graphical non-descendants given its graphical parents. A second assumption that is made for the sake of causal discovery is the *faithfulness condition*: all of the conditional independence relations in the probability distribution follow from the causal Markov condition. Based upon these two conditions, Spirtes et al. (2000) provide some algorithms (operationalised in a computer program called TETRAD) that identify the causal graph which has generated the data from tests on conditional independence relationships. Often this graph is not a unique DAG, but a set of Markov equivalent DAGs, i.e., a set of graphs which share the same conditional independence relations among the variables. Variants of these algorithms are given for environments where the possibility of latent variables is allowed (Spirtes et al., 2000, chap. 6). Richardson and Spirtes (1999) extend the procedure to situations involving

---

[6] I am not denying, however, that there are other econometric approaches that also perform very well in the task of mediating between deductivist and inductivist approaches. I am referring in particular to the London School of Economics approach to econometrics (Hendry, 1995) and to the "extreme bound analysis" of Leamer (1983). But in these two approaches the issue of causal inference is not as central as in graphical models.

[7] A graph can be thought as a pair (V, E), where V is a nonempty set of *vertices*, and E is a subset of the set $V \times V$ of ordered pair of vertices, called *edges*. For a more detailed graphical model terminology see Spirtes et al. (2000, p. 5-17).

cycles and feedbacks.

This opens up the possibility of a logic of scientific discovery, which was explicitly denied in the philosophy of science for many years, from Hempel onwards. However, I want to argue here that graphical models need not to be interpreted or used as instruments of pure inductive learning. To begin with, both the causal Markov and the faithfulness condition should be taken with caution, because, although in general statistical models for social sciences with a causal significance satisfy these conditions (Spirtes et al., 2000, p. 29), there are still several environments in which these conditions are usually violated.

In general, the causal Markov condition does not hold if variables relevant to the causal structure are not included in the set V of the vertices (although it is possible to test for latent variables), if probabilistic dependencies are drawn from non-homogenous populations, if variables are not properly distinct from one another, or if causality cannot be assumed to be local in time and space (for example in quantum mechanical experiments). In macroeconomics the problem is compounded by the problem of aggregation: causal structures may be effective at a low level of aggregation (at the micro level), but variables are measured at a high level of aggregation (at the macro level).

The faithfulness condition can also be thought as claiming that the probability distribution on V embodies only independence relations that can be represented in a causal graph (through the Markov condition), excluding independence relations that are sensitive to particular values of the parameters and vanish when such parameters are slightly modified. Pearl (2000, p. 48) calls this assumption *stability*, because it corresponds to assuming that all the independence relations remain invariant when the parameter values change. This means that external influence (exogenous shocks) will tend to change parameter values and not the causal structures (from which all the independence relations derive). In economics this concept recalls Simon's (1953) characterization of causal relations as invariant under interventions, and Frisch and Haavelmo's concept of "autonomy" or "structural invariance" (Aldrich, 1989).

Thus, it is important to stress the fact that causal Markov and faithfulness are *a priori* assumptions. In a macroeconometric framework, causal Markov and faithfulness condition should be taken as working assumptions. Indeed, it is important to be aware that the results may depend on the choice of variables, level of aggregation and presence of structural changes. Econometric tests are available for many of these specification issues (AIC criterion, Chow test, etc.) and should be taken into account before applying the algorithm. In other words, graphical causal models should be based on background knowledge based on independent statistical techniques (besides theoretical knowledge). Graphical models permit one to take into account the maximum amount of probabilistic information (partial correlations of all possible orders), which can be used to exclude false causal relations. Partial correlations, however, are never sufficient to isolate the unique true causal relations, except in very exceptional circumstances. Background knowledge has to be incorporated and this approach permits the use of background causal knowledge in a very efficient way.

The view I am proposing here is much in the spirit of the synthetic approach proposed by Williamson (2002). As Williamson (2002, p. 10) argues, "while the causal Markov condition may fail it remains a good default assumption, in the sense that if one knows of the causal relationships amongst a set of variables, and one knows of no counterexample to the causal Markov condition amongst those variables, then one's subjective probabilities ought to satisfy the condition."

4 Graphical models and structural VARs

In this section, I show how the task of mediating between a deductivist and an inductivist approach can be put forward, through graphical models, in the special context of structural VAR. In section (2.2), I have shown how the tension between a deductivist and inductivist approach emerges again in the identification of a structural VAR. I want to show here how graphical models can be useful in mediating between an inductivist and deductivist approach to impose the restrictions to identify a structural VAR. Recall that the problem of identification, in the VAR framework, consists in recovering the structural equation

(7) $\quad A_0 Y_t = A_1 Y_{t-1} + \ldots + A_m Y_{t-m} + \epsilon_t$

from the estimate of the reduced form equation

(8) $\quad Y_t = P_1 Y_{t-1} + \ldots + P_m Y_{t-m} + u_t,$

where $A_0 P_j = A_j$ (for $j = 1, \ldots, m$) and $A_0 u_t = \epsilon_t$. These systems of equations can be solved only by imposing restrictions on the matrix A_0. The elements of A_0, appropriately normalised, can be thought as the coefficient of l regression equations:

$u_{1t} = \alpha_{11} u_{2t} + \ldots + \alpha_{1(l-1)} u_{lt} + \epsilon_{1t}$
$u_{2t} = \alpha_{21} u_{1t} + \ldots + \alpha_{2(l-1)} u_{lt} + \epsilon_{2t}$
$\ldots$
$u_{lt} = \alpha_{l1} u_{1t} + \ldots + \alpha_{l(l-1)} u_{(l-1)t} + \epsilon_{lt},$

where some of the α's may be zero, but we do not know *a priori* which ones. But looking at the equation (7), it is straightforward to see that A_0 incorporates the structural relations, that is *causal* relations, among the contemporaneous elements of Y_t. Thus, there is an isomorphism between the causal relation among the residual variables $u_{1t}, \ldots, u_{lt}$ and the contemporaneous variable $y_{1t}, \ldots, y_{lt}$.

The idea of Swanson and Granger (1997), Reale and Wilson (2001), Blesser and Lee (2002), Demiralp and Hoover (2003), Moneta (2003) is to use graphical causal models to infer the causal relationships among the elements of u_t (equivalent to the causal relationships among the element of Y_t) from the estimate of vanishing partial correlations among u_t.[8] This allows the imposition of enough zero-restrictions on the elements of A_0 (i.e., on the α's) in order to get the model identified. A zero on A_0 corresponds to a lack of causality among two elements of u_t.

I will clarify this approach through an empirical example.[9] An important question in macroeconomics is which shocks are the main causes of income fluctuations. This is not only an important question per se, but it is crucial to assess a theoretical hypothesis, like, for example, the Real Business Cycle hypothesis, which claims that shocks to real variables (consumption, investment, income) are the dominant sources of income fluctuations and that shocks to nominal variables (money, interest rates) play an insignificant role in determining the long-run behaviour of real variables. To address this question, I estimate a VAR very similar to the one used by King et al. (1991). Let $Y_t = (C, I, M, Y, R, \Delta P)'$, where C denotes per capita consumption expenditure, I per capita investment, M the real balances, that is the ratio between money and price level, Y per capita gross national product, R nominal interest rate, and ΔP inflation. The data are six quarterly U.S. macro variables for the period 1947:2 to 1994:1 (188 observations).

A series of specification tests (cointegration, number of lags, structural change, etc.) confirmed the possibility of a stable causal structure for these years. Thus, assuming causal Markov and faithfulness condition, a modified version of the PC algorithm incorporated in TETRAD could be applied using as input the tests on vanishing partial correlations among the elements of u_t. The resulting graph is displayed in Figure 1.

---

[8]Swanson and Granger (1997) apply a technique which assumes the Markov condition, but not the Faithfulness condition; Reale and Wilson (2001) apply conditional independence graphs; Blesser and Lee (2002) and Demiralp and Hoover (2003) apply the PC algorithm incorporated in TETRAD; Moneta (2003) applies a modified version of the PC algorithm which is more severe in orienting edges.

[9]This empirical example is drawn from Moneta (2003). The reader is referred to this paper for more details.

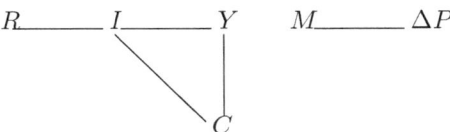

Figure 1. Output of the search algorithm.

Notice that the algorithm does not direct any causal relationship because the modification that I made to the PC algorithm rendered directing edges more severe. The set of DAGs for this pattern consists of 24 elements. Each of these 24 causal structures corresponds to overidentifying restrictions on the matrix A_0, i.e., restrictions such that the model has a number of known parameters (estimated coefficients and estimated covariance matrix) greater than the number of unknown parameters (parameters of the structural model). This constitutes an advantage with respect to the standard recursive VARs identified using the Choleski factorization of residuals covariance matrix (Sims, 1980), which are just-identified, because overidentified models can be tested using a χ^2 test statistic (see Doan, 2000). It turns out that some DAGs do not pass this test, in particular the DAGs which contain one or both of the following configurations: $R \to I \leftarrow Y$ and $R \to I \leftarrow C$. The number of DAGs ruled out is 8. Thus there are 16 DAGs left. The number of DAGs left is narrow enough to check if there are results about the effects of shocks on output (Y) fluctuations which are robust across the different specifications of the models. I will show the results imposing another *a priori* specification. Among the 16 models considered, four are consistent with the conjecture that interest rate and investment are leading indicator for output. Although this is an hypothesis which is well in tune with much economic theory and empirical stylised facts, it has not to be taken for granted: it is always possible to check whether the results change dropping this hypothesis. In Figure 2 two of the causal graphs for these four models are displayed. I call model 1 and model 3 the models corresponding to the causal graphs displayed in Figure 2. Model 2 and model 4 have causal graphs equal to model 1 and model 3, respectively, except that the causal relationship between M and ΔP runs in the opposite direction.

Figure 3 shows the calculations of the dynamic responses of output (impulse response functions) to the shocks to consumption, investment, money and interest rates for the 4 different model specifications. The results point out that not only shocks associated to real macroeconomic variables (out-

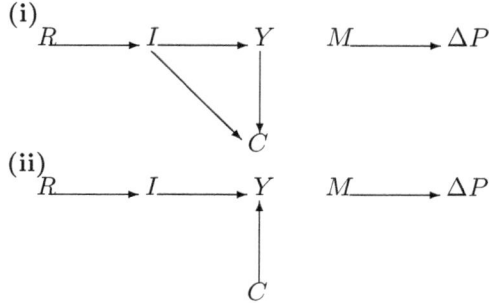

Figure 2. (i) Causal graph for model 1. (ii) Causal graph for model 3.

put, consumption and investment) but also shocks associated to nominal variables (money, inflation and interest rates) have a considerable effect on macroeconomic fluctuations (at all frequencies). This result shows how US data are not consistent with the Real Business Cycle hypothesis, which claims that a single productivity shock is driving output fluctuations. These general results are robust across different specifications of the other 12 models.[10]

5 Concluding remarks

The aim of this paper was to show how graphical models can help to approach the problem of causal inference mediating between deductive and inductive learning. Indeed, these techniques are very powerful in generating causal models starting from a probability distribution, but the general assumptions which permit them to work are not innocuous. I propose to use the causal Markov and faithfulness conditions as working assumptions to be used in a certain temporal window, when empirical evidence and theoretical

---

[10] These results are not reported here for limits of space. There are also several other tests that could be run to know how results are robust to change of number of lags (a main problem also in Granger-causality tests), significance level, and across sub-samples. However, a careful analysis of the epistemic virtues of robustness for each of this case has yet to be done for the methods presented in this paper. Another important issue is the exclusion of feedbacks and loops in the DAGs. This is a very useful simplification, but it is not always reliable in aggregated data. In another paper (Moneta, 2004) I have relaxed this restriction for a similar macroeconomic data set, but it remains an open question how two interpret similarities and differences between the results with and the results without the acycilicity condition.

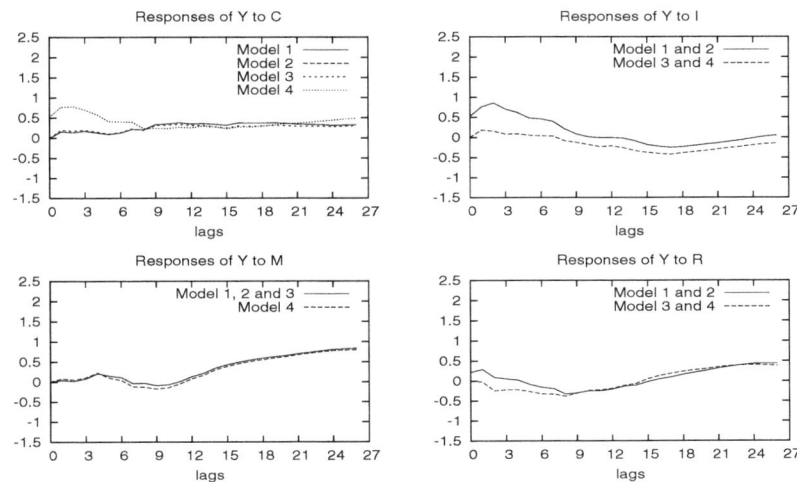

Figure 3. Impulse response functions of income to consumption, investment, money and interest rates shocks.

background knowledge are not at odds with the hypothesis of a stable causal structure generating the data. Thus, both the inductive stage carried over by the graphical algorithms, and the deductive stage are equally important. Thus, one should view the inferred causal structures as a set of hypotheses that has to be tested independently. Moreover, since the set of inferred causal relationships is only in exceptional cases a unique causal structure, additional information about the causal structure has to be derived from background or theoretical knowledge. The advantage of using graphical models is that they express such *a priori* assumptions in an explicit causal language, which helps in testing their validity.

Alessio Moneta
Max Planck Institute of Economics, Jena, Germany.
moneta@econ.mpg.de

BIBLIOGRAPHY

Aldrich, J. (1989). Autonomy. *Oxford Economic Papers*, 41:15–34.
Blesser, D. and Lee, S. (2002). Money and prices: US data 1869-1914 (a study with directed graphs). *Empirical Economics*, 27:427–446.

Demiralp, S. and Hoover, K. (2003). Searching for the causal structure of a vector autoregression. *Oxford Bulletin of Economics and Statistics*, 65:745–767.

Doan, T. A. (2000). *RATS Version 5, User's Guide*. Estima, Evanston, IL.

Forni, M. and Lippi, M. (1997). *Aggregation and the microfoundations of dynamic macroeconomics*. Clarendon Press, Oxford and New York.

Frisch, R. (1933). Editorial. *Econometrica*, 1:1–4.

Granger, C. (1969). Investigating causal relations by econometric models and cross-spectral methods. *Econometrica*, 37:424–438.

Granger, C. (1980). Testing for causality: a personal viewpoint. *Journal of Economic Dynamic and Control*, 2:329–352.

Haavelmo, T. (1944). The probability approach in econometrics. *Econometrica*, 12:1–115.

Hansen, L. and Sargent, T. (1980). Formulating and estimating dynamic linear rational expectations models. In Lucas, R. and Sargent, T., editors, *Rational expectations and econometric practice*, pages 295–320. Allen & Unwin, London.

Heckman, J. (2000). Causal parameters and policy analysis in economics: a twentieth century retrospective. *Quarterly Journal of Economics*, 115:45–97.

Hendry, D. (1995). *Dynamic Econometrics*. Oxford University Press, Oxford.

Hoover, K. (1994). Econometrics as observation: the Lucas critique and the nature of econometric inference. *Journal of Economic Methodology*, 1:65–80.

Hoover, K. (1995a). Facts and artifacts: calibration and the empirical assessment of real-business-cycle models. *Oxford Economic Papers*, 47:24–44.

Hoover, K. (1995b). *Macroeconometrics, developments, tensions, and prospects*. Kluwer, Boston, Dordrecht and London.

Hoover, K. (2006). The methodology of econometrics. In Mills, T. C. and Patterson, K., editors, *Palgrave Handbook of Econometrics, vol. 1, Econometric Theory*. Palgrave Macmillan.

Intriligator, M. (1983). Economic and econometric models. In Griliches, Z. and Intriligator, M. D., editors, *Handbook of econometrics*. Elsevier Science Publishers.

King, R., Plosser, C., Stock, J., and Watson, M. (1991). Stochastic trends and economic fluctuations. *American Economic Review*, 81:819–840.

Kirman, A. (1992). Whom or what does the representative agent represent? *Journal of Economic Perspectives*, 6:117–136.

Kydland, E. and Prescott, E. (1982). Time to build and aggregate fluctuations. *Econometrica*, 50:1345–1369.

Leamer, E. (1983). Let's take the con out of econometrics. *American Economic Review*, 73:31–43.

Lucas, R. (1976). Econometric policy evaluation: a critique. In *The Phillips curve and labor markets. Carnegie Rochester Conference Series on Public Policy*. North Holland.

Lucas, R. (1980). Methods and problems in business cycle theory. In Lucas, R., editor, *Studies in business-cycle theory*. Blackwell.

Moneta, A. (2003). Graphical models for structural vector autoregressions. *LEM Working Paper, Sant'Anna School of Advanced Studies, Pisa*, 03/07.

Moneta, A. (2004). Identification of monetary policy shocks: a graphical causal approach. *Notas Económicas*, 20:39–62.

Moneta, A. (2005). Causality in macroeconometrics: some considerations about reductionism and realism. *Journal of Economic Methodology*, 12:433–453.

Pearl, J. (2000). *Causality. Models, reasoning, and inference*. Cambridge University Press, Cambridge.

Reale, M. and Wilson, G. T. (2001). Identification of vector AR models with recursive structural errors using conditional independence graphs. *Statistical Methods and Applications*, 10:49–65.

Richardson, T. and Spirtes, P. (1999). Automated discovery of linear feedback models. In Glymour, C. and Cooper, G. F., editors, *Computation, causation, and discovery*. AAAI Press and The MIT Press.

Sargent, T. and Sims, C. (1977). Business cycle modeling without pretending to have too much a priori economic theory. In *New methods in business cycle research: proceeding from a conference*, Federal Reserve Bank of Minneapolis, Minneapolis, MN.

Sawyer, K., Beed, C., and Sankey, H. (1997). Underdetermination in economics. The Duhem-Quine thesis. *Economics and Philosophy*, 13:1–23.

Simon, H. (1953). Causal ordering and identifiability. In Hood, W. C. and Koopmans, T. C., editors, *Studies in econometric methods*. Wiley.

Sims, C. (1980). Macroeconomics and reality. *Econometrica*, 48:1–47.

Spirtes, P., Glymour, C., and Scheines, R. (2000). *Causation, prediction, and search*. The MIT Press, Cambridge, MA.

Spohn, W. (1984). Probabilistic causality: from Hume to Suppes via Granger. In Galavotti, M. C. and Gambetta, G., editors, *Causalità e modelli probabilistici*. CLUEB.

Stock, J. and Watson, M. (2001). Macroeconomics and reality. *Journal of Economic Perspectives*, 15:101–115.

Suppes, P. (1970). *A probabilistic theory of causality*. North Holland, Amsterdam.

Swanson, N. and Granger, C. (1997). Impulse response functions based on a causal approach to residual orthogonalization in vector autoregressions. *Journal of the American Statistical Association*, 92:357–367.

Williamson, J. (2002). Learning causal relationships. *Technical Report, Centre for Philosophy of Natural and Social Science, London School of Economics*, 02/02.

Williamson, J. (2005). *Bayesian nets and causality. Philosophical and computational foundations*. Oxford University Press, Oxford.

Causality in economics
STEPHEN LEROY

ABSTRACT. The initiation of quantitative analysis of formal structural models in the social sciences is generally attributed to the Cowles Commission economists in the 1950s (see Hood and Koopmans (1953) for a collection of some of the most important papers). Since then the new methods have been extended and refined and applied in the other social sciences. Graphical methods based on the Cowles developments are increasingly used in the social and biological sciences. The theme of this paper is that recent purported refinements of the Cowles analysis, particularly as it relates to representations of causal relations in formal models, departed from the Cowles analysis to a greater extent than is commonly realized. The suggestion is that the Cowles development as interpreted here compares favorably to more recent developments.

1 Causation

The term "structural" was given several distinct, though related, meanings by the Cowles economists, as has frequently been observed. At a minimum, the term refers to the distinction between the structural form and the reduced form of a model. In the structural form each internal variable is expressed as a function of some other internal variables and some external variables, whereas the reduced form referred to the solution of a model, in which each internal variable is expressed as a function of the external variables alone.

As the term implies, the structural form was viewed as more fundamental than the reduced form. It was seen as containing information not present in the reduced form, such as exclusion restrictions. For example, a structural supply-demand system might specify that one or several of the external variables affect demand but not supply, or vice-versa. These restrictions were used to identify structural coefficients.

A second meaning of "structural", one more basic than that discussed above, has to do with invariance under intervention. The Cowles economists implicitly distinguished two classes of hypothetical experiments: (1) determining the effects of routine realizations of external variables, and (2) analyzing changes in model structure. There is no formal justification for this distinction since hypothetical model shifts can always be parametrized, allowing the analyst to represent an intervention on the structure of a model via a hypothetical shift in an external variable. Thus in a model that is structural in this sense coefficients can be treated as if they were external variables: an intervention on one or several model coefficients by definition leaves other coefficients unchanged. The Cowles economists were often unclear as to when they were treating coefficients as external variables and when they were treating them as constants.

This conception of structural models has left its tracks in current practice in macroeconomics: the Lucas critique, with its assertion that policy change is properly modeled as an intervention on parameters rather than variables, is one example. The distinction between deep and shallow parameters is another.

In many analyses "structural" has a further meaning: a structural equation is one in which the right-hand side variables are causes of the left-hand side variable. In structural equations so defined the equals sign has the meaning of the assignment operator in computer languages, as Pearl (2000) observed (Pearl's work is discussed below). This is distinguished from its meaning in mathematics, under which equality by definition treats the right-hand side and left-hand side of an equation symmetrically. Graphical analyses of causation have heretofore been based on this causal interpretation of structural equations: the graphical representation of the equals sign is an arrow from the right-hand side variables to the left-hand side variable.

With a few exceptions, economists have been conspicuous in their absence from these developments in recent years. It is true that the topic of causation has been of passing interest to economists—witness Granger causality—but there is essentially no connection between the lines of inquiry that economists have pursued and the graphical methods used in the other disciplines. It is clear why economists have not adopted the new graphical methods: economic models use the equality symbol with its usual mathematical meaning, not with the meaning of the assignment operator.[1] Therefore economic models are not expressible as graphs, at least insofar as graphs are based on the causal interpretation of the equality symbol. That being so, economists' lack of interest in graphical analyses of causa-

---

[1] Some contemporary studies of causation in economic models carry over the interpretation of right-hand side variables as causes of the left-hand side variable. This is clearly so with Pearl (2000). In a supply-demand model Heckman (2004) justified putting quantity on the left-hand side and price on the right-hand side on the grounds that in competitive models individuals are modeled as price-takers. It is not clear to what extent Heckman's formal analysis depends on this interpretation. Pearl and Heckman's work is discussed below.

tion is not surprising. Further, economic models in the tradition of general equilibrium theory (including modern macroeconomics) make no use of the structural form/reduced form distinction.

These issues require more careful examination. We begin with Simon, whose analysis of causation is in fact completely different from that implicit in the graphical treatments. Contrary to the implication in the preceding paragraph and the presumption in many expositions of graphical analysis of causation, Simon's definition of causal orderings does not require a model that is structural in the sense that the equals sign is interpreted as causal.

1.1 Notation

One problem that the reader of the literature on causation encounters is that terminology is sometimes not clearly defined and, further, many of the terms are used with different meanings by different authors. To forestall confusion we begin by defining the terminology used here. It seems to me that these definitions are close to standard usage, insofar as standard usage exists, but some analysts appear to disagree.[2]

One essential distinction, stressed in logic and mathematics but often blurred in philosophical discourse and economic analysis, is that between variables and constants.[3] A *variable* is the argument or value of a function, while a *constant* is a standing for a number. By specifying that θ is a constant, the analyst stipulates that it makes no sense to consider alternative values of θ—doing so makes no more sense than asking what would happen in mathematics if π took on a value other than 3.14159. Variables, in turn, may be classified into *external* and *internal variables*. Internal variables are determined by the model; external variables are taken as model inputs. Therefore a model consists of a map from the space of external variables to that of internal variables. Of course, in some exercises involving formal models, such as data-description exercises, variables are not classified as between external and internal variables, making it impossible to discuss causation.[4] Since causation is the subject here, we will assume that the analyst is willing to make this categorization.

The terms "exogenous variable" and "endogenous variable" are often used with the same meaning as "external variable" and "internal variable", and the etymology of the former pair of terms supports this usage. However,

---

[2] For example, Hoover (2001b, p.171) referred to the paper preceding this one (LeRoy, 1995a), which has substantially the same terminology, as offering a "complex and difficult terminological landscape".

[3] As observed above, the Cowles economists were sometimes unclear as to whether model coefficients were to be interpreted as constants or external parameters.

[4] See Shafer (1996) for an extended formal discussion of causation that dispenses almost completely with the distinction between external and internal variables.

econometricians sometimes use the term "exogenous variable" with a different meaning: roughly, an exogenous variable is an observed variable of which an unobserved equation error is independent or mean-independent. Leamer (1985) itemized the various meanings of "exogenous" and "endogenous" in the economics and econometrics literature. We follow his recommendation that the terms "external variable" and "internal variable" be substituted to avoid confusion.

It is assumed that all interventions on a model—that is, all hypothetical experiments involving the model—can be characterized, either explicitly or implicitly, as interventions on the external variables. Direct interventions on the internal variables are inadmissible precisely because these variables are determined by the external variables. Thus hypothesized interventions on internal variables must be implicitly attributed to interventions on the external variables that determine them. Our analysis of causal orderings is predicated on this characterization of interventions.

It is assumed that external variables satisfy the "variation-free" condition: the external variables can be intervened on independently. If this condition fails, the interpretation is that there exists a functional relation linking the external variables. If so, that relation should be included in the model. The variation-free condition asserts that the model is "invariant under intervention", to use a phrase favored by the Cowles economists: a hypothesized change in one external variable leaves the other external variables unchanged, and does not otherwise alter the structure of the model.

In multidate models we distinguish between parameters and processes. A *parameter* is a variable which one wants to distinguish from a constant, while a *process* is a collection of variables, one for each element of some index set representing time. Equivalently, one could define a parameter as a process each element of which is assumed to take on the same value. In many discussions the terms "constant" and "variable" are used where we use "parameter" and "process", but this usage would invite confusion in the present paper because of the different definitions of constant and variable presented above. Parameters and processes, like variables, may be external or internal.

Observe that in this usage the term "variable" is used in static but not dynamic models, while the opposite is the case for the terms "parameter" and "process". In contrast, in much applied work involving static models the term "parameter" is used with the same meaning as "variable" under the above definition. Also, in multidate models the term "variable" is often used with the meaning of "process" as defined above. Generally there is no harm in either of these usages, but in the present context they would cause confusion: in static models variables as defined here are the same as parameters, so one of these terms should be deleted, while in multidate models we have already defined processes as collections of variables.

In some discussions the term "parameter" is used with a different meaning. For example, Hoover (2001b) defined a parameter as a variable that is subject to direct control (p. 61). This definition appears to coincide with our definition of an external variable (or an external process). In some of Hoover's applications parameters are assumed to take on different values at different dates, although Hoover did not time-subscript parameters or identify them as processes, as would seem to be appropriate at least according to the notation adopted here. The same practice is followed in many other sources (Engle et al. (1983) is an example). The merits of Hoover's terminology are not clear; however, our point here is to set out the notation of the present paper, not to argue against other possible choices.

Variables may be either observed by the analyst, or unobserved. External unobserved variables will be assigned probability distributions, and these will induce probability distributions on internal variables, both observed and unobserved (assuming that models have unique solutions, as we do throughout this paper). We will assume that constants are unobserved. In writing down uninterpreted models, the following notation is adopted:

external observed variables or processes	x
internal observed variables or processes	y
unobserved constants	$a, b, A, B, C, D, \alpha, \beta$
unobserved external variables or processes	u
external parameters	θ
internal parameters	ψ

(with interpreted models it is sometimes easier to depart from this notation so as to use variable names that evoke the meaning of the variable). Note that this classification is incomplete; other possibilities, such as internal unobserved processes (latent variables, in some characterizations) are deleted because they are not considered in this paper.

Below we will be considering the consequences of classifying terms as constants or coefficients. Now, as a logical matter we should be specifying a name for an entity that could be either a variable or a constant ("term" and "entity" are unsatisfactory), and we should also add a new symbol so as to avoid the need to write a or x. However, expanding the notation in this way is obviously unweildy. In dealing with models that are linear in variables we can use the term "coefficient", so as to be able to state that a linear model becomes bilinear if coeffients are treated as variables rather than constants. In some contexts below we will use the term "coefficient" in this sense even when the context does not require the limitation to linear/bilinear settings. No confusion should result.

1.2 Simon's definition

We begin with a (somewhat unconventional) review of Simon's definition of causation. Suppose that a model is representable as a linear operator from R^m, a space of external variables, into R^n, a space of internal variables. This operator is assumed to be representable by an $n \times m$ matrix C of constants:

(1) $\quad y = Cx.$

Here (1) is the *reduced form*.

Assume that (1) A is an $n \times n$ matrix of constants with zeros on the main diagonal, (2) B is an $n \times m$ matrix of constants, and (3) A and B satisfy

(2) $\quad (I - A)^{-1} B = C.$

Under these assumptions the model (1) can be written in the form

(3) $\quad y = Ay + Bx,$

as is readily verified by substituting the left-hand side of (2) for C in (1).

Simon defined causal orderings from (3): for two internal variables y_1 and y_2, y_1 causes y_2—denoted $y_1 \to y_2$—if y_1 appears in the block of equations that determine y_2, and also in a block of equations of lower order (see Simon (1953) for definitions of these terms). For example, in the model

$$y_1 = b_{11} x_1 + b_{12} x_2 \quad (4)$$
$$y_2 = a_{21} y_1 + b_{23} x_3, \quad (5)$$

we have $y_1 \to y_2$ because y_1 appears in equation (5), which determines y_2, but also in equation (4), which by itself constitutes a lower-order block. Formally, the causal ordering on Y, the set of internal variables, associated with a given structural model is a subset of $Y \times Y$; $y_1 \to y_2$ means that (y_1, y_2) is in the ordering.

In identifying the model (4)-(5) with the causal ordering $y_1 \to y_2$ we are implicitly assigning generic values to the coefficients. In special cases (for example, in the case $a_{21} = 0$ above) y_1 does not cause y_2. For reference below, note that the assumption that the coefficients are nonzero is not sufficient to assure the uniqueness of causal orderings; we will see that if coefficients obey certain restrictions, but restrictions that do not involve zero values, causal orderings are altered. Since these restrictions are nongeneric, assuming genericity assures uniqueness. This is demonstrated below. This qualification is not repeated below, but it is assumed.

There is a well-known difficulty with the above account of causation: algebraic operations on the equations of (3) can apparently alter causal orderings. For example, substituting (4) in (5) results in

(6) $\quad y_2 = a_{21}b_{11}x_1 + a_{21}b_{12}x_2 + b_{23}x_3.$

In the model (4), (6) neither y_1 nor y_2 causes the other according to Simon's definition. Different algebraic operations result in models in which y_1 and y_2 are simultaneously determined, or obey $y_2 \to y_1$, even though each of these models represents the same linear operator C. It appears as if apparently innocuous mathematical operations alter causal orderings.

This problem reflects the fact that the matrices A and B associated with a given C by (2) are not unique. Therefore on Simon's definition the causal ordering of the internal variables depends on which of an infinite number of pairs of matrices A and B satisfying (2) is chosen.

To avoid the problem posed by the apparent dependence of causal orderings on algebraic operations, we impose a further restriction: we require that *each equation contain at least one external variable not found in any other equation*. Hereafter we refer to this condition as the *exclusion condition*.[5] The exclusion condition rules out algebraic operations that involve more than one equation (because if the original model satisfies the exclusion condition, the modified model will not). For example, the model (4), (6) does not satisfy the exclusion condition: the external variables that appear in (4)—x_1 and x_2—also appear in (6).

Satisfaction of the exclusion condition requires that C have rank n. If the exclusion condition is satisfied and if in addition the model has the same number of internal as external variables, then causal orderings are unique. To see this, note that under the assumed condition C is nonsingular, so (1) can be solved to yield

(7) $\quad x = C^{-1}y.$

If D is defined by $D = I - C^{-1}$, where I is the identity matrix, this becomes

(8) $\quad y = Dy + x.$

Since D is unique in (8), it is clear that the causal ordering is unique.

If $m \geq n$, with any C there is always associated at least one pair A and B that satisfies the exclusion condition: the fact that C is of rank n implies that one can always find a square nonsingular matrix C_1 and a

---

[5]The exclusion condition is essentially the same as Hausman's *independence condition* (Hausman, 1998, p.64). See also Hoover (2001a, p.103ff).

matrix C_2 such that the external variables x can be partitioned (perhaps after reordering) into (x_1, x_2) and (1) can be written in the form

(9) $y = C_1 x_1 + C_2 x_2$.

Premultiplying by C_1^{-1} results in

(10) $C_1^{-1} y = x_1 + C_1^{-1} C_2 x_2$.

As before, if D is defined by $D = I - C_1^{-1}$, (10) can be written as

(11) $y = Dy + x_1 + C_1^{-1} C_2 x_2$.

Here each x_1 enters one and only one of the equations. The variables in x_2 can enter in any or all of the equations.

Fennell (2004) pointed out that if $m > n$, causal orderings under Simon's definition are not unique even if the exclusion condition is imposed. This is so because with $m > n$, different subsets of the external variables can be selected to satisfy the exclusion condition, and each choice implies a different causal ordering. To see this, consider the system

(12) $y_1 = b_{11} x_1 + b_{12} x_2$

(13) $y_2 = b_{22} x_2 + b_{23} x_3$,

in which y_1 and y_2 are not causally ordered. The exclusion condition is satisfied by the presence of x_1 in (12) and x_3 in (13). However, if (13) is solved for x_2 and the result is substituted in (12), we obtain

(14) $y_1 = b_{11} x_1 + a_{12} y_2 - (b_{12} b_{23}/b_{22}) x_3$.

(15) $y_2 = b_{22} x_2 + b_{23} x_3$,

where

(16) $a_{12} = b_{12}/b_{22}$.

In the system (14)-(15) the exclusion condition is again satisfied because of the exclusion of x_2 in (14) and x_1 in (15). In (14)-(15) we have $y_2 \to y_1$. Since (12)-(13) is mathematically equivalent to (14)-(15), it follows that causal orderings are not unique.

Despite this, causal orderings are unique generically: in (14)-(15) we have $y_2 \to y_1$, but that version of the model is nongeneric because of the restriction (16). Since we have already ruled out nongeneric special cases, it is seen that Fennell's observation about nonuniqueness of causal orderings when $m > n$ does not involve anything new.

It is noteworthy that assuming that a model satisfies the exclusion condition is weaker than assuming that it is structural in the sense that the equality symbol is asymmetric: imposition of the exclusion condition allows renormalization of individual equations (i.e., expressing them so that a different variable appears on the left-hand side), so it does not matter which variable is located on the left-hand side. Causal orderings as Simon defined them are not altered by such renormalizations. In contrast, under the causality definition based on the asymmetric interpretation of the equality symbol, renormalizations of individual equations result in a different model with a different causal ordering. That Simon's definition of causation does not rely on an unconventional interpretation of the equality symbol is an attractive feature of his treatment.

The foregoing discussion is very close to Simon's development. On a superficial comparison of the above discussion with Simon's paper, it appears that the exclusion condition has nothing to do with Simon's Section 6 discussion of when causal structures are "operationally meaningful". In fact, however, Simon's discussion is entirely consistent with the discussion here; the apparent differences are terminological.

Simon's Section 6 marks a change from the discussion that preceded it in his paper. Prior to that section Simon did not explicitly incorporate external variables in his discussion (except in Example 4.2), as that term is used here. His examples contained only variables x and constants a (or α). Simon's x corresponds to our y; his a (or α) corresponds to our x and a. Simon used the terms "exogenous variable" and "endogenous variable", but he assigned them a meaning that is derived from his definition of causal orderings: on Simon's usage if we have $y_1 \to y_2$, then y_1 is exogenous in the set of equations that determine y_2, and y_2 is endogenous in that set.

However, in Section 6 in dealing with the fact that algebraic operations can apparently alter causal orderings, Simon considered interventions in the a terms, implying that in that section he was viewing the a terms as variables that are external in the sense of this paper, as opposed to constants as in the earlier sections.

Simon did not distinguish between the coefficients and the intercept terms, implying that he was allowing for interventions in either. Here, in contrast, we are simplifying relative to Simon by maintaining the assumption that the coefficient terms are constants, so that they are not subject to intervention, implying that only the intercept terms are treated as external variables. Treating the coefficients as variables would convert what is a linear model into a bilinear model. Following Simon here would complicate the discussion unnecessarily (a bilinear model is considered briefly in Sec. 1.5).

For Simon, causal orderings are operationally meaningful only if the equations of a structural model have "individual identities". The equations of a structural model have "individual identities" insofar as interventions can be associated with particular equations or subsets of equations. In the terminology of the present paper, these interventions are associated with external variables. Therefore, translating into the terminology of the present paper, Simon's criterion for operational meaningfulness is that particular external variables be associated with particular equations. This corresponds exactly to our exclusion condition.

Simon stated this explicitly: "The causal relationships have operational meaning, then, to the extent that particular alterations or 'interventions' in the structure can be associated with specific complete subsets of equations" (p. 65). Continuing, "[w]e found that we could provide [a causal] ordering with an operational basis if we could associate with each equation of a structure a specific power of intervention, or 'direct control.' ... Hence, ... structural equations are equations that correspond to specified possibilities of intervention" (p. 66).

Simon's discussion would have been clearer if he had explicitly incorporated this idea in his definition of causal orderings, as we have, rather than implicitly attaching the relevant condition later as a condition for causal orderings to be operationally meaningful. This is, of course, a criticism of exposition, not substance.

Our simplification (relative to Simon) of treating coefficients as constants rather than external variables does not alter the substance of Simon's argument: it is easy to see from examination of examples that if $y_1 \rightarrow y_2$ when the coefficients are treated as constants, the same is true when the coefficients are treated as external variables.[6] Assuming that the model is linear in variables (a consequence of treating the coefficients in Simon's model as constants) limits the direct applicability of the analysis; contemporary models are likely to be nonlinear. Again, however, the analysis can be extended to the general case. The principal difference between the analysis of causation in linear vs. nonlinear models is that in the latter case causal orderings are no longer associated with constants measuring the strength of causal effects: in general the magnitude of a given change in the cause variable on the effect variable depends on the values of all external variables.

The theme of this paper is that Simon's analysis of causation differs in major respects from more recent treatments. At this stage we point out some of the distinguishing features of Simon's treatment. First, Simon made clear that he was analyzing causation in the context of a formal model, not

---

[6]However, the converse is not true: changing an external variable to a constant reduces the set of possible interventions, implying that y_1 may no longer cause y_2.

causation as it applies directly to reality or perceived reality. In contrast, in virtually all discussions in the philosophy literature, and in some in the economics literature, causation is discussed as a direct feature of reality. Second, under Simon's definition causality is not a matter of how a model is interpreted or applied: rather, the causal ordering implied by a model can be inferred unambiguously from its formal structure. Third, under Simon's treatment models are written in the usual form as maps from external to internal variables. As noted above, under some alternative treatments of causation the equals sign is interpreted as asymmetric, with cause variables on the right-hand side and effect variables on the left-hand side. In all three respects we follow Simon's lead in this paper.

These features of Simon's treatment of causality have the implication that some sentences that are customarily interpreted as causal do not satisfy the formal requirements for causation. For example, consider the statement "I drank too much yesterday (D); as a result I fell asleep while smoking (S), and my doing so caused my home to burn down (B)". Here D is clearly external, and it would be natural to diagram this sentence as "$D \to S \to B$". The problem is that the exclusion condition for "$S \to B$" fails: there does not exist an external variable that affects B but not S. Equivalently, S is determined in the same block of equations as B, implying that S and B are appropriately treated as simultaneous, not causally ordered. It follows that the inference that S caused B is a feature of the interpretation of the model, not its formal structure.

The fact that sometimes we are willing to infer causation in settings where Simon's formal analysis does not justify this inference does not reflect any shortcoming in Simon's treatment. Informal statements of causation, such as that just given, generally presume unstated background conditions ("the fire extinguisher did not work (E), so I could not put out the fire"). Explicit incorporation of variables representing background conditions generally allows satisfaction of the exclusion condition for a causal ordering. In this case E would be included as an external variable that appears in the external set of B but not that of S. Under this modification we would have $S \to B$[7].

1.3 Causality as sufficiency

Part of the reason Simon's characterization of causation is not much used currently is that Simon did not provide a clear explanation of what follows

---

[7]In this example causality is expressed as a relation among events rather than variables. The reader can supply the indicated modification of the formal structure set out above so as to deal with this case.

if one variable causes another.[8] What does the fact that the cause variable is determined in a lower-order subsystem relative to the effect variable have to do with causation? What is the content of "operationally meaningful" in this context, and what is the connection between this concept and the exclusion condition? What interventions are admissible if y_1 causes y_2, but not otherwise? What is the interpretation of these interventions?

The best way to supply intuitive content to causation is to consider simple examples. We will see that in some cases it is clear that causal statements are not appropriate, while in other cases it is equally clear that they are. Examination of the difference between these examples will suggest the general principle. This principle is stated more formally in the next subsection.

Consider the supply-demand model

$$q_s = a_{sp}p + b_{sw}w \tag{17}$$

$$q_d = a_{dp}p + b_{di}i \tag{18}$$

$$q_s = q_d = q, \tag{19}$$

where q_s is quantity supplied, q_d is quantity demanded, q is equilibrium quantity, i is income, p is price and w is weather. Here weather and income are external and the other variables are internal.

In the system (17)-(19) the question "What is the effect of weather on the equilibrium quantity?" is unambiguous: the effect can be directly calculated from the model. This is so because weather is external. However, if one were to ask "What is the effect of price on equilibrium quantity?" the appropriate response would be that the question is misposed. Price and quantity are both internal; they are simultaneously determined, and neither is causally prior to the other.

The reasoning here is worth elaborating. The assumed intervention results in the price changing from, say, p to $p + \Delta p$. The problem is to infer the effect of this intervention on q. The reason the question is ambiguous is that any of an infinite number of pairs of shifts in the external variables "weather" and "income" could have caused the assumed change in price, and these interventions map onto different values of q. Thus the reason the question is misposed is that it does not give enough information about the intervention being considered to allow a unique answer.

The suggestion is that causal statements involving internal variables as causes are ambiguous, and therefore inadmissible, *except when all the interventions consistent with a given change in the cause variable map onto*

---

[8]The material presented in this and the following subsections is drawn from LeRoy (1995a).

the same change in the effect variable. One is led to define two internal variables as causally ordered when the indicated condition is satisfied, and not otherwise.

Now consider the model

$$q_s = b_{sw}w + b_{sf}f \tag{20}$$
$$q_d = a_{dp}p + b_{di}i \tag{21}$$
$$q_s = q_d = q, \tag{22}$$

where f is fertilizer. Weather, fertilizer and income are the external variables. Here even though q is internal there is no problem with the assertion that q causes p. This is so because all the interventions in weather and fertilizer consistent with a given change in q map onto the same value of p, as the structure of the model makes obvious.

1.4 A formal ftatement

Let the *external set* X_j for a particular internal variable y_j be the minimal set of external variables such that y_j can be written as a function of X_j. Then the model (1) can be written in the form

(23) $\quad y_j = \beta_j \overline{X}_j \qquad j = 1, ..., n,$

where $\overline{X}_j$ is a vector of which the elements are the members of X_j, and β_j is a conformable vector of constants. Of course, β_j coincides with the j-th row of C with the zero elements deleted.

Suppose that $X_i \subset\subset X_j$, where $\subset\subset$ means "is a proper subset of". Hereafter we will call this the *subset condition*. Further, define $X_{j,i}$ as $X_j - X_i$ (i.e., as the set consisting of the elements of X_j that are not in X_i). Define $\overline{X}_{j,i}$ as a vector of which the elements are the members of $X_{j,i}$. Suppose in addition that there exists a scalar constant $\gamma_{j,i}$ and a vector of constants $\delta_{j,i}$ such that (23) can be written in the form

(24) $\quad y_j = \beta_j \overline{X}_j = \gamma_{j,i} y_i + \delta_{j,i} \overline{X}_{j,i}.$

Existence of $\gamma_{j,i}$ and $\delta_{j,i}$ with this property implies that all the interventions in X_i consistent with a given change in y_i have the same effect on y_j. Thus all the information relevant for y_j contained in X_i is summarized in y_i, so that even though many possible interventions in X_i could have caused the variation in y_i, each possible intervention has the same effect on y_j. When $\gamma_{j,i}$ and $\delta_{j,i}$ exist that satisfy the above property we will say that y_i is a *simple cause* of y_j, and will write $y_i \Rightarrow y_j$. Thus $y_i \Rightarrow y_j$ means that y_i is sufficient for X_i in the determination of y_j. We will call the condition that there exist $\gamma_{j,i}$ and $\delta_{j,i}$ with the properties just described the *sufficiency condition*. Thus we have $y_i \Rightarrow y_j$ if and only if both the subset condition and the sufficiency condition are satisfied.

In general, $X_i \subset\subset X_j$ does not imply existence of $\gamma_{j,i}$ and $\delta_{j,i}$ satisfying (24). Therefore it will not generally be the case that $X_i \subset\subset X_j$ implies $y_i \Rightarrow y_j$. However, if $X_i \subset\subset X_j$, there may exist some other internal variable y_k that satisfies the subset condition such that all the interventions in X_i consistent with a given change in y_i and a given value of y_k map onto the same value of y_j. Then we have *conditional causation*, indicated by $y_i \Rightarrow y_j | y_k$. Still more generally, the conditioning set may include several internal variables rather than just one, and may include one or more of the external variables.

Two conditions are required for conditional causation. First, y_i must be *variation-free*: if the variables held constant completely determine y_i, it makes no sense to talk about the effect of variations in y_i on y_j, ceteris paribus. For example, if the conditioning set includes all the external variables, no variation in y_i is possible. More precisely, the variation-free condition is satisfied for $y_i \Rightarrow y_j | y_k$ if the model permits independent variation in y_i and y_k without restricting y_j.

Second, if we are to have $y_i \Rightarrow y_j | y_k$, the conditioning set must be such as to ensure that all values of the external variables X_i consistent with a given change in y_i and a given level of y_k produce the same change in y_j. This requirement, which corresponds to the sufficiency requirement for simple causation, is needed to avoid ambiguity in the effect of variations in the external set for y_i on y_j.

As long as $X_i \subset\subset X_j$, there will always exist some subset (possibly the null set, if $y_i \Rightarrow y_j$) of the external and internal variables such that y_i causes y_j conditional on that set of variables. We will write $y_i \to y_j$ if either $y_i \Rightarrow y_j$ or $y_i \Rightarrow y_j | z_k$ for some scalar or vector z_k. Thus $y_i \to y_j$ under the definition just given is equivalent to $X_i \subset\subset X_j$.

If $y_i \to y_j$, there exists a set of conditioning variables z_k such that $y_i \Rightarrow y_j | z_k$, but that set is not necessarily unique. For example, in the model

$$y_1 = b_{11} x_1 + x_2 \tag{25}$$
$$y_2 = a_{21} y_1 + b_{21} x_1 + x_3, \tag{26}$$

where x_1, x_2 and x_3 are external variables, we have $y_1 \Rightarrow y_2 | x_1$, but also $y_1 \Rightarrow y_2 | x_2$. In the first case the intervention is on x_2, while in the second case it is on x_1. The coefficient associated with $y_1 \Rightarrow y_2 | x_1$ is clearly a_{21}. However, in the case of $y_1 \Rightarrow y_2 | x_2$ matters are more complicated: we must distinguish between the direct effect of x_1 on y_2, which has coefficient b_{21}, and its indirect effect. The indirect effect has coefficient a_{21} if the cause variable is identified with y_1, and $a_{21}b_{11}$ if it is identified with x_1.

Conditional causation may raise problems of interpretation. The indicated intervention requires a nonzero change in the variables in X_i, with the changes required to satisfy a linear relation so as to hold y_k constant. Existence of such functional relations among external variables appears to conflict with the assumption that the external variables are variation-free. If there exists a functional relation among the variables in X_i, then assuming that these variables are external is a misspecification. Thus the intervention is inappropriate to the assumed model.

In other cases this problem does not arise. For example, in the important case when the matrix A in the model (3) is triangular (not just block-triangular), each y_i causes y_j ($i < j$) conditional on $y_1, y_2, ..., y_{i-1}, y_{i+1}, ..., y_{j-1}$. However, the indicated intervention involves only one external variable, so there is no violation of the variation-free condition.

It is easily verified that the above definition of $y_i \to y_j$ coincides with Simon's definition: assuming that the exclusion condition is satisfied, y_i appears in the block of equations that determines y_j and also in a lower-order block if and only if $X_i \subset\subset X_j$. Thus $y_i \to y_j$ can refer to both conditional causation as defined here and Simon's definition of causation.

1.5 Causality and parameter interventions

Up to now we have taken coefficients to be constants rather than variables or parameters. This assumption was for convenience only. Many problems involving causation do not satisfy this restriction. For example, as soon as coefficients are treated as external variables rather than constants, models become bilinear rather than linear. In this subsection we make some observations about the consequences of treating coefficients as parameters in multidate models.

Neoclassical macroeconomists stress that analysis of macroeconomic policy changes requires identifying and estimating "deep parameters" (labeled here external parameters). This is correct if one is considering changes in policy regimes, and if one is modeling regime change by parameter interventions, as recommended by the Lucas critique (Lucas, 1976), at least on some readings. I have argued elsewhere (LeRoy, 1995b) that whether policy changes in dynamic models are appropriately modeled through interventions on external parameters or on external processes depends on the question being asked: if the intervention is intended to apply to the past as well as the present and future, parameter interventions are indicated. If only the present and/or future is to be affected, process interventions are indicated.

An important point is that if regime changes are modeled using process rather than parameter interventions, there is no need to model the dependence of internal parameters on external parameters. To see this, suppose that we have a model that is linear in variables if the coefficients are treated as constants. As observed above, if y_{1t} causes y_{2t} in this setting, then the same remains true when the coefficients are treated as parameters. This is so regardless of the causal ordering among parameters. The only difference between the two cases is that in the latter case the external parameters, or some subset of them, are included in the exogenous sets, but this change will not cause failure of the subset and sufficiency conditions, assuming that these are satisfied when coefficients are treated as constants. This point underscores the importance of Marschak (1953) observation that analysis of causation does not always require a complete characterization of a model's causal ordering.

Hoover (1991) pointed out that a potentially testable implication of causation is that interventions on the (external parameters that determine the) probability distribution of the cause variable should not affect the probability distribution of the effect variable conditional on the cause variable. Cartwright (1995, p.57), took issue with this assertion:

> If x causes y, then in a two-variable model $D(y|x)$ [the distribution of y conditional on x] measures the strength of x's effect on y [Cartwright's notation has been changed]. Clearly the question of the invariance of the strength of this influence across envisaged interventions in x is one of considerable interest in itself. But finding out the answer is not a test for causation, neither in the original situation nor in any of the new situations that might be created by intervention. Even if x causes y in the original situation and continues to do so across all the changes envisaged, there is in general no reason to think that interventions that change the distribution of x will not also affect the mechanism by which x brings about y, and hence also change the strength of x's influence on y.

The last sentence of this passage appears to be incorrect, at least under the implementation of causation analyzed in this paper. Cartwright is certainly correct that determining that a conditional distribution appears to be invariant over time does not constitute a test for causation. However, it is easy to show that if y_1 does cause y_2, then the distribution of y_2 conditional on y_1 will be invariant to interventions on X_1, as Hoover asserted. To see this, consider the model

(27) $\quad y_1 = x_1 + x_2 u_1$

(28) $\quad y_2 = x_3 + x_4 y_1 + x_5 u_2,$

where $x_1, ..., x_5$ are external variables, and u_1 and u_2 are independently distributed unobserved external variables. For concreteness we will take u_1

and u_2 to have zero mean and unit variance. Taking x_2 as a parameter rather than a constant allows the analyst to consider interventions on the standard deviation of y_1.

In the nonlinear model (27)-(28) y_1 causes y_2. To see this, note that we have that the external sets for y_1 and y_2 are

$$X_1 = (x_1, x_2, u_1) \qquad (29)$$
$$X_2 = (x_1, x_2, x_3, x_4, x_5, u_1, u_2), \qquad (30)$$

which satisfy the subset condition $X_1 \subset\subset X_2$. The sufficiency condition for $y_1 \Rightarrow y_2$ is also satisfied. An intervention on x_2 will affect the marginal distribution of y_1, but will not affect $D(y_2|y_1)$.

As Cartwright observed in the passage just quoted, it is easy to imagine models in which interventions on x_2 do affect the parameters of the conditional distribution of y_1. However, in such models by definition the conditions for $y_1 \Rightarrow y_2$ will fail, contrary to the assumption. It appears, then, that Hoover's assertion is correct.

2 Causality and identification

2.1 Identification and exclusion restrictions

We have seen that exclusion restrictions play a central role in determining causal orderings. As is well known, they also play a central role in determining whether the parameters in a model are identified (Fisher, 1976). Despite the common role of exclusion restrictions in determining causality and identification, the two are very different notions. Causation is an ordering on internal variables, whereas identification has to do with whether the econometrician can make inferences about parameter values from the (population) distribution of observed variables. Causation can be defined and analyzed without even specifying which variables are observed, and in fact this is exactly what we have done up to now. However, empirical estimation of causal parameters requires assumptions that assure identification, and this is different from causation. Whether parameters are identified depends not just on which variables are excludable from which equations, but also on which variables are observed and what distributional assumptions are imposed on those (external) variables that are not observed.

As just noted, discussion of identifiability requires specifying which variables are observed by the econometrician and which are not. Both internal and external variables can be either observed or unobserved. Probability distributions are assigned to unobserved external variables as part of model specification, and these induce distributions on both observed and unobserved internal variables. Henceforth we will use u to denote unobserved variables, and all other letters to denote observed variables.

To illustrate the difference between causality and identifiability, consider the supply-demand model

$$q = a_1 p + b_1 x_1 + b_2 x_2 + u_1 \tag{31}$$
$$q = a_2 p + b_3 x_1 + b_4 x_2 + u_2, \tag{32}$$

where x_1 and x_2 are external observed shift variables (weather, income and fertilizer played this role above) and a_1, a_2, b_1 and b_2 are constants. As it stands, none of the coefficients in this model are identified: observations on p, q, x_1 and x_2 cannot be used to estimate the coefficients. Under the restriction $b_2 = b_3 = 0$, all coefficients are identified, including a_1 and a_2, the price elasticities of demand and supply. However, these coefficients do not measure the strength of causal relations because p is not causally prior to q. We see that in general identification is an issue in estimating not just constants associated with causation, but also those associated with simultaneous determination of internal variables.

In bringing empirical evidence to bear on evaluating the causal relation (or lack thereof) between variables y_i and y_j in a given model, then, two questions must be distinguished. First, are the two variables causally ordered? If not, the question of identification of causal coefficients does not come up because there is no causal coefficient. If the two variables are causally ordered, then the associated coefficient is well defined conceptually, but it is not necessarily identified.

An assumption that plays a central role in assuring identifiability of causal coefficients is that *all unobserved external variables are (statistically) independent of observed external variables and of each other.* Henceforth we will call this the *independence assumption*.[9] The role of this requirement is to force the model-builder to state explicitly what is assumed about causation, rather than burying causation in uninterpreted correlations among external variables.

---

[9]The independence assumption is essentially the same as the principle of the common cause, which says that if two variables are correlated, then either one variable causes the other or the two variables have a common cause (Reichenbach, 1956). Here, however, independence is interpreted as a formal restriction on models rather than as a philosophical proposition. As such, the suitability of the assumption is evaluated according to whether it is analytically fruitful, rather than according to whether it is consistent with, for example, the correlation between British bread prices and the sea level in Venice (see Hoover (2003) for discussion).

To understand the role of the independence assumption, consider two observed variables y_1 and y_2. Knowledge of their joint probability distribution obviously does not allow any inference about which variable, if either, causes the other. However, the analyst can certainly define an unobserved variable u_1 and a constant a_{21} such that y_1 and y_2 satisfy

$$(33) \quad y_2 = a_{21} y_1 + u_1,$$

where the specified distribution of u_1 and the chosen value of a_{21} generate the joint probability distribution of y_1 and y_2. If the model-builder is willing to interpret y_1 and u_1 as external variables, then we have $y_1 \Rightarrow y_2$. Here, of course, the assumption that y_1 and u_1 are external variables is not implied by equation (33); it is a separate assumption. An infinite number of pairs of parameters a_{21} and random variables u_1 are available which generate a given distribution of y_1 and y_2, implying that a_{21} is unidentified.

Correspondingly, the model-builder can simply project y_2 onto y_1:

$$(34) \quad y_2 = c_{21} y_1 + u_2.$$

Here by construction u_2 is uncorrelated with y_1, and the random variable u_2 and parameter c_{21} are unique. However, in the absence of further assumptions this decomposition of y_2 into y_1 and c_{21} has nothing to do with causation; cause variables can be projected onto effect variables as well as vice-versa.

As is obvious, these two decompositions of y_2 into y_1 and an unobserved random variable are very different operations: the construction of u_1 and the assumption that it is external amount to assuming that $y_1 \Rightarrow y_2$, but the associated coefficient is unidentified. In contrast, the construction of u_2 guarantees its uncorrelatedness with y_1, but has nothing to do with causation. If we are willing to assume further that $u_1 = u_2$—or, equivalently, that y_1 and u_1 are uncorrelated, or that u_2 is external—we can interpret the coefficient c_{21} of the projection of y_2 on y_1 as a causal coefficient.

One implication of the independence assumption is that if one variable causes another, the two variables are necessarily correlated.[10] In some settings this may seem counterintuitive. Suppose that y_{1t} is generated according to

$$(35) \quad y_{1t} = \theta_{11} y_{1,t-1} + \theta_{12} y_{2t} + u_t,$$

---

[10] Of course, the converse is not true: if the external sets for two internal variables have a nonempty intersection, the independence assumption implies that two variables will be correlated. However, if neither external set is a subset of the other, then the two variables are not causally ordered, either unconditionally or conditionally.

where y_{2t} is a regulator, the behavior of which is generated by

$$y_{2t} = \theta_{21}y_{1,t-1}. \tag{36}$$

Then the behavior of y_{1t} follows

$$y_{1t} = (\theta_{11} + \theta_{12}\theta_{21})y_{1,t-1} + u_t, \tag{37}$$

where the external variables are $u_t, u_{t-1}, \ldots$ and y_0. The definitions of causation imply that we have $y_{1,t-1} \Rightarrow y_{1t}$ under the (heretofore unstated) assumption that $\theta_{11} + \theta_{12}\theta_{21} \neq 0$.

However, suppose that the regulator is operated by choosing c so as to minimize the unconditional variance of y_t. This results in $\theta_{11} + \theta_{12}\theta_{21} = 0$ or, equivalently, $\theta_{21} = -\theta_{11}/\theta_{12}$, implying that (37) becomes

$$y_{1t} = u_t. \tag{38}$$

In this special case the regulator y_{2t} is no longer correlated with y_{1t}, and we no longer have causation: $y_{1,t-1} \not\Rightarrow y_{1t}$. This result may seem to run counter to the ordinary-language usage of the term "causality", but it is an unavoidable consequence of the definitions.

Under any particular parametrization of a model, imposing the independence requirement is obviously very restrictive. However, imposing independence on some related parametrization of the model amounts only to requiring the modeler to state explicitly what he is or is not willing to assume about causation. This is not an unreasonable requirement insofar as the goal is to arrive at causal conclusions. For example, the modeler who is willing to assume $y_2 \Rightarrow y_1$ instead of $y_1 \Rightarrow y_2$ would generate y_1 from $y_1 = a_{12}y_2 + u_3$, with y_2 and u_3 assumed external and independent. Finally, if the model-builder were unwilling to assume that either y_1 or y_2 causes the other, he could specify the parametrization

$$y_1 = a_{12}y_2 + u_1 \tag{39}$$
$$y_2 = a_{21}y_1 + u_2. \tag{40}$$

Here u_1 and u_2 are independent by the independence assumption, but this implies neither $y_1 \Rightarrow y_2$ nor $y_2 \Rightarrow y_1$.

2.2 Causation and regression

In LeRoy (1995a) it was pointed out that, under the condition that all unobserved external variables are independently distributed, coefficients associated with causation can be estimated consistently by ordinary least squares.

This is so because under the stated restriction variables representing causes are statistically independent of error terms. It follows that econometric theory can sometimes be used to determine causation: knowledge that ordinary least squares results in inconsistency implies absence of (simple) causation. The example given to illustrate this point was the model

$$y_t = \theta_1 y_{t-1} + u_{1t} \tag{41}$$
$$u_{1t} = \theta_2 u_{1,t-1} + u_{2t}, \tag{42}$$

where the unobserved external variables $u_{2t}, u_{2,t-1}, ..., u_{21}, u_{20}$ and y_0 are assumed to be independent. When $\theta_2 \neq 0$ we have that y_{t-1} is correlated with u_{1t}, so a least-squares regression of y_t on y_{t-1} will not produce a consistent estimate of θ_1. From the stated result it follows that $y_{t-1} \nRightarrow y_t$. Checking, the subset condition for $y_{t-1} \Rightarrow y_t$ is satisfied, but the sufficiency condition fails: the elements of the external set for y_{t-1} (these are $u_{2,t-1}, ..., u_{20}, y_0$) affect y_t through u_{1t} as well as through y_{t-1}. Therefore we have $y_{t-1} \nRightarrow y_t$.

In response to this, Hoover (2001a) observed that applying the Koyck transformation results in

$$(43) \quad y_t = (\theta_1 + \theta_2) y_{t-1} - \theta_1 \theta_2 y_{t-2} + u_{2t},$$

in which the parameters $\rho + \lambda$ and $\rho\lambda$ are in fact estimated consistently by ordinary least squares. Further, separate consistent estimates of θ_1 and θ_2 are easily calculated from the estimated regression coefficients. Hoover appeared to view this result as raising questions about the validity of the inference that failure of ordinary least squares implies nonexistence of causation, although he did not point out an error in the reasoning or spell out the argument. In fact, the multiple regression of y_t on y_{t-1} and y_{t-2} is different from the univariate regression of y_t on y_{t-1}. The fact that ordinary least squares is valid in (43) suggests[11] that even though we do not have the simple causation $y_{t-1} \Rightarrow y_t$, we might have the conditional causation $y_{t-1} \Rightarrow y_t | y_{t-2}$, and it can be directly verified from the definition of conditional causation that this is the case.

3 Alternative treatments of causality

As noted in the introduction, many expositions of contemporary macroeconometric practice as it relates to causation give the impression that it is essentially a refinement and extension of the Cowles insights, particularly those of Simon. The essentials, we are told, were taken over as a whole;

---
[11] We use "suggests" rather than "implies" because the converse of the above result—that consistency of ordinary least squares implies causation—is not generally true.

subsequent developments added precision and amplified details, but did not affect the substance. On the contrary, contemporary discussions are much sketchier than the Cowles treatment (although the Cowles economists are not beyond criticism in this respect), and they generally differ in essential respects from the Cowles treatment. The reader who is persuaded by this assessment will be motivated to understand the differences in the various treatments so as to determine which line offers the best prospect for improving analytical practice. It will shortly be clear that the view here is that to the (considerable) extent that modern practice differs from the Cowles analysis, the differences do not represent clear improvements.

3.1 Sims-Granger

The Cowles Commission economists emphasized that in the absence of other identifying information, causal orderings are not empirically testable: two models with different causal orderings may be observationally equivalent (it remains true that, in conjunction with other restrictions, causal orderings may be overidentifying, hence testable). Subsequently various economists have challenged this dictum, apparently asserting that causation is directly testable. For example, Sims (1977, p.24) wrote:

> When, as econometricians estimating models ordinarily do, someone asserts that a particular variable or group of variables is strictly exogenous in a certain regression, that assertion is, in time series models, testable. "Exogeneity" here is given its standard econometrics textbook definition. Exogeneity tests are thus an easily applied test for specification error, powerful against the alternative that simultaneous-equations bias is present. The usefulness of these specification tests ought not to be controversial....

Despite the claim here that the term "exogeneity" has a standard meaning and the presumption in this literature that this meaning is closely related to that of causation, definitional issues come to the forefront. A process y_2 *Granger-causes* another process y_1 if lagged values of y_2 predict y_1 conditional on lagged values of y_1. If y_2 fails to Granger-cause y_1 then, according to Granger (1969), correlations between the two processes can be taken to represent the causal influence of y_1 on y_2. It shortly was made clear by a number of critics that this conception of causality bore no obvious relation to causality as defined either in ordinary language or in formal analysis.[12] However, the analysis was not as sharp as it might have been because of the lack of a suitable formal definition of causation to compare to Granger causality, or so it appears with hindsight. The definition of causation de-

---

[12]Recently, for example, the point was made in passing by Heckman (1999) in reviewing the Cowles contributions to econometric theory. Heckman's paper is discussed below.

veloped in this paper makes possible a precise comparison with Granger causality.

Suppose that processes y_1 and y_2 are generated by

$$y_{1t} = a_{12}y_{2t} + b_{11}y_{1,t-1} + b_{12}y_{2,t-1} + u_{1t} \qquad (44)$$

$$y_{2t} = a_{21}y_{1t} + b_{21}y_{1,t-1} + b_{22}y_{2,t-1} + u_{2t}, \qquad (45)$$

where the independence assumption is satisfied (so that the unobserved external variables u_{1t} and u_{2t} are uncorrelated with each other contemporaneously, with their own and each other's lagged values, and with the initial values y_{10} and y_{20}). This model is underidentified.

The reduced form of the model is

$$y_{1t} = c_{11}y_{1,t-1} + c_{12}y_{2,t-1} + u_{3t} \qquad (46)$$
$$y_{2t} = c_{21}y_{1,t-1} + c_{22}y_{2,t-1} + u_{4t}. \qquad (47)$$

From this it is clear that we have $y_{1,t-1} \Rightarrow y_{1,t}|y_{2,t-1}$, with associated parameter c_{11}. The other parameters c_{12}, c_{21} and c_{22} are associated with similar elements of the causal ordering. These parameters, of course, are identified. Note that these elements of the causal ordering obtain despite the lack of identifiability of the "structural form" (44)-(45), and whether or not either y_1 or y_2 Granger-causes the other.

Granger-causality is defined from the reduced form (46)-(47). We have that y_2 fails to Granger-cause y_1 if $c_{12} = 0$, where $c_{12} = (b_{12} + a_{12}b_{22})/(1 - a_{12}a_{21})$. If $c_{12} = 0$, $y_{1,t-1} \Rightarrow y_{1,t}|y_{2,t-1}$ simplifies to $y_{1,t-1} \Rightarrow y_{1,t}$, so Granger-causality is necessary and sufficient for $y_{1,t-1} \Rightarrow y_{1,t}$. However, we are interested in causation involving y_1 as a cause and y_2 as an effect, so the relation $y_{1,t-1} \Rightarrow y_{1,t}$ is not of much interest.

Under the restriction $a_{12} = 0$ the model generates $y_{1t} \Rightarrow y_{2t}|y_{1,t-1}, y_{2,t-1}$, which gives a precise sense in which the process y_1 is causally prior to the process y_2. The restriction $a_{12} = 0$, being just-identifying, by itself has no observed implications, and therefore is not testable in the absence of other restrictions. In particular, $c_{12} = 0$ is neither necessary nor sufficient for $a_{12} = 0$ ($b_{12} + a_{12}b_{22} = 0$ is neither necessary nor sufficient for $a_{12} = b_{12} = 0$), so Granger-noncausality is neither necessary nor sufficient for $y_{1t} \Rightarrow y_{2t}|y_{1,t-1}, y_{2,t-1}$.

Under the restriction $a_{12} = b_{12} = 0$ we have $y_{1t} \Rightarrow y_{2,t+j}|y_{1,t-1}$, $j = 1, 2, \ldots$. Granger-noncausality is necessary for $a_{12} = b_{12} = 0$, so $c_{12} \neq 0$ is evidence against $y_{1t} \Rightarrow y_{2,t+j}|y_{1,t-1}$, subject to the usual caveats about sampling error and the like, under the maintained assumptions of the model. However, Granger noncausality is not sufficient for $a_{12} = b_{12} = 0$, and it is easy to construct theoretical examples with "spurious exogeneity" (Granger-noncausality without $a_{12} = b_{12} = 0$). Sims (1977) expressed the view that spurious exogeneity is "unlikely". Since this opinion was rendered in the context of a model written in abstract form (like (44)-(45)), it is clear that he intended this judgment to apply to economic models in general. In conclusion, there are connections between Granger-noncausality and causality, but they are somewhat remote and not easily interpreted.

There remains the point that the causal elements discussed above involve conditional rather than simple causation. As such they involve interventions on subsets of the external variables subject to linear restrictions (see Section 1.4). We observed above that such interventions are not easily reconciled with the assumed variation-free status of the external variables. Thus the question of what economic meaning can be attached to the indicated interventions remains open. This point weakens still further the link between causality and Granger-causality.

3.2 Pearl

Pearl (2000) presented an alternative formalization of causality. He viewed his development as based on the Cowles analysis of the 1950s, particularly that of Simon (1953), as here. Pearl's view is that after a promising beginning during the Cowles years, social scientists lost touch with the idea of structural modeling and failed to develop the original formal analysis of causation. He criticized sharply the tendency of economists—as exemplified in this paper—to interpret the equals sign in (supposedly) structural models as having its usual mathematical meaning, rather than as directly representing causation. For Pearl the definition of structural models implies that the equals sign denotes causation. He also would reject the assertion in Section 1.2 that Simon's analysis of causation is relevant to current practice precisely because it does not depend on the interpretation of the equality sign as directly incorporating causation.

Under Pearl's interpretation of a structural model, each structural equation represents a distinct causal law for one of the internal variables. For Pearl interventions are analyzed by deleting the equation determining a particular internal variable and setting the value of that internal variable at a preassigned level. In this paper, in contrast, we have followed the current economics literature in modeling interventions by the straightforward device of simply specifying values for causal variables.

This is a distinction without a difference when the causal variable is external. When the cause variable is internal, however, Pearl's algorithm can lead to difficulties. The assumption that it makes sense to delete one or more of the structural equations and replace the value of the internal variable so determined by a constant without altering the other equations has been termed "modularity".[13] In special cases Pearl's assumption of modularity is satisfied, implying that his algorithm is valid even when the causal variable is internal. For example, modularity for all possible interventions on a given equation is satisfied if the external sets for the internal variables are disjoint. This property, however, is virtually never satisfied in economic models since each external variable typically affects equilibrium values of more than one internal variable. In fact, it is difficult to think of nontrivial models in any area of research in which the modularity assumption is satisfied (Cartwright, 2000). In any case, when modularity is satisfied the resulting causal ordering on internal variables is empty, so causal analysis is rendered trivial.

When modularity fails, Pearl's method of analyzing interventions is valid if the variable Pearl treats as a cause is in fact causally prior to the effect variable in the sense defined in this paper. If not, however, replacing the equation determining the purported cause variable with direct determination of the equilibrium value of that variable amounts to jettisoning the model in which the question at issue is inherently ambiguous—what is the effect of one variable on another?—in favor of a different model in which that question is unambiguous. There is no reason to presume that causal analysis based on the altered model has any relevance for the original model.

To get a clearer idea of the problems Pearl's algorithm entails when the requisite causal ordering fails, we consider Pearl's application in a supply-demand model like those analyzed above. Pearl (p. 215) wrote the model as follows:

(48) $q = a_{qp}p + b_{qi}i$

(49) $p = a_{pq}q + b_{pw}w$,

where we have deleted the error terms since they play no role in the analysis. Here (48) is a structural demand equation and (49) is a structural supply equation. As before, i is income and w is weather. These external variables enter the demand equation and the supply equation, respectively. This model conforms to Pearl's interpretation of the equations of structural

---

[13]This term was used by Cartwright and Reiss (2003), whose criticism of Pearl is similar to that presented here. Fennell (2006) argued against Pearl's sweeping assertion that structural models are inherently modular, and in particular against Pearl's attribution of modularity to Simon.

models as representing distinct causal laws, one for each internal variable; here price causes quantity in the demand equation, whereas quantity causes price in the supply equation. Economists will be puzzled by this asymmetric modeling of supply and demand; however, some such specification is required under Pearl's characterization of structural models.

Pearl noted that three queries can be distinguished:

1. What is the expected value of "the demand q" [quotation marks supplied] if the price is controlled at $p = p_0$?

2. What is the expected value of "the demand q" if the price is reported to be $p = p_0$?

3. Given that the current price is $p = p_0$, what would be the expected value of "the demand q" if we were to control the price at $p = p_1$?

Observe the syntax here: despite Pearl's terminology, the symbol q refers to (equilibrium) quantity, which equals quantity demanded and quantity supplied equivalently, as above. Pearl's use of the phrase "the demand q" reflects his specification that quantity is determined by price in the demand equation (48). In contrast, price determines quantity in the supply equation (49). One wonders whether Pearl would accept the question "What is the expected value of 'the supply q'if the price is controlled at $p = p_0$?" as being equivalent to question 1, on the grounds that quantity demand equals quantity supplied in equilibrium, or whether instead he would regard that question as inapplicable in the system (48)-(49), in which supply quantity is a cause of price, not an effect.

In a footnote Pearl reported that he has presented this model and these questions to well over one hundred econometrics students and faculty. He found that the respondents had no trouble answering 2, but only one person could solve 1, and none could solve 3. With the exception of the one respondent who could answer question 1 to Pearl's satisfaction, this is exactly the response pattern that one would hope for based on the analysis of this paper: if the unsatisfactory phrase "the demand q" is replaced by simply "q", the correct response to questions 1 and 3 is that they are ambiguous because price does not cause quantity in the system (48)-(49).

Under Pearl's algorithm, however, questions 1 and 3 are not ambiguous. They are answered by deleting (49) and replacing it with the equation $p = p_0$ or $p = p_1$, respectively. Thus the relevant causal parameter is a_{qp}; the supply elasticity a_{pq} by assumption plays no role.

These difficulties arise because, as we have seen, Pearl's representation of an economic model differs in key respects from the representation of a model which most economists would feel comfortable working with. In contrast to

Pearl's view, we have argued that defining "structural" models as models that directly encode causal ideas is a dead end.

3.3 Heckman

Heckman (1999) analyzed causality in the context of the standard supply-demand model, as here. The example Heckman used to analyze causation can be written as follows:

$$q_s = a_{sp}p_s + b_{sw}w \tag{50}$$
$$q_d = a_{dp}p_d + b_{di}i. \tag{51}$$

This structure is similar to (17)-(18) above except that the supply price p_s is distinguished from the demand price p_d. Because the two equations of this model have no common variables, they can be analyzed separately. Heckman did so: he interpreted a_{sp} as measuring the causal effect of p_s on q_s, just as b_{sw} measures the causal effect of w on q_s. The interpretation of the demand function is similar.

In characterizing equilibrium the equations for supply and demand are combined by appending the identities $p_s = p_d = p$ and $q_s = q_d = q$. Note the contrast between this treatment and that proposed here. We have analyzed causation from the system as a whole, whereas Heckman analyzed causation from each equation taken separately. Heckman was explicit about this:

> If prices are fixed outside of the market, say by a government pricing program, we can *hypothetically* vary p_d and p_s to obtain causal effects for (50) and (51) as partial derivatives or as finite differences of prices holding other factors constant (p. 10; emphasis in original).

Under the definitions proposed here, the parameter a_{sp} is not associated with causation because, with $p_s = p_d = p$ and $q_s = q_d = q$ added to the model, p does not cause q. Rather, these variables are determined simultaneously.

In this example Heckman treated p_s and p_d as external variables for the purpose of analyzing causation, even though p is internal when the equilibrium conditions are imposed. This treatment leads to puzzles. For example, suppose the equations are renormalized on prices rather than quantities (as observed in note 1, it is not clear that Heckman would accept the renormalized version of this equation as equivalent to the original version).

Then would the parameter $1/a_{sp}$ be interpretable as measuring the effect of q_s on p_s? Can a_{sp} measure the effect of p_s on q_s at the same time as

$1/a_{sp}$ measures the effect of q_s on p_s, or do we have to choose? Either way, under Heckman's treatment there appears to be no asymmetry involved with causation. Causal orderings are no longer orderings in the mathematical sense. In contrast, the treatment here is fully in the spirit of simultaneous-equations modeling; a supply-demand model with both price and quantity as internal variables is a different animal from a demand (or supply) equation with price taken as external, and one cannot substitute one for the other in analyzing causation.

4 Conclusion

In this paper we have presented a relatively detailed exposition of the received Cowles account of causation in social science models, together with a rationale for the Cowles treatment of causal orderings as stating conditions under which interventions are or are not unambiguous. Also, we have compared the Cowles analysis of causality with more recent discussions, concluding that the more recent discussions differ in essential respects both from the Cowles treatment and from each other. Considering that it is exactly the purpose of social science models to provide a framework for the disciplined analysis of causation, this is not a very satisfactory situation. One hopes that analysts with an interest in philosophical inquiry will try to pull together these various lines of thought in the analysis of causal structure.

Acknowledgments

I have received helpful comments from Nancy Cartwright, Daniel Hausman, Damien Fennell, Julian Reiss and two referees.

Stephen LeRoy
University of California, Santa Barbara, USA.
sleroy@econ.ucsb.edu

BIBLIOGRAPHY

Cartwright, N. (1995). Probabilities and experiments. *Journal of Econometrics*, 67:47–59.

Cartwright, N. (2000). Measuring causes: Invariance, modularity and the causal Markov condition. *Measurement in Physics and Economics discussion papers 10/2000*.

Cartwright, N. and Reiss, J. (2003). Uncertainty in economics: Evaluating policy counterfactuals. reproduced, CPNSS, London School of Economics.

Engle, R. F., Hendry, D. F., and Richard, J.-F. (1983). Exogeneity. *Econometrica*, 51:277–304.

Fennell, D. J. (2004). Comments on Stephen LeRoy's 'Causality in Economics'. reproduced, London School of Economics.

Fennell, D. J. (2006). Causality, mechanisms and modularity: Structural models in econometrics. reproduced, London School of Economics.
Fisher, F. M. (1976). *The Identification Problem in Econometrics*. Krieger, New York.
Granger, C. W. J. (1969). Investigating causal relations by econometric models and cross-spectral methods. *Econometrica*, 37:424–438.
Hausman, D. M. (1998). *Causal Asymmetries*. Cambridge U. P., Cambridge.
Heckman, J. J. (1999). Causal parameters and policy analysis in economics: A twentieth century retrospective. National Bureau of Economic Research.
Heckman, J. J. (2004). The scientific model of causality. Reproduced, University of Chicago.
Hood, W. C. and Koopmans, T. C., editors (1953). *Studies in Econometric Method*. John Wiley and Sons, Inc.
Hoover, K. D. (1991). The causal direction between money and prices. *Journal of Monetary Economics*.
Hoover, K. D. (2001a). *Causality in Macroeconomics*. Cambridge University Press, Cambridge.
Hoover, K. D. (2001b). *The Methodology of Empirical Macroeconomics*. Cambridge University Press, Cambridge.
Hoover, K. D. (2003). Nonstationary time series, cointegration, and the principle of the common cause. *British Journal of the Philosophy of Science*, 54:527–551.
Leamer, E. E. (1985). Vector autoregressions for causal inference? volume 22. Carnegie-Rochester Conference Series on Public Policy.
LeRoy, S. F. (1995a). Causal orderings. In Hoover, K. D., editor, *Macroeconometrics: Developments, Tensions and Prospects*. Kluwer Academic Publishers.
LeRoy, S. F. (1995b). On policy regimes. In Hoover, K. D., editor, *Macroeconometrics: Developments, Tensions and Prospects*. Kluwer Academic Publishers.
Lucas, R. E. (1976). Econometric policy evaluation: A critique. In Brunner, K. and Meltzer, A. H., editors, *Carnegie-Rochester Conference Series on Public Policy*. North-Holland.
Marschak, J. (1953). Economic measurement for policy and prediction. In Hood, W. C. and Koopmans, T. C., editors, *Studies in Econometric Method*. John Wiley and Sons, Inc.
Pearl, J. (2000). *Causality: Models, Reasoning and Inference*. Cambridge University Press, Cambridge.
Reichenbach, H. (1956). *The Direction of Time*. University of California Press, Berkeley.
Shafer, G. (1996). *The Art of Causal Conjecture*. Cambridge University Press, Cambridge.
Simon, H. A. (1953). Causal ordering and identifiability. In Hood, W. C. and Koopmans, T. C., editors, *Studies in Econometric Method*. John Wiley and Sons, Inc.
Sims, C. A. (1977). Exogeneity and causal ordering in macroeconomic models. In Sims, C. A., editor, *New Methods of Business Cycle Research: Proceedings from a Conference*. Federal Reserve Bank of Minneapolis.

Causality, mechanisms and modularity: structural models in econometrics

Damien Fennell

ABSTRACT. Structural equation models (SEMs) play a central role in econometrics. This paper begins by presenting a dispute between Stephen LeRoy and Judea Pearl over how structural models should be interpreted in econometrics. The dispute is about how interventions should be understood in these structural models. The dispute is over whether it should be assumed that the causal systems modelled are modular, that is, whether each effect can be intervened-to independently of its causes and the rest of the causal structure. This paper argues that modularity is not appropriate for understanding these econometric models. To do this, it first presents Herbert Simon's influential work on the semantics of econometric models to show that he does not assume modularity. The paper then adds to this by investigating how modularity relates to mechanisms modelled by economic models, and how these mechanisms are to be identified. This provides further reasons for not assuming modularity for structural models in econometrics.

1 Introduction

This paper investigates a dispute over how structural models in econometrics should be interpreted. In particular, it analyses how interventions should be understood for structural models in econometrics and whether, in making the structural content of structural models explicit, it should be assumed that the causal systems modelled are modular, that is whether it should be assumed that each effect can be intervened-to independently of its causes and the rest of the causal structure. After presenting the dispute, the paper presents influential work on the semantics of econometric models by Herbert Simon to argue that he does not assume modularity. The paper finishes with a brief, related analysis of how the issue of modularity relates to how one understands and identifies the mechanisms that economic models model. This provides further grounds for being suspicious of the appropriateness of modularity for structural models in econometrics.

2 A disagreement over causal interpretation

In his recent monograph *Causality* Pearl (2000) presents the following simple structural economic model which he has shown widely to econometrics students. The model is

$$q = b_1 p + d_1 i + u_1$$
$$p = b_2 q + d_2 w + u_2$$

where q denotes quantity demanded for a good, i denotes household income, p denotes the price of the good, w is the wage for producing the good while u_1 and u_2 are error terms denoting omitted factors. This is a simple supply and demand model of the type presented in basic economic textbooks. Pearl's interest is how such simultaneous structural models should be interpreted. In his book he sets out a careful and elaborate semantics for structural models which he applies to his example to answer the following questions.

1. What is the expected value of the demand Q if the price is *controlled* at $P = p_0$?

2. What is the expected value of the demand Q if the price is *reported to be* $P = p_0$?

3. Given that the current price is $P = p_0$, what would be the expected value of the demand Q if we *were to control* the price at $P = p_1$?

(Pearl, 2000, p.216, original emphases).

Pearl notes that econometrics students were able to answer the second question but almost all were unable to answer questions 1 and 3 (Pearl, 2000, p.216, n.10). Pearl takes this to be symptomatic that the original meaning of the structural models has become obscured and that the work of the founders of econometrics on the interpretation of structural equation models, such as that by Haavelmo, Koopmans and Simon has been forgotten. Part of Pearl's aim is to set out an explicit interpretation of structural models for econometrics (and other disciplines) and thus reverse the decline.

Pearl's semantics for such models is that the left hand (i.e., dependent) variable is to be read as directly caused by the right hand (i.e., independent) variables. So the first equation shows how quantity demanded is directly caused by price of the good, income and other omitted factors. Following Pearl, such equations denote mechanisms, in this case a demand mechanism relating the direct causes of demand to quantity demanded. Similarly, the second equation denotes a supply mechanism specifying the direct causes of the price. As such, an equation denotes an invariant relationship between the value of an effect and the values of its causes, that is, under some set of possible changes in the values of the causes the effect will take the value as specified by the equation.

In addition, Pearl adopts a particular position on how interventions in the system are to be modelled. In his view, interventions are made up of 'atomic interventions' in which an effect variable is set to a particular value by replacing its equation with an equality setting it to the intervened-to value while leaving other equations unchanged. As reflected in the absence of any change to any other equation, it is assumed that the intervention does not have any impact on any other part of the causal system. In addition, since one simply sets the variable to a particular value, replacing its original equation, no assumption is made that the intervention is carried out through the direct causes which the replaced equation modelled. In this way, Pearl's atomic interventions are 'surgical' in that they only affect the variable intervened to (and its effects in turn) and do so in a way that breaks any links with its causes, that is, the effect is changed independently of its direct causes, in the sense that this relationship from the direct causes and the effect is 'broken' by the intervention.

As Pearl notes, his semantics can be used to answer the three questions above. To answer the first, one uses the semantics for intervention into an effect variable. This requires that one replace the second equation by $p = p_0$. The intervention is assumed to have no impact on the distributions of the errors or other independent variables. Therefore calculating the expectation for Q in this revised model provides an answer to the first question. The second question (answerable by the econometrics students) is straightforward since one simply uses the distributions of the error terms and independent variables, and the two equations, to calculate the conditional expectation $E(Q|P = p_0)$. Whereas for the third question, one uses a revised model, that is where the second equation is replaced by $p = p_1$, and calculates the expectation of Q conditioned on $p = p_0$.

It is clear that a central element in Pearl's semantics is to use an equation-replacing method to represent interventions in which an effect variable is set to a particular value. Crucially it is also where interventions are involved in his questions that the econometrics students had trouble responding. Pearl takes this to be a sign that the distinct content of structural models has become obscured in econometrics and that the important work of the founders of econometrics has been forgotten. But is this the right interpretation? Perhaps econometricians simply have an alternative way of reading structural models like that presented by Pearl.

This, for instance, is the view of LeRoy (2004)[1] who takes issue with the way Pearl interprets structural models and offers an alternative interpretation as to why econometrics students were unable to answer Pearl's first and third questions above. LeRoy first criticises the syntax of Pearl's model, where in the supply equation the price is on the left hand side and on the right hand side in the demand equation. This asymmetric syntactic treatment reflects Pearl's asymmetric causal interpretation in which quantity demanded is caused by price and price is caused by quantity supplied. LeRoy notes that both the presentation and the associated interpretation is unorthodox and likely to confuse econometricians, who are more used to quantity and price being modelled symmetrically. In short, the econometric version of the Pearl's model would be

$$q = b_1 p + d_1 i + u_1$$
$$q = \frac{1}{b_2} p - \frac{d_2}{b_2} w + u'_2$$
$$where\ u'_2 = -\frac{u_2}{b_2}.$$

Of course, this version of the model is not compatible with Pearl's causal reading (of the left hand variable denoting the effect of the its direct causes on the right). In standard econometric readings this is an equilibrium model of supply and demand in which q and p denote respectively the equilibrium quantity transacted (i.e., that is supplied and demanded) and the equilibrium price at which the good is transacted. Under this reading, the equilibrium price and quantity are co-determined by the other, external (or exogenous) variables.[2] Indeed, this more standard reading is captured by LeRoy's alternative way of interpreting structural models in which neither price nor quantity is a cause of the other. For LeRoy, therefore, the econometrics students' answers are reassuring because the first and third questions posed by Pearl are ambiguous. For instance, in LeRoy's view, how p is set to p_0 (since the value taken by p is determined by i, w and the error terms) will have implications as to the (expected) value of q. The ambiguity is that unless further information is provided as to how p is set to p_0, there is no clear answer as to what will happen to q.

In any event, the aim here isn't to discuss in detail Pearl's or LeRoy's interpretations of simultaneous equation models, but to point out an important difference of opinion that shows where effort is required in clarifying structural models in econometrics. The above has shown that Pearl's method for interpreting structural equations models, by his own admission, does not seem to fit with the way econometrics students understand the

---

[1] For an updated and abridged version of this paper see LeRoy's paper in this volume.
[2] An external (exogenous) variable is one whose value is determined outside the model.

models. I think, in line with LeRoy, that it is rather uncharitable of Pearl to assume that the econometrics students are all ignorant of how to interpret structural models. Instead, I think a more charitable reading requires that one look more carefully at what interpretations of structural models might make sense of the econometrics students' puzzlement at Pearl's questions. Perhaps LeRoy's alternative reading is a more appropriate reading for econometric structural models after all?

One can further clarify the issue at stake here by noting the particular way Pearl models interventions using his equation-replacing method. These are surgical interventions that allow one to directly fix an effect variable without changing anything else. This property of causal systems in which every effect variable can be cleanly intervened to is known as *modularity*. To assume modularity is clearly a substantive constraint on causal systems and unsurprisingly there has been debate as to how widely it holds in causal systems. Specifically, Woodward (2003) and Hausman (1998) have adopted a position, like Pearl's, that modularity is a universal feature of causal systems. In sharp contrast, Cartwright (2001) has countered that modularity is not only not universal, but typically does not hold.

This raises an interesting question in relation to the economics example above. Pearl assumes modularity for causal systems and makes use of it in making counterfactual claims about the results of interventions for his economic model. Since these counterfactual claims are disputed by LeRoy, it raises a natural question: is modularity an important assumption for the causal systems modelled in econometrics? More specifically, if one follows Pearl, Woodward, Cartwright and others as reading structural equations as denoting mechanisms, should one expect the systems made up of economic mechanisms to be modular?

In the section that follows, I try to shed some light on this question. To do this, I go back to basics by presenting causal semantics set out by Simon (1953). The reason for looking at Simon's work is twofold. First, Simon is one of the founders of modern econometrics and his work on the causal semantics of structural models is not only highly developed, but it has been highly influential on all the discussants mentioned here, that is, Pearl, Woodward, Cartwright and LeRoy's. The second reason connects more directly with Pearl, who "wonders ... what has happened to [structural equation modelling] over the past 50 years, and why the basic (and still valid) teachings of Wright, Haavelmo, Koopmans, and Simon has been forgotten" (Pearl, 2000, p.137). It seems sensible therefore to try to clarify what the teachings of Simon were and whether he assumed modularity, since if he didn't, then this seems to raise serious questions over whether Pearl's modular approach is really the correct approach for making explicit the structural content of econometric models.

3 Simon's semantics for structural models

In his seminal paper, Simon (1953) presents a formal definition of causal order for structural equations models. To obtain the causal order, one first distinguishes between two types of variables in the model, internal and external.[3] The internal variables are those that are determined by the model (for example q and p in the example above), while the external variables (for example i and w) and the error terms have values that are taken as given, from outside the model. One then solves for the internal variables one-by-one using the fewest equations required to solve for them; this stipulates an order for the solution of the internal variables. Any variable used to solve for and solved for prior to another variable is defined to causally precede it. One variable directly causally precedes another if it causally precedes it and if it appears in the same equation as the other variable.[4] The resulting ordering among the variables is called the causal order.

Consider Pearl's earlier supply and demand example,[5] where one catego-

---

[3]The account given here is based on the more detailed analysis of Fennell (2005). Note that this version deviates slightly from that of Simon (1953) in one respect. The difference is that Simon bases his causal order on a distinction between coefficients and variables, rather than external and internal variables. My slight modification of Simon is motivated by an attempt to present the models, for which causal order is defined, in terms that fit more closely with econometric practice. Though I think the difference is not significant since it essentially amounts to difference in labelling, some might dispute this for the following reason. In Simon's paper he presents examples of sets of linear equations (like the example above) in which the coefficients in the equations are taken to denote directly controllable factors. In addition, in all the examples he presents in that paper, no coefficients are repeated. This implies that for each of these examples it is possible to intervene separately into each equation, and even more than that, separately wherever a coefficient appears. This property is very close to and in some ways stronger than modularity. However, I think that it would be a mistake to read Simon as making this assumption (that there are no repeated coefficients) generally because it is not an assumption that is made in his formal analysis and nor it is necessary for the formal analysis. In summary, I think that the lack of any repeated coefficients in the examples he presents is accidental, and given the lack of any explicit assumption to that effect in his formal analysis, the no repeated coefficients assumption should not be taken as a general commitment by Simon.

[4]This is a simplification, in fact the condition here is: if it appears in the same minimal set of equations used to solve for the variable. See Simon (1953) and Fennell (2005) for details.

[5]Though the equations are written in LeRoy's preferred form, the interpretation would be the same if they were written in Pearl's form. In other words, in Simon's interpretation the location of a variable with respect to the '=' in an equation is not significant.

rizes q and p as internal and i and w as external.

$$q = b_1 p + d_1 i + u_1 \ldots demand$$
$$q = \tfrac{1}{b_2} p - \tfrac{d_2}{b_2} w + u_2' \ldots supply$$

Here one constructs the causal order by noting that q and p can only be solved for together in terms of i and w and the error terms, using both equations. Moreover, since both i and w appear in an equation with q and p, both are direct causes of p and q. Also, since p and q are both determined together (in the same minimal set of equations) they are 'co-determined'. Thus, the causal order can be represented as in figure 1 (where the arrows denote direct causal precedence). Simon's causal order yields an intuitive result for the example since it makes explicit that income and production costs are direct causes of equilibrium price and quantity, while equilibrium price and quantity are co-determined, just what one would expect for an equilibrium model of supply and demand.

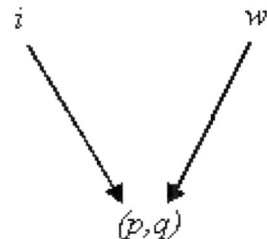

Figure 1. The Causal Order

Nevertheless, at this stage Simon's causal order is merely a formal relation among the variables defined from the functional form of the equations. It is simply a syntactic trick for deriving claims labelled as 'causal'.[6] Helpfully, however, Simon discusses how variables and equations should be interpreted. In his semantics, the external variables should be taken to denote factors that are directly controllable by an 'experimenter' or 'nature', and internal variables taken to denote factors that are indirectly controllable.[7] The

---

[6] Stated differently, the concern is that one follows Simon's particular solution procedure that yields an order which is then termed 'causal' by Simon, but why is it appropriate to call that particular solution order 'causal'?

[7] As shown below, proponents of modularity allow internal variables to have their

external variables are variation free, meaning they can take any value as a group as they can individually. Equations are taken to denote mechanisms and error terms are taken to denote the joint effect of omitted directly controllable factors in a mechanism. The core idea is that the experimenter or nature has direct access to the directly controllable factors and is free to change them. Changing these then has an impact on the other indirectly controllable factors in virtue of the mechanisms that connect the indirectly controllable factors to the directly controllable factors. Under this interpretation, the causal order arises from the joint action of mechanisms, and maps out how changes in a factor will 'in general'[8] lead to changes in other factors. Under this reading a cause 'in general' changes its effects, whereas changes in an effect need not be accompanied by changes in any one of its causes.[9]

4 Does Simon assume modularity?

In this section, I ask whether Simon adopts modularity. To do this, I compare and contrast Simon's analysis of an important example with that of James Woodward who explicitly uses modularity (in much the same way as Pearl) in his analysis of what is essentially the same example.

In Simon's paper there is an important section in which Simon uses his semantics to show how one can distinguish between two observationally equivalent systems of equations which have different causal orders. The following is an adapted version of the example (Simon, 1953, pp.24-25) presents.[10]

$$\begin{array}{ll} System\ 1 & System\ 2 \\ y_1 = ax + u_1 & y_1 = ax + u_1 \\ y_2 = bx + u_2 & y_2 = cx - y_1 + u_3 \end{array}$$

In the first system y_1 and y_2 are commonly caused by x whereas in

---

values set independently of the rest of the system in a 'direct' way. This suggests that proponents of modularity may disagree with indirect versus direct control distinction as applied to external and internal variables here. Though this suggest another siginficant difference between Simon's approach and those who adopt modularity, I leave its analysis as further work.

[8]The 'in general' here is meant in the colloquial sense of almost always. It is there to allow for the case where the impacts of causes cancel themselves out thus have no net impact on some of their effects.

[9]An exception is the case where an effect has only one cause.

[10]This system is different from the one Simon presents because my use of the internal-external variable distinction and the introduction of error terms. Also, it is put into a form that makes it easier to compare with Woodward's. The changes here do not affect the substance of the discussion.

the second y_1 is also a cause of y_2.[11] There is a problem of observational equivalence here because if one system holds it can be transformed into the system of the other form.[12] Therefore, both can be used to model the same observations. So how to distinguish between the two? In his discussion of the example, Simon notes that the following property can be used to distinguish the two systems. If the first system's causal order held then if one were to intervene by changing u_1 then only y_1 would change. In contrast, if the causal order in the second system held, then for almost all values of a and c,[13] a change in u_1 would change both y_1 and y_2. This gives a way, according to Simon, to operationalise causal order. The idea is to observe interventions such as these and to observe what variables change with which. So, for instance, if one were to change u_1 and only y_1 changed then this would suggest the first system and its causal order held rather than that of the second.

At this stage it may not be clear what Simon's example has to do with modularity. Yet Woodward (2003, p.330) uses essentially the same example with similar analysis to show how modularity of systems can be used to solve the observational equivalence problem. For Woodward, structural equations denote mechanisms where—like Pearl—the right hand variables in an equation denote the direct causes of the left hand variable. For Woodward, modularity of a system of equations causal system ensures it is possible to intervene into any effect-variable without changing any other equation in the system. Such interventions are modelled, like Pearl, by replacing the equation for the variable with an equality setting it to the intervened-to value.

So Woodward's interpretation of the systems

$$y_1 = ax + u_1 \qquad (1)$$
$$y_2 = bx + u_2 \qquad (2)$$

---

[11] Note that the causal order of these systems is the same read using the Pearl-Woodward approach and using the Simon approach.

[12] Provided one defines the coefficients and u's appropriately, for instance if one derives the second set of equations from the first, then one must define $c = a+b$ and $u_3 = u_1+u_2$ for mathematical equivalence. Such constraints don't help the observational equivalence problem, however, because typically the values of the coefficients are unknown and the u's are error terms denoting unknown omitted factors. Therefore, these constraints do not provide epistemic help in distinguishing between the two sets of equations (though in cases where there is some additional information about the coefficients or the error terms, the observational equivalence may be avoided).

[13] Though crucially if the two systems are mathematically equivalent then the change in u_1 would not change y_2. This is an important caveat which is discussed in more detail below.

and

$$y_1 = ax + u_1 \qquad (3)$$
$$y_2 = cx - y_1 + u_3 \qquad (4)$$

would be that to determine which system holds (and thus solve the observational equivalence problem) one can use knowledge as to which is modular. If the first system is modular, then it is possible to intervene in (1) to change y_1 without having to change x or u_1 and this does not change (2). As Woodward (2003, p.330) describes, such an intervention amounts to replacing equation (1) with $y_1 = y$.[14] Since equation (2) is unchanged then y_2 will not change under this intervention. Conversely, if the second system is modular then it is possible to intervene in (3), replacing it by $y_1 = y$, without violating (4), thus in this situation y_2 would change value. Since the same set of possible interventions, those that replace (1) or (3) by $y_1 = y$ lead to distinct consequences for y_2 depending on which system is modular, modularity can be used, as Woodward notes, to resolve the observational equivalence problem.

The similarity between Simon's and Woodward's analyses is obvious. The key difference is in how the intervention, which does the work of distinguishing the observationally equivalent systems, is conceived. Woodward follows Pearl in using equation replacing to model the interventions that are made possible by modularity, while Simon uses changes in error terms or external variables to model the interventions.

Returning to the question at hand, does the Simon approach assume modularity, like Woodward's? The similarity between the treatments makes it tempting to say 'yes', but I think this is mistaken. Though Simon's solution, like Woodward's, relies on intervening on a specific equation to see what changes, which suggests modularity.[15] Unlike Woodward, Simon puts an important caveat into his analysis. In his generalisation of the analysis of the example, Simon (1953, p.25) presents a theorem which states that for a system, interpreted with his causal order, changes in causes almost always lead to changes in their effects. The 'almost always' is an indication that there is a situation in which Simon's analysis will fail to lead to a discernible difference between the two systems. Such a situation can be

---

[14]The similarity of Woodward's modelling of an intervention to Pearl's is explicit here.

[15]It suggests modularity in the sense that each equation can be intervened to (in some sense) separately from the others.

seen if one considers the two systems once again.

$$
\begin{array}{ll}
System\ 1 & System\ 2 \\
y_1 = ax + u_1 & y_1 = ax + u_1 \\
y_2 = bx + u_2 & y_2 = cx - y_1 + u_3
\end{array}
$$

Suppose that $c = a + b$ and $u_3 = u_2 + u_1$, but that the second system is the correct system with correct causal order. In this case, y_1 causes y_2 but changes to y_1 by changing u_1 or x will have no net impact on y_2 because the direct impact of x or u_1 on y_2 and indirect impact via y_1 will cancel each other out. In this case the observed changes will be consistent with the causal order of the first system and observations will not clarify which system holds (i.e., the first system, or the second system where $c = a + b$ and $u_3 = u_2 + u_1$).[16]

To understand the connection with modularity, consider how Woodward's analysis yields a different result in the same case. For Woodward one can surgically intervene to change the first equation to $y_1 = y$ without having to intervene via x or u_1. In essence, Woodward's intervention 'breaks' the mechanism and replaces it. This allows y_1 to be changed independently of its explicitly modelled causes and thus enables the resulting impact of y_1 to be 'no longer cancelled out'. In this way, if the system is of the second form, one can intervene to fix y_1 and then infer from observations the second equation.

So, modularity as espoused by Woodward and Pearl takes a step beyond Simon's analysis of the two mathematically equivalent but causally distinct systems. I think that Simon's introduction of an explicit caveat to allow for the situation ruled out by modularity for Woodward gives us clear reason to believe that Simon does not assume modularity as it has been adopted by Pearl and Woodward.[17]

---

[16] Note that in my reading of the causal orders I here diverge from LeRoy's causal interpretation of the two systems of equations. In contrast, in LeRoy's interpretation (2004), y_1 and y_2 are causally unordered in both systems. For further information about the differences between my reading of Simon and LeRoy's, see Fennell (2005).

[17] Note that even if one disagrees with footnote 3 above and one claims that Simon, by not allowing repeated coefficients in his models, assumes a form of modularity the divergence set out here between Woodward and Simon's analyses remains. This is because of the difference in how the interventions are modelled. In Simon's case interventions (even clean surgical ones) come about via explicitly modelled variables, whereas in Woodward (and Pearl) one cuts away the explicitly modelled causes of an effect and sets that effect to some desired value independently.

5 Modularity and identifying mechanisms

There is a connection between modularity and how one identifies mechanisms which is not only important, since it highlights one of the main motivations for adopting modularity, but also because it provides further evidence that Simon, by adopting a non-modularity approach to mechanism identification, does not assume modularity.

One might wonder what one gains and loses by adopting modularity. There is one obvious epistemic limitation that comes with systems that are not modular—one loses the possibility of a clean intervention into any part of the causal system. In contrast in a modular system it is possible to fix an effect variable independently of its modelled causes. This provides obvious epistemic advantages. Crudely, one could wiggle each part of the system in turn to make inferences about individual causal relations.[18] As Nancy Cartwright puts it, modularity is an assumption which makes causal systems 'epistemically convenient' (Cartwright, 2001, p.66).[19] This is also a disadvantage, since it rules out modelling systems that are not epistemically convenient in this way.

Related to epistemic convenience is an advantage of modularity: it can be used as the basis of mechanism identification. For instance, one can follow Woodward (2002) and define a mechanism to be an equation which is invariant to some set of possible interventions in a modular system of equations.[20] Not only does this provide a reductive definition of mechanisms in terms of possible intervention, it has the epistemic advantage mentioned above. The advantage is if one observes a set of correlations which are adequately modelled by a class of (observationally equivalent) sets of functional relations, then provided one can intervene to fix the different effect variables, knowing that these are interventions that only fix the effect-variables independently of the rest of the causal system, then one can infer which of the sets of functional relations among the class of observationally equivalent

---

[18] Indeed, this is exactly what Woodward does to distinguish the two systems above.

[19] Cartwright (2001) also argues that one should not expect all causal systems to be epistemically convenient. For this reason she argues against those who build in such a concept into their causal systems, for example, Woodward, Pearl and Hausman.

[20] As far as I can tell from his brief discussion of the subject, Pearl does not adopt a Woodward view of mechanisms. For instance, he writes 'the world consists for a huge number of autonomous and invariant linkages or mechanisms, each corresponding to a physical process that constrains the behaviour of a relatively small group of variables' (Pearl, 2000, p.223, my emphasis). Statements like these (see Pearl (2000, pp.223-226)) suggest that Pearl associates mechanisms with the physical machinery that generate the invariant relations expressed in the structural equations, rather than equating the mechanisms with those invariant relations, as Woodward (2002) does. Therefore, unlike the modularity assumption, which is made by both Woodward and Pearl, the invariance view of mechanisms is attributed to Woodward only in the discussion that follows.

ones is the correct, causally accurate system. In other words, modularity allows one to define mechanisms in terms of possible interventions and has the epistemic advantage that provided one can knowingly carry out 'modular' interventions one can solve the observational equivalence problem by experimentation. Indeed, this is what is done in Woodward's analysis of the example above where, depending on which system holds, 'modular' interventions could be used to solve the observational equivalence problem.

In contrast it was seen above that Simon's approach did not make use of 'modular' interventions. This lead to the epistemic inconvenience of having an observational equivalence problem remain between the two systems, when the coefficients took on particular, inconvenient values. Hand in hand with this, one would expect Simon to adopt an alternative approach to mechanisms and to how one determines which mechanisms hold. This is indeed the case. In a later paper Simon and Rescher (1966, p.329) explicitly comment that what distinguishes two systems, such as the two in the example analysed above, is the assumption that each equation in the correct set of equations denotes a mechanism. Transforming such a system into a mathematically equivalent system with differing causal order scrambles up the mechanisms. Therefore, the transformed system does not have equations that denote mechanisms.

However, this move is purely conceptual and, as Simon and Rescher note, does not help with the problem of observational equivalence, since it leaves unaddressed the problem of how to work out which system has equations that correctly denote mechanisms. To address this, however, they do not appeal to sets of possible interventions like Woodward. Instead, Simon and Rescher give a brief discussion of some ways of making progress with this. They outline a 'principle of prepotence' and the 'empty world postulate'. The first assumes that systems of high energy will be causally independent of systems of low energy while the second assumes that there is no causal connection between two variables, unless there is a plausible mechanism by which one variable could affect the other. Now, it is debatable to what extent these two principles are acceptable and in what contexts. However, what is not debatable is that Simon's approach is completely different from Woodward's. Simon's approach is appealing to background assumptions about what is and is not causally connected to identify mechanisms. This is unlike Woodward who falls back on what relations are invariant to different kinds of interventions. In short, in the example above, where the two causal systems were indistinguishable by observation in Simon's analysis, these two systems would be distinguished by appeal to background knowledge as to the whether y_1 does cause y_2 or not. This appeals to information not explicitly modelled in the equations.[21]

In short, I think this analysis strengthens further the view that Simon's approach differs crucially from Pearl's and Woodward's by not assuming modularity. And, to return to the contentious example presented by Pearl, it suggests that LeRoy (and the confused econometrics students) may be closer to Simon than Pearl, since LeRoy (like Simon) treats interventions through the modelled variables.[22]

6 Modularity and econometrics

The discussion so far has briefly explored some of the work of Herbert Simon on how structural models in econometrics should be interpreted. It has presented some evidence to support the view that Simon does not adopt modularity. In as much as one is willing to take Simon's semantics as representative of the way econometricians interpret structural models—and LeRoy's reading of Pearl's example is consistent with this—then this seems to speak against using Pearl's or Woodward's semantics in econometrics. Of course this line of argument is open to criticism since it is still debatable that econometricians really do follow Simon's semantics. Nevertheless, there are independent reasons why econometricians might follow the Simon-approach to mechanisms rather than the modularity approach.

To see this, I consider a simple economic model and contrast what would be required to identify mechanisms following the Woodward approach and the Simon approach. Reconsider Pearl's contentious example:

$$q = b_1 p + d_1 i + u_1$$
$$p = b_2 q + d_2 w + u_2.$$

Following Woodward, these equations are invariant to a set of possible interventions in p, q, i and w. Moreover, by modularity there exist possible interventions by which q and p can be set independently of their causes as modelled in this system. Since for Woodward these invariance relations exhaust the mechanism, it seems that the only way to find out about such

---

[21] There is another way one can distinguish Woodward's approach from that of Simon's. Woodward (2002) aims to give a counterfactual account of causal relations (hence his counterfactual definition of a mechanism) whereas Simon and Rescher (1966) do the exact opposite. They explain causal counterfactuals using causal systems. Indeed, it seems that modularity is a key assumption in using counterfactuals to explain causal relations for Woodward, because it allows one to uniquely specify causal relations from a set of counterfactual claims about what happens under intervention.

[22] This should not be surprising since Simon's approach to structural modelling was historically much more influential than an alternative account, championed by Wold, which does explicitly use modularity in the equation-replacing way. See Strotz and Wold (1960) and Morgan (1991) for more on this.

mechanisms is to first find out functional relations that adequately describe the observed correlations between p, q, i and w and then to carry out some interventions to test that the functional relations remain suitably invariant. Any other approach, for instance, one that appeals to underlying causal reasons as to why one set of functional relations (rather than some observationally equivalent alternative) is the invariant one seems to fall back to a richer Simon-like approach. Yet consider what such an approach would require of the econometrician. They (the government etc.) would have to experiment with the economic system to test the invariance of the relations between quantity, price, income and so on. The problem with this is that econometricians simply do not have these possibilities for intervention. Of course, changes are sometimes observed that can be considered to be interventions, but these are usually too few and far between to provide a suitable epistemic basis for identifying mechanisms in line with the Woodward approach. It just seems much too ambitious to try to infer the correct causal system in this way.

In contrast, the Simon approach of appealing to a richer sense of mechanisms seems more practicable. This is because it does not tie the concept of mechanism so closely with intervention. Instead, it builds a justification for the invariance relations, assumed in the structural model, from background knowledge about the causal relations. So for instance, one would justify the simultaneous equations model (corresponding to Pearl's)

$$q = b_1 p + d_1 i + u_1 \ldots demand$$
$$q = \tfrac{1}{b_2} p - \tfrac{d_2}{b_2} w + u'_2 \ldots supply$$
$$where\ u'_2 = -\tfrac{u_2}{b_2}.$$

by appealing to theory as to what is and is not involved in the supply and demand mechanisms. So for instance, for the supply mechanism there might be an appeal to a belief that income is the most significant factor influencing aggregate demand for the market being modelled, because of a belief in consumer theory about the individuals in the market and beliefs that the relation continues to hold when the variables are aggregated. Likewise, there might be good reasons for believing wages to be the significant causal factor for supply.[23] In short, the appeal of the Simon approach, in contrast to that of Woodward, is that it allows an appeal to existing well-established theory and hypotheses about underlying mechanisms in postulating the invariant relations (the structural equations) which are to be parameterised from observation.

---

[23]There are further complexities here such as what justifies the assumption that the dynamic system can be modelled by static equilibrium relations. This was the subject of a fierce debate in econometrics in the mid-twentieth century, see Morgan (1991).

Of course, this advantage comes at a price, the problem is that the mechanisms postulated will be only as good as the theories and causal hypotheses that are used to support them, and there is a worry that the empirical claims of econometricians—if they follow the Simon path—become contingent on theories and hypotheses that may be disputable or contentious. Nevertheless, it has the advantage of being more practicable and being more transparent in that the form of the equations is connected with explicit theoretical claims as to what variables influence which.[24]

Moreover, I do not think it would be difficult to find evidence from econometric analysis that the standard way the functional forms are justified in structural models is by appeal to theory which incorporates a rich mechanistic story as to the causal paths by which one variable may affect others.[25] Though, to do a thorough review of the literature would require a separate study, if this is found to be the case, then this would add further support to the position that the Simon analysis is more appropriate for understanding structural econometrics than the Pearl or Woodward semantics which assume modularity.

7 Conclusion

In conclusion, this paper has attempted to show how the Pearl and Woodward approach of assuming modularity in interpreting structural equations is not that of Simon. This lends support to LeRoy's rejection of Pearl's analysis as to why econometrics students had trouble with the example they were presented. To the extent that one is willing to accept Simon's causal interpretation of structural equations as representative of econometric practice, the paper supports the view that modularity is not assumed in structural econometric modelling. That said, the argument presented here remains incomplete without further support for the claim that Simon's causal interpretation is indicative of how structural models are interpreted in econometrics. I think this argument can be made, though I leave it for another paper.

In addition, I have argued that an absence of modularity in causal semantics in structural econometrics makes sense given structural econometrics' reliance on rich background economic theory. This provides an alternative basis for distinguishing between distinct mechanisms. It (of course) doesn't address the related epistemic question of how we find out about mechanisms

---

[24]Unlike the Woodward position where the justification of the equation form would rely on claims about what sets of possible interventions the equations are invariant to, which would be more convoluted and more difficult to grasp and criticise.

[25]Though econometricians, not liking to be explicit in their use of causal language, will often talk about these using the language of exogeneity or predeterminedness.

and the invariant relations they support. The approach of Woodward has the advantage that it should be possible to identify a set of mechanisms by looking at covariances for a set of observables with respect to a large set of interventions. However, given the limited scope for intervention, it is questionable whether this approach is practicable for structural econometrics.

Acknowledgements

Many thanks to Nancy Cartwright, Stephen LeRoy, Julian Reiss and participants at the 'Causality and Probability in the Sciences' conference (Canterbury, June 2006) and the 'Plurality in Causality' workshop (Bologna, May 2006) for helpful comments I also want to acknowledge the AHRC 'Contingency and Dissent in Science' Project for supporting the research for this paper.

Damien Fennell
Centre for Philosophy of Natural and Social Science, London School of Economics, UK.
D.J.Fennell@lse.ac.uk

BIBLIOGRAPHY

Cartwright, N. (2001). Modularity it can—and it generally does—fail. In M.C. Galavotti, P. S. and Constantini, D., editors, *Stochastic Causality*, pages 65–84. CSLI Publications, Stanford, California, first (2001) edition.
Fennell, D. (2005). *A Philosophical Analysis of Causality in Econometrics*. PhD thesis, University of London, London, United Kingdom.
Hausman, D. (1998). *Causal Asymmetries*. Cambridge University Press, Cambridge, United Kingdom.
LeRoy, S. (2004). Causality in economics. *Causality: Metaphysics and Methods Technical Reports*, 20.
Morgan, M. (1991). The stamping out of process analysis in econometrics. In Blaug, M. and Marchi, N. D., editors, *Appraising Economic Theories*, pages 237–272. Elgar, Cheltenham, United Kingdom.
Pearl, J. (2000). *Causality: Models Reasoning and Inference*. Cambridge University Press, Cambridge, United Kingdom.
Simon, H. (1953). Causal ordering and identifiability. In Simon, H., editor, *Models of Man*. Wiley, New York, New York.
Simon, H. and Rescher, N. (1966). Cause and counterfactual. *Philosophy of Science*, 33:323–340.
Strotz, R. and Wold, H. (1960). Recursive vs. non recursive systems: An attempt at a synthesis. *Econometrica*, 28:417–427.
Woodward, J. (2002). What is a mechanism? A counterfactual account. *Philosophy of Science*, 69:S366–377.
Woodward, J. (2003). *Making Things Happen: A Theory of Causal Explanation*. Oxford University Press, Oxford, United Kingdom.

Time series, nonsense correlations and the principle of the common cause

JULIAN REISS

ABSTRACT. This paper examines some recent defences of the principle of the common cause (PCC) against Elliott Sober's famous counterexample. There are two lines of attack: attempts to defuse the counterexample, that is, to show that the scenario described by Sober only apparently conflicts with the PCC; and attempts to demonstrate that the counterexample has no practical consequences. I show in this paper that there are problems with both strategies. In response, I formulate an alternative version of the principle that avoids the known counterexamples and that makes its status as fallible epistemic principle explicit.

1 Introduction

Time series, that is, time-ordered sets of observations on a random variable or random variables, are of fundamental importance for empirical inferences in sciences as diverse as neurophysiology, climatology, epidemiology, astro- and geophysics and many of the social sciences. In this paper I shall argue that a number of particularities of time series pose serious difficulties for one of the most prominent kinds of account of causal inference: probabilistic theories. A core assumption of probabilistic theories is the principle of the common cause, according to which a correlation between two variables is indicative of a causal connection between these variables. "Nonsense correlations"—i.e., correlations that are artefacts of the statistical properties of the variables or that obtain for other non-causal reasons—pose an obvious problem for probabilistic theories.

Despite the recognition of the problem of nonsense correlations, probabilistic methods of causal inference have become increasingly popular in recent years. This has triggered some discussions about the seriousness of the problem. Some authors have tried to show that it is a mere pseudo-problem and that the principle can be saved once the notion of "correlation" is clarified. Others have accepted the genuineness of the problem but denied that it has serious methodological implications.

Here I shall show that the problem of nonsense correlations is both serious as well as ubiquitous in all areas of science where time-series matter. In response, I try to formulate a version of the principle of the common cause that avoids the nonsense correlation issue and, moreover, makes explicit its status as fallible epistemic principle. The lesson to draw from this discussion is that the methodological monism occasionally proposed by defenders of probabilistic theories of causal inference is mistaken: different kinds of situations require methods of inference to be tailored to the specifics of those situations if the researcher aims at inferring correct causal claims.

Let us start with some essential definitions. Time series $X = \{x_1, x_2, ..., x_T\}$ are time-ordered sets of observations on quantitative characteristics of an individual or a structure such as a socio-economic system. At each point in time, the observations are assumed to be drawn from a probability distribution $P_t(X)$.

It is important to distinguish a time series from the stochastic process that generates it. The stochastic process is the world line of the persisting object (a die, a socio-economic structure) itself whereas the time series records measurements or observations on the process made through (usually identical intervals in) time. Quantitative characteristics of an object can assume different values at different points in time and at each point are assumed to be drawn from a probability distribution $f_t(x) = P_t(X = x)$ that satisfies the usual axioms. X is thus a *random variable*. I will represent a variable by a capital letter X and a value of a variable by a small letter x.

2 The PCC, British bread prices and Venetian sea levels

The principle of the common cause (PCC) lies at the heart of many accounts of probabilistic causation (cf., Pearl (2000); Reichenbach (1956); Salmon (1994); Spirtes et al. (2000); Suppes (1970)).[1] Simplifying slightly, it can be stated as follows:

PCC. If two random variables X, Y are probabilistically dependent, then either X causes Y, Y causes X or X and Y are the joint effects of a common cause Z.[2]

---

[1] More recent accounts adopt the related causal Markov condition (CMC) as core principle. Since in the two-variable case the PCC can be shown to follow from the CMC, a counterexample to the PCC is a counterexample to the CMC, too. I will thus not consider it separately here.

[2] In most formulations the PCC contains also the screening-off condition, which states that the (full) common cause Z screens off the dependence between X and Y: $P(Y|X, Z) = P(Y|Z)$. The screening-off condition is controversial itself, and since my discussion focuses on violations of the first part of the PCC, I omit it here.

Two variables X and Y are probabilistically dependent just in case $P(XY) \neq P(X)P(Y)$. A situation in which a probabilistic dependence between two variables is due to non-causal facts about the properties of the variables constitutes an obvious violation of the PCC. The first systematic discussion of the problem is due to G. Udny Yule (Yule, 1926). Among philosopher's of science, a counterexample introduced by Elliott Sober has been widely discussed (Sober (2001, p.322), Sober (1987, pp.161-2)). In this example, X = sea levels in Venice and Y = cost of bread in Britain. Sober assumes the two variables to increase monotonically in time (Sober, 1987, p.334):

Year (t)	British Bread Prices (Y)	Venetian Sea Levels (X)
1	4	22
2	5	23
3	6	24
4	10	25
5	14	28
6	15	29
7	19	30
8	20	31

An intuitive test for whether two variables are probabilistically dependent is asking whether observing one variable is informative about the likely value the other variable will take. This criterion is clearly fulfilled in this case: a higher observed Venetian sea level allows us to infer higher British bread prices and vice versa. And yet, the two variables are *ex hypothesi* not causally connected.

In principle there are two strategies for saving the PCC in the light of Sober's argument. One could, first, argue that the scenario merely *appears* to be a counterexample to the PCC. Though the data Sober provides makes the underlying variables *look* probabilistically dependent, in fact they are not. We make a fallacious inference from sample statistics to population statistics if we used the PCC in this case rather than a fallacious *causal* inference. A second strategy is to argue that the PCC is indeed violated in the Sober scenario but that it is possible (and indeed, required) to prepare the data prior to analysis in such a way as to avoid the violation of the PCC. I will consider each line of response in turn. (The two strategies can be regarded as complements and used jointly; I'll separate them analytically and examine them one by one.)

3 Fallacious statistical inference

In a recent article (Hoover, 2003), Kevin Hoover argues that although Sober's scenario appears to violate some formulations of the PCC, it does not constitute a counterexample to the spirit of Reichenbach's original idea, which was: "If an improbable coincidence has occurred, there must exist a common cause" (Reichenbach, 1956, p.156). To Hoover, understanding the correct meaning of the word "improbable" is essential here: it means that the observed coincidence must be something out of the ordinary, something unexpected in order to be evidence for an underlying causal connection. That the car that just passes by is a green Volvo, built in 1990, is in some sense improbable—out of all cars that could have driven by, why should it be exactly this green Volvo?—but it is nothing out of the ordinary. If, by contrast, all the members of a theatre troupe develop identical symptoms of food poisoning after a common meal in the theatre refectory, something out of the ordinary has happened.

In order to flesh out the meaning of "improbably" more formally, Hoover distinguishes between two stages of inference from observations to underlying causal relations. In the first stage, statistical inference, the reasoning proceeds from observed sample frequencies to underlying probabilities. In the second, from probabilities to causal relations. The PCC pertains to the second step: it says that once can infer from a fact about probabilities— the probabilistic dependence between two variables—to a fact about causal relations—either one variable causes another or there exists a common case. Sober mistakenly infers from facts about sample frequencies that the two series are probabilistically dependent, which, thus Hoover, they are not.

Statistical inference, according to Hoover, is always conducted against a probability model, that is, a hypothesis about the stochastic process responsible for the generation of the data; that model is accepted, which is most likely to be true, given the data (Hoover, 2003, pp.530f.). Claiming that Sober's data violate the PCC makes a fallacious inference at this stage. There is no doubt that the two variables are associated at the level of the sample. That can be readily verified by calculating the sample correlation coefficient:

(1) $$r_{XY} = \frac{\sum (x_i - \bar{x})(y_i - \bar{y})}{\sqrt{\sum (x_i - \bar{x})^2 \sum (y_i - \bar{y})^2}},$$

where a bar above a variable denotes the sample mean. For the data Sober provides, $r_{XY} = .99$. But, says Hoover, we cannot readily take this as evidence that the underlying population correlation:

(2) $$\rho_{XY} = \frac{E[(x_i - \mu_X)(y_i - \mu_Y)]}{\sqrt{E(x_i - \mu_X)^2 E(y_i - \mu_Y)^2}} = \frac{\sigma_{XY}}{\sigma_X \sigma_Y},$$

where E denotes the expected value of the expression in parentheses, the μ's are the population means of X and Y, σ_{XY} denotes the covariance between X and Y and σ_X and σ_Y are the standard deviations of X and Y, is positive too. "Correlation" and "probabilistic dependence" are related but not strictly equivalent concepts. If $P(XY) = P(X)P(Y)$, then $\rho_{XY} = 0$ (if X and Y are independent, then they are uncorrelated) but the reverse is not necessarily the case.[3]).

In order to understand the details of this argument, a number of concepts from the analysis of time series has to be introduced (cf. Hamilton (1994, pp.45f). for the following definitions). The jth autocovariance of a variable Y_t of some process (denoted γ_{jt}) is defined as:

(3) $\quad \gamma_{jt} = E[(Y_t - \mu_t)(Y_{t-j} - \mu_{t-j})],$

In other words, the jth autocovariance of Y_t is the covariance of Y_t and Y_{t-j}. Further, if neither the mean nor the autocovariances of Y_t depend on time t, then the process of Y_t is said to be covariance- or weakly stationary:

$$E(Y_t) = \mu \qquad \text{for all } t \qquad (4)$$
$$E[(Y_t - \mu)(Y_{t-j} - \mu)] = \gamma_j \qquad \text{for all } t \text{ and any } j. \qquad (5)$$

A time series is said to be *strictly stationary* if, for any values $j_1, j_2, ..., j_n$, the joint distribution of $(Y_t, Y_{t+j_1}, Y_{t+j_2}, ..., Y_{t+j_n})$, depends only on the intervals separating the dates (i.e., the j's) and not the date t itself. Sober's series are non-stationary if only because the mean of the process increases monotonically with every observation.

A time series can be non-stationary in several ways. For example, it can be stationary around a deterministic trend, as in:

(6) $\quad Y_t = \delta t + \epsilon_t,$

where $\epsilon_t \sim N[0, \sigma_\epsilon]$. Such a series is called "trend-stationary". Another form of non-stationarity obtains when past errors accumulate, as in:

(7) $\quad Y_t = Y_{t-1} + \epsilon_t,$

with ϵ as before. A series such as (7) is called "integrated". Let the difference operator be $\Delta X_t = X_t - X_{t-1}$. The difference operator transforms variables measured in levels into variables measured in changes and can be

---

[3]Some distributions may have expectations such that $\sigma_{XY} = E(XY) - E(X)E(Y) = 0$ even though $P(XY) \neq P(X)P(Y)$. For the bivariate normal, the concepts are equivalent though

applied several times: $\Delta^2 X_t = \Delta(\Delta X_t)$. If differencing d times renders an integrated series stationary, it is said to be integrated of order d. More formally, (p.543, emphasis original):

> Let d be the minimum integer such that $\{\Delta^d X_t\}$ is weakly stationary. Then $\{X_t\}$ is said to be integrated of order d, which is notated I(d). (By convention, a stationary time series is notated as I(0).)

Processes such as (7) are I(1) or integrated of order 1 and also called unit-root processes (because the coefficient on Y_t is unity) or random walks.

Now, Sober makes a mistake in applying the PCC to the data series he provides because inferring from a sample correlation to a probabilistic dependence means that one takes the most likely data-generating process to be stationary.[4] However, given the data, the most likely data-generating process is non-stationary, for example, trend-stationary or integrated. But that means that a sample correlation or association is no evidence for an underlying probabilistic dependence. And if the two variables are associated yet not probabilistically dependent, the antecedent of the PCC is not satisfied, hence the principle does not apply.

How do we know whether two non-stationary time series are probabilistically dependent? If the series are trend-stationary, Hoover says (p.541): "Principle (P) [the PCC] would naturally be applied to the stationary components of a pair of trend-stationary series". For integrated series, the test is slightly more complicated. If we have two distinct I(1) processes, a linear combination is usually I(1) too. However, in some cases a linear combination of two I(1) series can be stationary. Then the series is said to be "co-integrated" (p.545):

> Two time series $\{X_t\}$ and $\{Y_t\}$ are cointegrated if, and only if, each is I(1) and a linear combination $\{X_t - \beta_0 - \beta_1 Y_t\}$, where $\beta_1 \neq 0$, is I(0). (Paraphrased from Hamilton (1994), p.571.)

In turn, evidence for two series being co-integrated constitutes evidence for their probabilistic dependence (p.547). His reformulated principle reads as follows (*cf.* p.548):[5]

PCC* If variables X and Y are probabilistically dependent (for instance, they are *each stationary or trend-stationary and correlated with each other* or *each integrated and cointegrated with each other*), then either

---

[4]In fact, it needs to be ergodic. But most series that are stationary are also ergodic and vice versa, so the exact details are of no concern here. For a discussion, see Hamilton (1994), Ch. 19.

[5]I changed Hoover's wording slightly to make it consistent with the remainder of this paper but without, I hope, distorting his intentions.

X causes Y or Y causes X, or X and Y are joint effects of a common cause.

Hoover thus provides an elegant solution to the difficulty raised by Sober's scenario. Nevertheless I would like to suggest an alternative. My worry is the following: defining the PCC in terms of facts about probabilities rather than sample frequencies deprives the principle of much of its inferential power and to some extent betrays the motivation behind it. Let me explain what I mean by this.

We can understand the PCC (as stated) either in a metaphysical reading or in an epistemic reading. As a metaphysical principle, it would help to *define* the notion of causation.[6] The problem with the metaphysical reading is that, Sober's scenario aside, there are a number of clear-cut counterexamples, such as:[7]

- *Colliders.* When two variables cause an effect, the two can be probabilistically dependent conditional on the effect even though they are unconditionally independent.

- *Mixing.* When populations from different probability distributions are mixed, dependencies can arise even though the homogenous populations are probabilistically independent (see for instance Spirtes et al. (2000)).

- *Laws of coexistence.* Frank Arntzenius has pointed out that many laws of physics can be read as laws of coexistence without posing the need for a causal explanation. He mentions Maxwell's equations, Newtonian gravity, the Pauli exclusion principle and relativistic laws of coexistence (Arntzenius, 2005).

- ...

The PCC thus cannot serve as a metaphysical principle in a definition of causation. Indeed, few philosophers would attempt to define causation in such a way today. It is rather used as an epistemic principle for causal inference. As such, however, the PCC as formulated is both too strong and too weak.

---

[6]The PCC by itself obviously gives only a necessary, not a sufficient condition for causation. The reverse condition, in some versions called "faithfulness", is less plausible than the PCC, and I won't discuss it here.

[7]Hoover is fully aware that there are situations in which the PCC does not hold. See for example Hoover (2001).

Using probabilistic dependencies rather than empirical correlations deprives the principle of much of its inferential power. One reason is that probability distributions do not always exist (as Hoover is aware, see pp.531f.). They arise rather in fairly special circumstances, in what Ian Hacking called a "chance set-up" or Nancy Cartwright's "nomological machines" (Hacking, 1965; Cartwright, 1999). A chance set-up or nomological machine is essentially a persistent structure that can operate unimpededly and thus allow the generation of probabilities. But there is no reason to restrict the PCC to such special situations. Indeed, at least in some of the examples that are usually given to motivate the principle no such arrangement seems to be in place: the theatre troupe (we suspect a common cause to be responsible for the simultaneous appearance of symptoms of food poisoning in all the members of a theatre troupe after taking a joint meal completely independently of whether or not they regularly eat in the same place or such poisonings occur regularly); two students handing in the exact same term paper; correlations between phenotypic traits in evolutionary biology.

Second, using the PCC as formulated above presupposes that statistical inference is always prior to causal inference (perhaps in a temporal sense, but definitely in an epistemic sense: we need to know probabilities in order to use the principle for inferring causal relations). But such neat division into two stages of inference, and such that one is prior to the other, is not always possible and surely not always the most efficient way to do things. Although I don't think many would disagree (and I know Hoover would not disagree), I would like to point out that background knowledge, including *causal* background knowledge, often plays a role in inferring probabilities. In this sense knowledge about probabilities cannot be prior to knowledge about causal relations. Judging whether or not probabilities exists is a case in point: we can determine whether a chance-set up or nomological machine is in place on the basis of the causal knowledge about the situation. We can use that kind of knowledge for determining the potentially relevant variables. And we can use that kind of knowledge in order to determine whether or not time series are stationary.

Suppose we would like to determine the causal efficacy of a training programme and therefore need to determine average test results X. Our data are $X = (x_1 = 153, x_2 = 157, x_3 = 161, x_4 = 168, x_5 = 175, x_6 = 183)$. Suppose further that a greater subscript means "measured later in time" and thus that the data constitute a time series. It is obvious that whether or not we can use this time series for certain kinds of inferences depends on how the measurements were made. If, for example, the data are the results of a test the students sat on a specific date, and the different times are the times at which we finished marking the test (say, at $t_1 = 10:25$,

$t_2 = 10{:}27$, $t_3 = 10{:}29$ etc.) we are entitled to draw an average over time and use it for causal inference. We simply happened to mark the test of the worst-performing student first, then that of the second-worst performing student and so on. Whether data are arranged in this order or in another order does not make a difference. If, however, the relevant times are $t_1 = 2000, t_2 = 2001, t_3 = 2002$ *etc.* and data record a student's performance on subsequent (though identical) tests, we have to respect the order and drawing an average would not be meaningful. This is because, as we know, students' performance may change over time and the effectiveness of training programmes may accumulate. But there is nothing in the data themselves that tells us this. Moreover, causal background assumptions will often solve the statistical and causal inferential problem at once. If we judge on their basis that sea levels and bread prices cannot be causally connected, it simply does not matter whether they are correlated a sample or population level.

Using causal background knowledge for statistical inference would only be a problem if the contention was that knowledge about probabilities is *always* prior to knowledge about causes (say, because we wanted to use probabilities for a *definition* of causation). But we could subscribe to a more modest claim: there are situations in which after having successfully used our (causal and non-causal) background knowledge for statistical inference, we can use the PCC for further causal inference. Of course, I agree. But even in such situations using the PCC may be unnecessary and cumbersome. Often, our causal background knowledge will allow us to make the causal inference without worrying about probabilities. To use an example of Hoover's (p.547), it is a very unlikely coincidence that his daughter should have been born on the day the Challenger space shuttle blew up. But are we looking for a common cause here? Of course not. Our causal background knowledge tells us immediately that this is a mere coincidence (rather than a genuine co-occurrence of events that warrants the search for causal relations). We do not detour via judgements about probabilities in such cases.

The principle as stated by PCC* is also too weak because the problems for the metaphysical reading of course double up as problems for the epistemic reading—unless one qualifies the latter. Colliders for instance are a serious problem for practical causal inference because we often collect data with a specific purpose in mind. But this may mean that everybody in that population has a specific characteristic, say Z. Now, if X and Y are both causes of Z, they are probabilistically dependent *conditional upon* Z even though (let us suppose) X and Y are unconditionally independent. The problem is that often we may not notice that we sampled only members of the populations in which Z is present. Importantly, this is a problem at the level of populations, not samples. And: it may obtain for stationary as well as non-stationary variables. Similarly, problems regarding heterogeneous populations are a serious problem for practical inferences.

Before presenting my own proposal for a reformulated PCC, let me discuss the second strategy to deal with Sober's scenario, data preparation.

4 Data preparation

In a discussion note on Hoover's paper Daniel Steel disagrees with Hoover's analysis that the Sober scenario is a problem for statistical rather than causal inference. With Spirtes, Glymour and Scheines he thinks that the problem in Sober's case is just a special case of mixing. He provides the following argument (Steel, 2003). Central to his demonstration is the so-called mixing theorem, which can be applied to time series (p.310). For the simple case of $T = 2$ it reads:

Mixing Theorem. Let $P(XY) = nP_1(XY) + mP_2(XY)$, where n and m are real numbers greater than zero such that $n+m = 1$. Let $P_1(XY) = P_1(X)P_1(Y)$ and $P_2(XY) = P_2(X)P_2(Y)$. Then $P(XY) = P(X)P(Y)$ if and only if

$$P_2(X)P_2(Y) + P_1(X)P_1(Y) = P_1(X)P_2(Y) + P_2(X)P_1(Y)$$

An important corollary is the following:

Corollary. Let $P(XY) = nP_1(XY) + mP_2(XY)$, where n and m are real numbers greater than zero such that $n + m = 1$. Let $P_1(XY) = P_1(X)P_1(Y)$ and $P_2(XY) = P_2(X)P_2(Y)$. Then $P(XY) \neq P(X)P(Y)$ if and only if $P_1(X) \neq P_2(X)$ and $P_1(Y) \neq P_2(Y)$.

As briefly mentioned above this shows that dependencies can arise whenever populations from different probability distributions are mixed, even though the homogenous subpopulations are independent. It is important to note that what matters here is that the mixed population is *probabilistically* heterogeneous independently of whether or not it is *causally* heterogeneous. Some arguments to the effect that this case isn't one of mixing because two processes may be causally identical (because time is not a cause) and yet the problem associated with the Sober scenario arises are somewhat beside the point (this argument has been made for instance by Cartwright (2001); Sober (2001)).

What is wrong with the Spirtes, Glymour and Scheines/Steel proposal in my view is that it suggests the wrong methodological lessons. The natural response to a problem posed by the heterogeneity of populations is to require that populations be made homogenous prior to analysis. We can, for example, regard each x_t (for all $t = 1, 2, ..., T$) in the time series as drawn from a different variable X_t (for all $t = 1, 2, ..., T$) and analyse only contemporaneous probabilistic relationships.[8] Alternatively we can condition each variable on its past and analyse the probabilistic relations between the conditional variables $X_t|X_{t-1}$ etc.[9] Another alternative would be to difference the data prior to analysis or to use another preparation method in order to homogenise the variables. This seems to be exactly what the proponents of this solution have in mind. Clark Glymour, for one, says (Glymour, 1999, pp.73f., emphasis added):

> Applying the program [that incorporates a version of the PCC as core assumption] to real data requires a lot of adaptation to particular circumstances: variables must often be transformed to better approximate normal distributions, decisions made about modeling with discrete or continuous variables, *data must be differenced to remove auto-correlation*, and on and on.

Similarly, Steel writes (Steel, 2003, p.134):

> The above discussion illustrates how researchers interested in drawing conclusions from statistical data can design their investigation so that counter-examples like Sobers are not a concern. For instance, if the series is non-stationary but transformable into a stationary one via differentiating with respect to time, then differentiate. Then PCC can be invoked without concern for the difficulty illustrated by the Venice-Britain example.

The idea seems to be that data can always be suitably prepared (by, say, conditioning on time or on series' past or by prior differencing or detrending) before using the PCC for analysis. Thus, we can reformulate the PCC as follows:

PCC** If two *suitably prepared* random variables X, Y are probabilistically dependent, then either X causes Y, Y causes X or X and Y are the joint effects of a common cause Z.

Unfortunately, data preparation does too much and too little at the same time. Regarding only contemporaneous statistical relations, conditioning on the past of variables and differencing, detrending *et al.* all result in the loss of important long-run information that a prudent statistician should make use of (see for instance Hendry (1995, Sect. 7.4). The statistical concepts of cointegration, which Hoover discusses at length, and co-breaking (see

---

[8] An anonymous referee, for example, wrote: "What two variables [this relates to my X and Y]? What we have is two times series: $X_t, X_{t-1}, X_{t-2}, ..., X_{t-n}$ and $Y_t, Y_{t-1}, Y_{t-2}, ..., Y_{t-n}$. There are $2n+2$ variables!"

[9] This seems to be suggested by Frank Arntzenius, see Arntzenius (2005), section 2.3.

for instance Hendry and Mizon (1978)) were developed specifically in order to deal with Sober-like situations while retaining the long-run information contained in the time series. Consider the following. Unit-roots processes are sometimes said to be subject to *stochastic* shifts because the error term accumulates over time. By contrast, when *deterministic* shifts occur, parameters of a process such as its mean, variance or trend change. In recent econometrics, methods have been developed that exploit information about common shifts in two or more series for causal inference. Through differencing, however, this information can be lost. Consider the following series:

$$X_t = \alpha_X + \beta_X \delta t + \epsilon_{X_t} \tag{8}$$
$$Y_t = \alpha_Y + \beta_Y \delta t + \epsilon_{Y_t}, \tag{9}$$

where the δ denotes a common trend. When there are shifts in the value of that coefficient, these common "breaks" can be detected by statistical methods. The breaks will disappear, however, when the series are differenced. In other words, if the non-stationarity of a series is due to shifts in deterministic coefficients, series should not be differenced prior to causal analysis (even though such series may be, as Steel demands, non-stationary and transformable into a stationary one via differencing). The point is that prior data preparation of the kind discussed here ignores that we are dealing with continuous and persisting processes here, and not merely with contemporaneous events or changes, and the analysis methods we use should reflect just that (Hoover, 2001, p.165).

While data preparation will often result in information loss, it may not solve the problem either. Differencing can remove some sources of nonsense correlation but is often inapplicable. Of course, it is an analytical truth that differencing removes unit roots. If we have two independent random walks:

$$X_t = X_{t-1} + \epsilon_{X_t} \tag{10}$$
$$Y_t = Y_{t-1} + \epsilon_{Y_t}, \tag{11}$$

then a regression of the differenced series such as

(12) $\Delta Y_t = \beta \Delta X_t + \nu_t$

will, correctly, find a zero regression coefficient. But unit roots constitute only one source of non-stationarity, and non-stationarity is only one source of nonsense correlation.[10] Differencing is ineffective when nonsense correlation arise in stationary series. That is, even stationary time series can

---

[10]On the pitfalls of differencing see the papers Courakis (1978); Hendry and Mizon (1978); in defence of *a priori* differencing, see Williams (1978).

appear correlated even though they are *ex hypothesi* causally independent. Indeed, in his original article on the problem Yule did not discuss the problem in the context of non-stationarity. For example, for two series of the form:

$$X_t = \theta_X X_{t-1} + \epsilon_{X_t} \tag{13}$$
$$Y_t = \theta_Y Y_{t-1} + \epsilon_{Y_t}, \tag{14}$$

where $|\theta_i| < 1$ and the ϵ_{it}'s are i.i.d. and zero mean one can show that nonsense correlations obtain regularly (Granger, Hyung and Jeon, p.899). As one can see easily, the mathematical form of the differenced series is exactly the same as that of the original series. Although in this context, the problem arises to a somewhat lesser extent—when θ is 0.75, significant correlations obtain in about 30% of the cases—the problem does obtain despite the fact that the series are stationary. Something similar happens when time series are moving averages, as in the following:

$$X_t = \sum_{j=0}^{k} e_{X,t-j} \tag{15}$$
$$Y_t = \sum_{j=0}^{k} e_{Y,t-j} \tag{16}$$

Even if k is only 5, nonsense correlation results obtain in about a third of cases (Granger, Hyung and Jeon, p.902).

Lastly, serial correlation may persist even after differencing, and in fact will in general persist. In his 2001 paper, Sober discusses an example from evolutionary biology in which similar developmental sequences evolve independently in two lineages (pp.335ff.). This is also common in time-series analysis. Series that are integrated of orders higher than 1 will have to be differenced several times before achieving stationarity. There are series that are fractionally integrated (i.e., they have a non-integer order of integration), which can lead to spurious results whenever the orders of integration sum up to more than 0.5 (Tsay and Chung, 2000). Such fractional orders of integration can obviously not be removed by differencing.

Other *a priori* data preparation methods do not fare better. An alternative method to remove non-stationarity is detrending, i.e., subtracting a linear deterministic trend before the analysing data. However, detrending too can yield spurious results, namely when the processes are unit roots (see Hendry (1995, Sect. 4.3)). Lesson: there are no data preparation methods that can be used prior to systematic statistical (and therefore causal, see above) analysis. Hence, PCC**, too, fails.

5 Non-statistical sources of nonsense correlations

There are also various non-statistical sources of nonsense correlations. For example, when variables are connected because of logical, mathematical and conceptual links, they may be correlated but the correlation is not due to a causal connection. Non-causal physical laws may provide yet another source of correlation. These problems are not peculiar to time series and they are well recognised in the literature (for a valuable overview, see Williams (1978, pp.52ff.)). For the sake of completeness, let us briefly consider each issue in turn.

When variables have logical links, correlations are not indicative of causal connections. "Day" may be perfectly negatively correlated with "Night" but this is because of a logical, not a causal relation. A variable and a function thereof are highly correlated. Time-series econometrics often uses logarithms of variables such as money, income and prices. The logarithms are correlated with the original variables but not for causal reasons. More serious are conceptual links between variables. Many economic variables are linked because they have interdependent measurement procedures. For example, the measurement of some variables is based on the same national accounts.

There may be other non-causal constraints between variables, for example, when two variables are related by budget constraints. Consumption (C) will be correlated with savings (S) not because they are causally related but because a third variable, income (I) constrains them by the mathematical relation $C + S = I$. Problems of this kind and conceptual relations can be very serious in time-series analysis, especially in social science applications.

6 Evidence, eliminative induction and the PCC

Clearly, there is a core of truth in the PCC. But what is it? In some cases it does indeed work, just think of Salmon's famous examples: the theatre company all of whose members out of a sudden get violently ill; the students who hand in the exact same term paper; the twin quasars (Salmon (1994, p.158f.)). Can we formulate the principle in a way as to avoid the counterexamples discussed here?

The core of it, in my view, is that an empirical or sample correlation between variables sometimes provides evidence for the hypothesis that these variables are causally connected. In many cases, of course, the correlation arises for reasons other than causal connectedness: the sample is small; there is selection bias; there are logical, mathematical or conceptual relations between them; they are generated by unit-root processes. If we can rule out these non-causal accounts for the correlation, then the causal account is probably true.

In order to formalise this idea to some degree, recall Patrick Suppes' probabilistic theory of causation (Suppes, 1970). He defined an event A as a *prima facie* cause of another event B if and only if A precedes B and

$P(AB) > P(B)$. Not every *prima facie* cause is, however, also a genuine cause. Thus he defined as *spurious* cause an event A that is a *prima facie* cause of event B but such that there is a partition π prior to A such that every element C_i in that partition renders A and B probabilistically independent: $P(B|A, C_i) = P(B|C_i)$ (for all i). A genuine cause is a *prima facie* cause that is not spurious.

Analogously, we can define e as *prima facie* evidence for hypothesis h if and only if e stands in an appropriate relationship with h. In the present case, the appropriate relationship is explanatory: e is *prima facie* evidence for h if and only if h, if true, explains e: a causal relation between two variables (whether direct or due to a common cause) explains the correlation between the variables.[11] e is spurious evidence for h if and only if e is *prima facie evidence* and h is explained by an alternative hypothesis h_i^a. If e is *prima facie* evidence and not spurious, it is genuine evidence.

Thus the PCC reads:

PCC*** The proposition e = "Random variables X and Y are (sample or empirically) correlated" is *prima facie* evidence for the hypothesis h = "X and Y are causally connected". If all alternative hypotheses h_i^a (*e.g.*, "the correlation is due to sampling error", "the correlation is due to the data-generating processes for X and Y being non-stationary", "X and Y are logically, conceptually or mathematically related") can be ruled out, then e is genuine evidence for h.

There are various advantages of this formulation. First, and foremost, it is very explicitly formulated as *epistemic* principle. There is no way to misread the principle as saying that all correlations must have causal explanations for instance. Unlike previous versions, this formulation makes evident that the principle is fallible. It nowhere says that the evidence is conclusive or that the evidence entails its hypothesis. Evidence provides a reason to believe, not metaphysical certainty. Second, it makes evident that causal inference is context-dependent. I haven't said a lot about where the alternative hypotheses h_i^a come from. Usually our background knowledge about the situation under investigation will determine what plausible candidates there are, and how much effort needs to be taken to rule them out. Third, unlike Hoover's PCC*, PCC*** is very widely applicable. In particular, it is applicable to cases of empirical correlation where no probability

---

[11]There are also cases where evidence explains the hypothesis for which it is evidence, and where a third statement c explains both evidence and hypothesis (see Achinstein (2001)). Moreover there are cases where the evidential relationship is not explanatory (for a discussion, see Reiss (2007, Ch. 1)). My definitions are thus intended to apply only to the present case.

distributions exist or where statistical inference may be difficult or cumbersome. Fourth, unlike Steel's PCC**, it does not rely on data-preparation techniques that frequently do more harm than good.

There may be a residual worry that I throw out the baby with the bath water. What happens if one of the alternative hypotheses h_i^a is true but X and Y are causally connected nevertheless? But this isn't a problem as I make no suggestion to the effect that empirical correlations can be the *only* evidence for causal connections. Of course, this is absurd. When samples are small and statistical tests lack power, one can try to physically investigate the units. When time series are non-stationary, one can try to exploit "structural breaks" in the series for causal inference and use tests of the kind Hoover (Hoover, 2001) and David Hendry and his collaborators (e.g. Hendry and Massmann (forthcoming)) have developed. In other cases we may be able to improve the quality of the data and thus ameliorate problems. In yet other cases we may be able to use one of a host of qualitative methods such as ethnographic methods. Violations of the PCC are only a problem if one thinks that all correlations need a causal explanation and if one thinks that the principle is the only or only appropriate or most important or "core" principle of causal inference. But it isn't. It is one of many such principles and has it's own advantages and drawbacks. What is important, though, is to keep its limitations in mind.

Thus let me end with an irony. What kinds of systems do we have good reason to believe that most or all of the alternative hypotheses are false? As mentioned above, shifts that render a time series non-stationary can be of two kinds: "stochastic" and "deterministic". Stochastic shifts obtain when error terms accumulate; deterministic shifts, when deterministic parameters (such as coefficients on trends) change. Systems where neither kind of change is likely are systems that lack internal dynamics and that are shielded from outside influences or "closed". Moreover we want to rule out chance associations and thus require that the static and closed system persists for a while so that sample sizes are sufficient. We also want to make sure that the populations in the system are homogeneous, that variables are well-measured and so on. Now, aren't these characteristics the characteristics of experimental systems? But if they are, why do we need the PCC to draw causal conclusions?

Acknowledgements

I'd like to thank Nancy Cartwright, Damien Fennell, David Hendry, Elliott Sober, three anonymous referees as well as audiences in Kent and Vancouver for helpful suggestions. Special thanks go to Kevin Hoover who provided extensive and extremely valuable comments on several drafts of this paper.

Some disagreements remain but the paper profited enormously from the discussion.

Julian Reiss
Department of Logic and Philosophy of Science, Complutense University, Madrid, Spain.
and
Centre for Philosophy of Natural and Social Science, London School of Economics, UK.
jreiss@filos.ucm.es

BIBLIOGRAPHY

Achinstein, P. (2001). *The Book of Evidence*. Oxford University Press, Oxford.
Arntzenius, F. (2005). Reichenbach's principle of the common cause. In Zalta, E., editor, *Stanford Encyclopedia of Philosophy*. CSLI, Stanford.
Cartwright, N. (1999). *The Dapppled World*. Cambridge University Press, Cambridge.
Cartwright, N. (2001). What's wrong with Bayes' nets? *Monist*, 84 (2):242–64.
Courakis, A. (1978). Serial correlation and a Bank of England study of the demand for money: An exercise in measurement without theory. *Economic Journal*, 88:537–48.
Glymour, C. (1999). Rabbit hunting. *Synthese*, 121:55–78.
Granger, C., Hyung, N. and Jeon, Y. (2001). Spurious regressions with stationary series. *Applied Economics*, 33:899–904.
Hacking, I. (1965). *The Logic of Statistical Inference*. Cambridge University Press, Cambridge.
Hamilton, J. (1994). *Time-Series Analysis*. Princeton University Press, Princeton (NJ).
Hendry, D. (1995). *Dynamic Econometrics*. Oxford University Press, Oxford.
Hendry, D. and Massmann, M. (forthcoming). Co-breaking: Recent advances and a synopsis of the literature. *Journal of Business and Economic Statistics*.
Hendry, D. and Mizon, G. (1978). Serial correlation as a convenient simplification, not a nuisance: A comment on a study of the demand for money by the Bank of England. *Economic Journal*, 88:549–63.
Hoover, K. (2001). *Causality in Macroeconomics*. Cambridge University Press, Cambridge.
Hoover, K. (2003). Nonstationary time-series, cointegration, and the principle of the common cause. *British Journal for the Philosophy of Science*, 54:527–51.
Pearl, J. (1994). *Causation: Models, Reasoning and Inference*. Cambridge University Press, Cambridge.
Reichenbach, H. (1956). *The Direction of Time*. University of California Press, Berkeley (CA).
Reiss, J. (2007). *Error in Economics: Towards a More Evidence-Based Methodology*. Routledge, London.
Salmon, W. (1994). *Scientific Explanation and the Causal Structure of the World*. Princeton University Press, Princeton.
Sober, E. (1987). The principle of the common cause. In Fetzer, J., editor, *Probability and Causality: Essays in Honor of Wesley Salmon*, pages 211–28. Reidel, Dordrecht.
Sober, E. (2001). Venetian sea levels, British bread prices, and the principle of the common cause. *British Journal for the Philosophy of Science*, 52:331–46.
Spirtes, P., Glymour, C. and Scheines, R. (2000). *Causation, Prediction, and Search*. MIT Press, Cambridge (MA), 2nd. edition.

Steel, D. (2003). Making time stand still: A response to Sober's counter-example to the principle of the common cause. *British Journal for the Philosophy of Science*, 54:309–17.

Suppes, P. (1970). *A Probabilistic Theory of Causality*. North-Holland, Amsterdam.

Tsay, W.-J. and Chung, C.-F. (2000). The spurious regression of fractionally integrated processes. *Journal of Econometrics*, 96:155–82.

Williams, D. (1978). Estimating in levels or first differences: A defence of the method used for certain demand-for-money equations. *Economic Journal*, 88:564–68.

Yule, G. U. (1926). Why do we sometimes get nonsense-correlations between time series? *Journal of the Royal Statistical Society*, 89:1–64.

Conceptual tools for causal analysis in the social sciences

Erik Weber

ABSTRACT. This paper has three aims. I want to show that social scientists must deal with at least three kinds of causal relations. My second aim is to argue that the first kind is well understood, but the second and third kind not. Finally, I also want to show that by adapting Wesley Salmons causal mechanical approach, we can get a better grip and the two latter kinds.

1 Introduction

1.1 Let me start with some general reflections. Philosophers who want to study causation can do at least two different things. On the one hand, they can do *conceptual analysis*: they can develop a definition of causation that adequately represents our everyday causal talk (and can be used to revise our everyday causal talk). On the other hand, they can try to develop a set of causal concepts that helps scientists to achieve their aims.[1]

Since I don't intend to do conceptual analysis here, I will only briefly describe my position on that topic, without an elaborate argument. The reason why I want to go into this briefly is that I don't want the second type of project (which is the kind of project I will undertake) to be confused with conceptual analysis. The aims of the two projects, as stated above, are clearly different. The way one has to argue in conceptual analysis is also different from the way one has to argue in my type of project. These differences are often neglected in the philosophical analysis of causation. I want to pay some attention to them in order to prevent confusion and misunderstanding of my project.

I am a *conceptual pluralist* about causation: I think it is impossible to have a single definition of causation that adequately represents all our "causal talk". Suppose we try to define causation and come up with a definition of this form:

---
[1] A third kind of project is the metaphysical project of describing what causation is in the world. I will not discuss that here.

C causes E if and only if (conditions X_1, X_2, ..., X_n are satisfied) or (conditions Y_1, Y_2..., Y_n are satisfied) or ... or (conditions Z_1, Z_2, ..., Z_n are satisfied).

In its strongest form, conceptual causal pluralism claims that there is no condition that occurs in all the disjuncts of an adequate definition of causation (i.e., a definition that represents all our intuitions). A weaker form is the claim that what is common in all disjuncts is very unspecific (e.g., a cause does not come after its effect).

Let me illustrate this with an example of Ned Hall:

> Suzy and Billy have grown up, just in time to get involved in World War III. Suzy is piloting a bomber on a mission to blow up an enemy target, and Billy is piloting a fighter as her lone escort. Along comes an enemy fighter plane, piloted by Enemy. Sharp-eyed Billy spots Enemy, zooms in, pulls the trigger, and Enemy's plane goes down in flames. Suzy's mission is undisturbed, and the bombing takes place as planned. If Billy hadn't pulled the trigger, Enemy would have eluded him and shot down Suzy, and the bombing would not have happened. (Hall, 2004, p. 241)

In this example, the effect counterfactually depends on the cause, but there is no mechanism linking cause and effect. In other cases, e.g., a firing squad, there is a causal mechanism but no counterfactual dependence (because there is simultaneous overdetermination). Ned Hall concludes from this:

> Events can stand in one kind of causal relation – dependence – for the explication of which the counterfactual analysis is perfectly suited ... . And they can stand in an entirely different kind of causal relation – production – which requires an entirely different kind of causal analysis... . (Hall, 2004, p. 226)

I agree with this, though I would rather say "almost entirely different", because there might be a time order (Hall excludes backward causation in both cases, so there is a tiny similarity between the two causal relations).

The second type of project that philosophers of causation can set themselves belongs to the philosophy of science (rather than to the philosophy of language). It consists in developing an adequate set of causal concepts which helps scientists to achieve their aims. As already mentioned, this paper fits in this perspective. From this perspective, examples like Hall's are irrelevant, because they relate to everyday causal talk and our intuitions about them. My aim is not to represent the way scientists talk about causation, but to equip them with an adequate conceptual apparatus, so they can do a better job.

1.2 Projects like mine have to start with a view on the aims of science. My view is something like Hempel (1965, p. 333):

> Among the many factors that have prompted and sustained inquiry in the diverse fields of empirical science, two enduring human concerns have provided the principle stimulus for man's scientific efforts.
> One of them is of a practical nature. Man wants not only to survive in the world, but also to improve his strategic position in it. This makes it important for him to find reliable ways of foreseeing changes in his environment and, if possible, controlling them to his advantage. The formulation of laws and theories that permit the prediction of future occurrences are among the proudest achievements of empirical science; and the extent to which they answer man's quest for foresight and control is indicated by the vast scope of their practical applications, which range from astronomic predictions to meteorological, demographic, and economic forecasts, and from physico-chemical and biological technology to psychological and social control.
> The second motive for man's scientific quest is independent of such practical concerns; it lies in his sheer intellectual curiosity, in his deep and persistent desire to know and to understand himself and his world. So strong, indeed is this urge that in the absence of more reliable knowledge, myths are often invoked to fill the gap. But in time, many such myths give way to scientific conceptions of the what and why of empirical phenomena.

I believe that every scientific discipline needs a set of causal concepts in order to provide causal and other (e.g., functional) explanations, and in order to improve our interventions in the natural and social world. But for the sake of argument, I will assume that science has only a practical aim. I do this in order to make my conclusions acceptable for people who have a more narrow view on the aims of science.

1.3 The aims of my paper are (i) to show that social scientists must deal with at least three kinds of causal relations, (ii) that the first kind is well understood, but the second and third kind are not, and (iii) to argue that by adapting Wesley Salmon's causal mechanical approach, we can get a better grip on the two latter kinds. In Section 2, I will give some examples of research relating to the first type of causal relation, and link them to probabilistic concepts of causation. In Section 3 and 4, I will argue that social scientists need other types of causation and adapt the causal mechanical approach in such way that it can explicate these types of causation.

2 Probabilistic causation in the social sciences

2.1 I start with some examples. In a book on ethical problems in the social sciences, Paul Davidson Reynolds discusses an experiment which investigates the effects of negative income tax:[2]

> The research involved the examination of the effects of different negative income tax plans (direct cash payments) to "guarantee" a predetermined minimum household income: partial reductions in payments occurred if household earnings increased. The basic question was the extent of labor-force participation of individuals in households with a guaranteed income—i.e., would they work less? The study also estimated the costs of a guaranteed income program if adopted as the major welfare strategy for the nation. The initial study involved 1,400 families in five cities in the New Jersey-Pennsylvania area randomly assigned to one of the eight plans (negative income tax schedules) or to a control group (families receiving no guaranteed income). (Reynolds, 1982, p. 36)

The eight plans differed in the amount of money that was given if there was no other income, and in the reductions in payment that occurred when there was another income. But in each plan the reductions were only partial. The aim of the study was to determine whether the advantages of a guaranteed income plan (administrative simplicity, dignity, equity, ...) were or were not outweighed by a possible disadvantage, viz. reduced labour-force participation.

My second example is taken from *Investigating the Social World*, a textbook on methodology for the social sciences by Russell K. Schutt:[3]

> (Police assigned the 330 domestic assault cases in the study to one of three conditions: to result in an arrest, to result in an order that the offending spouse leave the house for eight hours, or to result in some type of verbal advice by the police officers.) Police officers responding to the domestic assault cases were not allowed to choose which "treatment" to apply (except in extreme cases, such as when there was severe injury or when the spouse

---

[2] Reynold's source is Kershaw (1972).
[3] Shutt's source is Sherman and Berk (1984).

demanded that an arrest be made). Instead the treatments were assigned by police in random order, according to the color of the next report form on a pad that had been prepared by the researchers. (Schutt, 1996, p. 48-49)

The main aim of this experiment was to determine whether arrest has an effect on recidivism: does arrest reduce the risk of repeat offenses?

My third example is also taken from a methodological textbook (*Research Methods in the Social and Behavioral Sciences* by Russell A. Jones), more specifically from the chapter on simulations. In a simulation we experiment with real people in an artificial context. The individuals are subjected to certain inputs, and we observe their behaviour. There is always a qualitative gap (called 'simulation gap') between the simulation context and the real world. This gap relates to the effect variable that is being studied: in a simulation the effect is always in some sense "not real". Let us look at some examples. Quite a lot of research has been done with mock juries (see Jones, 1996, p. 310-313). Suppose we want to know whether sex, age, or occupation of members of a jury in a criminal court have an influence on the verdict. For a defence lawyer it would be useful to know whether women are more lenient (i.e., declare less people guilty and/or propose less severe punishments) than men. If so, a lawyer could try to get as many women as possible in the jury. Randomised field experiments are not possible here on moral grounds (if there is a difference, some accused are better of than others). This is why simulations have been performed with respect to these research questions. The experimental subjects were asked to read the (real) evidence for and against (real) defendants. Then they had to propose a verdict (guilty or not guilty) and (if guilty) a punishment. So the experimenters tried to stay as close as possible to the real world, with one exception: the decisions of the experimental subjects did not have an effect on the defendants. The fate of the defendants was decided by a real court.

Mock jury experiments (as these kind of simulations are called) have also been used to investigate some fundamental characteristics of legal systems. For instance, it has been established that jurors are better in disregarding illegally obtained evidence (evidence that is mentioned, but which the judge instructs them to disregard) if there is deliberation among jury members (as opposed to a jury system in which the verdict is determined by the individual decisions of the jury members, which they take without deliberating with other members).

What do these examples have in common? First, the claims are about causal relations in populations, not about individuals. Second, they have policy relevance because of a difference between the purported cause variable and the purported effect variable. The cause variable is something over which policy makers have direct control: the rules of the social security system (first example), the operating rules of a police force (second example), the rules in a legal system (third example). The effect variable is something which policy makers cannot or do not want to control in a similar direct way: we don't want forced labour (first example), one cannot supervise all offenders permanently (second example), one should not force the jurors to make a certain verdict (third example). So policy makers are interested in indirect ways to change the effect variables.

Causal claims with these characteristics can be adequately conceptualised by probabilistic notions at the population level. In the literature we can find two types of probabilistic theories: the so called "average effect theories" and the "context unanimity theories". I will present them in 2.2 and 2.3 and compare them in 2.4.

2.2 Versions of the average effect theory can be found in e.g., chapter 9 of Dupré (1993) and chapter 9 of Hausman (1998). I will use the version that can be found in chapter 7 of Giere (1997). Though the theory can be extended to other types of variables, Giere considers only binary variables. So in the following, **C** is a variable with two values (C and $\neg C$); the same for **E** (values E and $\neg E$). The crucial definitions for causation in populations are:

> C is a *positive causal factor* for E in the population U whenever $P_X(E)$ *is greater than* $P_K(E)$.
> C is a *negative causal factor* for E in the population U whenever $P_X(E)$ *is less than* $P_K(E)$.
> C is *causally irrelevant* for **E** in the population U whenever $P_X(E)$ *is equal to* $P_K(E)$. (Giere, 1997, p. 204)

X is the hypothetical population which is identical to U, except that each individual exhibits the value C of the causal variable C. K is the analogous hypothetical population in which all individuals exhibit $\neg C$. For instance, if we claim that smoking is a positive causal factor for lung cancer, we claim that, if everyone smoked, there would be more people affected by lung cancers than if no one smoked.

Causal factors in individuals are defined as follows

> C is a *positive causal factor* (deterministic) for E in an individual, I, characterized by residual state, S, if in I, C produces E and $\neg C$ produces $\neg E$.
> C is a *negative causal factor*(deterministic) for E in an individual, I, characterized by residual state, S, if in I, C produces $\neg E$ and $\neg C$ produces E. (Giere, 1997, p. 200)

> If C is neither a positive nor a negative causal factor for E in I, given S, then we say that the variable C is causally irrelevant for E in I, given S. (Giere, 1997, p. 201)

The residual state S refers to all other characteristics of the individual besides the cause and effect variable.

In order to understand the difference between the average effect theory and the contextual unanimity theory, it is important to see that Giere's population claims "always average over individuals and, therefore ignore what might be important differences among individuals" (Giere, 1997, p. 204-205). Giere discusses only one specific case: if in U there are some individuals for which C is a positive causal factor for E, and an equal number for which C is a negative causal factor, C is causally irrelevant for E in U. More generally, Giere's definitions leave open the possibility that a population contains individuals for which C is a positive causal factor for E, as well as individuals for which C is a negative causal factor for E. If the first subpopulation is larger, C will be a positive causal factor in E. If the second subpopulation is larger, C will be a negative causal factor.

In the following chapter of his book, Giere links his definitions to randomised experimental designs; he shows that the relative frequency of E in the experimental group of a randomised experiment is a reliable estimate of $P_X(E)$, and the relative frequency of E in the control group a reliable estimate of $P_K(E)$. This means that statistically significant differences between experimental and control groups are good evidence for causal claims about populations. Giere also discusses possible dangers (placebo effect, observer bias, ...) and goes into the relative advantages and disadvantages of randomised experiments, prospective designs and retrospective designs. What is important is that experimental methods enable us to determine whether a causal relationship holds in a population even if there is no way to determine causal relations at the individual level.

2.3 The first version of the context unanimity theory can be found in Cartwright (1979). More recent versions can be found in Humphreys (1989) and Eells (1991). I use the latter one. In chapter 2 of his book, Eells gives the following example:

> To use an example of Cartwright's (1979), ingesting an acid poison (X) is causally positive for death (Y) when no alkali poison has been ingested ($\neg F$), but when an alkali poison has been ingested (F), the ingestion of an acid poison is causally negative for death. I will argue that in a case like this it is best to deny that X is a positive causal factor for Y, even if, overall

> (for the population is a whole), the probability of death when an acid poison has been ingested is greater than the probability of death when no acid poison has been ingested (that is, even if $Pr(Y|X) > Pr(Y|\neg X)$). I will argue that it is best in this case to say that X is causally *mixed* for Y, and despite the *overall* or *average* probability increase, X is nevertheless not a positive causal factor for Y in the population as a whole. (Eells, 1991, p. 58)

What he does with this example fits in his conceptual scheme:

> Then we say that X is a *positive causal factor* for Y if and only if, *for each i*, $Pr(Y|K_i \& X) > Pr(Y|K_i \& \neg X)$. *Negative causal factorhood* and *causal neutrality* are defined by changing the "always rises" (>) idea to "always lowers" (<) and "always leaves unchanged" (=), respectively. The idea that the inequality or equality must hold for *each* of the background contexts K_i is sometimes called the condition of *contextual unanimity*, or *context unanimity*. ... Note that these three relations of positive, negative and neutral causal factorhood are not exhaustive of the possible causal significance that a factor X can have for a factor Y: There remains the possibility of various kinds of *mixed* causal relevance, corresponding to various ways in which unanimity can fail. (Eells, 1991, p. 86-87)

2.4 The great advantage of causes in the sense of the context unanimity theory is their strength: the causal tendency cannot be reversed (from positive to negative) or annihilated (from positive or negative to causally neutral) in a subpopulation. Knowing such causes is certainly an advantage from the point of view of policy. However, as John Dupré has pointed out, there is an epistemological problem:

> ...it should be noted that for someone who believes that the unanimity condition is an essential part of an account of (probabilistic) causation, the importance attaching to controlled experiments in science is quite mysterious. For a controlled experiment tells us only about average effects, and nothing about the various positive and negative effects that may contribute to that average. Consequently they can never provide us with grounds for believing that the experimental treatment causes the effect in question in every possible causal context, and for a proponent of the unanimity thesis they cannot, therefore, provide any information about causality. (Dupré, 1993, p. 202)

There is an important exception to what Dupré says, viz. deterministic cases. But that will be of little comfort to the social scientist, who almost never knows all the causes of a phenomenon. So I think we have to choose for the average effect theory: it represents the information that social scientists can and want to provide when doing the kind of research exemplified by the examples of section 2.1.

3 Causation at the individual level

3.1 The first question to be answered here is: do social scientists need a type of causation besides the well understood probabilistic one? I think the answer is positive, because social scientists should seek knowledge about social mechanisms. With social mechanism, I mean this:

> Social mechanisms in particular are usually thought of as complexes of interactions among individuals that underlie and account for aggregate social regularities. (Steel, 2004, p. 57-58)

In my view, one of the reasons why social scientists should investigate social mechanisms is that they are indispensable for explaining social regularities (including causal claims at the population level). But I promised to base my argument on a more narrow view on the aims of science, one that does not include explanation as an aim of science. Even then social mechanisms are important, for at least two reasons:

(1) As is clearly shown in Steel (2004), social mechanisms can play an important role in solving the problem of confounders (and thus in distinguishing real causal relations from spurious correlations in the absence of randomised experiments).

(2) There are cases in which social mechanisms are the only kind of evidence we have for accepting causal claims at the aggregate level (see DeMey and Weber, 2003).

In other words: mechanisms are important as part of the evidence for causal claims at the aggregate level.

Granted that social scientists need social mechanisms, the question arises what kind of causation is involved. In 3.2 and 3.3 I will argue that a constructivist reconstrual of the concept of *causal interaction*, as it was defined by Wesley Salmon in his book *Scientific Explanation and the Causal Structure of the World* (1984), can help us a great deal in answering this question. In section 4 I will show why causal interaction is not enough: there is more in social mechanisms than causal interactions.

3.2 The concept of *causal interaction* was introduced by Salmon in order to capture what he calls the innovative aspect of causation (as opposed to the conservative aspect, for which he developed the concept of *causal process*; cfr. Section 4). I will adopt a definition that is very close to Salmon's original definition:

(CI) There is a causal interaction between objects x and y at time t if and only if

(1) there is an intersection between x and y at t (i.e., they are in adjacent or identical spatial regions at t),

(2) x exhibits a characteristic P' in an interval immediately before t, but a modified characteristic P immediately after t,

(3) y exhibits a characteristic Q' in an interval immediately before t, but a modified characteristic Q immediately after t,

(4) x would have had P' immediately after t if the intersection would not have occurred, and

(5) y would have had Q' immediately after t if the intersection would not have occurred.

An object can be anything in the ontology of science (e.g., atoms, photons,...) or common sense (humans, chairs, trees,...). This definition incorporates the basic ideas of Salmon. The main difference is that, according to my definition, interactions occur between two objects. In Salmon's definition, an interaction is something that happens between two processes (see Salmon, 1984, p. 171). This modification was suggested in Dowe (1992) (I don't agree with the other modifications that Dowe suggests; so this is the only change I want to make; see section 5.2 for some comments on this). The modification is not substantial (processes are world-lines of objects, i.e., collections of points on a space-time diagram that represents the history of an object).

Because I stick close to Salmon's original definition, I can borrow his examples. Collision is the prototype of causal interaction: the momentum of each object is changed, this change would not have occurred without the collision, and the new momentum is preserved in an interval immediately after the collision. When a white light pulse goes through a piece of red glass, this intersection is also a causal interaction: the light pulse becomes and remains red, while the filter undergoes an increase in energy because it absorbs some of the light. The glass retains some of the energy for some time beyond the actual moment of interaction. As an example of an intersection which is not a causal interaction, we consider two spots of light, one red and the other green, that are projected on a white screen. The red spot moves diagonally across the screen from the lower left-hand corner to the upper right-hand corner, while the green spot moves from the lower right-hand corner to the upper left-hand corner. The spots meet momentarily at the centre of the screen. At that moment, a yellow spot appears, but each spot resumes its former colour as soon as it leaves the region of intersection. No modification of colour persists beyond the intersection, so no causal interaction has occurred.

3.3 Though I think that Salmon defined an interesting concept, I also think that he did not realise what the preconditions of its use were. Suppose we want to make a claim about a causal interaction, of the following form:

(CCI) There was a causal interaction between x and y at time t, in which x acquired characteristic P and lost characteristic P', and in which y acquired characteristic Q and lost characteristic Q'.

Making claims about causal interaction presupposes a frame of reference that settles the level of description, the spatial scale and the timescale that are to be used. The level of description determines the kinds of system we talk about (e.g., individuals or groups of individuals, macroscopic objects or elementary particles). The spatial scale determines the smallest unit of distance, and thus determines whether two systems are or are not in adjacent spatial regions (they are if the distance between them is smaller than the smallest unit of distance). Likewise, the timescale determines the smallest unit of time we will use, and thus allows us to distinguish between "sudden changes" as they occur in interactions, and slower evolution: we have a sudden change if and only if the change takes place in a period of time that is smaller than the smallest unit of time.

Let me clarify this by means of a series of examples. Consider a group of people in a seminar room. There is a speaker that tells his audience things that are really new to them. The seminar lasts 59 minutes. Now take the following frame of reference:

> Objects = common sense macro objects
> Space = rooms and multiples of them (floors, buildings)
> Time = 1 hour and multiples (days, weeks,...)

In this frame of reference, a set of interactions has occurred: the speaker and each member of his audience were in adjacent spatial regions (because they were in the same room), and a sudden change has occurred (they learned something new within 1 hour).

Contrast this with a different frame of reference:

Objects = common sense macro objects
Space = 1 mm distance and multiples
Time = 5 seconds and multiples

In this frame of reference, the seminar does not constitute a causal interaction because the distances are too big and the changes are too slow. However, someone inoculating me to protect me against some disease would be causally interacting with me: there is less than 1 mm distance between my body and the needle of the syringe, and there is a sudden change in my body (within 5 seconds, it contains a fluid it did not contain before the interaction).

If we modify the last clause into:

Time = 0.5 seconds and multiples

the inoculation is not a causal interaction any more (because the change is too slow).

In this modified frame of reference, collisions between two billiard balls still constitute causal interactions. However, if we take smaller units of space and time, these collisions cease to be causal interactions.

I think we can draw three lessons from these examples:

(1) Salmon's concept of causal interaction is a "skeleton concept": it cannot be applied to empirical phenomena until we supplement it with a frame of reference as outlined above.

(2) Scientists in a given discipline should find the appropriate frame of reference for their domain, a frame of reference that "makes a difference" (a frame in which some phenomena constitute causal interactions, while others don't).

(3) There are no causal interactions in the world: if something is a causal interaction given a frame of reference, refining the frame of reference is sufficient to ensure that the phenomenon fails to satisfy the conditions.

These three lessons together constitute what I meant with the preconditions of the use of the concept of causal interaction. Salmon, who was a realist, would probably not have liked the constructivist flavour of these preconditions. But there is no way to avoid them: they are a consequence of the vagueness of certain words in the definition.

The constructivist preconditions may be bad news for causal realists, they are good news for our project: they entail that Salmon's definition is a polyvalent one that can be applied in all areas of science, including the social sciences. So we have a concept that captures the interactions between individuals that occurred in our definition of social mechanisms.

4 Spontaneous preservation

4.1 We have defined social mechanisms as complexes of interactions among individuals. There is no trace of further types of causation in our definition. However, Salmon has argued that causal mechanisms are more than complexes of causal interactions. In his view, causation also has a *conservative* aspect: properties acquired in causal interactions are often spontaneously preserved. This kind of causation is also involved in social mechanisms, as is clear from the following example.

During World War II, the American government asked the famous sociologist Robert Merton to analyse the success and failure of propaganda campaigns (see Merton, 1957, chap. 14). In his analysis, Merton takes individual propaganda documents (e.g., a movie, a pamphlet, a radio speech) as units. Each propaganda document has a specific aim, viz. to convince the reader/viewer/listener to adopt a specific role in the war machine. One of Merton's most interesting examples is a pamphlet that was meant to convince Afro-Americans to join the army. The pamphlet tried to reach this aim by communicating two messages:

(1) while Afro-Americans still suffer from discrimination, great progress has been made; and

(2) these attainments are threatened if the Nazis win the war.

The pamphlet failed because the authors tried to communicate the second message mainly by text items. The—mostly lower educated—Afro-Americans did not read the text, they just looked at the pictures. Most of the pictures related to the first theme: they featured Afro-Americans that occupied important positions in the American society. The result was that their self-confidence increased, but they did not conclude from the pamphlet that their attainments were in danger. Hence, they did not volunteer for military service.

Merton showed that the authors were wrong about the kind of changes the pamphlet would cause (i.e., about the causal interactions that would take place). But there is no problem with the other underlying assumption, viz. that once someone is convinced of something, this opinion is spontaneously preserved till another relevant causal interaction occurs. More generally, social scientists often assume that the opinions people acquire through causal interactions persist spontaneously (some for a short time, some for a longer time). We don't need constant causal interactions with other people to maintain our opinions.

4.2 The conclusion we can draw from 4.1 is that, in order to describe social mechanisms, we need causal claims of the following form:

(PSP) Property P was spontaneously preserved in object x during the period $[t, t']$.

I define spontaneous preservation as follows:

(SP) Property P is spontaneously preserved in object x during the period $[t, t']$ if and only if x has property P at t, t' and all times between, and one of the following conditions are satisfied:

(1) in the period $[t, t']$ there is no intersection between x and another object, or

(2) there are intersections between x and other objects in the period $[t, t']$, but even without these intersections x would have property P throughout the interval $[t, t']$.

The concept of spontaneous preservation is inspired by Salmon's *causal processes*. I will first clarify the relationship between these concepts and then explain why I do not simply use Salmon's concept.

Salmon divides processes (world lines of objects) into causal processes and pseudo-processes. Causal processes are capable of transmitting marks, pseudo-processes cannot transmit marks. Mark transmission is defined by Salmon as follows:

> Let P be a process that, in the absence of interactions with other processes, would remain uniform with respect to characteristic Q, which it would manifest consistently over an interval that includes both of the space-time points A and B ($A \neq B$). Then a *mark* (consisting of a modification of Q into Q'), which has been introduced into process P by means of a single local interaction at point A, is *transmitted* to point B if P manifests the modification Q' at B and at all stages of the process between A and B without additional interventions. (Salmon, 1984, p. 148)

Salmon mentions material objects and electromagnetic waves as examples of causal processes. This is quite strange: a process is a world line of an object, so it is very awkward to call some objects causal processes. We have to make a clear distinction between objects and world lines of objects. If we make this distinction, we can also distinguish between objects that have the capacity to transmit certain modifications of their structure to other spatiotemporal regions (like e.g., material objects) and world lines of such objects (= causal processes). The movement of a material object is a process (world line of an object). Moreover, it is a causal process: the underlying object has a capacity to transmit marks. But the material object itself is not a causal process, since it is not a process. The movement of an object is a causal process, but the moving object itself is not. The relation between spontaneous preservation and causal processes is now obvious. Spontaneous preservation requires an object that has the capacity to transmit marks: spontaneous preservation means that this capacity was realised for some characteristic P during some interval. If x spontaneously preserves P in the period $[t, t']$, the series of states of x in this period is a causal process in which the capacity of transmitting marks was effectively realised with respect to P.

The reason why I introduce the concept of spontaneous preservation, is explanatory relevance. As is argued in Hitchcock (1995), claiming that an object had the *capacity* to preserve *some* property spontaneously has no explanatory value. In explanations, we have to claim that a *specific* property *really* was spontaneously preserved.

One brief but important remark must be made. Like the concept of causal interaction, my concept of spontaneous preservation must be interpreted in a constructivist way. There is always an underlying reference frame which specifies the objects and the time and space scales. For, instance, if the objects are human individuals, then a person that has no contact with anybody else for one week, satisfies clause (1) of definition (SP), even if he eats, drinks and interacts in various other ways with his non-human biological environment.

5 Conclusion and topics for further research

5.1 I have argued that social scientists need at least three causal concepts, and I have tried to explicate them. So I am not only a conceptual pluralist, but also pluralist with respect to causation in the social sciences. The first concept (Section 2) was located at the population level, while the other ones (Section 3 and 4) are located at the individual level. This is the first distinction on which my pluralism was based. The second distinction is that between innovation and preservation.

5.2 I conclude this paper with some possibilities for further research. First, I think that most of what I say about the social sciences can be extrapolated to the biomedical sciences. Consider an example of Patrick Suppes in his seminal book *A Probabilistic Theory of Causality*. The example is based on a study of the efficacy of inoculation against cholera (Suppes, 1970, p. 12-13). Only 1 percent of the individuals that were inoculated (3/279) were attacked by cholera, while the mean probability of attack in the whole population was 9 percent (69/818). As in the social science examples of 2.1, the causal claim investigated here is located at the population level and has policy relevance. I think it is possible to argue that the biomedical sciences need biological mechanisms (cfr. the argument for social mechanisms in 3.1). However, exploring the similarities between biomedical and social sciences is beyond the scope of this paper.

A second follow-up question is: what about physics? The ideas Wesley Salmon presented have been criticized and elaborated by many people. Let us focus on Phil Dowe, who has suggested a definition of causal interaction that is much simpler than my definition (and Salmon's original one):

> A *causal interaction* is an intersection of world lines that involves exchange of a conserved quantity. (Dowe, 1992, p. 210) & (Dowe, 2000, p. 90)

As already mentioned, a *world line* is the collection of points on a spacetime diagram representing the history of an object. An *object* can be anything found in the ontology of science (particles, waves, fields,...) or common sense (chairs, buildings, people,...). In his Salmon (1994, p. 304) says that this definition is "acceptable as it stands". The main difference between (CI) and Dowe's definition is that the counterfactuals that occur in (CI) have been substituted for the requirement that a conserved quantity is exchanged.

Whatever the merits of Dowe's definition may be, it is not applicable in the social sciences (nor in the biomedical sciences) because there are no conservation laws in these domains. But even if we restrict ourselves to physics, I think there is a problem of vagueness and arbitrariness. Dowe defines conserved quantities as follows:

> A *conserved quantity* is any quantity universally conserved according to current scientific theories. (Dowe, 1992, p. 210)

> A *conserved quantity* is any quantity that is governed by a conservation law, and current scientific theory is our best guide as to what these are. (Dowe, 1992, p. 91)

The vagueness lies in the fact that the concept of conservation law is not defined (can it be defined without bringing in counterfactuals again?), there is arbitrariness in the reference to current scientific theories. So in the area of physics there is also work to be done. But that falls outside the scope of this paper.

Acknowledgments

I thank Jeroen Van Bouwel and the two anonymous referees for their comments on earlier drafts of this paper.

Erik Weber
Centre for Logic and Philosophy of Science, Ghent University, Belgium.
erik.weber@ugent.be

BIBLIOGRAPHY

Cartwright, N. (1979). Causal laws and effective strategies. *Nous*, 13:419–437.

DeMey, T. and Weber, E. (2003). Explanation and thought experiments in history. *History and Theory*, 42:28–38.

Dowe, P. (1992). Wesley Salmon's *Process Theory of Causality and the Conserved Quantity Theory*. *Philosophy of Science*, 59:195–216.

Dowe, P. (1995). Causality and conserved quantities: a reply to Salmon. *Philosophy of Science*, 62:321–333.

Dowe, P. (2000). *Physical Causation*. Cambridge University Press, Cambridge.

Dupré, J. (1993). *The Disorder of Things*. Harvard University Press, Cambridge & London.

Eells, E. (1991). *Probabilistic Causality*. Cambridge University Press, Cambridge.

Giere, R. (1997). *Understanding Scientific Reasoning (4th edition)*. Harcourt College Publishers, Fort Worth.

Hall, N. (2004). Two concepts of causation. In Collins J., H. N. and A., P. L., editors, *Causation and Counterfactuals*, pages 225–276. The MIT Press, MA.

Hausman, D. (1998). *Causal Asymmetries*. Cambridge University Press, Cambridge.

Hempel, C. (1965). Aspects of scientific explanation. In *Aspects of Scientific Explanation and other Essays in the Philosophy of Science*, pages 331–496. Free Press, New York.

Hitchcock, C. R. (1995). Salmon on explanatory relevance. *Philosophy of Science*, 62:304–320.

Humphreys, P. (1989). *The Chances of Explanation*. Princeton University Press, Princeton.

Jones, R. A. (1996). *Research Methods in the Social and Behavioral Sciences*. Sinauer Associates, Sunderland, Massachusetts.

Kershaw, D. N. (1972). A negative income tax experiment. *Scientific American*, 227(4):19–25.

Merton, R. (1957). *Social Theory and Social Structure*. The Free Press, Glencoe, Illinois.

Reynolds, P. D. (1982). *Ethics and Social Science Research*. Prentice-Hall, Englewood Cliffs, New Jersey.

Salmon, W. (1984). *Scientific Explanation and the Causal Structure of the World*. Princeton University Press, Princeton, New Jersey.

Salmon, W. (1994). Causality without counterfactuals. *Philosophy of Science*, 61:297–312.

Salmon, W. (1997). Causality and explanation: a reply to two critiques. *Philosophy of Science*, 64:461–477.

Schutt, R. K. (1996). *Investigating the Social World. The Process and Practice of Research*. Pine Forge Press, Thousand Oaks, California.

Sherman, L. W. and Berk, R. A. (1984). The specific deterrent effects of arrest for domestic assault. *American Sociological Review*, 49:261–272.

Steel, D. (2004). Social mechanisms and causal inference. *Philosophy of the Social Sciences*, 34:55–78.

Suppes, P. (1970). *A Probabilistic Theory of Causality*. North-Holland Publishing Company, Amsterdam.

PART IV

CAUSALITY AND PROBABILITY IN THE BIOMEDICAL SCIENCES

Interpreting probability in causal models for cancer

FEDERICA RUSSO AND JON WILLIAMSON

ABSTRACT. How should probabilities be interpreted in causal models in the social and health sciences? In this paper we take a step towards answering this question by investigating the case of cancer in epidemiology and arguing that the objective Bayesian interpretation is most appropriate in this domain.

After introducing the problem in §1 and giving an overview of causal analysis in the social and health sciences in §2, in §3 we present the cancer case study in some detail. In §4 we discuss the importance of correctly interpreting probability. Then, in §5, we put forward some desiderata that an interpretation of probability ought to satisfy; two Bayesian interpretations of probability come out well according to these desiderata. In §6 we go further by showing how the full-blown objectivity of objective Bayesianism is needed for the practice of cancer treatment. Finally we discuss the ramifications of this conclusion for the social and health sciences in §7.

1 Introduction

Whilst it might seem uncontroversial that the health sciences search for causes—that is, for causes of disease and for effective treatments—the causal perspective is less obvious in the social sciences, perhaps because it is apparently harder to glean general laws in the social sciences than in other sciences. Thus the search for causes in the social sciences is often perceived to be a vain enterprise and it is often thought that social studies merely describe the phenomena.

On the other hand an explicit causal perspective can already be found in pioneering works of Adolphe Quetelet and Emile Durkheim in demography and sociology respectively, and the social sciences have taken a significant step in *quantitative causal analysis* by following Sewall Wright's path analysis, which was first applied in population genetics. Subsequent developments of path analysis—e.g., structural models, covariance structure models and multilevel analysis—have the merit of making the concept of cause operational by introducing causal relations into the framework of statistical modelling. However, these developments in causal modelling leave a number of conceptual issues unanswered: for instance the question of how probability should be interpreted in probabilistic causal models.

In the philosophy of probability many interpretations have been proposed and crucial objections raised. For instance, it has been argued that the frequency interpretation does not make sense in the single-case, that subjectivist accounts lead to arbitrariness in probability assignments, and that logical interpretations, though suited to gambling situations, are of scarce applicability in science.

In this paper we raise the problem of the interpretation of probability within a specific context: causal models in cancer epidemiology.[1] This is motivated by the thought that competing interpretations are not right or wrong, but that they are better or worse suited to particular contexts and the demands we make of them. To this end, we first introduce causal analysis in the social and health sciences and then present the case of cancer epidemiology in some detail. We pay particular attention to explaining different possible meanings of probabilistic statements in this context and the importance of choosing one interpretation of probability over another. We then argue that any satisfactory interpretation of probability should satisfy five desiderata; this narrows down the choice to the frequency interpretation twinned with an empirically-based subjective interpretation or with an objective Bayesian interpretation. We go on to argue that the probabilities in causal models in cancer epidemiology should be given a frequency-cum-objective-Bayesian interpretation; the main reason for this choice is the need to cope with two different types of probabilistic inference, generic and single-case.

2 Causality in the social and health sciences

Different social sciences study society from different perspectives. Sociology studies the structure and development of human society, demography attends to the vital statistics of populations, economics studies the management of goods and services, epidemiology studies the distribution of disease in human populations and the factors determining that distribution, etc. In spite of these differences, the social and health sciences share a common objective: to understand, predict and intervene on society. Knowledge of

---

[1] We are fairly liberal as to which models count as *causal*. Arguably, a model is causal if its relationships are interpreted causally or put to causal use. Thus associational and regression models, which tend not be explicitly causal, would count as causal for us if the relationships in the model are interpreted causally (e.g., if an association between smoking and cancer is interpreted as supporting the claim that smoking causes cancer) or used as a basis for intervention (e.g., by banning tobacco advertising).

causes is required to achieve this common goal; such knowledge provides an explanation of social phenomena as well as of individual behaviour.

Whilst social scientists have been looking for social causes, health scientists traditionally focus on the biological causes of diseases. For instance, in *Le Suicide* Emile Durkheim studies suicide as a social phenomenon and consequently looks for the social causes of the suicide rate in the population. Modern medicine, since the work of Claude Bernard (1813-1878), tries to identify physiological mechanisms active in living beings. The Henle-Koch Postulates, formulated by Robert Koch in 1882, provided criteria for judgements about the presence of micro-organisms as causes of disease (Koch, 1882). For instance, the current theory of carcinogenesis involves a particular molecular mechanism. Under normal circumstances the growth of cells is controlled accurately by inherited mechanisms and stimulated or inhibited as required. Cancer occurs when genetic alterations (mutations) disturb the normal regulation of a cell and cause it to erroneously multiply—with potentially fatal consequences. Mutations like this trigger cell growth either directly or indirectly by disrupting the mechanisms responsible for limiting cell division. On the other hand, the social sciences try to understand the variety of ways in which the population is exposed to carcinogenic substances, according to levels of education, economic status, type of occupation, etc.

The association between diseases and various socio-economic factors is the object of research in several disciplines. In particular, epidemiology typically tries to single out individual genetic factors, biological factors, and environmental risk factors. Studies have recently focused on the role of neighbourhood environment (see Pickett and Pearl (2001) for a literature review). Neighbourhood environment affects health through the availability and accessibility of health services, infrastructure deprivation, stress, lack of social support, and so on. An established tradition in sociology successfully studies the impact of neighbourhood environment on sociological outcomes such as educational attainment and labour market opportunity by taking advantage of multilevel analysis. Epidemiology picks up this tradition and uses multilevel analysis to examine group level effects on individual health. Thus, although for long time the social and health sciences have trodden quite different and independent paths, a new perspective integrating both approaches is now seeming to emerge.

The case of epidemiology is of particular interest since it has been argued that an integration of the social science approach and the health science approach will get at a better understanding of causal relations (see Susser and Susser (1996); Vineis et al. (2004); Weed (2000) and references therein). Some time ago the same methodological turn was also advanced in demography (Mosley and Chen, 1984) for studying the phenomenon of child mortality in developing countries: social and economic determinants exert an impact on mortality through biological mechanisms. Hence, Mosley and Chen's approach—which now constitutes the received view in the specific field of child mortality in developing countries—incorporates both social and biological variables and integrates research methods proper to the two domains. Because of this move to consider both biological and social factors when explaining a given phenomenon, we see epidemiology as paradigmatic of the social and health sciences.

Epidemiology, being interested in the biological and social determinants of diseases and in their distribution on the population, has cognitive and practical goals. The cognitive goal concerns the aetiology of diseases, and the practical goal concerns the implications of such causal knowledge on policy making and also on causal attribution and diagnosis in particular individuals.

In §3 we shall focus our attention on the case of cancer epidemiology. This will help us make two major points in the paper. First, epidemiology is concerned with two different types of causal inferences. One is *generic* and concerns the population as a whole, and the other is *single-case* and concerns particular individuals.[2] Second, both causal inferences are essentially probabilistic; consequently we raise the question of which interpretation of probability best fits these probabilistic inferences. We shall argue that probabilities have to be interpreted according to an objective Bayesian approach in the single case and according to a frequentist approach in generic inferences.

3 Case study: cancer

In the late '50s, when he was a scientific consultant for the Tobacco Manufacturers Standing Committee, Sir Ronald Fisher advanced that the correlation between smoking and lung cancer was due to an unknown genotype influencing both smoking behaviour and the predisposition to lung cancer, thus casting doubt on the hypothesis of a direct causal link from smoking to lung cancer.[3] The primary intent of Fisher was to point to the well known fact that correlation is not causation and that alternative explanations—

---

[2]There is a subtlety here: the distinction between population-level and individual-level (commonly drawn with regard to causal claims) does not quite correspond to the distinction (commonly drawn with regard to interpretations of probability) between the generic and the single case. This can be seen from the following claims. In general, inequality (population-level, generic) causes deterioration in health (population-level, generic). The inequality of her compatriots (population-level but single-case) is a cause of Naomi's deterioration in health (individual-level, single-case) (Glymour, 2003). To simplify matters we shall restrict our attention to the generic / single-case distinction in this paper.

[3](Fisher, 1957, 1958).

such as the one of an unknown genotype—could equally well account for statistical correlations.

Since then, cancer research has gone a long way. The link between smoking and cancer has been investigated under almost every possible angle. For instance, scientists try to establish the effect of tobacco control programs on declines in smoking and heart disease mortality (Barnoya and Glantz, 2004). In this type of study the causal effectiveness of tobacco consumption on lung cancer is implicitly assumed. However, if we can ascribe a decrease in mortality rates to a decrease in smoking, this will provide further epidemiological evidence. Fisher's hypothesis of a gene regulating cancer predisposition has not been dismissed, however. For instance, a recent study (Hwang et al., 2003) tried to assess the causal role of the p53 germline. Results indicate that a cancer predisposition due to a p53 mutation is significantly increased by cigarette smoking.

Specific studies concerning particular populations or subpopulations, such as those mentioned above, contribute to the mapping of the aetiology of cancer. From an epistemological viewpoint, the problem is how to gather together knowledge acquired in those specific studies in order to attain general epidemiological knowledge abut cancer. Consider for instance the two following papers: Lagiou et al. (2005); Vineis et al. (2004). These articles intend to provide a rather complete overview of the present state of affairs. That is, the aim is to summarize various epidemiological evidence on tobacco and cancer coming from specialised studies.

The mapping of the aetiology of cancer bears on several questions. One is the question of which types of cancer are due to tobacco consumption. There are several: lung, nasal cavity, stomach, liver, kidney, uterine cervix.[4] Another is the question of which biological factors are carcinogenic (e.g., hepatitis B and C virus, helicobacter pylori, human papilloma viruses, etc.) and which cancer sites are associated with these factors. A third question concerns the occupational chemicals that produce cancer; for instance, arsenic, asbestos, benzene, hair dyes, painting materials, soot. From a medical viewpoint it is important to understand which cancer sites are most often associated with exposure to these substances; on the other hand from a demographic viewpoint it is also important to figure out which parts of the population, in terms of social class or occupation, are more exposed to those carcinogenic substances and what the intensity of exposure is. A large part of medical and epidemiological research in cancer aetiology also focusses on

---

[4]Although epidemiological studies have variably shown positive, inverse or null associations between cigarette smoking and breast cancer, experimental studies indicate that tobacco smoke contains potential human breast carcinogens. See Terry and Rohan (2002).

the genetic factors predisposing or preventing cancer. Dietary behaviour, lifestyle factors (e.g., passive smoking, consumption of alcoholic beverages, ultraviolet radiation), and socio-demographic characteristics related to cancer are likewise investigated.

It is worth pointing out that results of particular studies do not automatically count as epidemiological evidence. Their soundness is evaluated according to specific criteria. The International Agency for Research on Cancer (IARC) classifies evidence of carcinogenicity into four categories:[5] sufficient, limited, inadequate and evidence suggesting lack of carcinogenicity. We have sufficient evidence when a positive relationship has been observed between the exposure and cancer in studies in which chance, bias and confounding could be ruled out with reasonable confidence. Evidence is limited if the working group considers the association credible, but chance, bias and confounding could not be ruled out with reasonable confidence. Evidence is inadequate if the available studies are of insufficient quality, consistency or statistical power to permit a conclusion about the presence or absence of a causal relation. Finally, evidence suggests lack of carcinogenicity when several studies are consistent in not showing a positive association between exposure to the agent and any studied cancer at any observed level of exposure.[6]

If this tells us something about the kind of evidence (biological and social) used to support causal statements and about the criteria used for evaluating those results, it doesn't say anything about the very concept of causation underlying causal analysis in cancer epidemiology.

In the following, we shall not review the extensive epidemiological literature to come up with an inventory of different concepts of cause. Such an overview of different concepts of cause in epidemiology is offered in recent publications (see Parascandola and Weed (2001) and references therein); virtues, faults and applicability of various concepts—production, necessary and sufficient, sufficient-component, counterfactual, probabilistic—are presented there. We won't even enter the debate about whether a probabilistic concept of causality does a better job than a deterministic one (although this seems to be the most recent point of view emerging, see Parascandola and Weed (2001); Vineis et al. (2004)). Instead, we shall focus on causal models and inferences in cancer epidemiology where, as matter of fact, causal relations are probabilistically characterized. It is worth noting, however, that a probabilistic methodology and epistemology do not necessarily im-

---

[5]See Vineis et al. (2004, p. 100) and Lagiou et al. (2005, p. 569), and the IARC web site http://monographs.iarc.fr/ENG/Preamble/index.php.

[6]Of course, evidence of lack of carcinogenicity requires that studies meet to a sufficient degree the standards of design, and in particular that bias, confounding and missclassification be ruled out with a reasonable degree of certainty.

ply that the *concept* of cause—i.e., the cause in the metaphysical sense—is itself probabilistic.[7] Consequently, "probabilistic" has to be understood in a non-metaphysical sense—i.e., as just referring to the use of probabilistic models for causal analysis in cancer epidemiology.

As mentioned above, there is now unquestionable evidence that tobacco is a powerful carcinogenic substance that can cause cancer in many different organs. It is also commonly agreed that while tobacco consumption raises the probability of developing cancer, tobacco consumption—whether active or passive—is not a sufficient cause. In other words, what scientists seek to establish is the extent to which smoking increases the probability of developing cancer, or the extent to which exposure to a carcinogenic substance such as asbestos influences cancer rates, or the extent to which particular dietary habits prevent—i.e., lower the probability of developing—cancer. Thus, causation of cancer is conceptualized in a probabilistic sense involving statistical terms and procedures. Whether studies are experimental or observational, the goal in both cases is to reduce uncertainty, by performing as many studies as possible to generate sensible summary statistics, and by reducing confounding and bias. Different statistical models, ranging from multiple regression analysis to structural modelling, are used to accomplish this task.

It is not hard to see that these probabilistic models only allow for probabilistic inferences. But what types of inferences are we concerned with? The first type of inference, which we shall call *generic* inference, aims at establishing whether or not a factor is a cause of disease by deciding whether, roughly, alterations in the frequency or intensity of this factor are accompanied by alterations in the frequency or intensity of disease. This corresponds to the naive causal statement 'smoking causes lung cancer.' The second type of inference, which we shall call *single-case* inference, is instead concerned with particular individuals. For instance, exposure to a known cause of cancer implies that this individual is now more likely to develop cancer. Assessing single-case probabilities for *particular* individuals is a real worry for practicians, and in fact one goal of Evidence Based Medicine is to provide guidelines to tackle particular situations.[8] Parascandola and Weed (2001, p. 908) echo Cox's, Holland's and Olsen's criticisms of probabilistic accounts: a probabilistic theory of causation, based on statistical inequalities, is inadequate since it leaves unclear what it means for smoking to raise the probability of an individual developing lung cancer. They argue that in this respect counterfactuals help. Instead, we argue that the problem is

---

[7]Parascandola and Weed (2001, p. 906) make this point but don't develop it further.
[8]See for instance online resources of the Centre for Health Evidence, http://www.cche.net/usersguides/.

not the inadequacy of probabilistic theories but rather a neglected aspect of them: the distinction between the generic and the single-case.

Lagiou et al. (2005, p. 569) seem to have grasped the importance of such a distinction, for they claim that although criteria such as those mentioned above are surely important for discerning causal association from non-causal association, they do not allow one to separate the different issues posed by (i) the results of a single study, (ii) the results of several studies, and (iii) the likelihood of causation in a particular individual. They acknowledge that an individual study (what they call level I) does not allow one to infer causation but can provide evidence when we already have several studies (level II) pointing to the same causal relation. Level II, however, is not sufficient to establish a causal link between an individual's exposure and disease. What it is still possible to do, they claim, is to infer from level II that the specific individual's illness was more likely than not caused by the specified exposure. This highlights the levels of causation and the different types of inferences we make either concerning the population as a whole or concerning particular individuals.

The discussion, however, is not pushed further. In particular, only to recognise different types of inferences is not enough: those inferences are probabilistic and consequently raise the problem of how probability has to be interpreted. This is the question we turn to next.[9]

4 Interpreting probability

In the philosophy of probability there is a wide-ranging debate about interpretation. Indeed, discussions about the *meaning* of probability began as early as the first formulations of probability theory. We direct the reader to Hacking (1975) and Gillies (2000) for interesting historical overviews.

As we have seen in §3, causal inferences in cancer epidemiology are essentially probabilistic. A standard objection to probabilistic theories of causation is that the claim 'smoking raises the probability of lung cancer' is ambiguous for it might be interpreted in different ways. It might say that within the population, the proportion of those who develop lung cancer is greater amongst smokers. It might also say that if a *particular patient* smokes, then it's more likely that she will develop lung cancer. This ambiguity motivates a distinction between different levels of causation and consequently between different types of causal inferences: generic and single-case. Those claims state, in different terms, a probabilistic relation between

---

[9]Inferences in cancer epidemiology, whether generic or single-case, are also based on the computation of relative risks and odds ratios. In the appendix we briefly address the problem of correctly interpreting these measures and argue that the generic/single-case distinction is again illuminating.

smoking and lung cancer. But what does probability mean in these cases? Is there a unique interpretation fitting the two examples equally well? Or should we attach a different meaning to probability depending on the claim at stake?

It is not hard to see that those claims state quite different things. A generic causal claim posits an average causal relation, which is supposedly valid for the majority of individuals exposed within the population. The adoption of a probabilistic framework entails, by itself, that such a causal claim is not a universal and necessary law, and, consequently, that not every individual in the population who instantiates the cause will instantiate the effect. In fact, as is well known, some smokers never develop cancer, and some non-smokers instead do. This leads us to ask what the meaning of a single-case causal claim is. Two meanings ought to be distinguished. In one case we wish to make a prediction: your smoking now makes you more likely to develop cancer in the future. A second interpretation is instead retrospective: it is likely that smoking caused you to develop lung cancer.

It is then apparent that the meaning of probability for generic causal claims is connected with frequency of occurrence, whereas for single-case causal claims the meaning is closer to something like belief or credence about what will happen or happened. A competing interpretation for the single-case is single-case chance. However, as we shall see, single-case chance raises problems of epistemic access.

The overview we offer next is meant to see whether any of the leading interpretations provide sensible meaning both for generic and single-case claims. However, we shall see that none of these interpretation succeeds. In §5 we argue that a frequency interpretation is needed for the generic and that a Bayesian approach is needed for the single case. In §6 we will argue that practical considerations motivate choosing the objective Bayesian approach over the empirically-based Bayesian approach.

We will now sketch very briefly the main features of the four leading contenders: (i) the classical and logical interpretations; (ii) the physical interpretations: frequency and propensity; (iii) the subjective interpretation; (iv) the objective Bayesian interpretation.

The classical interpretation of probability defines probability as the ratio between the number of favourable cases and the number of all equipossible cases. The easiest way to grasp the meaning of favourable and equipossible cases is to think of dice. The six sides of a die constitute the probability space—i.e., the six possible outcomes. Assuming that the die is not biased, the six sides are all equipossible—i.e., they all have the same probability of occurring uppermost. The favourable cases constitute the event we are interested in. For instance, the probability that an even number will result is given by the number of favourable cases (sides 2, 4 and 6) over the total number of the equipossible cases (for an unbiased die: 1, 2, 3, 4, 5 and 6) which gives $\frac{3}{6}$, i.e., $\frac{1}{2}$.

First developed by Laplace (1814), a similar interpretation was also proposed by Pascal. Probability values are assigned in the absence of any evidence—the probability of an event is simply the fraction of the total number of possibilities in which the event occurs. The notion of equipossibility is expressed by what Keynes called the Principle of Indifference. This principle states that whenever there is no evidence favouring one possibility over another, these possibilities have the same probability. The classical interpretation seems especially well suited to games of chance, although it is sometimes objected that this interpretation suffers the problem of circularity, for equipossible means equiprobable, hence 'probable' is not properly explicated. A second traditional objection is that the classical interpretation is of scarce applicability in science; in fact, adopting the classical interpretation we have no meaningful way to express knowledge of the population probabilistically, nor to evaluate individual hypotheses.

A generalization of the classical interpretation is the so-called logical interpretation, advanced by Keynes (1921); Jeffreys (1939); Carnap (1950). This interpretation depends on the Principle of Indifference and thus rests on the idea that probabilities can be determined a priori by an examination of the space of possibilities, but only when no knowledge indicating unequal probabilities is available. The main aim of the logical interpretation is to provide an account, as general as possible, of the degree of support or confirmation that a piece of evidence e confers upon a given hypothesis h. In Carnap's notation, the c-function precisely expresses this idea.[10]

According to the physical view, probability values are quantitative expressions of some feature of the world, not of our knowledge or beliefs. The physical view is typically taken to encompass the frequency and the propensity interpretations. A simple version of frequentism, due to Venn (1866), states that the probability of an attribute A in a finite reference class B is the relative frequency of the actual occurrence of A within B. Further developments of frequentism are due to von Mises (1928) and Reichenbach (1935), who consider infinite reference classes and identify probabilities with the limiting relative frequencies of events or attributes therein. This second sort of frequentism is also advocated by Salmon (1967). Limiting relative

---

[10] As well as appealing to symmetry or indifference, Carnap (1950, §§41–42) bases probability values upon knowledge of physical probabilities. He says, 'in these cases the probability is determined with the help of a given frequency and its value is either equal or close to that of the frequency' (Carnap, 1950, §42B). Consequently, Carnap's development of the logical interpretation might be classified alongside the objective Bayesian approach, which we will shortly introduce.

frequencies serve, in his approach, to determine the probability of evidence. Note that frequentism interprets generic probabilities; in order to associate a frequency with a single case, a unique reference class must be associated with the single case; that this can not be done in general is known as the *reference class problem*.

The propensity interpretation is also located in the physical realm, since probability is 'in the world', so to speak, rather than 'in our heads'—as it is in the subjectivist approach and the classical and logical interpretations. Probability is here conceived as a physical propensity, or disposition, or tendency, of a given type of physical situation to yield an outcome of a certain kind, or to yield a long run relative frequency of such an outcome. The propensity interpretation was advanced by Popper (1957, 1959), who was motivated by the desire to make sense of single-case probabilities, for instance in quantum mechanics.

In the subjective interpretation probabilities are quantitative expressions of an agent's opinion, or degree of belief, or epistemic attitude, or something similar. First advances are due to Ramsey (1926) and de Finetti (1937). De Finetti's viewpoint is paradigmatic of personalistic approaches, for he firmly stated that probability does not exist (in the physical sense), and that it is possible to reconstruct and deduce probability theory just relying on the subjectivist interpretation (de Finetti, 1993, pp. 248 ff). In subjectivist approaches, also called subjective Bayesian, probabilities are analyzed in terms of betting behaviour. Probabilities are identified with the announcement of the betting odds that a rational agent is willing to accept. A Dutch book (against an agent) is a series of bets, each acceptable to the agent, but which collectively guarantee her loss, whatever happens. Two Dutch book theorems then follow. (i) If an agent's degrees of belief violate the probability calculus, then she is liable to a Dutch book, and, conversely, (ii) if an agent's degrees of belief conform to the probability calculus, then no Dutch book can be made against her. A series of bets is called *coherent* if it is not susceptible to a Dutch book. In subjectivist approaches, adherence to the probability calculus is taken to be a necessary and sufficient condition for rationality.

It is typically objected that this personalistic account leads to arbitrariness, that is, it is too subjective. In fact, two agents with exactly the same evidence may assign different probability values to the same event and be equally rational, provided that they do not violate the probability calculus. (It is worth pointing out that de Finetti's betting interpretation derives probabilities from utilities and rational preferences. The intimate link between utility of outcomes and probabilities is even more prominent in the approaches of Savage (1954) and Jeffrey (1965). The main idea of the utility interpretation is that probabilities and utilities can be derived from preferences among options that are constrained by certain putative consistency principles.)

A solution to the objection of arbitrariness is attempted by empirically-based subjectivism and also by the objective Bayesian interpretation of probability. These interpretations require that further constraints beyond coherence be satisfied before an agent's degrees of belief can be considered rational. The former approach, empirically-based subjectivism, was advocated e.g., by Salmon (1967, 1990). This approach requires that empirical information constrain degrees of belief: if, for instance, an agent knows that 60% of people with a certain type of cancer recover, knows that a particular patient has this type of cancer, but knows nothing else pertinent, then she should believe that the patient will recover to degree 0.6. The latter approach, objective Bayesianism, was put forward by Jaynes (1957) and goes beyond empirically-based subjectivism. According to this view, lack of knowledge should also constrain degrees of belief: in the absence of evidence the agent should be as equivocal as possible, e.g., if the agent does not have any knowledge at all pertinent to a cancer patient then she should believe that the patient will recover to degree 0.5; if her knowledge constrains her degree of belief to fall in the interval $[0.6, 0.8]$ then she should chose the point that most equivocates between recovery and non-recovery, i.e., 0.6.[11] Thus both information and lack of information about the world should be taken into account in shaping epistemic probabilities. Information-theoretic considerations motivate the use of entropy as a measure of the extent to which a probability function equivocates; consequently Jaynes put forward the *maximum entropy principle*, which provides a formal framework for objective Bayesianism: the agent's belief function should be a probability function, from all those that satisfy constraints imposed by evidence, that has maximum entropy. In this framework, on a finite domain an agent's background knowledge fully determines the degrees of belief that she ought to adopt.[12]

Note that degree-of-belief interpretations—including subjectivism, empirically-based subjectivism, and objective Bayesianism—interpret single-case rather than generic probabilities: degrees of belief are associated with

---

[11] Note that the interval $[0.6, 0.8]$ is not a confidence interval. If a study indicates that 70% of people recover and provides a confidence interval $[0.6, 0.8]$, then the best (albeit defeasible) evidence is that 70% of people recover and the agent should simply set her degree of belief in recovery to 0.7. Rather, the interval constraint might be generated by two studies, one of which finds a 60% recovery rate, the other of which finds 80%, and neither of which is to be preferred over the other (on the grounds of sample size, specificity etc.)—then it is reasonable to place one's degree of belief somewhere in the ordered interval generated by the frequencies.

[12] (Williamson, 2005, §5.3). Note that this is not necessarily the case on infinite domains—see Williamson (2006c, §19).

betting, and a bet in a generic outcome makes little sense.

In §5 we will argue that generic causal claims demand a frequency interpretation, while single-case claims require an empirically-based or objective Bayesian interpretation. In §6 will go further in arguing that an objective Bayesian interpretation should be chosen.

5 Desiderata

In this section we shall put forward some requirements that an interpretation of probability should meet.

A philosophical theory of probability should:

Objectivity: account for the objectivity of probability,

Calculi: explain how we reason about probability,

Epistemology: explain how we can know about probability,

Variety: cope with the full variety of probabilistic claims that we make,

Parsimony: be ontologically parsimonious.

We shall discuss each of these desiderata in turn, paying special attention to the application to causal models of cancer.[13]

Objectivity

Many applications of probability invoke a notion of probability that is objective in a logical sense: there is a fact of the matter as to what the probabilities are; if two agents disagree about a probability, at least one of them must be wrong.[14] For example, the probability that a patient's breast cancer will recur after treatment is supposed to depend on features of the cancer (e.g., whether it is metastatic, whether it is HER2 positive, its ER status), of the treatment, and of the patient. It is not simply a matter of personal opinion: if two prognostic probabilities differ, at least one of them must be wrong. A philosophical theory of probability should yield a notion of probability that is objective in this logical sense—otherwise it is not meeting the demands of these particular applications.

Clearly the subjective interpretation of probability suffers in this respect. According to the subjective theory, probabilities are degrees of belief and one can adopt any prior probabilities one likes as one's degrees of belief. According to the subjective theory, then, one agent can give probability

---

[13]See Williamson (2006a) for discussion of similar desiderata as requirements of a philosophical theory of causality.

[14]Logical objectivity contrasts with the ontological sense of objectivity: probabilities are ontologically objective if they exist as physical entities.

0.9 to the patient's breast cancer recurring, another agent with the same knowledge of the situation can give probability 0 to the same event, and neither agent can be considered wrong.[15] Empirically-based subjectivism also suffers, but to a lesser extent: if frequencies are known then probability assignments are not arbitrary, but where frequencies are not known an agent can choose her degrees of belief arbitrarily. The classical and logical interpretations can also suffer at the hands of Objectivity, since different agents can construe different partitions of events as equipossible.

In contrast, frequency, propensity and objective Bayesian interpretations all yield objective probabilities of varying forms. A frequency is objectively determined by a reference class; a propensity is objectively determined by the history of the universe up to the present time; under objective Bayesianism a probability is objectively determined by an agent's knowledge. Thus these interpretations fare better with respect to this desideratum.

Calculi

Probabilities are manipulated and inferences are drawn from them by means of the probability calculus. This mathematical apparatus, based on axioms put forward by Kolmogorov (1933), has by now become well entrenched. Consequently a philosophical theory of probability should yield a notion that satisfies the axioms of probability. Otherwise it is not a theory of probability, but a theory of something else.

Some theories suffer in this respect. According to some accounts, probabilities are not real numbers but are intervals of numbers, pairs of real numbers, or qualitative entities.[16] According to other accounts probabilities satisfy some axioms but not others—the frequency theory of von Mises (1928), for instance, does not satisfy the axiom of countable additivity; the propensity theory has problems with conditional probabilities.[17] With respect to this desideratum, then, degree-of-belief interpretations (subjectivism, empirically-based subjectivism, and objective Bayesianism) fare better than these other approaches.

---

[15] Proponents of the subjective account tend to respond in two ways: by saying that the subjective theory can account for objectivity in the long run as different agents' beliefs converge to frequencies, and by saying that there is no further objectivity to be found. While the first claim is notoriously problematic (Williamson, 2005, §2.8), the second claim is simply dangerous. If the subjectivist has no knowledge that bears on recurrence of breast cancer and awards a degree of belief 0.9, instead of the objectively-determined middling value 0.5, then she may initiate unnecessarily aggressive treatment rather than collect further evidence—see §6. Thus there is further objectivity—derived from the need to equivocate in the absence of evidence—that the subjective account ignores.

[16] See e.g., Keynes (1921); Kyburg Jr (2003); Walley (1991).

[17] (Humphreys, 2004)

Epistemology

We come to know about probabilities in various ways: we measure population frequencies, we appeal to symmetry arguments or scientific theories, we make educated guesses, we derive some probabilities from others using the probability calculus. A philosophical theory of probability should explain how we can use such techniques to discover probabilities. If the theory rejects some of these techniques it should say where they go wrong and why they are apparently successful.

This desideratum is a stumbling block for several theories. The classical, logical and subjective theories can not account for the widespread use of frequencies, while the frequency theory can not explain how degrees of belief can offer access to probabilities. The propensity theory is oft criticised for being metaphysical: it connects probability with scientific theories and even degrees of belief,[18] but struggles to identify a precise link with frequency. However the empirically-based subjectivist and objective Bayesian approaches allow background knowledge of any form—frequencies, symmetries, scientific theories included—to constrain an agent's rational degrees of belief; by design these theories admit a variety of sources of probabilistic knowledge.

Variety

Probabilistic claims are extremely varied. For instance, claims are made about single-case probabilities (e.g., the probability that a particular patient's cancer will recur) and generic probabilities (e.g., the probability of recurrence among those who receive radiotherapy). Moreover, probabilities are attached to a variety of entities, including events, sets, variables, sentences, propositions and hypotheses. A philosophical theory of causality should be able to cope with this variety—it should account for each use of probability, or, if some uses are to be viewed as illegitimate, it should say how such uses should be eliminated in favour of the legitimate uses. Otherwise, the theory is at best a partial theory, a theory of *some* of the uses of probability.

This desideratum is a problem for many of the interpretations of probability. The frequency and propensity theories can not ascribe a probability to a given hypothesis, but only yield the probability of observing a sample if the hypothesis is true—on this point see Courgeau (2004). Moreover, the frequency theory is a generic theory; it views single-case probabilities as illegitimate but provides no means of interpreting single-case claims in terms of frequencies. If single-case probabilities are to be abandoned the theory really ought to explain why their use, if so erroneous, is apparently

---
[18] (Lewis, 1980)

so successful. Other interpretations ascribe probabilities to single cases, and do not provide a means for interpreting population-level probabilities. It seems that pluralism is the only option: have one interpretation for the single-case and another for generic claims. But then work needs to be done to explain why we apparently have a single concept of probability when in fact there are at least two. The empirically-based subjectivist or the objective Bayesian route is perhaps most attractive here: use frequencies in the generic case, and use these frequencies to constrain single-case degrees of belief. The two notions of probability, frequency and degree of belief, are tightly connected under these accounts and do not seem so disparate after all.

Parsimony

Arguably a philosophical theory of probability should not make unwarranted ontological commitments: if one can reduce probabilities to something else in one's ontology then one should do that rather than take probabilities as primitive. This is just Ockham's razor; it may be viewed as a methodological or psychological requirement and as such subsidiary to the other desiderata.

Parsimony tells against the propensity interpretation, which usually takes probabilities to be primitive. Keynes (1921) in his development of the logical view also takes probabilities to be primitive. In contrast a frequency is a feature of a sequence of observed outcomes and so presumably reducible to entities already in a natural ontology. Similarly rational degrees of belief, the entities of the subjective, the empirically-based subjective and the objective Bayesian interpretations, will already be included in an ontology and do not count as an ontological extra. Of course all this depends on ontology; an ontology that contains only propensities may be more parsimonious than one that contains rational degrees of belief among other things.

We see then that these desiderata help to isolate a viable interpretation of probability. The propensity theory falls foul of Calculi, Epistemology, Variety and Parsimony; the frequency theory of Calculi, Epistemology and Variety; the classical, logical and subjective theories of Objectivity, Epistemology and Variety. The empirically-based subjective theory does well, though perhaps suffers with respect to Objectivity. The objective Bayesian interpretation seems to offer the most promise, when twinned with a frequency interpretation of generic probabilities. This combination is a particularly attractive way of interpreting probability in causal models for cancer: crucially, perhaps, the epistemology desideratum is satisfied—we can know about generic probabilities as well as single-case probabilities; moreover this combination makes sense both of generic causal inferences where probabilities can be interpreted as frequencies, and of single-case causal inferences where objective Bayesian probabilities are determined by empirical knowledge including frequencies.

6 Objective Bayesianism

Clearly frequencies are required to make sense of generic probabilities, e.g., the probability of surviving more than five years given metastatic breast cancer is 0.4. We have also suggested that degrees of belief constrained by frequencies should be used to make sense of the single case: if one knows only the aforementioned generic probability then one should believe that Audrey, who has metastatic breast cancer, will survive more than five years, to extent 0.4. This ties the two levels together in a natural way: generic knowledge yields predictions about the single case.

As yet though, this leaves two interpretations of probability for the single case. First, we have empirically-based subjectivism: an agent's degrees of belief ought to be constrained by knowledge of frequencies; in the absence of such knowledge they may be chosen arbitrarily. The second alternative is objective Bayesianism: an agent's degrees of belief ought to be constrained by knowledge of frequencies; in the absence of such knowledge they should be as equivocal as possible. In our view the latter approach should be adopted, as we shall now explain.

The main reason for preferring objective Bayesianism over the empirically-based subjective theory is that objective Bayesianism is on average more cautious when it comes to risky decisions. In cancer applications, single-case probabilities are used to make treatment decisions.[19] For example, if the probability of recurrence in a particular patient is very high, aggressive treatments might be used; if the probability of recurrence is very low then no further treatment is given; otherwise more evidence is garnered and non-aggressive treatments are given. Now suppose empirical evidence forces degree of belief in recurrence to lie between 0 and 0.4, say. Under the empirically-based subjective theory, an agent is free to choose any degree of belief within this interval $[0, 0.4]$. So the agent may set degree of belief 0, which will trigger abandonment of treatment. But under objective Bayesianism, the agent must choose the most equivocal—i.e., middling— degree of belief from this interval. So she must have degree of belief 0.4, which may trigger the collection of more evidence in order to reach a firmer opinion, and may trigger a non-aggressive treatment in the meantime. In general, high-risk decisions tend to be triggered by high or low degrees of belief; the objective Bayesian protocol ensures that such drastic actions only

---

[19] See e.g., Williams and Williamson (2006).

get taken if there is sufficient empirical evidence to force the extreme probabilities required to trigger them. If evidence is lacking then more middling probabilities must be adopted so that less risky actions can be initiated and further evidence can be collected.[20]

Tim McGrew has posed the following objection to this argument:

> Suppose that someone was diagnosed with cancer, underwent treatment T_1, and is being assessed for the success of T_1. Unless there is good evidence that T_1 was successful, a more radical treatment T_2 is indicated. Empirical information indicates a probability interval $[0.6, 0.95]$ for the success of T_1. It would appear that objective Bayesianism requires us to adopt the probability 0.6 that T_1 was successful, which may lie within the zone that triggers aggressive treatment T_2. There may be a good answer to this worry, but it does, *prima facie*, cast doubt on the idea that objective Bayesianism always errs on the side of caution, where caution can be equated with a preference for the less radical treatment over the more radical.[21]

In response, the first thing to note is that the most equivocal probability, in this case 0.6, will not always be the most cautious—this depends on the particular decision structure. At best we can say that *on average* with respect to risky decisions the maximum entropy approach is the more cautious policy—see Williamson (2006b, §8) for a full discussion of this point. We can see that equivocal probabilities tend to be more cautious when we see how decision scenarios arise. Consider the above scenario. Here the decision rule is that T_2 will be instigated unless the probability of the success of T_1 is greater than some threshold. Now a middling probability such as 0.6 indicates a lack of evidence and will normally trigger the collection of further evidence to try to shift the probability to one extreme or the other. In this case, however, collecting further evidence is not an option; this suggests that the costs and risks associated with collecting further evidence are greater than those associated with the treatment T_2, for otherwise the decision structure would be different. T_2 is triggered by 0.6 because, aggressive though it is, this treatment is the least risky action available. So even in this example, the objective Bayesian degree of belief is most cautious. The fact is that when we set up decision protocols, the risky actions tend to be triggered by extreme probabilities and conversely equivocal probabilities tend to trigger the less risky actions; if a middling degree of belief triggers a

---

[20]This argument is presented in detail in Williamson (2006b).

[21]Personal communication. We are very grateful to Tim McGrew for this point and other insightful comments.

risky action then that is because it is the least risky action overall. We must not equate caution with a preference for the less radical treatment over the more radical, but with the choice of a less risky action over a more risky action (in the absence of evidence that warrants the more risky action). Objective Bayesianism is then the more cautious policy, on average.

In sum, when we take the applications of single-case probabilities into account—in particular the application to cancer treatment—it becomes clear that these probabilities are a guide to action. Some actions are more drastic than others, and objective Bayesianism ensures that such actions are not embarked upon lightly.

7 Conclusion

In this paper we have argued that epidemiology can be seen as paradigmatic of the methodology of the social and health sciences. As far as causal analysis is concerned, we considered cancer epidemiology for two reasons: causal models for cancer (i) try to take into account both socio-economic and biological factors, and (ii) are essentially probabilistic.

The analysis of causal models and of the probabilistic inferences they induce suggests that two levels of inference ought to be distinguished: generic and single-case. Generic causal statements aim at positing average causal relations, for instance by claiming that tobacco consumption is a powerful carcinogen or that healthy dietary habits can possibly prevent cancer. Single-case causal statements tend to concern, instead, particular individuals. They are used for diagnosis or for causal attribution, for instance to assess the probability of recurrence of breast cancer in a particular patient.

As a matter of fact, both generic and single-case inferences are probabilistic and therefore raise the problem of how probability should be interpreted. An overview of the leading interpretations—classical and logical, frequency, propensity, subjective, empirically-based subjective, and objective Bayesian—shows that if we want to make sense of probabilistic statements at both levels we have to opt for a pluralist interpretation: the frequency interpretation is most appropriate for generic causal claims, and the single case demands a Bayesian interpretation in which probabilities are thought of as degrees of belief constrained by frequencies.

Moreover, cancer epidemiology appears to be of particular interest because it is both concerned with gaining general causal knowledge—e.g., about cancer aetiology—and with applying such general knowledge to a particular individual. This is not true of all social and health sciences. Many disciplines are more concerned with the general level and only indirectly with the single-case or the other way around. Demography, for instance, studies migration behaviour of populations but it is not directly interested in the probability that a particular individual will migrate. On the other hand, it is oft said that in medicine there are not illnesses but only ill persons to cure. Physicians, then, are more concerned with single-case probabilities than with frequencies of disease.

Though not directly concerned with individuals, social sciences such as sociology, demography or economics do have a bearing on the individual since their results orient and guide public policies, for instance to reduce unemployment or to discourage tobacco consumption. On the other hand, to correctly assign single-case probabilities, physicians do need to take generic probabilities into account. Thus, we'd better have a unified account of the interpretation of probability that makes sense both at the generic level and at the single-case. Such an account—we have argued—is objective Bayesianism twinned with the frequency interpretation.

Acknowledgments

This research forms a part of the project *Causality and probability in the social and health sciences*. We are grateful to the British Academy and to the FSR (Université catholique de Louvain) for funding this project. We are also grateful to two referees for very helpful comments.

Federica Russo
Philosophy, University of Kent, UK.
f.russo@kent.ac.uk

Jon Williamson
Philosophy, University of Kent, UK.
j.williamson@kent.ac.uk

A Probabilities, Odds and Risks

The medical and epidemiological sciences often summarise results—e.g., of logistic regressions or of meta-analyses—by means of risks and odds. Although these are associational measures, arguably they have causal import insofar as they provide evidence for a generic causal claim—e.g., smoking causes lung cancer—or inform single-case inferences—e.g., predicting the survival time of a particular individual—on the basis of these measures.

In several types of biomedical research, for instance case-control, cohort, cross-sectional or experimental studies, risks and odds are used to quantify the strength of the relation between two binary variables: a particular outcome (disease) and presence of factor (exposure). Those results are customarily presented in 2x2 contingency tables.

Let E and D denote two binary or dichotomous variables, each having only two possible levels. For the explanatory variable (exposure) E: *exposed, unexposed*, and for the outcome (disease) D: *yes, no*. Results in the table are presented as counts of observations at each level. A 2x2 contingency table has thus 4 cells. The table below shows the general layout of a 2x2 contingency table.

	Disease	
Exposure	Yes	No
Exposed	n_{11}	n_{12}
	p_{11}	p_{12}
Unexposed	n_{21}	n_{22}
	p_{21}	p_{22}

The notation n_{ij} refers to the *number* of subjects observed in the corresponding cell, i.e., to the number of observations in the i-th row ($i = 1, 2$) and j-th column ($j = 1, 2$); the notation p_{ij} refers to the *proportion* of subjects observed in the corresponding cell, where $p_{ij} = n_{ij}/n$, n being the total number of observed subjects. The notation $P(E)$ and $P(D)$ will refer to the marginal probabilities of exposure and disease, respectively.

With this data we can compute relative risks, odds, odds ratios and estimate probabilities.

Relative risk

The relative risk (RR) is defined as the ratio of risk in the exposed and unexposed group:

$$\frac{n_{11}/n}{n_{21}/n} = \frac{p_{11}}{p_{21}}$$

Thus RR measures the *incidence* of the disease. $RR > 1.0$ indicates that the risk of disease is increased when the risk factor (exposure) is present; $RR < 1.0$ indicates that the risk of disease is decreased when the risk factor is present, i.e., the factor is a protective factor or preventative. RR can also be given a definition in terms of conditional probabilities:

$$\frac{P(D|E)}{P(D|\neg E)}.$$

Odds ratio

The odds ratio (OR) is another way to compare proportions in a 2x2 contingency table. OR is computed from odds: it is the ratio of the odds of disease in the exposed group and the odds of disease in the unexposed group. The odds of an outcome is equal to the probability that the outcome does occur divided by the probability that the outcome does not occur. In a 2x2 contingency table, the probability of an outcome is equal to the number of times the outcome is observed divided by the total observations.

$$OR = \frac{Odds_{exp}}{Odds_{unexp}}$$

$$Odds_{exp} = \frac{n_{11}/(n_{11}+n_{12})}{n_{12}/(n_{11}+n_{12})} = \frac{n_{11}}{n_{12}}$$

where $n_{11}/(n_{11}+n_{12})$ is the probability that the disease occurs in the exposed group and $n_{12}/(n_{11}+n_{12})$ is the probability that the disease does not occur in the exposed group. In terms of conditional probabilities,

$$Odds_{exp} = \frac{P(D|E)}{P(\neg D|E)}.$$

Similarly,

$$Odds_{unexp} = \frac{n_{21}/(n_{21}+n_{22})}{n_{22}/(n_{21}+n_{22})} = \frac{n_{21}}{n_{22}},$$

where $n_{21}/(n_{21}+n_{22})$ is the probability that the disease occurs in the unexposed group and $n_{22}/(n_{21}+n_{22})$ is the probability that disease does not occur in the unexposed group. Again, in terms of conditional probabilities,

$$Odds_{unexp} = \frac{P(D|\neg E)}{P(\neg D|\neg E)}$$

OR can now be computed as

$$\frac{n_{11}/n_{21}}{n_{12}/n_{22}} = \frac{n_{11}n_{22}}{n_{12}n_{21}}.$$

or, equivalently

$$\frac{P(D|E)}{P(\neg D|E)} \times \frac{P(\neg D|\neg E)}{P(D|\neg E)}.$$

It is worth noting that the odds ratio of exposure $OR = \frac{Odds_{exp}}{Odds_{unexp}}$ is equal to the odds ratio of disease $OR = \frac{Odds_{Dyes}}{Odds_{Dno}}$. There is a mathematical relation between odds and probabilities:

$$P(D|E) = \frac{Odds_{exp}}{1+Odds_{exp}}$$

and

$$Odds_{exp} = \frac{P(D|E)}{1-P(D|E)}.$$

Interpreting RR, OR and probabilities

Although calculations are fairly easy, the interpretation appears to be more tricky. For instance, Sistrom and Garvan (2004, p. 16) claim: '... an RR equal to 2.0 means that an exposed person is twice as likely to have an adverse outcome as one who is not exposed, and an RR of 0.5 means that an exposed person is half as likely to have the outcome.' Or (ibidem): 'Odds and probabilities are different ways of expressing the chance that an outcome may occur.' Here there is a tension between a generic and single-case interpretation of RR, OR and probabilities.

Similarly, Bland and Altman (2000) on the one hand explain the odds and odds ratios as means to compare *groups*, but, on the other, in giving an example, they talk in terms of individuals: 'The probability that a child with eczema will also have hay fever is estimated by the proportion 141/561 (25.1%). ... Similarly, for children without eczema the probability of having hay fever is estimated by 928/14453 (6.4%).'

Two remarks are in order. Firstly, this last quotation may look puzzling unless we make clear that the 'child' that Bland and Altman refer to is a *statistical individual*, i.e., an individual randomly sampled from the population. In this case the probability may be construed as generic rather than single-case.

Secondly, as shown above, the calculation of risks and odds involves *proportions*—i.e., the numbers of subjects who got/didn't get the disease and were/were not exposed to the factor compared to the whole population. Because calculation involves proportions, RR and OR have a natural *generic* interpretation and do not make sense in the single case. Consequently, the corresponding probabilities need a frequentist interpretation. However, if the definitions in terms of conditional probabilities are preferred instead, one might argue that these probabilities are not the frequencies drawn from the 2x2 table but subjective probabilities. If so, then risks and odds are all single-case, referring to a single individual, and they do not say anything about the population. But of course this view is prone to the objections raised in §5. From a normative point of view, rational degrees of belief should be based on empirical evidence such as frequencies, and otherwise maximally equivocal. Thus an *objective Bayesian* interpretation ought to be preferred instead for the single case. A further advantage of the objective Bayesian interpretation is that if a contingency table based on frequencies is incomplete, one can generate a complete contingency table by filling in missing values with objective Bayesian probabilities.

BIBLIOGRAPHY

Barnoya, J. and Glantz, S. (2004). Association of the tobacco control program with declines in lung cancer incidence. *Cancer Causes and Control*, 15:689–695.

Bland, J. M. and Altman, D. G. (2000). The odds ratio. *British Medical Journal*, 320:1468.

Carnap, R. (1950). *Logical foundations of probability*. Routledge and Kegan Paul, London.

Courgeau, D. (2004). Probabilité, démographie et sciences sociales. *Mathematics and Social Sciences*, 167:27–50.

de Finetti, B. (1937). Foresight. its logical laws, its subjective sources. In Kyburg, H. E. and Smokler, H. E., editors, *Studies in subjective probability*, pages 53–118. Robert E. Krieger Publishing Company, Huntington, New York, second (1980) edition.

de Finetti, B. (1993). *Induction and Probability*. D. Montanari and D. Cocchi, editors. CLUEB, Bologna.

Fisher, R. (1957). Alleged dangers of cigarette smoking. *British Medical Journal*, 2:297–298.

Fisher, R. (1958). Lung cancer and cigarettes. *Nature*, 182.

Gillies, D. (2000). *Philosophical theories of probability*. Routledge, London and New York.

Glymour, B. (2003). On the metaphysics of probabilistic causation: lessons from social epidemiology. *Philosophy of Science*, 70:1413–1423.

Hacking, I. (1975). *The emergence of probability*. Cambridge University Press, Cambridge.

Humphreys, P. (2004). Some considerations on conditional chances. *British Journal for the Philosophy of Science*, 55:667–680.

Hwang, S.-J., Shu-Chung, C. L., Lonzano, G., Amos, C., Gu, X., and Strong, L. (2003). Lung cancer risk in germline p53 mutation carriers: association between an inherited cancer predisposition, cigarette smoking, and cancer risk. *Human Genetics*, 113:238–243.

Jaynes, E. T. (1957). Information theory and statistical mechanics. *The Physical Review*, 106(4):620–630.

Jeffrey, R. (1965). *The logic of decision*. University of Chicago Press, Chicago IL, second (1983) edition.

Jeffreys, H. (1939). *Theory of Probability*. Clarendon Press, Oxford, third (1961) edition.

Keynes, J. M. (1921). *A treatise on probability*. Macmillan (1948), London.

Koch, R. (1882). Über die Ätiologie der Tuberkulose. In *Verhandlungen des Kongresses für Innere Medizin*, Erster Kongress, Wiesbaden.

Kolmogorov, A. N. (1933). *The foundations of the theory of probability*. Chelsea Publishing Company (1950), New York.

Kyburg Jr, H. E. (2003). Are there degrees of belief? *Journal of Applied Logic*, 1:139–149.

Lagiou, P., Adam, H.-O., and Trichopoulos, D. (2005). Causality in cancer epidemiology. *European Journal of Epidemiology*, 20:565–574.

Lewis, D. K. (1980). A subjectivist's guide to objective chance. In *Philosophical papers*, volume 2, pages 83–132. Oxford University Press (1986), Oxford.

Mosley, W. and Chen, L. (1984). An analytical framework for the study of child survival in developing countries. *Population and Development Review*, 10:25–45. Supplement: Child Survival: Strategies for Research.

Parascandola, M. and Weed, D. (2001). Causation in epidemiology. *Journal of Epidemiology and Community Health*, 55:905–912.

Pickett, K. and Pearl, M. (2001). Multilevel analysis of neighbourhood socioeconomic context and and health outcomes: a critical review. *Journal of Epidemiology and Community Health*, 55:111–122.

Popper, K. R. (1957). The propensity interpretation of the calculus of probability and the quantum theory. In Körner, S., editor, *Observation and Interpretation*, volume 9, pages 65–70. Butterworths.
Popper, K. R. (1959). The propensity interpretation of probability. *British Journal for the Philosophy of Science*, 10:25–42.
Ramsey, F. P. (1926). Truth and probability. In Kyburg, H. E. and Smokler, H. E., editors, *Studies in subjective probability*, pages 23–52. Robert E. Krieger Publishing Company, Huntington, New York, second (1980) edition.
Reichenbach, H. (1935). *The theory of probability: an inquiry into the logical and mathematical foundations of the calculus of probability*. University of California Press (1949), Berkeley and Los Angeles. Trans. Ernest H. Hutten and Maria Reichenbach.
Salmon, W. C. (1967). *Foundations of Scientific Inference*. University of Pittsburgh Press, Pittsburgh.
Salmon, W. C. (1990). Rationality and objectivity in science, or Tom Kuhn meets Tom Bayes. In Wade Savage, C., editor, *Scientific theories*, pages 175–204. University of Minnesota Press, Minneapolis. Minnesota Studies in the Philosophy of Science 14.
Savage, L. (1954). *The Foundations of Statistics*. John Wiley.
Sistrom, C. L. and Garvan, C. W. (2004). Proportions, odds, and risk. *Radiology*, 230:12–19.
Susser, M. and Susser, E. (1996). Choosing a future for epidemiology II: from black box to chinese box and ecoepidemiology. *American Journal of Public Health*, 86:674–677.
Terry, P. and Rohan, T. (2002). Cigarette smoking and the risk of breast cancer in women: a review of the literature. *Cancer Epidemiology, Biomarkers and Prevention*, 11:953–971.
Venn, J. (1866). *Logic of chance: an essay on the foundations and province of the theory of probability*. Macmillan, London.
Vineis, P. et al. (2004). Tobacco and cancer: recent epidemiological evidence. *Journal of the National Cancer Institute*, 96(2):99–106.
von Mises, R. (1928). *Probability, statistics and truth*. Allen and Unwin, London, second (1957) edition.
Walley, P. (1991). *Statistical reasoning with imprecise probabilities*. Chapman and Hall, London.
Weed, D. (2000). Interpreting epidemiological evidence: how meta-analysis and causal inference methods are related. *International Journal of Epidemiology*, 29:387–390.
Williams, M. and Williamson, J. (2006). Combining argumentation and Bayesian nets for breast cancer prognosis. *Journal of Logic, Language and Information*.
Williamson, J. (2005). *Bayesian nets and causality: philosophical and computational foundations*. Oxford University Press, Oxford.
Williamson, J. (2006a). Dispositional versus epistemic causality. *Minds and Machines*.
Williamson, J. (2006b). Motivating objective Bayesianism: from empirical constraints to objective probabilities. In Harper, W. L. and Wheeler, G. R., editors, *Probability and Inference: Essays in Honour of Henry E. Kyburg Jr*. College Publications, London.
Williamson, J. (2006c). Philosophies of probability: objective Bayesianism and its challenges. In Irvine, A., editor, *Handbook of the philosophy of mathematics*. Elsevier.

Galton's blinding glasses. Modern statistics hiding causal structure in early theories of inheritance

BERT LEURIDAN

ABSTRACT. Probability and statistics play an important role in contemporary philosophy of causality. They are viewed as glasses through which we can see or detect causal relations. However, they may sometimes act as blinding glasses, as I will argue in this paper. In the 19th century, Francis Galton tried to statistically analyze hereditary phenomena. Although he was a far better statistician than Gregor Mendel, his biological theory turned out to be less fruitful. This was no sheer accident. His knowledge of statistics generated two explananda (unknown to Mendel) which in turn generated constraints for any theory of heredity. These constraints misguided Galtons search for the causal mechanism of inheritance. This is not just an interesting case for philosophers and historians of science. Notwithstanding the progress made by statitics, Galtons problem is still relevant today. In the final section, I briefly explore the implications for statistical techniques such as structural equation modelling.

1 The problem: probability and statistics as a tool for discovering causal patterns

More than a century ago, biological knowledge of the mechanism of heredity was rather scarce. In that context, the influential scientist and statistician Francis Galton tried to analyze hereditary phenomena statistically. He discovered interesting phenomenological regularities and posited a theoretical or causal mechanism of hereditary transmission to explain them. But, as I will argue, his causal ideas were perniciously biased by the statistical techniques he used.

Nowadays it is commonplace to attribute to probability theory and statistical inference a central place in the philosophy of causality. Several authors, such as Pearl and Spirtes, Glymour & Scheines (Pearl, 2000; Spirtes et al., 2000), have built theories of causal inference and causal discovery on the basis of knowledge of (conditional) (in)dependence relations between variables, together with some graph theoretical theorems and a set of assumptions. Their theories are tightly linked with contemporary statistical techniques such as structural equation modelling (SEM).[1] To put it metaphorically: *probability and statistics are viewed as glasses through which we can see or detect causal relations.*

This raises a problem. If I succeed in showing that Galton's knowledge of statistics impeded him from successfully developing a biological theory of inheritance (and thus acted as *blinding glasses*), it could be called into question whether contemporary statistical techniques are neutral with respect to their domain of application (i.e., with respect to the theory being developed or tested).

So here's the plan. In sections 2 and 3, I will discuss an important methodological difference between Mendel and Galton (the role of statistics in their scientific research) and I will shortly glance through the main characteristics of Mendel's theory.[2] Then, in section 4, I will show how probability and statistics generated two *explananda* which in their turn generated constraints for any theory of inheritance. In sections 4.1 and 5, I'll present Galton's *explanans* (his theory of heredity) and discuss the differences with Mendel's view on the matter. After parrying two possible counter-arguments to my reasoning (6.1, 6.2), I will summarize what this case-study shows us regarding the neutrality or non-neutrality of statistics in the work of Galton. Finally, in section 7, I will analyze the consequences for contemporary statistical techniques like SEM.

2   Gregor Mendel and Francis Galton: two different scientists

In the second half of the 19th century Francis Galton studied the processes of heredity. He started his research by considering 'hereditary genius' (Galton, 1869), but soon he turned to more easily observable characteristics. In *Natural Inheritance* (Galton, 1889) he bundled the results of more than twenty years of research on this topic. One of the most interesting aspects of

---

[1]'Structural equation modelling' refers to many related techniques. Other, more or less equivalent labels are: 'covariance structure analysis', 'covariance structure modeling,' and 'analysis of covariance structures'. The term 'causal modeling' happens to be somewhat dated. (Kline, 2005, p.9)

[2]My treatment of Mendel's theory of inheritance will closely follow the 'textbook' interpretation of his work, i.e., the interpretation that was formed in the beginning of the twentieth century by people like Bateson (1902) and that can still be found in modern genetics textbooks (Klug and Cummings, 1997). This interpretation in many ways strongly differs from Mendel's original theory, which was finely discussed by Onno G. Meijer (Meijer, 1983) and which I present in my (Leuridan, 200+). This choice is legitimate, since the biometricians rivalled textbook-Mendel, not the original one (cf. *infra*, p. 245).

Galton's work was the fact that he used and developed several modern statistical techniques, some of which are still used today (e.g., linear regression, which was mathematically developed by e.g., Karl Pearson).

At the time Galton started to work on the problem of heredity, Gregor Mendel had just finished a long series of experiments with pea plants (*P. sativum*). In 1866 he wrote his *Versuche über Pflanzenhybriden* (Mendel, 1933)[3] in which he meticulously presented his theory of inheritance. As this paper met little or no response in the biological community at the time, his ideas remained silent until they were rediscovered by Carl Correns (1900) and Hugo de Vries (1900).[4] In the meantime, Galton independently developed his biometric *theory of ancestral inheritance*. After 1900, these theories would become vehement rivals and it would take several years or even decades before the dispute between the Mendelians (e.g., Bateson, 1902) and the biometricians (e.g., Karl Pearson, but also Weldon, 1902) was settled.

Contrary to Galton, Mendel made little use of statistical theory. He inferred in a rather intuitive way from particular observations to general regularities.[5] So, given the current dominant role of statistics in the special sciences, shouldn't we expect Galton to have found the most 'true' regularities? Of course, every one knows that Mendel is still considered the founding father of genetics, while Galton's name is now only associated with dubious disciplines such as phrenology and eugenics.

3   Mendel's theory of inheritance

All of the phenotypic characteristics that Mendel observed in *P. sativum* were qualitative and discrete. He studied seven pairs of contrasting traits, such as seed shape (round or wrinkled) and stem length (tall or dwarf). His *explanandum* consisted of some very straightforward empirical regularities (phenotypic distributions). The task was to explain why after crossing dwarf plants with true-breeding tall plants, all the off-spring in the first filial generation was tall; or why, after selfing this off-spring, 75% of the second filial generation was tall, while 25% was dwarf (see figure 1).

Mendel's *explanans* was based on a causal mechanism invoking material bearers of hereditary traits. In each pair of contrasting traits, one is dominant and the other is recessive (round is dominant to wrinkled, tall is

---

[3] This was translated as *Experiments on Plant Hybrids* (see Stern and Sherwood, 1966, pp.1–48).

[4] For a nice demystification of this mysterious rediscovery and of the preceding neglect, see Meijer (1983, sections 1 and 2) and also Darden (1991, section 4.3.).

[5] This statement is perhaps somewhat unfair and should be nuanced. Mendel *was* trained mathematically and statistically (see Meijer, 1983, p.128), but not in the 'new' tradition of Quetelet, Galton, etc.

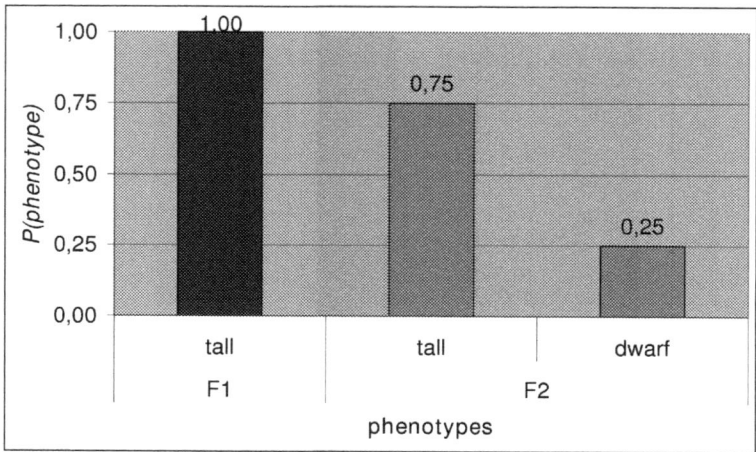

Figure 1. Mendel's explananda illustrated. The F1-generation resulted from crossing dwarf plants with true-breeding tall plants. The F2-generation resulted from selfing F1.

dominant to dwarf). These traits are caused and carried over from parental plants to filial plants by *unit factors* ('*Factoren*', see Mendel, 1933, p.24). A unit factor either codes for the dominant trait or for the recessive trait. They occur *pairwise* in the individual pea plants, but singly in the gametes.[6] Roughly half of the gametes carries (a copy of) the first factor. The other half carries (a copy of) the second factor. Later this would be called the '*principle of segregation*'. As would be emphasized by William Bateson in the beginning of the 20th century, no gamete carries an intermediate factor—this was dubbed '*the purity of the gametes*' in Bateson (1902, p.108).[7]

A last point to be noted with respect to Mendel's theory is that the 'genotype'[8] of a pea plant is *screened off* or *d-separated* by the set of gametes that produced it (cf. Pearl, 2000, p.16). Once it is established what unit factors are carried by the germinal cell and the pollen cell, the origin of

---

[6] At least, this is what classical Mendelism has taught us. However, there is every indication that according to Mendel factors occur pairwise in heterozygotes, but singly in homozygotes (Meijer, 1983; Leuridan, 200+); cf. footnote 2.

[7] I will not discuss the 'principle of independent assortment', as Galton did not explicitly treat multi-hybrid crosses.

[8] I apologize for anachronistically using terms like 'genotype', 'phenotype' and their derivatives, but I think it's worth the trouble so as to avoid unnecessary circumlocution.

these gametes plays no further role. Remote ancestry has no influence, conditional on the gametes.

To summarize, these are the main features of Mendel's theory that we should bear in mind in the following sections and which we should oppose to Galton's. Traits are grouped in pairs, in which one is dominant, the other recessive. They are caused and transmitted by unit factors that occur pairwise in individual organisms. Each gamete, however, contains only one factor, according to the principle of segregation. Moreover, the gametes are pure (they never carry intermediate factors), and they *screen off* the resulting individual from its ancestry.

4 Statistics generating two *explananda* for Galton's theory of heredity

4.1 Preview of Galton's theory of *ancestral inheritance*

Like Mendel, Galton invoked bearers of hereditary traits to set up a causal mechanism of heredity.[9] He called them 'elements' or 'particles'.[10] But contrary to Mendel's unit factors, these did not occur in pairs. In each individual organism, an *indefinite* or *incalculable number of elements* responsible for the same phenotypic trait is present.

From a modern perspective, this may look weird, as chromosomes occur pairwise in diploid organisms. At the time Mendel and Galton developed their views on the mechanism of inheritance, however, cytological constraints were rather poor (Darden, 1991, chapter 7). This should be borne in mind if we want to assess the merits of Galton's work.

Mendel had a relatively clear view on the transmission of unit factors and on the possible genetic make-up of the gametes.[11] By contrast, the genetic make-up of gametes or individuals was never treated concretely in Galton's texts and the gametes played no inferential or predictive role in his theory. Although he stated, in the beginning of his *Natural Inheritance*, that "there is no direct hereditary relation between the personal parents and the personal child" and that "the main line of hereditary connection unites the sets of elements out of which the personal parents had been

---

[9]I have found no explicit discussion of the concept of 'causation' either by Mendel or by Galton. Their writings, however, bespeak that both had a 'mechanistic' view of causation.

Note that in this respect they strongly differed from e.g., Pearson who had a Machian view on causation in which, first, causation is defined as perfect correlation and in which, secondly, the existence of causal relations in the empirical world is excluded *a priori* (Pearson (1900, chapter IV) or (1911, chapter V)).

[10]Galton also used the word 'element' to refer to the phenotypic traits themselves, instead of the particles that caused them (see section 5.1).

[11]See 'The reproductive cells of hybrids' (in Stern and Sherwood, 1966, pp.23–32).

evolved with the set out of which the personal child was evolved" (both are quoted from Galton, 1889, 19), he directly predicted the traits of issue from the traits of its ancestors. These ancestors not only included the parental generation, but also the grand-parents, great-grand-parents, ... Galton was convinced that all ancestry may in principle have an influence on the set of elements from which the organism is built, and thus on its phenotype.[12] The influence is neither screened off by the parents, nor by their gametes. Therefore, Galton's theory can be called the *theory of ancestral inheritance*.

This suffices to show that the causal mechanisms proposed by Mendel and by Galton were really different. In the following sections, I will show in what ways this difference can (partly) be explained by laying bare the role played by Galton's statistical knowledge. His statistics generated two *explananda* which in their turn imposed influential constraints on any would-be *explanans*.[13]

4.2 The first *explanandum*: the normal distribution

Contrary to Mendel, Galton mostly studied continuous traits. In his *Natural Inheritance*, he paid a lot of attention to the schemes of distribution and the schemes of frequency of e.g., human strength, stature, span of arms, weight, breathing capacity, etc. (Galton, 1889, pp.35–50 and 200). All these characteristics, Galton noted, are normally distributed.

In the 19th century, the normal distribution played a very important role, not only in astronomy, but also in the biological and the social sciences. Adolphe Quetelet, the Belgian statistician and sociologist, discovered that measurements of phenomena such as human stature, birth ratios and crime rates were all normally distributed. The term 'Quetelismus' refers, then, to the exaggeration of the dominance of the normal distribution, i.e., to the view that "*all* naturally occurring distributions of properly collected and sorted data follow a normal curve" (Stigler, 1986, p.201, my italics; see also 203–205). Galton explicitly acknowledged the influence of Quetelet and stated that the latter introduced the idea that the Law of Error (i.e., the normal distribution) might be applicable to human measures (e.g., Galton (1877, p.493) and (1889, 54–55)).

Galton knew several ways to represent distributions of data. One way was to represent them graphically. Another way was to cite a series of eleven percentiles (the 5th, 10th, 20th, ..., 80th, 90th and 95th). But the most economical method, applicable in case the data were normally distributed (as seemed mostly the case), was to cite just two numbers: M and Q. M was the mean or median. Q he called the 'Prob. Deviation' and it was defined

---

[12]Cf. *infra*, p. 254.

[13]In fact, Galton's statistics generated at least three explananda (see footnote 31).

as one half of the interquartile range: $Q = \frac{1}{2}(Q_2 - Q_1)$, where Q_2 and Q_1 are the third and the first quartile respectively. It conveyed the dispersion of the distribution and thus played a role similar to the standard deviation σ.[14] Once M and Q were known, all percentiles could be calculated and the scheme of distribution could be drawn.

The normal distribution of Human Stature played a tremendously important role in Galton's biometrical work. He wrote:

> In particular, the agreement of the Curve of Stature with the Normal Curve is very fair, and forms a mainstay of my inquiry into the laws of Natural Inheritance. (Galton, 1889, p.57)

Several sets of data, collected by Galton himself, indicated that the median male stature P was 68.25 inch (Galton used the symbol P to refer to the median in the context of Stature) and that $Q = 1.7$ inch.

Obviously, the normal distribution generated a major constraint on Galton's theory of heredity. Whatever mechanism one was to propose to explain the processes of inheritance, it had to be able to explain why inherited traits are normally distributed. As Galton stated in 1877,

> The conclusion is of the greatest importance to our problem. It is, that the processes of heredity must work harmoniously with the law of deviation, and be themselves in some sense conformable to it. (Galton, 1877, p.512)

What causes variables to be normally distributed? In 1810, Pierre Simon Laplace first introduced what was later to be called the *'Hypothesis of Elementary Errors'*: that the joint action of a multitude of independent 'errors' produces a normal distribution (Stigler, 1986, p.201–202). Laplace's main topic of interest was the distribution of astronomical observations,[15] but his ideas forcefully influenced Quetelet and later also Galton. Galton wrote:

> The Law of Error finds a footing wherever the individual peculiarities are wholly due to the combined influence of a multitude of "accidents" [...]. (Galton, 1889, p.55)

So the constraint imposed by the normal distribution was the following: Galton had to introduce a *'variety of petty influences'* in his biological theory (Galton, 1889, pp.16–17).

---

[14]The standard deviation σ is by definition larger that Q, since $M \pm \sigma$ includes about 68% of the observations, while $M \pm Q$ includes only 50% of them.

[15]If every measurement is the aggregate of many independent component measurements, each of them subject to small errors, the normal distribution of astronomical data is explained.

4.3 The second *explanandum*: Regression towards the mean

Perhaps the most important one of Galton's contributions to modern statistical theory was his concept of '*regression towards the mean*'. It would later result in the theory of linear regression (elaborated by Galton's protégé Karl Pearson). What exactly did Galton mean by this regression?

In the 1860's, he first tackled the topic of inheritance by studying human genius or talent and the way it was distributed within families. In *Hereditary Genius* (1869), Galton observed that, generally speaking, the relatives of gifted men (such as Johann Sebastian Bach or Jacob Bernoulli) are gifted too, but less so. And the more remote a relative is, the less he is talented (Stigler, 2002, pp.176–177). In other words, Galton found a 'regression towards mediocrity'.

Of course, genius or talent is not easily observable, let alone measurable, but other characteristics are.[16] In 1884, Galton gathered hundreds or even thousands of records about the human population, called the *R.F.F. Data* (*Record of Family Faculties*). They comprised information about the Stature, Eye-colour, Temper, Artistic Faculty, ... of whole families (spanning several generations). The major part of *Natural Inheritance* concerned Human Stature, and the main research question was whether it is inheritable or not, i.e., whether the offspring of tall people is—on average—tall, or not.[17] But what do we mean by 'tall'?

As I stated earlier, Galton knew that male stature was normally distributed with $P = 68\frac{1}{4}$ inch and $Q = 1.7$ inch (see page 248). Women are slightly smaller than men, but this difference disappears if their statures are multiplied or transmuted by 1.08. These figures suggested a very straightforward criterion for 'tallness'. People are tall *iff* they have a stature larger than P. Human stature could be written as the sum of two components:

(1) $\text{Stature} = P \pm D$,

in which D is the individual's deviation from the mean (Galton, 1889, pp.51–52 and Chapter VII). So the sign and size of D indicated whether and to what extent an individual was tall or small and Galton's research question could be rephrased as follows: does the issue of large parents with a large D itself also have a large D?

The stature of the issue of unlike parents does not depend on the specific statures of the father and of the mother, as Galton's data revealed. It

---

[16] Galton took great pains, however, to classify men according to their talent, both on basis of their 'reputation' and on basis of their 'natural ability' (Galton, 1869, pp.37–38).

[17] The *R.F.F. Data* contained the statures of 205 pairs of parents and their 930 adult children. Other data sets also reported on Stature: the *Special Data* covered about 783 brothers from 295 families and the *Measures at the Anthropometric Laboratory* consisted of about 10,000 data (Galton, 1889, pp.71–82).

depends only on their *average* stature. Therefore, Galton introduced the concept of the 'Mid-Parent', which is defined as "an ideal person of composite sex, whose Stature is half way between the Stature of the father and the transmuted Stature of the mother." (Galton, 1889, p.87). The Mid-Parental Statures are normally distributed with P = $68\frac{1}{4}$ inch and Q = 1.21 inch.[18]

(2) $\quad$ Stature Mid-Parent $= \dfrac{\text{Stat.Father + Transm.Stat.Mother}}{2}$,

The *R.F.F. Data* revealed that the off-spring of tall Mid-Parents is on average taller than P and that the relation between the Mid-Parental Stature and the *average* Stature of the Son is stable (where the Son refers to both the sons and the transmuted daughters). If the Mid-Parental deviation is D, then the filial deviation is on average $\frac{2}{3}$ D:

> I call this ratio of 2 to 3 the ratio of "Filial Regression." It is the proportion in which the Son is, on the average, less exceptional than his Mid-Parent. (Galton, 1889, p.97)

So Galton could now describe with numerical precision what he had discovered years before: the inheritance of characteristics is subject to 'regression towards the mean'.

(3) $\quad$ Stat.Son $= P \pm \dfrac{2}{3}$ D, where D is the Mid-Parent's deviation.

Take for example a Mid-Parent that is very tall, say 71.25 inch (D = 3 inch). Equation 3 predicts and figure 2 illustrates that its issue will, on average, be 70.25 inch (D = 2 inch = $\frac{2}{3} \times 3$ inch).[19]

So here we are presented with a second set of constraints. Every theory of heredity should be able to explain this filial regression. In the following section, I will show how Galton's theory of Ancestral Inheritance did this.[20]

---

[18] Note that 1.21 is a theoretically predicted number, as $\frac{1.7}{\sqrt{2}} = 1.21$. According to the *R.F.F. Data*, the Mid-Parental Q is 1.19. Galton considered the agreement between these numbers to be excellent (Galton, 1889, pp.92–94 and 208).

[19] Figures 1 and 2 nicely illustrate the difference qua explananda between Mendel's theory and Galton's.

[20] 'Filial Regression' was not the only kind of regression that Galton discovered (Galton, 1889, pp.99–110). Related notions which will prove to be relevant in the following sections are 'Mid-Parental Regression' (equation 4), 'Parental Regression' (equation 5) and 'Fraternal Regression' (equation 6):

(4) $\quad$ Stat.Mid-Parent $= P \pm \dfrac{1}{3}$ D, where D is the deviation of the Son,

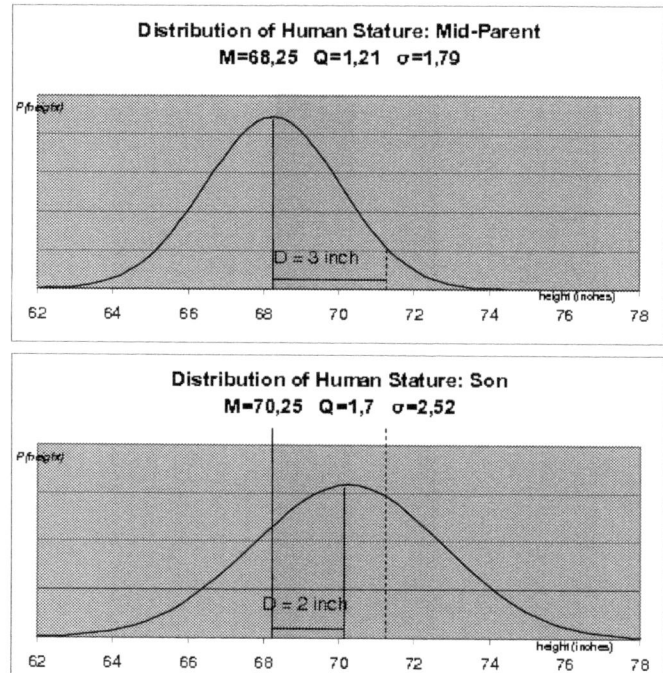

Figure 2. Filial Regression illustrated

5 Galton's theory of ancestral inheritance as an *explanans*

5.1 Particulate Inheritance and the Hypothesis of Elementary Errors

Mendel's causal mechanism responsible for the processes of heredity consisted of unit factors in pairs. In Galton's theory, the situation was less clear-cut. He called it the theory of *'particulate inheritance'* and used the words *'elements'* and *'particles'* several times. But sometimes these elements seemed to denote (elements of) phenotypic traits, other times they might have referred to carriers of hereditary traits. Nonetheless, as Galton was influenced by August Weismann's theory of the germ-plasm and Charles Darwin's concept of Pangenesis, a 'material' interpretation of the elements or particles is certainly justified (Galton, 1889, pp.7–9 and 192–193).

Galton's hereditary particles are transmitted from parents to offspring,[21] but in principle every ancestor may contribute to an individual's elements. So the parents don't screen off the offspring from the rest of its ancestry. The influence of remote ancestry is obviously smaller than the parental influence. Nevertheless, it exists and played an important role in Galton's predictive inferences. The separate contribution of each ancestor follows a very simple rule, which would later be called the *'Law of Ancestral Heredity'* by Karl Pearson:[22]

(5) $\quad$ Stat.Son $= P \pm \dfrac{1}{3} D$, where D is the deviation of one of his Parents,

(6) $\quad$ Stat.Brother $= P \pm \dfrac{2}{3} D$, where D is the deviation of a known man.

---

[21] Since phenotypic traits sometimes seem to skip a generation, he distinguished between *personal* elements (causing the traits they code for to be present), and *dormant* or *latent* elements ('unused' elements, having no phenotypic influence). Prima facie, this strongly resembles the Mendelian distinction between dominant and recessive traits or factors. There is an important difference, however. The relation of dominance/recessiveness is fixed for each pair of contrasting traits (round seed shape is always dominant to wrinkled seed shape in *Pisum*). By contrast, Galtonian elements that are latent in one organism can be personal in another. Galton had no definite answer to the question of what determined whether an element would be latent or personal. He thought there were three possible answers:

> [...] first, that in which each element selects its most suitable immediate neighbourhood, in accordance with the guiding idea in Darwin's theory of Pangenesis; secondly, that of more or less general co-ordination of the influences exerted on each element, not only by its immediate neighbours, but by many or most of the others as well; finally, that of accident or chance [...]. (Galton, 1889, p.19)

[22] Galton explicitly discussed the validity of this 'Law of Ancestral Heredity' for the inheritance of personal elements. At the end of his *Natural Inheritance* he hypothesized

[...] the influence, pure and simple, of the Mid-Parent may be taken as $\frac{1}{2}$, and that of the Mid-Grand-Parent as $\frac{1}{4}$, and so on. Consequently the influence of the individual Parent would be $\frac{1}{4}$, and of the individual Grand-Parent $\frac{1}{16}$, and so on. (Galton, 1889, p.136)

Taken together, the set of all ancestors fully determines the Son's set of elements, as

$$(7) \quad (2 \times \frac{1}{4}) + (4 \times \frac{1}{16}) + (8 \times \frac{1}{64}) + \ldots = 1.$$

At first glance, this picture seems paradoxical. Although all hereditary influence passes through the parents, there is still room for the influence of the grand-parents, great-grand-parents, etc. This semblance of paradox is dissolved if we distinguish between *personal allowance* and *ancestral allowance*. Personal allowance is the allowance 'pure and simple' and it is governed by the Law of Ancestral Heredity (so that e.g., the father's personal allowance is $\frac{1}{4}$). The ancestral allowance comprises all influence that is just passed through an ancestor. An individual's total allowance thus is the sum of its personal and its ancestral allowance.[23]

From this picture it is easy to explain why characteristics like human stature are normally distributed. In principle, an infinite number of ancestors influences the set of hereditary particles of a man. And his 'genotype' consists of an indefinite or incalculable number of elements. This makes sure that the inheritance of traits is determined by a *variety of petty influences* (Galton, 1889, pp.16–17).[24] If it is assumed that these are to some degree independent of one another (as Galton did), a physical basis is provided for the Hypothesis of Elementary Errors. So the theory of Ancestral Heredity is capable of explaining the normal distribution of phenotypic traits (Galton, 1889, pp.84–85).

---

that it would also apply to the latent elements (Galton, 1889, pp.187–191).

[23] Galton did not use the concepts 'personal allowance' and 'ancestral allowance' in *Natural Inheritance*. They appear one time in his "The Average Contribution of each several Ancestor to the total Heritage of the Offspring" (Galton, 1897, p.441).

[24] Galton incorporated two more sources of 'petty influences' in his theory. First, whether or not an element will be personal or dormant depends on very numerous influences (Galton, 1889, p.22). Secondly, a trait such as Human Stature is not one element, but "a sum of the accumulated lengths or thicknesses of more than a hundred bodily parts" (Galton, 1889, pp.83–84), and each element or length of a body part is subject to errors or environmental effects.

5.2 The law of ancestral heredity, dilution and taxation

Now what is the cause of Filial Regression (equation 3)? Why is it that offspring tends to be more mediocre than its parents? Galton proposed an answer in his paper "Regression towards mediocrity in hereditary stature" (Galton, 1886), which he recapitulated in *Natural Inheritance* (Galton, 1889).

Suppose some Mid-Parent has Stature $P \pm D$ and call D her peculiarity. From equation (4), Galton stated, it follows that the peculiarity of the Mid-Grandparent is $\frac{1}{3}D$, that of the Mid-Great-Grand-Parent $\frac{1}{9}D$, etc. If each generation would contribute its whole peculiarity, we should expect the Son to inherit $D(1 + \frac{1}{3} + \frac{1}{9} + \&c.) = D\frac{3}{2}$. This contradicts the expected Filial Regression of $\frac{2}{3}$ (Galton, 1889, p.134).

So Galton considered the possibility that the bequests of the successive generations are somehow taxed or diminished. His data did not allow the direct measurement the size of this tax, but he had two limiting hypotheses. On the one hand, if the bequest of every generation is taxed just once, the tax rate has to be $\frac{4}{9}$.[25] On the other hand, if the tax is repeated at each successive transmission, the rate should be $\frac{6}{11}$.[26] Galton's data did not allow to choose between these hypotheses. But as the both values differed but slightly from $\frac{1}{2}$, he decided that this would be a very good approximation (Galton, 1889, pp.134–136).

Galton's reasoning lacked logical rigour and can be challenged from several sides.[27] Nevertheless, one should not consider it as a totally *ad hoc* hypothesis. Both in 1886 and in 1889, Galton concluded that, as the tax rate should be estimated to be $\frac{1}{2}$, the Mid-Parent contributes half of his peculiarity, the Mid-Grand-Parent one quarter, etc. That is, he combined it with the Law of Ancestral Heredity (see page 254).[28] In 1897, Galton published an extra argument for this law. If each ancestor *may* contribute to the heritage of the offspring ('as is shown by observation'), if remote ancestry contributes less than near ancestry ('as is well known'), if the contribution of the parents to the children is the same as that of the grand-parents to the parents, etc. ('as is reasonable to believe'), and if the total amount contributed equals 1 ('as is necessarily the case'), then only the series of $\frac{1}{2} + (\frac{1}{2})^2 + (\frac{1}{2})^3 + etc.$ can describe the share of the Mid-Parent, the Mid-Grand-Parent, etc. (Galton, 1897, p.403). Michael Bulmer deems it very

---

[25] $D\frac{2}{3} = D\frac{3}{2} \times \frac{4}{9}$
[26] $D\frac{2}{3} = 1D \times \frac{6}{11} + \frac{1}{3}D \times (\frac{6}{11})^2 + \frac{1}{9}D \times (\frac{6}{11})^3 + etc.$
[27] For a crushing discussion of Galton's derivation, see Bulmer (1998) and Bulmer (2003, p.243).
[28] Note that the Law of Ancestral Heredity matches only the second of both limiting hypotheses, namely that the tax is repeated at each successive transmission.

plausible that Galton had this argument in mind in 1886 (Bulmer, 2003, p.246).[29] If he is right, as I think he is, Galton's choice for the tax rate of $\frac{1}{2}$ was not totally *ad hoc*.[30]

So now we see how Galton's theory explains regression towards the mean. Since the peculiarity D of the Mid-Parent is mixed with the smaller peculiarities of more remote ancestry, the Son's deviation from P is smaller than D. Galton used a very powerful metaphor to illustrate this:

> [The] effect resembles that of pouring a measure of water into a vessel of wine. The wine is diluted to a constant fraction of its alcoholic strength, whatever that strength may have been. (Galton, 1889, p.105)

The exceptionality of the parents is diluted by the mediocrity of the rest of the ancestry (hence, I call this the *Dilution Theory*), so that their offspring is more mediocre too. But, as we have seen, the Dilution Theory needed to be completed with the concept of *Taxation* to get the correct ratio of Filial Regression.

6 Two possible objections and a conclusion

Before I turn to the conclusion, I want to anticipate two possible objections to my thesis that statistics played a blinding role in the development of Galton's theory of heredity. I will show that neither the observational nature of the data on Human Stature, nor the continuous or blending nature of the observed characteristics can be cited as an alternative explanation for Galton's failure to arrive at the 'right' theory of heredity.

6.1 Observational versus experimental data

One of Mendel's major merits was that he paid a lot of attention to his experimental set-up. By carefully selecting a well-suited organism (pea plants) and manageable pairs of opposing characteristics, and by meticulously planning the right monohybrid as well as multihybrid crosses, he was able to confirm his causal theory of inheritance—a theory that still is considered the basis of modern genetics, although it has been subject to a vast amount of changes, specifications and additions (see Darden, 1991).

It is certainly true that Galton preferred observational data about the human population. He preferred data about humans because he considered

---

[29] In the bibliography of Bulmer (2003), however, this article is wrongly dated to 1885.

[30] Bulmer's conviction is based on the following quote by Galton: "These and the foregoing considerations were referred to when saying that the law might be inferred with considerable assurance *à priori* [...]." (Galton, 1897, p.403) Galton had indeed stated in 1886 that his law might have been deductively foreseen (Galton, 1886, p.253).

them more interesting or relevant. He was forced back to observational data because experimenting on human beings would have been (and still is!) rather problematic.

In the 1870's, however, after having published *Hereditary Genius* (1869), but before collecting the *R.F.F. Data*, Galton did experiment with plants (*sweet peas*, not to be confused with *P. sativum*). He weighted thousands of seeds to determine their size and then selected several sets for planting. Each set consisted of seventy seeds, divided in seven packets of ten seeds of exactly the same weight $(K, L, \ldots, Q)$. The K-class contained very heavy seeds, L the next heaviest, and so on. He sent these sets to his friends throughout the United Kingdom and asked to plant them according to very minute instructions and to collect the produce of each class separately. Seven experiments (with in total $7 \times 7 \times 10 = 490$ parental seeds) succeeded (see Galton (1877, pp.512–514) and (1889, pp.79–82 and 225–226)).

The experimental data showed that large seeds beget large seeds. But, as was the case with hereditary genius, '*Reversion*' could be observed (the label 'regression towards the mean' was not yet introduced in 1877). In 1877, he gave no exact value for the regression coefficient, stating only that it is constant. In "Regression towards mediocrity in hereditary stature", which dates nine years later, this lacuna is removed:

> It will be seen that for each increase of one unit on the part of the parent seed, there is a mean increase of only one-third of a unit in the filial seed; and again that the mean filial seed resembles the parental when the latter is about 15.5 hundredths of an inch in diameter. (Galton (1886, p.259); see also (1889, p.225))

This suffices to show that the difference between Mendel's and Galton's scientific practice should not be sought in the experimental data of the former and the observational data of the latter.

6.2 Alternative inheritance versus blended inheritance

Can't we explain the difference between Galton and Mendel by looking at the variables, i.e., the phenotypic traits, they studied? After all, Human Stature and the Size of sweet peas are continuous variables, while Mendel observed pairs of discrete, opposing characteristics.

It is certainly true that Galton paid a lot of attention to continuous traits and it is equally true that that was the best way to discover regression-phenomena.[31] In *Natural Inheritance*, however, he took great pains to argue that the theory of Ancestral Inheritance could encompass the transmission

---
[31] As I phrase it here, it looks as if Galton did not come upon the phenomenon of

of both *blended heritages* and *alternative heritages* (Galton, 1889, pp.12–14 and 138–153). To prove this point, Galton studied the transmission Eye-Colour:

> If notwithstanding this two-fold difference between the qualities of Stature and Eye-colour, the shares of hereditary contribution from the various ancestors are alike in the two cases, as I shall show they are, we may with some confidence expect that the law by which those hereditary contributions are found to be governed, may be widely, and perhaps universally applicable. (Galton, 1889, p.139)

How could the Law of Ancestral Heredity be used to predict the distribution of Eye-colours in issue, conditional on the Eye-colours of its parents, grand-parents, etc.? Galton distinguished between three types of Eye-Colour: light, hazel and dark, and then treated the problem as if it concerned Stature.

Suppose you want to predict the stature of some man, S, but that you only have information about one of his parents, F (having peculiarity D). By equation (2), both parents of F have on average the peculiarity $\frac{1}{3}D$, while his grand-parents (i.e., the great-grand-parents of S) have $\frac{1}{9}D$, ...From the Law of Ancestral Heredity it follows that F transmits only $\frac{1}{4}$ of his peculiarity to S; his parents transmit $\frac{1}{16}D$, etc. So the total calculable or predictable heritage that is transmitted through F is[32]

$$D\{1 \times 1 \times \frac{1}{4} + 2 \times \frac{1}{3} \times \frac{1}{16} + 4 \times \frac{1}{9} \times \frac{1}{64} + \&c.\} \approx D \times 0.30,$$

consisting of F's known personal allowance (0.25 D) and its predictable ancestral allowance $((0.30 - 0.25)D = 0.05\ D)$. By analogy, two parents have

---

regression by merely analyzing his data, but that he actively sought for it. In fact, this is true. Galton needed 'Regression towards the Mean' to explain another statistically inspired *explanandum*.

In 1877, Galton published "Typical Laws of Heredity" (Galton, 1877). One of the main explananda in that paper was the fact that the distribution of characteristics (specifically the Size of sweet peas) remained more or less constant in each successive generation. Using the terminology of 1889, it was to be explained why both its M and Q remained constant.

In the absence of regression, the dispersion Q would increase from generation to generation. The offspring of tall men would on average be as tall as its parents, some of it would even be taller; in the next generation, even taller issue would result, ... But this would contradict Galton's data about sweat peas, as well as the findings of Quetelet, Galton and others on human characteristics (Venn, 1889, p.415). Regression was the perfect candidate to solve this problem, as it would act as a counterbalance to this dispersive tendency.

[32] See Galton (1889, p.148–149). Note that Galton's formula on p. 149 contains a printing error.

a known total allowance of 0.60, "leaving an indeterminate residue of 0.40 due to the influence on ancestry about whom nothing is either known or implied" (Galton, 1889, p.149). This residue is a direct consequence of the fact that the ancestral influence is not screened off by the parents or the parental gametes in Galton's theory.

These results can be easily re-interpreted in the context of Eye-colour. We only need to interpret the personal allowance or ancestral allowance as fractions of the total number of children in a family that will inherit some specific trait (Galton, 1889, p.149–150). If a parent has dark eyes, that will cause 30% of his children to have dark eyes. Two dark-eyed parents will cause 60% of their children to be dark-eyed. Of the residue, 40% in this case, Galton assigns 28% to dark eyes and 12% to light eyes, proportionally to their overall ratio in the population. This gives rise to the following table, from which it is easy to predict the distribution Eye-colours in issue, conditional on the Eye-colours of its parents and grand-parents.[33]

Contribution to the heritage from each	Data limited to the eye-colours of the					
	2 parents		4 grand-parents		2 parents and 4 grand-parents	
	I.		II.		III.	
	Light	Dark	Light	Dark	Light	Dark
Light-eyed parent	.30				.25	
Hazel-eyed parent	.20	.10			.16	.09
Dark-eyed parent		.30				.25
Light-eyed grandparent			.16		.08	
Hazel-eyed grandparent			.10	.06	.05	.03
Dark-eyed grandparent				.16		.08
Residue, rateably assigned	.28	.12	.25	.11	.12	.06

For example, from the premise that in a family there are two light-eyed parents, three light-eyed grand-parents and one hazel-eyed grandparent, you can calculate that on average 91% of the children will have light eyes. The rest, 9%, will be dark-eyed.[34]

Thus we can conclude that, even if Galton paid heavy attention to continuous variables, he also tried to incorporate the transmission of discrete characteristics.

---

[33] Note that only the distribution of dark eyes and of light eyes is calculated. This table is reproduced from Galton (1889, p.213).

[34] $2 \times 0.25 + 3 \times 0.08 + 1 \times 0.05 + 0.12 = 0.91$,
$2 \times 0.00 + 3 \times 0.00 + 1 \times 0.03 + 0.06 = 0.09$ (see also Galton, 1889, p.215, Table 19).

6.3 Conclusion: the neutrality of Galton's Statistics

I have shown that, in the second half of the 19th century, Galton's views on the mechanism of heredity were so much constrained by his statistical explananda, that he failed to discover the Mendelian scheme.[35] From this it follows that statistics *can* bias scientific research. Moreover, Galton's case can be supplemented with examples from outside biology. In his very interesting book, *The Taming of Chance* (1990), Ian Hacking has argued that the practice of descriptive statistics by the bureaucracies of nation states affected the way people conceived of society, of other people and of themselves. So statistics have biased people's world views and they have (partly) constrained theory development in the social sciences too.[36] So clearly, probability and statistics were *not* always *neutral* scientific tools.

7 Consequences for contemporary statistics

In the literature on causal discovery, probability and statistics play an important role, as several flourishing research programmes use conditional (in)dependence relations as indicators of the presence and absence of causal relations and maintain strong ties to statistical techniques such as SEM (Pearl, 2000; Spirtes et al., 2000). In the SEM-literature, however, one is frequently warned not to draw causal inferences from structural models too quickly. Causal inference is only justified on basis of models that fit the data well. But good model fit is not enough. It may indicate that the model accurately reflects reality, but not necessarily so. It leaves open the possibility that the model is equivalent to one that corresponds to reality but itself is incorrect, or that it fits the data from a nonrepresentative sample but has poor fit in the population, or that it has so many parameters that it cannot have poor fit (Kline, 2005, p.321). In addition to fitting the data well, the model should also correctly describe the causal relations between its variables.

A SEM-model has to be specified independently before it is confronted with the data. In the SEM-literature, heavy stress is laid on the need for reliable background knowledge or theory in this process.[37] But this

---

[35] By this I of course not mean that Mendel's theory was either unconditionally true or that it is still used in its original form. I only mean that he laid the fruitful basis of modern genetics, even if the development of genetics involved a lot of changes to his basic tenets (cf. Darden, 1991).

[36] In the *The Taming of Chance*, Hacking also discussed Galton's work. His interpretation of the relation between the normal distribution, regression towards the mean, and Galton's hereditary mechanism, however, strongly differs from what I have argued in this paper (Hacking, 1990, chapter 21). I plan to elaborate the differences between his interpretation and mine in the near future.

[37] The following quote illustrates this and shows at the same time that, contrary to what

presupposes that this background knowledge is not perniciously biased by statistics itself.

So what about the neutrality of contemporary statistics? I would like to make two statements in this respect. First of all, we should not throw out the baby with the bathwater. One of statistics' fundamental strengths is that it has much thought for the presuppositions and internal limitations of each of the techniques developed (presuppositions with respect to the distribution of the data or the relation between variables, robustness against missing values and outliers, requisites concerning sample sizes, etc.). Add to this that tests have been developed for most of these presuppositions (tests which were not available to Francis Galton). Secondly, however, we should not be blindly optimistic. Galton's problem still is relevant today. Specifying a SEM-model involves that we prestructure our domain of interest. Examples of such prestructuring can be found at many places in Kline's introductory work (Kline, 2005). SEM-modeling involves marking out variables, specifying their possible values and developing and including methods for measuring them. It involves fixing covariance relations and perhaps imposing constraints on (covariances between) disturbance variables. It involves specifying the directionalities of presumed causal effects ...

This shows that, in practice, SEM (and like methods) should be used very carefully, i.e., one should always bare in mind the possibly *blinding* influence of the techniques themselves.

Acknowledgments

I would like to thank Erik Weber, Joke Meheus and the other members of the Centre for Logic and Philosophy of Science (Ghent University) for their helpful comments and I would especially like to thank both anonymous referees for their very stimulating suggestions and their critical questions.

Bert Leuridan
Centre for Logic and Philosophy of Science, Ghent University, Belgium.
Bert.Leuridan@Ugent.be

Pearl has argued, the exclusion of causal interpretations from SEM cannot be completely put down to SEM-practitioners seeking respectability by keeping causal assumptions implicit, or to the unsuitability of algebraic language to making or expressing causal assumptions (Pearl, 2000, pp.137–138).

> It is only from a solid base of knowledge about theory and research that one can even begin to address [the] requirements for inferring causation from correlation. Although facility with the statistical details of SEM is essential, it is not a substitute for what could be called wisdom about one's research area. (Kline, 2005, p.95)

BIBLIOGRAPHY

Bateson, W. (1902). *Mendel's Principles of Heredity. A Defense.* Cambridge University Press, London.

Bulmer, M. (1998). Galton's Law of Ancestral Heredity. *Heredity*, 81:579–585.

Bulmer, M. (2003). *Francis Galton. Pioneer of Heredity and Biometry.* The John Hopkins University Press, Baltimore, London.

Correns, C. (1900). G. Mendel's Laws concerning the behavior of progeny of varietal hybrids. In Stern and Sherwood (1966), pages 119–132.

Darden, L. (1991). *Theory Change in Science: Strategies from Mendelian Genetics.* Oxford University Press, Oxford.

de Vries, H. (1900). The law of segregation of hybrids. In Stern and Sherwood (1966), pages 107–117.

Galton, F. (1877). Typical laws of heredity. *Nature*, 15:492–5, 512–4, 532–3.

Galton, F. (1886). Regression towards mediocrity in hereditary stature. *Journal of the Anthropological Institute*, 15:246–263.

Galton, F. (1889). *Natural Inheritance.* Macmillan and Co., London.

Galton, F. (1892 (1869)). *Hereditary Genius.* Macmillan and Co., London, 2nd edition.

Galton, F. (1897). The average contribution of each several ancestor to the total heritage of the offspring. *Proceedings of the Royal Society*, LXI:401–413.

Hacking, I. (1990). *The Taming of Chance.* Cambridge University Press, Cambridge.

Kline, R. B. (2005). *Principles and Practice of Structural Equation Modeling.* The Guilford Presss, New York, 2nd edition.

Klug, W. and Cummings, M. (1997). *Concepts of Genetics.* Prentice-Hall, New Jersey, 5th edition.

Leuridan, B. (200+). Mendel's Bayesian net or the causal structure of classical genetics. unpublished manuscript.

Meijer, O. G. (1983). The essence of Mendel's discovery. In Orel, V. and Matalová, A., editors, *Gregor Mendel and the Foundation of Genetics*, pages 123–178. The Mendelianum of the Moravian Museum, Brno.

Mendel, G. (1933). *Versuche über Pflanzenhybriden.* Ostwald's Klassiker der Exakten Wissenschaften. Akademische Verlagsgesellschaft, Leipzig.

Pearl, J. (2000). *Causality. Models, Reasoning, and Inference.* Cambridge University Press, Cambridge.

Pearson, K. (1900). *The Grammar of Science.* Adam and Charles Black, London, 2nd edition.

Pearson, K. (1957 (1911)). *The Grammar of Science.* Meridian Books, New York, reissue of the 3rd edition.

Spirtes, P., Glymour, C., and Scheines, R. (2000). *Causation, Prediction, and Search.* MIT Press, Cambridge, Massachusetts.

Stern, C. and Sherwood, E. R., editors (1966). *The Origin of Genetics: A Mendel Source Book.* Freeman, San Francisco.

Stigler, S. M. (1986). *The History of Statistics. The Measurement of Uncertainty before 1900.* The Belknap Press of Harvard University Press, Cambridge (Mass.).

Stigler, S. M. (2002). *Statistics on the Table. The History of Statistical Concepts and Methods.* Harvard University Press, Cambridge (Mass.).

Venn, J. (1889). Critical notices. *Mind*, XIV(55):414–420.

Weldon, W. (1902). Mendels laws of alternative inheritance in peas. *Biometrika*, 1:228–254.

Mendelian randomisation: why epidemiology needs a formal language for causality

VANESSA DIDELEZ AND NUALA A. SHEEHAN

ABSTRACT. For ethical or practical reasons, randomised controlled trials are not always an option to test epidemiological hypotheses. Epidemiologists are consequently faced with the problem of how to make causal inferences from observational data, particularly when confounding is present and not fully understood. The method of instrumental variables can be exploited for this purpose in a process known as Mendelian randomisation. However, the approach has not been developed to deal satisfactorily with a binary outcome variable in the presence of confounding. This has not been properly understood in the medical literature. We show that by defining the problem using a formal causal language, the difficulties can be identified and misinterpretations avoided.

1 Introduction

Detection and assessment of the effect of some modifiable risk factor on a disease with view to informing public health intervention policies are of fundamental concern in epidemiology. For example, it is now well established that the risk of neural tube defects, such as spina bifida, can be greatly reduced by periconceptual maternal folate supplementation (MRC Vitamin Study Research Group, 1991; Czeizel and Dudás, 1992; Scholl and Johnson, 2000). A simple public health intervention of adding folic acid to flour and bread has been reported to have reduced the risk by 30-50% in the USA and Canada. The House of Commons Hansard Debates of October 19 2005 recommended this strategy for the UK which, if implemented, would be the first mandatory fortification of food in the UK since the compulsory addition of calcium, iron and vitamins B1 and B2 to flour after the second World War.

Clearly, it is important to have solid evidence that such a public health intervention will have an effect. The problem faced by epidemiologists is that an observed association or correlation between a risk factor and a disease does not necessarily mean that the risk factor is *causal* for the disease, and if the relationship is not causal, the prescribed intervention will be useless. Inferring causality from observational data is difficult as it is not always clear which of two associated variables is the cause, or which the effect. For example, sick people may change their diets or other aspects of their lifestyle (reverse causation). On the other hand, both disease and exposure levels may be associated purely through another possibly unmeasured factor such as smoking (confounding). In randomised controlled trials (RCTs), the random assignment of "treatment" levels to "experimental units" (Fisher, 1926) essentially renders reverse causation and confounding implausible, but such trials are neither ethical nor practical for many exposures of epidemiological interest like smoking, exercise regimes and alcohol consumption, to name but a few (though some attempts at such trials have been made but inevitably suffer from compliance and other problems).

In situations where randomisation *is* possible, epidemiological studies have been severely criticised for the large numbers of reported associations that have been interpreted as causal and have failed to be replicated in large-scale follow-up RCTs. For example, early observational findings suggesting that increased dietary intake of the anti-oxidant vitamin beta-carotene reduces the risk of smoking-related cancers (Peto et al., 1981) were negated by subsequent RCT findings. (Alpha-Tocopherol, Beta Carotene Cancer Prevention Study Group, 1994). Since only candidate causes with the strongest observational support are evaluated in RCTs, we can only assume that many reported associations, as yet untested, are even less likely to be causal (Davey Smith et al., 2005). Confounding is usually the main reason for such spurious findings as reverse causation can often be ruled out by the underlying biology. It is sometimes possible to control for confounding but in general it is difficult to know whether all the relevant confounders (or a sufficient subset of these) have been accounted for. Furthermore, confounding in these applications is usually due to social, behavioural or physiological factors which are difficult to control for and particularly difficult to measure accurately. Epidemiological exposures are also prone to reporting bias. Heavy drinkers, for example, will often under-estimate their alcohol intake.

There is hence a need to infer causality from observational data in the presence of confounding that cannot be controlled for because it is not fully understood. A possible approach in this situation is based on the method of instrumental variables (Bowden and Turkington, 1984; Angrist et al., 1996; Greenland, 2000; Pearl, 2000) which is known under the name of *'Mendelian randomisation'* [1] if the instrument is a genetic predisposition (Davey Smith

---

[1] The term 'Mendelian randomisation' seems to have become a fixed expression in the

and Ebrahim, 2003; Katan, 2004; Thomas and Conti, 2004). For example, observational studies have indicated that elevated plasma homocysteine levels are associated with increased risk of coronary heart disease (CHD) (Ford et al., 2002) but this effect is suspected to be heavily confounded by the usual factors such as smoking and socioeconomic status (Davey Smith and Ebrahim, 2003). RCTs have confirmed that homocysteine levels can be reduced substantially with a small increase in folate consumption (Homocysteine Lowering Trialists' Collaboration, 1998). However, the T allele of the MTHFR gene is known to be associated with higher homocysteine levels than the more common C allele and thus mimics the effect of low folate intake. In the absence of a definitive folate trial, the causal effect of homocysteine levels (and hence of folate intake) on CHD can be investigated by examining the association of the MTHFR genotype with CHD instead. The former association is affected by confounding but the latter can often be assumed to be free of confounding since alleles are assigned randomly from the two copies of the parents and so causality can be inferred. If the relationship between homocysteine and CHD is truly causal, adding a given quantity of folate to flour would also be worthwhile as a public health intervention to reduce CHD risk in the general population.

The practical difficulties typically encountered when inferring causation from observational data are compounded by the theoretical problem of expressing causal aims and methods in a mathematical language. Causal vocabulary features regularly in the epidemiological literature but this is often accompanied by standard regression methods that do not justify any causal conclusions. Despite recently proposed advances towards a formal causal framework for epidemiological applications (Greenland et al., 1999; Robins, 2001; Hernán, 2004; Hernán et al., 2004) such frameworks are not very widely adopted in general and in particular, are not reflected in the Mendelian randomisation literature at all.

The purpose of this article is to show that a formal, mathematically precise, causal framework is required for Mendelian randomisation applications. It is necessary, firstly, to state precisely what the quantity (parameter) of interest is (e.g., the amount by which CHD risk is reduced from adding folate to flour) and secondly, to formalise how associational findings and causal implications are related in order to obtain an estimate for this particular parameter. Failure to adopt a formal approach has led to misconceptions in the medical literature.

---

literature, but note that this is not a randomisation by study design and hence not fully comparable to a RCT. It has been suggested that 'Mendelian deconfounding' would be a better term (Tobin et al., 2004).

2 A formal language for causality

The medical literature often employs causal vocabulary loosely to express something that is more than association between potential risk factors and their effects. Underlying knowledge about the biology of the problem may indeed allow one to deduce the direction of an observed association and "causal pathways" for disease are familiar terms in the epidemiological literature (see Stanley et al. (2000) for example). The central argument in the present paper is that it is imperative to formally differentiate, with appropriate mathematical notation, between association and causation in order to be explicit about what can be inferred about causality from an observational study. Even the term "causal effect" is used loosely in practice and can mean different things in different settings. We present three approaches to defining a formal language for causality, each of which uses specific mathematical notation to represent that we are interested in interventions such as the public health intervention of adding folic acid (folate) to flour.

2.1 Interventions

As in Pearl (1995); Lauritzen (2000); Dawid (2002, 2003) we regard causal inference to be about the effect of intervening in a given system. For the applications we are considering, this would typically be the motivation for investigating a causal effect. There are many other notions of causality such as used in a courtroom for retrospective assignment of guilt, but we will not consider any other interpretations here. Let X be the cause under investigation and Y the response. In epidemiological applications, X would be the intermediate phenotype (homocysteine level) and and Y would be the disease status (CHD). We focus on the question of whether intervening on X has an effect on Y. By intervening on X, we mean that we can set X (or more generally its distribution) to any value we choose without affecting the distributions of the remaining variables in the system other than through the resulting changes in X. This is clearly an idealistic situation and not always easily justified for the examples of public health interventions given above. For example, increasing dietary folate will not determine a specific homocysteine level. However, a causal analysis can be used to generate hypotheses that can afterwards be investigated by controlled randomised trials where applicable. Moreover, if a risk factor is found to be causal in the above sense, different types of intervention can then be explored.

2.2 Three definitions of causal effect

Roughly speaking, the causal effect contrasts the effects of different interventions in X on the outcome Y, in some sense or another. We will now present three different approaches to doing this.

Pearl's do(·) operator

Pearl (2000) suggests the notation $P(Y|do(X = x))$ to distinguish between conditioning on intervention in X and 'ordinary' observational conditioning $P(Y|X = x)$ which is sometimes denoted by $P(Y|see(X = x))$ to make the distinction clearer. The former reflects how the distribution of Y should be modified given the information that X has been 'forced' to take on the value x by some external intervention, whereas the latter reflects how the distribution of Y should be modified when we have simply observed $X = x$. The average causal effect (ACE) is then defined as the difference in expectations under different settings of X:

$$ACE(x_1, x_2) = E(Y|do(X = x_1)) - E(Y|do(X = x_2)) \qquad (1)$$

where x_2 is often chosen to represent some baseline value. In particular, X is regarded as causal for Y if the average causal effect (1) is non-zero for some values x_1, x_2 with $x_1 \neq x_2$.

Regime indicators

This approach goes back to Pearl (1993) and has been further advocated by Lauritzen (2000); Dawid (2002, 2003); Dawid and Didelez (2005). It is based on an indicator F_X assuming values in $\mathcal{X} \cup \emptyset$ with $F_X = x$ if X is being set to the value x by external intervention and $F_X = \emptyset$ (or 'F_X is idle') indicates that X is allowed to arise 'naturally'. Observe that

$$P(Y|do(X = x)) = P(Y|X = x, F_X = x)$$

and

$$P(Y|see(X = x)) = P(Y|X = x, F_X = \emptyset).$$

Due to the deterministic relationship between $F_X = x$ and X we have that $P(Y|X = x, F_X = x) = P(Y|F_X = x)$, or that X is independent of any other variable given $F_X = x$.[2] The advantage of including an intervention indicator like F_X is that the intervention is made explicit and, as we will see later, can be represented visually in a graph. Besides, while the properties of the $do(\cdot)$ operator need to be formulated in separate 'axioms' (see Pearl, 2000, Section 3.4), F_X can be treated as a decision variable that is conditioned on and the properties of conditional independence can be applied (Dawid, 1979) to yield Pearl's axioms. In this context it should be noted that because F_X is a decision variable, it must always be in the conditioning set of a conditional probability, and when we write $Y \perp\!\!\!\perp F_X | X$, for example,

---

[2] Note that regime indicators can also be used for non–deterministic, i.e., random regimes, where the value to be assigned is drawn from a distribution. We do not go into more details here, but see Dawid and Didelez (2005).

we mean that $P(Y|X = x, F_X = \emptyset) = P(Y|X = x, F_X = x)$ for all $x \in \mathcal{X}$. This intervention variable also permits generalisation of causal inference to other types of intervention. For instance, interventions in X that depend on variables that have been observed before the intervention took place can be considered to reflect, for example, that the dosage of a drug should be different for different sexes and age groups. In such more general situations the type of intervention would be specified via the conditional distribution $P(X|C, F_X)$ where C represents some covariates that are taken into account by the intervention. Conditional interventions such as these are difficult to describe with the $do(\cdot)$ operator.

It is straightforward to reformulate the ACE using the regime indicator:

$$ACE(x_1, x_2) = E(Y|F_X = x_1) - E(Y|F_X = x_2). \qquad (2)$$

Counterfactuals

A philosophically quite different approach to causality is based on counterfactual variables (Neyman, 1923; Rubin, 1974, 1978; Robins, 1989). Here, Y_{x_1} represents the outcome if a subject is set to the value $X = x_1$ whereas Y_{x_2} is the outcome if the same subject is set to the value $X = x_2$. The variables are counterfactual because they can never both be observed together. With this notation one may define the *individual causal effect* (ICE) as

$$ICE(x_1, x_2) = Y_{x_1} - Y_{x_2}. \qquad (3)$$

The ACE is now expressed as

$$ACE(x_1, x_2) = E(Y_{x_1} - Y_{x_2}). \qquad (4)$$

Since this can be rewritten as $E(Y_{x_1}) - E(Y_{x_2})$ we can see that under certain assumptions we do not need to observe Y_{x_1} and Y_{x_2} together in order to make inference about the average causal effect.

In epidemiological applications, the ICE would represent the difference, say, in CHD status of an individual who starts off with high homocysteine levels as opposed to the same individual starting off with low levels, were both observable at the same time. This example demonstrates that, even though it might appear that counterfactuals do not require explicit specification of an intervention, they are only well defined when an intervention is implicit. How could the homocysteine level of a given individual be different from what it actually was? (See Hernán (2004) for further discussion.) Ideally, the risk factor (or potential cause), X, should be a variable on which subjects could be randomised (Rubin, 1974). Hence, all three approaches are only meaningful in the context of interventions that are actually feasible. The implications, for our purposes, are that some concrete public health intervention should be aimed at.

Comments

One could argue informally that the intervention distribution $P(Y|do(X = x))$, or $P(Y|F_X = x)$, corresponds to the distribution of the counterfactual Y_x, but we recommend some caution against this as Y under $do(X = x)$ is still not the same thing as Y_x. In the case of the former, $P(Y|F_X = x)$, we consider *one* variable Y describing the outcome of interest, the distribution of which has to be modified according to what we condition on. In the case of the counterfactual, Y_x, we consider *a number* of potential outcome variables in parallel that have possibly different distributions and, in particular, have a joint distribution. In addition, we note that the ICE being a comparison between two *values* rather than two distributions, does not have a counterpart in the other two frameworks. We will not give any further consideration to the ICE here as it represents a quantity that can never be observed, even in principle. Methods that claim to identify the ICE typically make strong untestable assumptions. Furthermore, it would seem that while the ACE is of obvious public health interest, the ICE might be more of medical/clinical interest. However, our aim is not to discuss the advantages and disadvantages of the three approaches. (For further discussion, see Dawid (2000) and the discussion of that paper, and also Dawid (2007) in this volume.) We wish, rather, to point out that such formal frameworks for causal inference exist and should be used more widely in epidemiological reasoning, especially for Mendelian randomisation applications. We would like to emphasise that it is neither possible to express the desired aim nor quantify the effect of a public health intervention with the "usual" conditional probabilities as the intervention creates a situation that is different from the purely observational one. Some additional 'ingredient' is required to clarify that a public health intervention will be applied to everyone (in a targeted population) and that inferences about this new situation are of interest. Only if this distinction is clearly made, can we work out the precise conditions that will allow such inference from the available data. We will mainly use the approach that includes an intervention indicator F_X but will also demonstrate how some things can be expressed in the other frameworks.

2.3 Identifiability

A causal parameter is identifiable if we can show that it can be estimated consistently from data under the conditions of how those data were obtained (e.g., randomised trial, case-control study, cohort study etc.). Mathematically, this amounts to being able to express the parameter in terms that do not involve the intervention (i.e., without $do(\cdot)$, F_X or counterfactuals) by using 'observational' terms only. These can then be estimated from data. As noted earlier and detailed in the following section, the distribution under intervention is not necessarily the same as the observational distribution because of confounding, for example. In cases when confounding is well understood, it can be shown that the intervention distribution can mathematically be re-expressed in observational terms and can hence be estimated from the observed data by adjusting for certain variables (Section 3.2) (Pearl, 1995, 2000; Lauritzen, 2000; Dawid, 2002). The instrumental variable technique on which Mendelian randomisation is based, permits a different way of identifying causal parameters when the confounding is poorly understood.

3 The Issue of confounding

Inferring causality from observational data is complicated by problems that may induce different types of bias. Here, we focus on the problem of confounding, as this is what Mendelian randomisation attempts to circumvent, and show how at least some of the concepts can be formally clarified.

3.1 What is confounding?

We will not attempt to provide a formal definition of confounding here. This issue is addressed in almost every textbook on epidemiology with varying degrees of clarity (see Rothman and Greenland (1998), for example, and for a discussion within the framework of causal graphs see Pearl (2000, chapter 6)). We will, however, highlight a few central aspects that will be relevant later.

Confounding could be said to be present whenever $P(Y|X=x, F_X=\emptyset)$ is not equal to $P(Y|F_X=x)$ or similarly if $P(Y|X=x)$ is not $P(Y_x)$. This dual notation reflects the common phrase "correlation is not causation". The well known implication is that a typical model for the regression of Y on X does not necessarily give us any information about the ACE. However, this problem could also be due to reverse causation or time trends which are typically not regarded as confounding but as separate mechanisms inducing bias. A common explanation of confounding is that there exists a variable (or set of variables) C that 'affects' both X and Y. As is well understood in epidemiology (Weinberg, 1993), a crucial implication is that C is not "on the causal pathway" between X and Y. This is important since we do not want to adjust for such variables as the true effect of X on Y could be hidden. Apart from being quite vague, such a requirement is difficult if not impossible to verify from observational data and conditional (in)dependencies since no testable implications arise in either scenario and all variables could be mutually dependent.

Our causal framework allows a formal definition of the requirement that

C should not be "on the causal pathway" as follows:

$$C \perp\!\!\!\perp F_X \qquad (5)$$

i.e., C is not affected by whether or not we intervene in X. This amounts to saying that we expect the marginal distribution of C to be the same in an observational study as in an RCT, for instance, where X has been randomly allocated. In the counterfactual framework the condition analogous to (5) is that there is no counterfactual version of C (i.e., there are no C_{x_1} and C_{x_2}), which implies that it is not affected by an intervention in X (see Dawid's contribution in this volume).

It is perhaps helpful at this point to demonstrate why $P(Y|X=x, F_X=\emptyset)$ and $P(Y|F_X=x)$ are not necessarily the same in the presence of confounding—and with the notion of interventions this can easily be formalised. Consider the above situation where C satisfies (5). Since

$$P(Y, X, C|F_X) = P(Y|X, C, F_X)P(X|C, F_X)P(C|F_X)$$

we have from (5) that $P(C|F_X) = P(C)$. Furthermore, under intervention, $F_X = x$, we have that $P(X = x'|C, F_X = x) = \mathbf{1}\{x' = x\}$, where $\mathbf{1}\{\cdot\}$ denotes the indicator function. Hence

$$\begin{aligned} P(Y, X=x, C|F_X=x) &= P(Y|X=x, C, F_X=x)P(C), \text{ and} \\ P(Y, X=x', C|F_X=x) &= 0, \text{ whenever } x' \neq x. \end{aligned}$$

On the other hand, under the observational regime, $F_X = \emptyset$,

$$P(Y, X, C|F_X = \emptyset) = P(Y|X, C, F_X = \emptyset)P(X|C, F_X = \emptyset)P(C).$$

Even if we are willing to assume that $P(Y|X = x, C, F_X = x) = P(Y|X = x, C, F_X = \emptyset)$ (i.e., that $Y \perp\!\!\!\perp F_X | (X, C)$), we can see immediately that the two expressions differ by the factor $P(X|C, F_X = \emptyset)$. This factor reflects that in the observational case, X is informative for C whereas in the intervention case it is not, and this can induce bias if C is predictive for Y. This is crucial to the understanding of the difference between the intervention situation that we are interested in for causal inference, and the observational situation that the data represent.

3.2 Adjusting for confounding

Confounding can be described in a purely operational manner by showing how one can adjust, or control, for it. Consider a variable, or set of variables C, such that

(i) $C \perp\!\!\!\perp F_X$, i.e., C is "not on the causal pathway", and

(ii) $Y \perp\!\!\!\perp F_X | (X, C)$, i.e., once we know C and X the distribution of Y is independent of how X was generated.

Then C is called 'a sufficient set of covariates' for identifying the ACE (Dawid, in this volume, calls it an 'unconfounder'). More precisely, we can show the following:

$$P(Y = y | F_X = x) = \sum_c P(Y = y | C = c, F_X = x) P(C = c | F_X = x)$$

$$\stackrel{(i)}{=} \sum_c P(Y = y | X = x, C = c, F_X = x) P(C = c | F_X = \emptyset)$$

$$\stackrel{(ii)}{=} \sum_c P(Y = y | X = x, C = c, F_X = \emptyset) P(C = c | F_X = \emptyset) \quad (6)$$

(Pearl, 1995; Dawid, 2002). Now all quantities are observational and the ACE can be calculated by substituting x_1 and x_2, i.e., the causal effect is identifiable from data on X, C, Y on imposition of an additional positivity condition (Dawid, 2002).

Within the counterfactual framework, the corresponding assumptions are:

(i*) $X \perp\!\!\!\perp Y_x | C$, the "no–unmeasured confounder" assumption, and

(ii*) $Y_x = Y$ if $X = x$, the "consistency" assumption.

The distribution of Y_x can then be identified as

$$P(Y_x = y) = \sum_c P(Y_x = y | C = c) P(C = c)$$

$$\stackrel{(i^*)}{=} \sum_c P(Y_x = y | X = x, C = c) P(C = c)$$

$$\stackrel{(ii^*)}{=} \sum_c P(Y = y | X = x, C = c) P(C = c),$$

where again, all distributions in the last line are 'counterfactual–free' and can hence be estimated from observational data.

Comments

We would like to point out that it is important to state explicitly the conditions that enable adjustment for confounding and that, as before, this requires a formal framework for interventions: otherwise it is difficult to express why the adjustment (6) gives us a desirable quantity. The conditions

(i) and (ii) (or (i*) and (ii*)) state precisely what is required to connect the observational data situation to the interventional situation that we are actually interested in. If we do adjust for confounding in the above way we have to justify these conditions based on background knowledge or prior studies for any given data situation. In addition, we need to be able to identify, observe and measure a sufficient set C in a way that ensures that these conditions are satisfied. In practice this may be very difficult; there are many ways to measure smoking behaviour or alcohol intake, for example, and such factors are prone to measurement error and recording bias.

3.3 Confounding in linear models

We now address a very popular class of models, linear models without interactions. The assumption of such a model class means that very strong restrictions regarding the shape of (causal) relations are imposed. In practice, these are often not justifiable, but some of the basic concepts explained so far can be illustrated within this simplistic setting. In the following, omission of F_X from the conditioning set implies $F_X = \emptyset$.

Assume that Y is continuous and that the causal dependence of Y on X is linear (possibly after suitable transformations), i.e.,

$$E(Y|F_X = x) = \alpha + \beta x.$$

In this case, the average causal effect is $\beta(x_1 - x_2)$ and can be summarised simply by β which is now interpreted as the average effect on Y of increasing X by one unit through some intervention. Similarly, we can make the assumption that $E(Y_x) = \alpha + \beta x$ and obtain the same ACE in the counterfactual framework. In contrast, a standard linear regression models

$$E(Y|X = x, F_X = \emptyset) = \tilde{\alpha} + \tilde{\beta} x$$

and there is a priori no reason to assume that $\tilde{\beta} = \beta$ as discussed in previous sections. In rare cases one may be able to justify $Y \perp\!\!\!\perp F_X \mid X$ implying $E(Y|X = x, F_X = \emptyset) = E(Y|X = x, F_X = x)$ so that in this case $\tilde{\beta} = \beta$ holds. However, this relation usually cannot be assumed in the presence of confounding and the following adjustment is necessary.

Now assume the situation where we include an additional variable (or set of variables) C to adjust for confounding, i.e., C is not on the causal pathway in the sense of (i) and also satisfies (ii). We still assume linear models for all (observational) relationships:

$$\begin{aligned} E(Y|X = x, C = c) &= \alpha + \beta_1 x + \beta_2 c \quad \text{and} \\ E(X|C = c) &= \gamma + \delta c, \end{aligned} \quad (7)$$

with both X and Y having constant (possibly different) conditional variances. In addition we assume that the first expectation is the same if we intervene in X:

$$E(Y|C = c, F_X = x) = \alpha + \beta_1 x + \beta_2 c.$$

This reflects assumption (ii): given X and C, the distribution of Y, and hence its expectation, is the same regardless of how X was generated. From the above, we have that

$$\begin{aligned}
E(Y|F_X = x) &= E_{C|F_X=x}E(Y|C, F_X = x) \\
&\stackrel{(i)}{=} E_C E(Y|C, F_X = x) \\
&\stackrel{(ii)}{=} \alpha + \beta_1 x + \beta_2 \mu_C \\
&= \alpha^* + \beta_1 x,
\end{aligned}$$

where $\mu_C = E(C)$ and using obvious notation for iterated conditional expectation. Hence

$$ACE(x_1, x_2) = \beta_1(x_1 - x_2)$$

and so β_1 is the causal parameter of interest.

A regression of Y on X alone corresponds to

$$\begin{aligned}
E(Y|X = x, F_X = \emptyset) &= E_{C|X=x,F_X=\emptyset}E(Y|X = x, C, F_X = \emptyset) \\
&= \alpha + \beta_1 x + \beta_2 \mu_{C|x},
\end{aligned}$$

where $\mu_{C|x} = E(C|X = x, F_X = \emptyset)$ is typically not constant in x and, in particular, is not equal to μ_C due to the dependence between X and C in the observational regime. Hence β_1 cannot be identified from a regression of Y on X alone. However, as we have assumed that C is sufficient for adjustment we can use our adjustment formula (6) to obtain

$$\begin{aligned}
E(Y|F_X = x) &= \sum_c E(Y|X = x, C = c)P(C = c) \\
&= \sum_c (\alpha + \beta_1 x + \beta_2 c)P(C = c) \\
&= \alpha + \beta_1 x + \beta_2 \mu_C
\end{aligned}$$

as desired. Hence, if we have data on X, Y and C we can estimate β_1 from a linear regression of Y on X and C.

Note that if X is binary the ACE is unique (up to its sign) given by ACE(1,0) but in the more general cases of more than two categories of X and/or nonlinear dependency the average causal effect is not necessarily summarised by a single parameter.

4 Formal graphical representation

It is useful to introduce a formal graphical representation so that existing substantive background knowledge can be formally encoded and conditions such as (i) and (ii) can be verified visually. We use directed acyclic graphs (DAGs) to represent conditional independencies among a set of joint variables in the following way. A graph is denoted by $\mathcal{G}$ and consists of nodes and directed edges. Every node of the graph represents a variable and these can be linked by directed edges which we represent by arrows ($\longrightarrow$). If $a \longrightarrow b$ we say that a is a *parent* of b and b is a *child* of a. If $a \longrightarrow \cdots \longrightarrow b$ then a is an *ancestor* of b and b is a *descendant* of a. A cycle occurs when a node a is its own ancestor or descendant meaning that there exists an unbroken sequence of directed edges leading from a back to itself. DAGs have no such cycles. All the conditional independencies represented in the graph can be derived from the *Markov properties* of the graph by which every node is independent of all its non-descendants given its parents (Pearl, 1988; Cowell et al., 1999). Furthermore, these Markov properties are equivalent to a *factorisation* of the joint distribution. By this we mean that if $X_1, \ldots, X_K$ are the variables represented by the nodes of the graph and pa(i) denotes the set of parents of X_i in the graph, the above Markov properties hold if and only if

$$P(X_1, \ldots, X_K) = \prod_{i=1}^{K} P(X_i | X_{\text{pa}(i)}). \tag{8}$$

Some of the nodes can be decision variables, but these would typically not have any parents and would always be conditioned on. Also note that the requirement that the distribution of X given $F_X = x$ be degenerate at x ($x \neq \emptyset$) is not explicitly displayed in the graph and still has to be introduced as an implicit, externally specified, constraint (Dawid, 2002).

The graph in Figure 1 represents a situation where the assumptions (i) and (ii) of Section 3.2 are satisfied and the ACE can be identified by adjusting for C. We can see that $C \perp\!\!\!\perp F_X$ because they are non–descendants of each other and have no parents and $Y \perp\!\!\!\perp F_X | (X, C)$ because F_X is a non–descendant of Y and (X, C) are the parents of Y. The graph induces the following factorisation

$$P(Y, X, C | F_X) = P(Y|X,C)P(X|C,F_X)P(C).$$

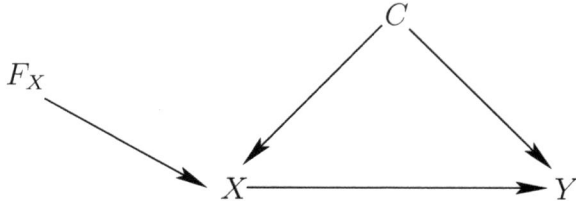

Figure 1. A situation where the set C is sufficient for identification of the ACE.

Depending on what value F_X takes, the distribution $P(X|C, F_X)$ is either observational or a one–point distribution on the value x. In order to read off the other conditional independencies that are implicit in the factorisation, we can use either the *moralisation* criterion (Lauritzen et al., 1990; Cowell et al., 1999) or alternatively, the method of d–separation (Pearl, 1988; Verma and Pearl, 1990). The former constructs an undirected (moral) graph $\mathcal{G}^m$ such that conditional independencies correspond to path separation in this undirected graph. The moral graph is constructed by adding an undirected edge between any two unconnected parents of a common child and removing the directions from all remaining edges. Any separation in the resulting undirected graph corresponds to a conditional independence in the underlying probabilistic model, where we say that "C separates A and B" if every (undirected) path between A and B contains nodes in C. These could equivalently be derived from the factorisation (8) but the graphical manipulations are often easier to carry out than the algebraic ones.

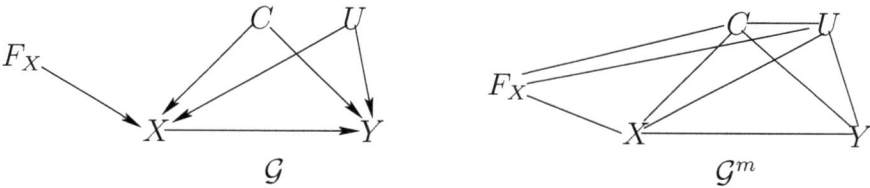

Figure 2. A graph $\mathcal{G}$ and corresponding moral graph $\mathcal{G}^m$ where C is not sufficient to identify the causal effect.

Consider the example in Figure 2. In such a situation, C is not sufficient to identify the causal effect as can be seen from the moral graph on the right: Y and F_X are not separated by $\{X, C\}$ because there is a path Y—U—F_X linking them. Hence the condition $Y \perp\!\!\!\perp F_X | (X, C)$ is not satisfied.

However, the conditions $Y \perp\!\!\!\perp F_X|(X,C,U)$ and $(C,U) \perp\!\!\!\perp F_X$ hold implying that if U were observable, an adjustment with regard to both (U,C) would be possible and yield valid causal inference.

Figure 3, on the other hand, depicts a situation where each of C or D alone is sufficient for adjustment. Here we have that $C \perp\!\!\!\perp F_X$ and $D \perp\!\!\!\perp F_X$, as the two variables in each statement are non–descendants of each other, and both $Y \perp\!\!\!\perp F_X|(X,C)$ and $Y \perp\!\!\!\perp F_X|(X,D)$, as can be seen from the corresponding moral graph $\mathcal{G}^m$. This situation illustrates the claim by Dawid (2002) that we do not need to adjust for 'all confounders' in the following sense. According to standard definitions of confounding that do not use a formal causal framework, each of C and D in Figure 3 are confounders in that they are not 'on the causal pathway' and they are both 'affecting' X and predicting Y. However, as we have shown, it is sufficient to adjust for only one of them in order to estimate the ACE.

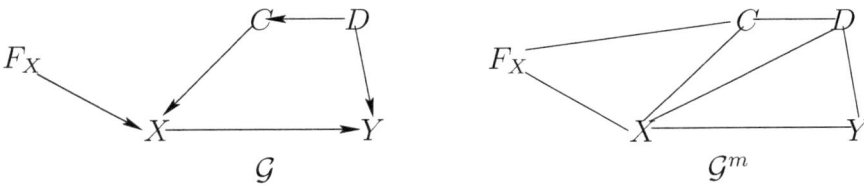

Figure 3. A graph $\mathcal{G}$ and corresponding moral graph $\mathcal{G}^m$ indicating that C or D are each on their own sufficient for adjustment.

5 Mendelian randomisation

'Mendelian randomisation' denotes the random assortment of genes from parents to offspring that occurs during gamete formation and conception. This 'randomness' can be exploited to test for, or estimate, the causal effect of an intermediate phenotype that has a genetic component on a disease in situations where confounding between the phenotype and the disease status is believed to be likely and is not fully understood (Davey Smith and Ebrahim, 2003; Katan, 2004; Thomas and Conti, 2004). There are other uses of the method including the provision of information about alternative biological pathways to a disease (Davey Smith and Ebrahim, 2004; Davey Smith et al., 2005). However, our focus here will be on its use to test for and estimate the causal effect of an intermediate phenotype on a disease in order to inform public health interventions.

5.1 The basic idea

The notion of Mendelian randomisation that we will use derives from an idea put forth by Katan (1986). In the mid-1980s, there was much debate over the direction of an observed association between low serum cholesterol levels and cancer. The hypothesis of interest was that low serum cholesterol increases the risk of cancer but it is also plausible that hidden tumours lower cholesterol in future cancer patients or other lifestyle factors affect both cholesterol levels and cancer risk (Katan, 2004). Katan noted that people with the rare genetic disease abetalipoproteinaemia, resulting in extremely low serum cholesterol levels, do not seem especially predisposed to getting cancer prematurely. It was known that the apolipoprotein E (APOE) gene is associated with cholesterol levels and that the E2 allele relates to lower levels than either E3 or E4. Crucially, by Mendel's Second Law (the law of assortment), E2 carriers should be no different from other genotypes in socioeconomic position, lifestyle and all other respects (this can be violated for various reasons and should always be checked in the light of background knowledge). Katan reasoned that if low serum cholesterol level is really a risk factor for cancer, then patients should have more E2 alleles and controls should have more E3 and E4 alleles. Otherwise, APOE alleles should be equally distributed across both groups.

The causality of the low cholesterol-cancer association was disproved by the subsequent large statin trials primarily concerned with the effects of *high* cholesterol levels on CHD risk (Scandinavian Simvastin Survival Study (4S), 1994; Heart Protection Study Collaborative Group, 2002), but the idea has been applied several times since and is what is now understood as *Mendelian randomisation* in the epidemiological literature. Katan's original idea was centred around hypothesis testing to confirm or disprove causality but the method is also used to estimate the size of the effect of the phenotype on the disease (Minelli et al., 2004) and, indeed, to compare this estimate with that obtained from observational studies in order to assess the extent to which confounding has been accounted for. Essentially, this approach exploits the idea that a genotype affecting the phenotype of interest, and thus indirectly affecting the disease status, is assigned randomly at meiosis, given the parents' genotype, independently of any possible confounding factors. It is well known in the econometrics and causal literatures (Bowden and Turkington, 1984) that these properties define an *instrumental* variable but additional fairly strong assumptions are required for unique identification of the causal effect of the phenotype on the disease status. These additional assumptions can take the form of *linearity* and *additivity* assumptions for all dependencies, as are typically assumed in econometrics applications but could also be assumptions about the *compliance* behaviour of subjects under study, as are often made in the context of randomised trials with incomplete compliance (Angrist et al., 1996).

5.2 Instrumental variables

We will present the basic properties that characterise an instrumental variable in terms of conditional independence statements. These conditions have been given in many different forms, using intervention indicators (Dawid, 2003) or counterfactuals (Greenland, 2000; Angrist et al., 1996; Robins, 1997) or linear structural equations (Goldberger, 1972; Pearl, 2000) and a comparison of some of these can be found in Galles and Pearl (1998). The conditions we give below are common to most instrumental variable methods but on their own they do not necessarily allow for identification of the ACE as we will discuss more fully in the following sections. For now, we will focus on these core assumptions and illustrate their meaning.

Core conditions

Let X and Y be defined as above with the causal effect of X (e.g., homocysteine level) on Y (e.g., CHD) being of primary interest and F_X being the intervention indicator. Furthermore, let G be the variable that we want to use as the instrument (the MTHFR genotype in our case).

The following 'core conditions' that G has to satisfy (e.g., Greenland (2000); Dawid (2003)) assume the existence of a concrete, although possibly unobservable, variable (or set of variables) U^3 such that, under the observational regime i.e., under $F_X = \emptyset$,

1. $G \perp\!\!\!\perp U$, i.e. G must be (marginally) independent of U;

2. $G \not\perp\!\!\!\perp X$, i.e., G must not be (marginally) independent of X; and

3. $Y \perp\!\!\!\perp G \mid (X, U)$, i.e., conditionally on X and U, the instrument and the response are independent.

These alone do not allow us to infer anything about the intervention situation, i.e., about quantities under $F_X = x$. Hence, we need to supplement them with suitable further assumptions. We note that such extra assumptions are only implicit in the counterfactual approach of Greenland (2000) and also that of Pearl (2000). To motivate these additional assumptions, note that the conditional independencies of condition 1 and 3 are equivalent

---

[3]Note that Dawid (2003) points out that "there is no compelling reason to posit the existence of such an unobserved variable. To make this assumption is to say something non–trivial about how the world is. And even if it can be assumed to exist, there is no reason why the variable U should be essentially unique."

to the factorisation

$$P(Y, X, U, G | F_X = \emptyset) =$$
$$P(Y|X, U, F_X = \emptyset) P(X|U, G, F_X = \emptyset) P(U|F_X = \emptyset) P(G|F_X = \emptyset).$$

We now assume that if we change to the interventional setting, only the factor $P(X|U, G, F_X = \emptyset)$ changes to $P(X = x'|U, G, F_X = x) = I\{x' = x\}$ while the other terms remain the same. This reflects the crucial assumption that an intervention in X is possible without affecting the generation of the remaining variables in the system. Clearly, this would be quite idealistic in many circumstances. More formally it means that our intervention is such that

$$(G, U) \perp\!\!\!\perp F_X \text{ and } Y \perp\!\!\!\perp F_X | (X, U) \qquad (9)$$

i.e., G and U are 'not on the causal pathway' from X to Y, and U would be sufficient for identifying the causal effect by adjustment if it were observable. As it is not we might call it a sufficient concomitant (Dawid, 2003). The distribution under intervention in X hence satisfies the following factorisation,

$$P(Y, X = x, U, G | F_X = x) =$$
$$P(Y|X = x, U, F_X = \emptyset) P(U|F_X = \emptyset) P(G|F_X = \emptyset),$$

where $P(Y, X = x', U, G | F_X = x) = 0$ if $x' \neq x$.[4]

Just as in the case of the assumptions required for confounder adjustment (Section 3.2), these conditions essentially have to be justified by subject matter background knowledge. Conditions 1–3 could be tested if U were observed, but otherwise they do not imply any testable independencies regarding the instrument G. In particular, they do not imply that G and Y are independent either marginally or conditionally on X alone (as has been assumed by Thomas and Conti (2004); Thompson et al. (2003) and implied by Foster (1997)). Moreover, conditions (9) must also be justified by background knowledge and will depend on what kind of actual intervention is being contemplated. Of course, U can be empty indicating that there is no need to adjust for confounding and hence no need to use an instrumental variable if X and Y can be simultaneously observed.

The typical Mendelian randomisation setting, where G corresponds to the genotype for phenotype X usually provides very detailed biological background knowledge to verify conditions 1–3. We know that genes are randomly assigned at meiosis and can therefore be reasonably assumed not to

---

[4]One could tentatively formulate alternative conditions to 1–3 avoiding the assumption of the existence of such a U as: 1A. $G \perp\!\!\!\perp F_X$, 2A. $G \not\!\perp\!\!\!\perp X$, 3A. $Y \perp\!\!\!\perp G | F_X = x$. This has not been considered yet, but see Pearl (2000, p.248).

be associated with anything concerning lifestyle factors confounding the relationship between X and Y, thus supporting condition 1. Condition 2 is typically verified by the fact that we only use genes with well-understood biological function in these applications and likewise, the biology can also help to exclude the possibility that G is affected by X (reverse causation), although the direction of this association is not crucial. Likewise, the existence of alternative biological pathways between G and Y other than through X can often be ruled out thus supporting condition 3. The assumption that we can intervene in the phenotype, setting it to a fixed value so that none of the other variables are affected, is generally more problematic in terms of justification.

Graphical representation

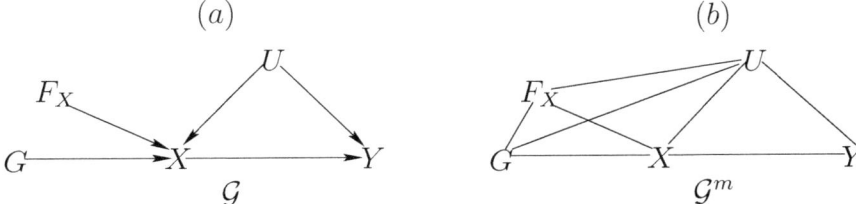

Figure 4. A graph $\mathcal{G}$ and corresponding moral graph $\mathcal{G}^m$ indicating that the core conditions for G to be an instrument are satisfied.

Figure 4 (a) shows a DAG involving G, X, Y and U that satisfies the core conditions 1–3, where the moral graph (b) in particular shows that condition 3 holds as Y and G are separated by (X, U) despite the moral edges that have to be added. In addition, by including the node F_X in the way shown, we ensure (9). The conditional independence restrictions imposed by the graph in Figure 4 (a) are equivalent to a factorisation of the joint density in the following way:

$$P(Y, X, U, G | F_X) = P(Y|U, X) P(X|U, G, F_X) P(U) P(G). \qquad (10)$$

From this, or from the moral graph in Figure 4 (b), it can be seen (by integrating out Y and conditioning on X) that $G \not\!\perp U | (X, F_X = \emptyset)$, for instance. Similarly, by integrating out X and conditioning on Y, we have that $G \not\!\perp U | (Y, F_X = \emptyset)$ despite $P(G, U) = P(G) P(U)$ (under either, $F_X = x$ or $F_X = \emptyset$). This is the so-called *selection effect* whereby two variables such as G and U, which are marginally independent, may become dependent once we condition on a common descendant. The selection effect is particularly relevant to case–control data when everything is conditional on the outcome Y. Hence the additional (moral) edge between G and U in the moral graph.

5.3 Linear no–interaction models

As mentioned above, the core conditions of Section 5.2 alone are not sufficient to allow us to obtain the causal effect in the presence of confounding for which we cannot adjust. More assumptions have to be made, typically with regard to the parametric shape of the relationships amongst the variables. In this section we continue with the simplistic situation described in Section 3.3 where a potential confounder is added, but this time we call it U to emphasise that it is unobservable.

As in Section 3.3 (with U replacing C), our model assumptions are that $U \perp\!\!\!\perp F_X$, i.e., we assume (9), which as explained earlier (cf. e.g., equation (5)) can be regarded as stating that U is not on the "causal pathway" from X to Y, and

$$E(Y|X=x, U=u, F_X=\emptyset) = \alpha + \beta_1 x + \beta_2 u \quad \text{and}$$
$$E(X|G=g, U=u, F_X=\emptyset) = \gamma + \delta_1 g + \delta_2 u.$$

In addition we assume that the first expectation is the same if we intervene in X i.e.,

$$E(Y|U=u, F_X=x) = \alpha + \beta_1 x + \beta_2 u.$$

reflecting assumption (9), i.e., that $Y \perp\!\!\!\perp F_X | (X, U)$. As before, β_1 is the causal parameter of interest here.

As we cannot adjust for U and a regression of Y on X alone does not yield the correct parameter (as shown in Section 3.3), we instead consider a regression of Y on G alone based on observational data, i.e., under $F_X = \emptyset$ (omitted from the conditioning sets below for brevity). This corresponds to

$$\begin{aligned} E(Y|G=g) &= E_{(X,U)|G=g} E(Y|X, U, G=g) \\ &= E_{U|G=g} E_{X|U,G=g} E(Y|X, U) \quad \text{since} \quad Y \perp\!\!\!\perp G | (X, U) \text{ by 3} \\ &= E_U E_{X|U,G=g} E(Y|X, U) \quad \text{since} \quad U \perp\!\!\!\perp G \text{ by 1} \\ &= E_U(\alpha + \beta_1(\gamma + \delta_1 g + \delta_2 U) + \beta_2 U) \\ &= \alpha + \beta_1 \gamma + \beta_1 \delta_1 g + (\beta_1 \delta_2 + \beta_2)\mu_U \\ &= \alpha^* + \beta_1 \delta_1 g. \end{aligned}$$

Hence, the coefficient of G in a regression of Y on G is $\beta_1 \delta_1$.

Furthermore, a regression of X on G alone based on observational data corresponds to

$$\begin{aligned}E(X|G=g) &= E_{U|G=g}E(X|G=g,U)\\&= E_U E(X|G=g,U)\\&= \gamma + \delta_1 g + \delta_2 \mu_U,\end{aligned}$$

so the coefficient of G in this regression is δ_1. Thus the causal parameter of interest, β_1, can be estimated consistently from the ratio of these two regression coefficients where the requirement $\delta_1 \neq 0$ is ensured by core condition 2. Note that the previous argument and model assumptions can easily be generalised to the case where X and G are binary. The case where Y is binary is more difficult and addressed below.

We should emphasise that, given that we are using the IV approach in the first place because confounding is not sufficiently understood, it seems unrealistic to believe that one would be willing to make such strong assumptions about U as are required for the above, in particular with regard to the parametric shape of the dependence of Y on U. Note that generalisations to the non–linear case have been developed in the econometrics literature but cannot necessarily be used for the present purpose as they are often targeted at situations with measurement errors (Amemiya, 1974; Hansen and Singleton, 1982).

5.4 More realistic situations

The instrumental variable approach for linear models without interactions as described above is well known from econometric theory. However, it is only of limited value for typical epidemiological applications where the primary aim of an investigation into the causal effect of a risk factor on a disease is to inform public health interventions . Firstly, the response Y is often a binary variable. Secondly, the data often arise from case–control studies with retrospective sampling (i.e., conditional on Y). Based on the above framework for causal inference we have shown (Didelez and Sheehan, 2005) that:

- If the core conditions 1–3 are satisfied, a test for *no causal effect* of the intermediate phenotype on the disease can be performed by testing that $Y \perp\!\!\!\perp G$, regardless of whether the data have been collected retrospectively and regardless of how the relevant variables have been measured. This was in fact the original idea of Katan (1986).

- If G, X and Y are all binary, the well-known ratio estimate derived in Section 5.3 cannot be applied. In fact, it is not straightforward to even specify the causal parameter in the latter case and the case-control scenario is further complicated by the fact that only odds ratios can be used. This has often been overlooked in the epidemiological literature.

Instead, bounds for the ACE can be derived as in Robins (1989); Manski (1990); Balke and Pearl (1994); Lauritzen (2000) and without assumptions about counterfactuals as in Dawid (2003). These can be modified to account for the case–control situation when gene frequencies $P(G)$ are available (Didelez and Sheehan, 2005) and can also be used as a rough test to rule out poor instruments. (Pearl, 2000) The calculations become computationally expensive when some variables have more than two categories and are intractable for continuous variables.

- An approximate check for confounding can be carried out when Y is binary and X continuous as in many epidemiological applications.

5.5 Problems with mendelian randomisation

The limitations of Mendelian randomisation, from the perspective of complicating features leading to poor estimation of the required genotype-phenotype and genotype-disease associations, have been discussed in detail in several places in the literature (Davey Smith and Ebrahim, 2003, 2004; Thomas and Conti, 2004; Davey Smith et al., 2005; Nitsch et al., 2006). However, existing approaches to testing and estimating the causal effect have not been formally challenged. Background knowledge is always required to verify untestable assumptions in order to make causal inferences. Mendelian randomisation applications have an advantage in that substantial biological background information can frequently be exploited in order to check that a particular genotype satisfies the conditions for an instrumental variable. It is unlikely that our simple model of Figure 4 will pertain, in practice. The common complex diseases that are of most interest from a public health perspective are generally multifactorial in nature and the definition of disease outcome itself is often ambiguous. As before, we can use directed acyclic graphs to represent the conditional dependencies that we believe are implied by the underlying biology and check the core conditions visually. We will illustrate this with a few examples below. Note however that, even if the core conditions would appear to be satisfied, the additional parametric assumptions that permit estimation of the causal effect of interest may not be justifiable.

Linkage disequilibrium refers to an association between alleles at different loci across the population and can be due either to tight *linkage* (i.e., because the loci are physically close on the chromosome and thus tend to be inherited together) or to other reasons such as natural selection, assortative mating,

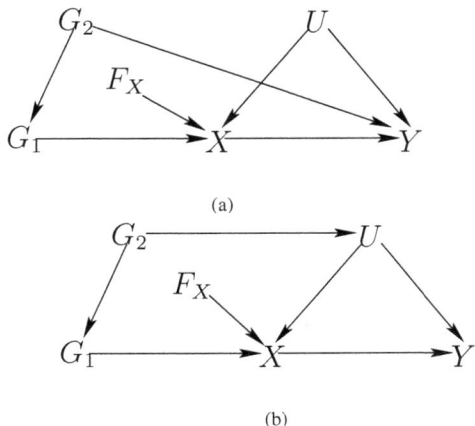

Figure 5. The instrument G_1 is in linkage disequilibrium with G_2 influencing Y, directly as in (a), or influences Y indirectly via the confounder U, as in (b).

and migration—see for example Lynch and Walsh (1998). If our chosen instrument G_1 is in linkage disequilibrium with another gene G_2 which has a *direct* influence on the disease Y, condition 3, $Y \perp\!\!\!\perp G_1 | (X, U)$, might be violated as shown in Figure 5 (a). If G_2 affects Y indirectly via a route other than through X, (Figure 5 (b)), condition 1, $G \perp\!\!\!\perp U$ might be violated. Note that even if the conditions appear to hold, linkage disequilibrium can cause attenuation of the genotype–phenotype association leading to poor inference.

Pleiotropy is the phenomenon whereby a single gene may influence several traits. If the chosen instrument G is associated with another intermediate phenotype which is also associated with the disease Y (Figure 6 (a)), condition 3 $Y \perp\!\!\!\perp G | (X_1, U)$, is again violated. As before, the association of X_2 with Y can also be via U (e.g., pleiotropic effects might influence consumption of tobacco or alcohol, for example) to violate condition 1 as in Figure 6 (b).

Population Stratification, referring to the co–existence of different disease rates and allele frequencies within subgroups of individuals, could lead to an association between the two at the population level which in turn can result in confounding of the genotype–disease association. That the disease

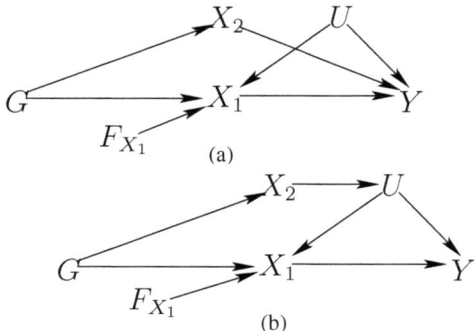

Figure 6. G has pleiotropic effects X_1 and X_2 where (a) both have a direct effect on the outcome Y of interest, or (b) X_1 has a direct effect but X_2 has an indirect effect via the confounder U.

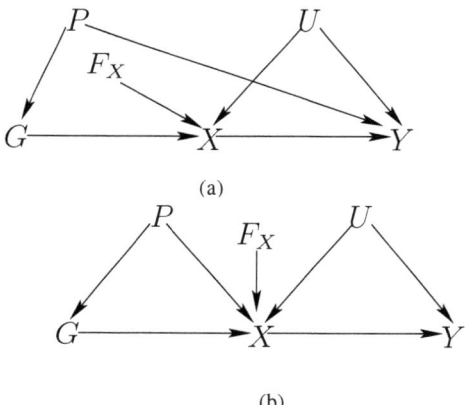

Figure 7. Two examples of population stratification where one of the conditions for G to be an instrument is violated (a) and all conditions are satisfied (b).

rates and allele frequencies are different for different population strata is represented in Figure 7 (a) by the arrows from P into G as well as into Y. We see from this graph that condition 3, $Y \perp\!\!\!\perp G_1|(X,U)$, is again violated: we need to condition on the population subgroup as well. However, if population stratification causes an association between allele frequencies and phenotype levels, as in Figure 7 (b), all conditions for G to be an instrument are still satisfied, and, in this situation, the $G - X$ association may in fact be strengthened, as a result.

6   Summary and discussion

The aim of this article is to justify our opinion that epidemiology in general, and the applications of Mendelian randomisation in particular, can benefit greatly from a formal framework for causal inference. We have presented three possible approaches. The first two, the $do(\cdot)$–operator and the intervention indicator F_X are very similar, with F_X being more general and explicit. The third, based on counterfactuals, is philosophically quite different especially with respect to the type of assumptions required. We clearly lean towards the first two. Note that the counterfactual language has been used very successfully to solve many intricate questions of causal inference, especially in the field of epidemiology, although it is still not widely adopted. Another formal approach that we have not considered here is provided by functional models (see Pearl, 2000, chapters 5–10) which include linear structural equation models (Haavelmo, 1943; Goldberger, 1972) as a special case. These are closely related to counterfactual reasoning and discussed in Dawid (2002), for example. Note that the models we considered in Sections 3.3 and 5.3 can be regarded as (partial) sturctural mean models in the sense that they provide equations for the mean of the response variable that remain stable under interventions in X, but this alone does not permit construction of the counterfactual responses for individuals.

We strongly advocate the adoption of a formal approach to causal inference in epidemiological applications—whichever of the above is favoured. Such a framework allows explicit specification of what the causal aim under investigation actually is: this cannot usually be expressed in a purely probabilistic way with 'normal' conditional probabilities. In the typical Mendelian randomisation setting, the causal aim would be some public health intervention and thus it would seem particularly desirable to make such interventions an integral part of the statistical model via decision variables such as F_X. For one thing, such an explicit representation can aid communication between the biologist and the statistician. Moreover, a formal language of causality allows explicit formulation of the conditions under which the target of inference can be attained. This, again, seems important with regard to facilitating the discussions and investigations that statisticians and subject matter experts have to carry out *together* when assessing whether the necessary conditions are satisfied and hence whether the ensuing causal conclusions are valid. Recall, for example, that our core condition 3 has been misinterpreted in the literature due to reliance on imprecise verbal descriptions. A particular advantage of the F_X indicator is that we can express the assumption that the core conditions 1–3 hold under $F_X = \emptyset$ and that *in addition* to these core conditions, (9) has to hold in order to specify which variables remain unaffected by the particular intervention that is being contemplated. The latter condition is implicit in Pearl's $do(\cdot)$ formalism and in the counterfactual approach.

We argue and confirm that Mendelian randomisation can often be reasonably assumed to satisfy the instrumental variable conditions. Moreover, subject matter background knowledge can be encoded using directed acyclic graphs to facilitate verification of the core conditions by visual inspection, and violations such as those of Section 5.5 can easily be identified. But there is also reason for some concern as to whether the public health interventions that can be carried out in practice satisfy conditions such as (9). For example, we cannot expect to be able to *fix* homocysteine at a prespecified level for the whole population. We can fix the amount of folic acid added to flour, but the amount of bread people eat and the amount by which folic acid affects the homocysteine level of individuals will vary. Such assumptions hence have to be scrutinised within the context of a specific application and with a concrete intervention in mind. Further research is required to address extending the notion of interventions in this respect.

On a more technical level, we showed in Didelez and Sheehan (2005) that under the above conditions, testing for a phenotype-disease causal effect by testing for a genotype-disease association is reasonable in all cases for practical purposes and that for calculation of the average causal effect, one must rely on additional strong parametric assumptions such as linearity and no interactions. In Sections 3.3 and 5.3 of the present article we have considered the linear case in some more detail in order to exemplify the use of the intervention indicator within this simple and well–known model class. But we emphasise that this model class is typically not useful for Mendelian randomisation settings, where non–linear models are called for and interactions may play a role. We then have to deal with even more technical problems that have yet to be formally tackled, For example, in the non–linear / interaction case, even the specification of the causal parameter is not obvious and determination of its relationship to the relevant regression parameters is not straightforward. "There is, in fact, no agreed upon generalisation of instrumental variables to non-linear systems" (see Pearl (2000, p.248)). However, the technical issues cannot be satisfactorily addressed if the causal questions are not clearly stated in the first place. We argue that this cannot be achieved consistently without adopting some formal framework for causality.

Acknowledgments

The authors acknowledge research support from the Centre for Advanced Study of the Norwegian Academy of Science and Letters.

Vanessa Didelez
Department of Statistical Sciences, University College London, UK.
vanessa@stats.ucl.ac.uk

Nuala A. Sheehan
Departments of Health Sciences and Genetics, University of Leicester, UK.
nas11@leicester.ac.uk

BIBLIOGRAPHY

Alpha-Tocopherol, Beta Carotene Cancer Prevention Study Group (1994). The effect of vitamin E and beta carotene on the incidence of lung cancer and other cancers in male smokers. *New England Journal of Medicine*, 330:1029–1035.

Amemiya, T. (1974). The nonlinear two-stage least-squares estimator. *Journal of Econometrics*, 2:105–110.

Angrist, J., Imbens, G., and Rubin, D. (1996). Identification of causal effects using instrumental variables. *Journal of the American Statistical Association*, 91(434):444–455.

Balke, A. A. and Pearl, J. (1994). Counterfactual probabilities: Computational methods, bounds and applications. In Mantaras, R. and Poole, D., editors, *Proceedings of the 10th Conference on Uncertainty in Artificial Inteligence*, pages 46–54.

Bowden, R. and Turkington, D. (1984). *Instrumental Variables*. Cambridge University Press, Cambridge, U.K.

Cowell, R. G., Dawid, A. P., Lauritzen, S. L., and Spiegelhalter, D. J. (1999). *Probabilistic Networks and Expert Systems*. Statistics for Engineering and Information Science. Springer-Verlag, New York, Inc.

Czeizel, A. E. and Dudás, I. (1992). Prevention of the first occurence of neural-tube defects by periconceptional vitamin supplementation. *New England Journal of Medicine*, 327:1832–35.

Davey Smith, G. and Ebrahim, S. (2003). Mendelian randomization: can genetic epidemiology contribute to understanding environmental determinants of disease? *International Journal of Epidemiology*, 32:1–22.

Davey Smith, G. and Ebrahim, S. (2004). Mendelian randomization: prospects, potentials, and limitations. *International Journal of Epidemiology*, 33:30–42.

Davey Smith, G., Ebrahim, S., Lewis, S., Hansell, A., Palmer, L., and Burton, P. (2005). Genetic epidemiology and public health: hope, hype, and future prospects. *Lancet*, 366:1484–1498. (a).

Dawid, A. P. (1979). Conditional independence in statistical theory (with Discussion). *Journal of the Royal Statistical Society, Series B*, 41:1–31.

Dawid, A. P. (2000). Causal inference without counterfactuals. *Journal of the American Statistical Association*, 95:407–448.

Dawid, A. P. (2002). Influence diagrams for causal modelling and inference. *International Statistical Review*, 70:161–189.

Dawid, A. P. (2003). Causal inference using influence diagrams: the problem of partial compliance. In Green, P. J., Hjort, N. L., and Richardson, S., editors, *Highly Structured Stochastic Systems*, pages 45–81. Oxford University Press, Oxford, UK.

Dawid, A. P. and Didelez, V. (2005). Identifying the consequences of dynamic treatment strategies. Research Report 262, Department of Statistical Sciences, University College London.

Didelez, V. and Sheehan, N. A. (2005). Mendelian randomisation and instrumental variables: what can and what can't be done. Technical Report 05-02, Department of Health Sciences, University of Leicester. (http://www.homepages.ucl.ac.uk/~ucakvdi/vlon.html).

Fisher, R. (1926). *The Design of Experiments*. Oliver & Boyd, Edinburgh, 1st edition.

Ford, E., Smith, S., Stroup, D., Steinberg, K., Mueller, P., and Thacker, S. (2002). Homocysteine and cardiovascular disease: a systematic review of the evidence with special emphasis on case-control studies and nested case-control studies. *International Journal of Epidemiology*, 31:59–70.

Foster, E. (1997). Instrumental variables for logistic regression: An illustration. *Social Science Research*, 26:487–504.

Galles, D. and Pearl, J. (1998). An axiomatic characterization of causal counterfactuals. *Foundations of Science*, 3:151–182.

Goldberger, A. (1972). Structural equation methods in the social sciences. *Econometrica*, 40:979–1001.

Greenland, S. (2000). An introduction to instrumental variables for epidemiologists. *International Journal of Epidemiology*, 29:722–729.

Greenland, S., Pearl, J., and Robins, J. M. (1999). Causal diagrams for epidemiologic research. *Epidemiology*, 10:37–48.

Haavelmo, T. (1943). The statistical implications of a system of simultaneous equations. *Econometrica*, 11:1–12.

Hansen, L. and Singleton, R. (1982). Generalized instrumental variable estimation of non-linear rational expectation models. *Econometrica*, 50:1269–1286.

Heart Protection Study Collaborative Group (2002). MRC/BHF heart protection study of cholesterol lowering with simvastin in 20,536 high-risk individuals. *Lancet*, 360:7–22.

Hernán, M. (2004). A definition of causal effect for epidemiologic research. *Journal of Epidemiology and Community Health*, 58:265–271.

Hernán, M., Hernández-Diaz, S., and Robins, J. (2004). A structural approach to selection bias. *Epidemiology*, 15:615–625.

Homocysteine Lowering Trialists' Collaboration (1998). Lowering blood homocysteine with folic acid based supplements: meta-analysis of randomized controlled trials. *BMJ*, 316:894–898.

Katan, M. B. (1986). Apolipoprotein E isoforms, serum cholesterol, and cancer. *Lancet*, i:507–508.

Katan, M. B. (2004). Commentary: Mendelian randomization, 18 years on. *International Journal of Epidemiology*, 33:10–11.

Lauritzen, S., Dawid, A., Larsen, B., and Leimer, H. (1990). Independence properties of directed Markov fields. *Networks*, 20:491–505.

Lauritzen, S. L. (2000). Causal inference from graphical models. In Barndorff-Nielsen, O. E., Cox, D. R., and Kluppelberg, C., editors, *Complex Stochastic Systems*, chapter 2, pages 63–107. Chapman & Hall.

Lynch, M. and Walsh, B. (1998). *Genetics and Analysis of Quantitative Traits*. Sinauer Associates Inc., USA.

Manski, C. F. (1990). Nonparametric bounds on treatment effects. *American Economic Review, Papers and Proceedings*, 80:319–323.

Minelli, C., Thompson, J., Tobin, M., and Abrams, K. (2004). An integrated approach to the Meta-Analysis of genetic association studies using Mendelian randomization. *American Journal of Epidemiology*, 160:445–452.

MRC Vitamin Study Research Group (1991). Prevention of neural tube defects: Results of the medical research council vitamin study. *Lancet*, 338:131–137.

Neyman, J. (1923). Sur les applications de la thar des probabilités aux experiences agariales: essay des principles. *Statistical Science*, 5:463–472. 1990. Excerpts reprinted in English. (D.Dabrowska and T.Speed, Trans.).

Nitsch, D., Molokhia, M., Smeeth, L., DeStavola, B. L., Whittaker, J. C., and Leon, D. A. (2006). Limits to causal inference based on Mendelian randomization: a comparison with randomised controlled trials. *American Journal of Epidemiology*, 163:397–403.

Pearl, J. (1988). *Probabilistic Reasoning in Intelligent Systems*. Morgan Kaufmann, San Mateo, CA.

Pearl, J. (1993). Graphical models, causality and intervention. *Statistical Science*, 8:266–269.

Pearl, J. (1995). Causal diagrams for empirical research. *Biometrika*, 82:669–710.

Pearl, J. (2000). *Causality*. Cambridge University Press.

Peto, R., Doll, R., Buckley, J., and Sporn, M. (1981). Can dietary beta-carotene materially reduce human cancer rates? *Nature*, 290:201–208.

Robins, J. (1989). The analysis of randomized and nonrandomized aids treatment trials using a new approach to causal inference in longitudinal studies. In Sechrest, L., Freeman, H., and Mulley, A., editors, *Health Service Research Methodology. A Focus on AIDS*, pages 113–159. U.S. Public Health Service, Washington, D.C.

Robins, J. (1997). Causal inference from complex longitudinal data. In Berkane, M., editor, *Latent Variable Modeling with Applications to Causality*, pages 69–117. Springer-Verlag, New York.

Robins, J. (2001). Data, design and background knowledge in etiologic inference. *Epidemiology*, 11:313–320.

Rothman, K. and Greenland, S. (1998). *Modern Epidemiology*. Lippincott-Raven, Philadelphia, 2nd edition.

Rubin, D. (1974). Estimating causal effects of treatments in randomized and nonrandomized studies. *Journal of Educational Psychology*, 66:688–701.

Rubin, D. (1978). Bayesian inference for causal effects: the role of randomization. *Annals of Statistics*, 6:34–58.

Scandinavian Simvastin Survival Study (4S) (1994). Randomised trial of cholesterol lowering in 4444 patients with coronary heart disease. *Lancet*, 344:1383–1389.

Scholl, T. O. and Johnson, W. G. (2000). Folic acid: influence on the outcome of pregnancy. *American Journal of Clinical Nutrition*, 71 (Suppl.):1295S–3035.

Stanley, F., Blair, E., and Alberman, E. (2000). *Cerebral Palsies: Epidemiology and Causal Pathways*. Mac Keith Press, London, U.K.

Thomas, D. and Conti, D. (2004). Commentary: The concept of "Mendelian randomization". *International Journal of Epidemiology*, 33:21–25.

Thompson, J., Tobin, M., and Minelli, C. (2003). On the accuracy of the effect of phenotype on disease derived from Mendelian randomisation studies. Genetic Epidemiology Technical Report 2003/GE1, Centre for Biostatistics, Department of Health Sciences, University of Leicester, (http://www.prw.le.ac.uk/research/HCG/getechrep.html).

Tobin, M., Minelli, C., Burton, P., and Thompson, J. (2004). Commentary: Development of Mendelian randomization: from hypothesis test to "Mendelian deconfounding". *International Journal of Epidemiology*, 33:26–29.

Verma, T. and Pearl, J. (1990). Causal networks: semantics and expressiveness. In Bonisonne, P., Henrion, M., Kanal, L., and Lemmer, J., editors, *Uncertainty in Artificial Interrligence 6*, pages 255–268. North-Holland, Amsterdam.

Weinberg, C. (1993). Towards a clearer definition of confounding. *American Journal of Epidemiology*, 137:1–8.

PART V

CAUSALITY AND PROBABILITY

The causal roots of probability
MARIANNE BELIS

ABSTRACT. Causality and probability are two models that account for the happening of phenomena. I argue that an objective probability exists in the case of the unique uncertain event and that this probability can be defined in the frame of a causal process. If the process is repeatable, then the probability value comes out in the long run in the same way in which a signal embedded in noise is recovered through multiple transmissions. Repetitive or mass phenomena are beneficial only because they are a sound, and sometimes the only, way to assess the value of objective probability, whose evaluation in the single case is generally difficult. Mass phenomena are not, instead, a condition of existence for objective probability. Some practical examples are given in the paper.

1 Introduction

One of the obvious features of our world is the interdependence of the objects and phenomena whose specific properties interact leading to observable changes. The mechanism which governs changes cannot be easily discovered because of the huge number of entities which interact simultaneously, with different intensities and speeds. Moreover, some of these entities are hidden because of our limited means of exploring the environment.

A general model of change is the causal model. It aims to model the very mechanism of change whatever the concrete features of the interacting entities. To explain the existence of an observable entity we look for its cause and to predict the occurrence of a future entity we look effects. Cause and effect are two observable entities from an endless chain of phenomena we deliberately truncate to obtain a causal connection. The causal model encompasses the very foundations of our knowledge. Widespread in everyday life, it represents the keystone of scientific reasoning.

Finding a general model of causality is difficult because of the different behaviours observed when a set of active entities is put to work. Some sets of causes always produce the same effect when they are assembled, but other sets exhibit a puzzling variability. The need to cope with this variability motivated the search for a measure of the uncertainty which it implies. This is a challenging task, as it is difficult to separate the subjective uncertainty that is due to the lack of knowledge from the objective indeterminacy that exists in reality.

Probability solved the problem of measuring uncertainty, without being concerned with the underlying causal connection responsible for the uncertain behaviour. A solid theoretical foundation reinforced the status of the probabilistic concept, which was developed far beyond its initial purpose, giving rise to the impressive corpus of statistics. Later, probability was used to define causality, giving rise to the theory of probabilistic causality.

Obviously, probability is a powerful tool to represent uncertain causal connections, as probability can be determined empirically while causality can be only hypothesized. Nevertheless the resort to empirical probabilities for defining causality shifts attention away from the singular causal connection, thanks to which, a certain pattern of regularity of the outcomes emerge. However, modern philosophical thought begins to be interested by the singular causal connection, whose modelling requires both physical and metaphysical concepts.

Philosophers of science looked for deeper explanations of causality. This inevitably led them to the heart of the causal process, the singular event. The objective probability of the singular event also gained the attention of physicists working on thermodynamics and quantum mechanics. As philosophical and scientific thought progressed, the streams of thought began to merge, illuminating new aspects of the relationship between causality and probability.

The objective of this paper is to continue this trend using new vistas opened in philosophy of science. The causal structure of random events is analysed, emphasizing the specific features that distinguish them from deterministic causal connections. Eventually, probability is derived from the causal structure of the singular event, leading to a *causal definition of probability*.

2 Singular causes

The Humean dogma of "regularities of associations" dominated Western philosophy for centuries. According to Hume (1739, p.199):

> ...all our reasoning concerning causes and effects are derived from nothing but custom; and that belief is more properly an act of the sensitive, than of a cogitative part of our nature.... The essence of things, their inner being escape us... All we know are sensations.

To the scientist of the 21st century, these assertions look strange. It is perhaps true that we ignore the "essence of things" (in fact, what does this mean exactly?) but scientific experiments helped to discover most of their features. Interactions are more and more controlled and used for our purposes. It is not by simple "associations" that we have walked on the moon, built nuclear plants, or performed in vitro fertilization. As Cartwright (1989, p.92) rightly noted:

> It does not take a lifetime of associations to convince a reasonable person of electromagnetic induction; Oersted's single experiment was quite sufficient... arbitrary regularities do not amount to causal connections... information about singular causes is vital.

What is essential in a causal connection and cannot be expressed by a mathematical relation is its productive or genetic character. As Bunge (1963, p.45) rightly remarked:

> Productivity is as absent from mathematics as from the Humean formula of causation... The Humean formula... accounts neither for the uniqueness nor for the *genetic* character of the relation between C and E. It does not convey the productivity or efficacy of the causation: it does not, in short, say that the effect happens to be *produced* by the cause, but only that it is regularly conjoined to it.

Some causal connections show a remarkable stability if reproduced repeatedly in the same conditions. That's why "regular associations" were used for a long time, and possibly even today. They have predictive force, which helps us to make decisions. But this is only *a proof of the validity of the causal connection at the singular level*, a proof that "the generic and the singular are inextricably intertwined... singular facts are not reducible to generic ones" (Cartwright, 1989, p.95). How else could regular associations exist if a causal connection did not work at the singular level? Regular associations are nothing but the long run behaviour of the singular causal connection. If we do not understand the latter, we cannot explain the former.

Moreover, the generic level alone cannot be a proof of the existence of the causal connection since there are many types of regular associations which are not causal as, for example, the events produced by a common cause. The barometer is not the cause of the storm though it fulfills Hume's conditions: spatio-temporal contiguity and regular association. By considering only regular associations in detecting causal connections one could fall into serious errors. I agree with Cartwright (1989, p.95) that "Hume's picture is exactly upside down".

Recently, an "anti-Humean" trend of thought began to emerge, looking for the very features of the singular causal connection, beyond "regular associations". According to Salmon (1984, p.146-147):

> A causal process is one that transmits energy, as well as information and causal influence. The propagation of causal influence by means of causal processes constitute I believe the mysterious connection between cause and effect which Hume sought.

In the case of random processes which do not exhibit a constant conjunction, but only a long-run pattern of regularity, the connection between cause and effect at the singular level is even more mysterious. Why don't the same causes produce the same effects each time they are assembled?

This puzzling variability was explained by the presence of minute influences which reinforce or weaken the main causal connection. These influences are generally not noticeable by a human observer, and so his lack of knowledge became the explanation of this kind of phenomena. Clothed with Laplace's authority, the argument looked so appealing that it shifted the attention away from the objective features of random phenomena. Determinism was saved thanks to subjectivism.

In fact, in order to find the objective features of indeterminate causal connections we have to separate ontology from epistemology, that is, determinism from prediction and indeterminism from unpredictability (Earman, 1986).

"Patterns of regularities" drew human attention because they were the only way to evaluate random behaviour. Like "constant conjunction" they provide an empirical way to manage the becoming of things. To explain the pattern of regularities of random processes we have to look for the genuine structure of the individual connection and to find out its specific features. As it will be shown later, the presence of minute influences alone (whether known or unknown) cannot justify the variability of the individual results. A special structure is responsible for this indeterminate behaviour which has an objective character at the individual level (Belis, 1995).

There was much debate about how to justify the assertion that an objective probability exists for a single event. The genuine features of the singular indeterminate connection must be unveiled in order to provide a justification. A thorough exploration of that claim was impeded by the difficulties of evaluating these features and of testing the probability of the single case against frequencies, which is seldom possible except in a few particular cases as the urn paradigm. Because of these *epistemic* difficulties, the probability of the single case was relegated to the subjective realm where it was evaluated either by means of a bet or, in some simple cases, by means of objective data. But these epistemic difficulties ought not to distract us from the singular connection, even when only its long-range behaviour is perceptible.

Nancy Cartwright (1989) convincingly argued that the single case is basic for understanding causality while Popper' propensities paved the way for taking into account the objective probability of the single case. In the following section we focus on these two concepts and we propose a relation between them.

3 Propensities and capacities

Propensity is an old concept. It goes back to Aristotle who conceived it as a potentiality inherent in the individual things, a general concept with universal value. However, Popper's interpretation of propensities (Popper, 1959, p.37) differs from Aristotle's:

> ...they cannot, as Aristotle thought, be inherent in the individual things... but in something a little more abstract, though physically real: they are relational properties of the experimental arrangement of the conditions we intend to keep constant during repetition.

Popper claimed that these relational properties are propensities to realize singular events. He looked for arguments in favour of an objective interpretation of the probability of the single case which he rightly considered fundamental. Popper didn't make precise how the "set of conditions" generates the corresponding propensity and what these conditions are. He conceived of propensities as generalized causes, that is he conceived of causes as propensities of measure one.

We consider that propensities depend on all the active factors which influence an outcome in both deterministic and random processes. The active factors include the so called causes and conditions whose dividing line is sometimes difficult to establish. To make this influence explicit, we shall include in our model the concept of *capacity* recently introduced by Nancy Cartwright (1989) in her analysis of the concept of causation. She focused on the single case which she rightly considers basic for the scientific image of the world.

The concept of capacity is a scientific tool for exploring the mysterious causal connection, providing strong arguments for the fundamental character of the singular cause. Cartwright claimed that individual objects have

properties and when interacting with other objects, these properties give rise to **capacities** whose strengths are observable and measurable.

Propensities and capacities are steps through which changes take place, enabling the transformation of cause into effect. We consider that a capacity refers to how an individual property may influence a change, while propensity includes the action of several capacities to produce a change. In an idealized case where the change is due to a unique cause, the capacity is the same as the propensity.

EXAMPLE 1 (The growth of a plant). The causal structure of this process is composed of *active factors* as for example: C_1 = seed, C_2 = humidity, C_3 = warmth, C_4 = fertilisers etc. Each has specific properties which entail various capacities to make the plant grow with various strengths. Some interaction between capacities could also exist. The combination of the various capacities leads to a *propensity* of the plant to grow, which is the result of all these contributions. This hypothetical disposition, tendency, or force results from the capacities of the active entities. Even if a process proceeds from known and measurable properties, the combination of these properties is generally hidden from our investigation.

Propensities lack the scientific flavour of capacities because they are a global concept, including the action of several capacities with various strengths and as such cannot be generally measured directly. How to capture the internal force which brings about a new plant? We have scientific methods to measure humidity, temperature and the quality of the soil, but the way and the proportion in which all these active factors combine for producing the plant is Nature's secret as it takes place at deep systemic levels, at the heart of the causal process.

The difficulties, even the impossibility, of directly measuring propensity form the main argument against this concept. New scientific concepts arouse distrust unless they permit measurement and verification by experience. However, contemporary science embraces a lot of concepts which cannot be apprehended directly. Concepts like force, gravitational field, electromagnetic waves, black holes, etc. contribute hugely to our ability to model the world. We rely on them because they are useful to explain experimental data.

Propensity is a hybrid concept. It is *metaphysical* in the sense that it cannot be directly perceived or measured, but it is *physical* in the sense of being the result of a combination of perceptible and measurable entities represented by the causal strengths. Propensity is a hypothesis which accounts for and makes intelligible the observed transformation of a group of causes into an effect. It is the bridge between individual capacities and observable change, in other words, between the inputs and the output of the black box inside which a causal process proceeds. Propensity's metaphysical character prevents direct measurement, but its relation with causal strengths sometimes permits indirect evaluation.

Popper himself was aware of this double character when he wrote (*ibid*) that he proposed a new physical hypothesis (or perhaps a metaphysical hypothesis). Moreover, he provided several versions of propensities, opening the door for a variety of interpretations. I agree with Fetzer (1988) when he associates propensities to a complete set of (nomically and/or causally) relevant conditions. Gillies (2000) observed that the complete list of relevant conditions might often be difficult to formulate and hard to test, thereby rendering the corresponding propensities metaphysical rather scientific.

Indeed, these difficulties exist. It is hard to take into account all the active entities which participate in a causal process, either because they cannot be measured or simply because we are unaware of their existence. Consequently, if a propensity cannot be verified by repetition, if it cannot be measured, then it is "metaphysical". Does that mean that we have to get rid of it? I do not think so. Even if propensities are not directly measurable and their physical existence is in doubt, they play an important role in explaining some experimental data which otherwise are mysterious. Propensities can complete a scientific model, which becomes a hybrid, but which also becomes more explicit and coherent.

We can ask further whether propensity really exists. Perhaps it is only our mind which synthesized the perceptible causal influences into a combination responsible for the observed effect. If so, then propensity would only satisfy our need for coherence, and this alone could not justify asserting that propensity exists. As will be shown later (§9) propensity proves to be essential in revealing the ontological roots of probability and in justifying claims for the existence of the objective probability of the singular event. It is a hidden entity which comes out through stable frequencies in the long run.

In defence of metaphysics, I argue that it is justified if it helps us to have a deeper insight into real-world phenomena, if metaphysical entities explain experimental data. The dogma of "direct measurement" ought not to prevent us from looking deeper into the heart of phenomena with our minds, if instruments are useless. Scientific models sometimes need help from metaphysics.

In the same vein, Tom Settle (1973, p.120) wrote:

> ...for generations we have tried in science to do without ontology (or metaphysics) despite the deep roots of science in ontology. The consequent divorce of science and philosophy has weakened

both. Perhaps the introduction of propensity theory, particularly of propensity theory of probability may help build the more fruitful relation between philosophy and science for which many of us hope.

In the following, we shall try to explain the relation between propensity and the causal strengths which compose it, providing a causal definition of propensity.

Let us consider a causal structure with $C_1, C_2, \ldots, C_n$ active factors (causes and conditions) which gives rise to $e_1, e_2, \ldots, e_m$ outcomes (or effects). Each of the active factors has a capacity of influencing the various outcomes with a certain strength which derives from its specific properties.

The propensity of an outcome to occur is due to multiple influences coming from various capacities of the causal structure. Some of them support it, others weaken it. This entails positive and negative strengths. The way in which they combine is difficult to know and to model (often impossible). Are the capacities independent? Is their action linear or non-linear? In its more general form the propensity of an effect e_j can be viewed as a function of all the actions it undergoes from the capacities of the causal structure:

$$\mathrm{Prop}(e_j) = F(c_{1j}, c_{2j}, \ldots, c_{nj})$$

where c_{ij} is the strength of the action of capacity C_i on effect e_j.

The function $F()$ has the following properties:

- $F()$ increases with positive elementary strengths and decreases with negative ones.

- $F()$ can take positive, null, or negative values. A positive value represents a support given to the respective outcome. A negative or a null value means that the outcome cannot occur. Negative or null values occur because of the combination of positive and negative elementary strengths.

In view of the difficulty to explicit the function $F()$, the propensity of e_j can be evaluated, in its simplest form, by the *algebraic sum* of all the strengths exerted upon it:

$$\mathrm{Prop}(e_j) = \sum_{i=1}^{n} c_{ij}$$

Among the objections raised against the concept of propensity is the difficulty or impossibility of discovering relevant propensities, since they are usually unmeasurable. The relation above, which connects causal strengths to propensity, transfers this objection to the underlying causal connections: how are causal connections discovered in the world? That question has sparked long-lasting debates among contending philosophical and scientific viewpoints.

Causal connections have been discovered by scientific means in laboratory and by experience in every day life. The development of the scientific thought submitted the observed regularities to rigorous and methodical experiments. The *frame* problem as it is called in artificial intelligence reminds us that it is practically impossible to describe all the conditions in which a change might take place and all possible consequences of a given action.

We have to limit the amount of both, by focusing on some few entities whose participation in the change is important. But how to separate what is important from the host of minor or irrelevant entities which are present? This cannot be done by passive observation. A passive observer is unable to distinguish efficient connections from simple associations. Active experiments are needed in which induced changes are controlled and measured. Effective strategies (Cartwright, 1983) and controllability (Nakashima and Osawa, 1997) enable one to separate causal connections from simple associations.

4 Deterministic and random phenomena

In the following the causal structure of deterministic and random processes will be analysed putting into fore the specificity of each of them.

Quantitative and qualitative changes → Continuous and discontinuous behaviour

Entities of the real world have various qualitative states. A state is characterised by various parameters whose values range between some limits. To exceed these limits amounts to a qualitative transformation (change of state). The accumulation of small quantitative changes leading to a sudden qualitative change of state is one of the basic principles of Hegelian dialectics. Here are some examples.

Mercury is in liquid state if its internal temperature ranges between -39 °C and +357 °C. Below -39 °C it becomes solid, and above +357 °C it becomes gaseous. At these values the behaviour of the mercury is *discontinuous*, as a little variation of the temperature entails a qualitative change of state. Between these temperatures and outside them, the behaviour is *continuous*, as the internal temperature of mercury increases quantitatively with the temperature of the source, while its phase is preserved.

Many examples of continuous behaviour alternating with discontinuous behaviour exist in the world. Abundant rain makes the level of water in a lake grow continuously. At a certain level, the dam breaks, that is, a change of state occurs. The same with the continuous accumulation of snow which suddenly releases an avalanche, the progressive deposition of cholesterol which finally causes thrombosis, or the slow rise of social discontent which at a certain level ushers in a revolution. The evolution of the above phenomena takes place in the continuous domain. Discontinuities represent only a limit, or particular points of behaviour. The behaviour of the process close to these points is *locally indeterminate* being controlled by small influences which can entail a change of state.

The threshold of qualitative transformation

We have seen that for mercury, a new qualitative state occurs when its inner temperature reaches a certain critical value. Such a critical level exists in every causal structure whose outcomes represent new qualitative states of the world. If the process has many active entities whose capacities exert various influences on the outcome, it is the sum of these influences that is the propensity (§3) which has to reach the critical level, specific to each causal process. If the negative elementary influences are important or the positive ones too weak, this critical level will never be reached and the effect will not occur.

We call *threshold of qualitative transformation*s the critical value of an inner parameter which determines a change of state. The nature and the level of this threshold is specific to each process: a certain quantity of snow releases an avalanche, a certain number of points makes a team win the match, a certain level of anger makes someone lose his temper, a certain number of votes decides the election etc.

Constant, variable and instantaneous propensities

In the real world we seldom find causal processes whose active entities have only constant strengths. This is because our world is dynamic, with multiple interactions coming from various hierarchical levels. Macroscopic entities can be influenced by molecular and atomic processes or by other processes belonging to the same systemic level. All these influences can be more or less variable with respect to the duration of the causal process.

An active entity is *constant* if the strength of its capacity does not vary during the process and its repetitions; otherwise it is *variable*. Variable entities arise inside or outside the causal process and their strengths can have regular or irregular variations. We focus on irregular variations with a *zero mean value*.

Constant and variable strengths combine together, giving rise to a propensity which is no longer constant but has an *instantaneous* value which can be greater or smaller than the constant one, depending upon the signs of the variable strengths which combine with it. As the irregular variations have zero mean value, the mean value of the instantaneous propensity obtained in the long run equals its constant value.

The variations of the instantaneous propensity have different consequences on the outcome of a process, depending upon the domain of behaviour of the process: continuous or discrete. In the continuous domain, the instantaneous variations of the propensity entail only quantitative variations of some parameters of the issue, the qualitative state being preserved. As these variations are generally small (sometimes imperceptible), the outcome is the same for an external observer. Imagine that in the continuous domain of variation of the mercury the temperature of the source varies from $+35\,°C$ to $35.2\,°C$. This entails a proportional increase in the inner temperature of the mercury, with no perceptible consequences: it is still in liquid state.

In the case of a discontinuous domain of behaviour, the situation is completely different. If the value of the constant propensity is near the threshold of qualitative transformation, the smallest variations of the instantaneous propensity can help the internal parameter to reach the threshold, thus entailing a qualitative change of state which is generally observable. In the case of the mercury, if the internal temperature is a little smaller than the threshold, the change can occur at any time, because of little and uncontrollable variations of the external temperature which superpose to the constant one. The mercury changes its state in an *indeterminate* way.

It follows that the same imperceptible variations have very different consequences on the outcome of a process, depending upon its domain of behaviour.

In the continuous domain the variation's existence has no visible effect, entailing only small quantitative changes in the parameters of the effect. It is the realm of *deterministic phenomena*.

In the discontinuous domain, the action of small variations has a dramatic influence leading to a qualitative change of state. It is the realm of *random phenomena*.

What makes the difference between deterministic and random phenomena is not the existence of small, often imperceptible, variations. These variations exist in both. What makes the difference is the domain of behaviour, continuous or discontinuous. This is an objective characteristic, independent of the existence of irregular events and of the knowledge we have about them. The absence of knowledge that is unpredictability is only an *epistemic consequence* of the objective indeterminacy. The Laplacian demon doesn't change the nature of the world; he only has a better insight about it than we do.

5 The objective features of the concept of chance

Generally, a causal structure is able to produce not only one, but several possible effects. These effects may occur simultaneously (compatible effects) or they may be mutually exclusive. An example of the first case is a medicine whose principal effect occurs simultaneously with some secondary ones, qualitatively different.

Many real world processes belong to the second case when only a single outcome from among many alternatives is possible. Examples: an election, the result of a match, the weather forecast, the draw of a ball from an urn, the toss of a coin, all are causal processes with many possible outcomes (or only two: the effect occurs or not), of which only a single one may occur. As it can be noticed all these cases concern causal processes with a *discontinuous* behaviour, leading to qualitatively different outcomes. As none of these outcomes is determinate we say that each has a *chance* of occurrence.

It has to be stressed that the indeterminacy of the outcomes is not related —for the moment— to some human uncertainty. No observer is present to watch the result. If some mechanical device extracts a ball from an urn, each time another ball could be drawn. If a horse race could be performed in "exactly the same conditions" the race can be won however by different horses. The human consequence of indeterminacy that is uncertainty and its measure will be analysed in the following section. For the moment we want to discern the features of the objective chance which favours more or less each outcome.

Suppose a causal process which comprises 'n' active entities endowed with individual capacities, generates 'm' possible mutually exclusive effects. Each effect is supported by a constant propensity resulting from the contribution of a part (or even all) of the existing capacities. The causal structure is then able to generate 'm' *constant* propensities, having different or equal values and leading to 'm' outcomes.

Constant propensities represent the *potentialities* existing inside the causal structure for producing the various outcomes. Variable influences (causal strengths) originating from inside or outside the process, transform these constant propensities in *instantaneous propensities* having variable values, greater or smaller than the constant ones, depending upon the sign of the variable strengths.

As the various possible effects are mutually exclusive, only one of the instantaneous propensities will succeed in producing its outcome. A true competition arises between the instantaneous propensities, a kind of "struggle for life" each one supporting its own effect. The one that reaches the threshold of qualitative transformation succeeds to realize its corresponding outcome, thus blocking the occurrence of the others.

Depending upon their irregular support, at each repetition of the process another instantaneous propensity wins the competition, whence the variability of the results. However, as expected, in the long run the stronger propensities will be realized more often as they require only little support from the irregular variations (or no support at all), while a small constant propensity needs to be reinforced by strong irregular variations in order to reach the threshold and realize its outcome. The outcomes with strong constant propensities have a greater facility to occur, a greater *chance* so that they will occur more often if the same process is repeated. Nevertheless, it is possible that at times, a small constant propensity, reinforced by strong positive irregular influences, may succeed in realizing its outcome.

The chance of occurrence of an undetermined outcome is represented by the value of its constant propensity.

In other words, chance represents the stable determination of an event. Depending on its value, the constant propensity needs more or less support from irregular events in order to produce its associated outcome. In some cases the value of the constant propensity alone is sufficient to reach the threshold and produce its effect. However negative irregular variations can prevent its action so that another outcome (with a smaller propensity but a favourable irregular influence) will have its place. That is why a horse with less stable qualities, i.e., less constant propensity or less "chance" can win the race.

Another example is the urn with balls of various colours. The stable propensities are provided by the number of balls of each colour. The more balls of a certain colour, the greater the chance for a ball of that colour to be extracted. But the threshold of qualitative transformation cannot be reached unless an extraction occurs, whether by mechanical or human means. The means for extraction provide the variable propensity, which adds to the constant propensities, helping one of the balls to reach the threshold. If the constant propensity of the white is greater (more numerous), white will be extracted more often. They have a greater chance.

6   The human model of chance

Human beings often confront events whose indeterminacy engenders uncertainty, which complicates their decision making. So, the human mind has always had a vital interest in foretelling the future. The road from the Delphic Oracle to probabilities was long and riddled with fallacies. It is only a few centuries ago that a theoretical framework and a numerical measure were developed in connection with games of chance, giving rise to the concept of probability.

In spite of its rigorous mathematical foundation, probability describes various aspects of reality only with difficulty, giving rise to various, sometimes discordant, interpretations. The major difficulty arises because of the complex nature of uncertainty which mixes human lack of knowledge with objective indeterminacy, whose existence has been denied for a long while.

After centuries of trials and confusion, objective or aleatory probability, which is a feature of the world, was separated from subjective or epistemic probability, which is a feature of our knowledge or belief. Objective probability was measured by frequencies, a choice which excludes the objective single case.

However, in everyday life, we confront unrepeatable indeterminate events. The frequency method cannot be used then, and chances must be evaluated with respect to the singular event only. To cope with this kind of situation, a subjective probability was defined as a "degree of rational belief" which can be measured through a bet. One might wonder how this degree of belief is built and whether it has any connection with objective reality.

As Popper (1959, p.29) ironically observed:

> ...what a gambler or a 'rational better' tries to find out, in order to bet upon it, is always and invariably the *objective* propensities, the *objective* odds of the event: thus the man who bets on horses is anxious to get more information about horses – rather than information about his own state of belief, or about the logical force of his total information in his possession.

Another approach towards the single case proposes that the values of the subjective probability be objectively based on the observable entities of the process. Surely, as the information about these conditions is incomplete, the objectively-based subjective probability differs from the objective probability of the single case. Moreover, one can ask how this subjective probability is to be evaluated on the basis of objective data.

According the causal scheme presented above in connection with propensities, the evaluation of the subjective probability of the singular event implies gathering information about the process which produces the uncertain event in order to detect all the constant active factors, their capacities and their strengths as well as the propensities which result. As already mentioned, the evaluation of propensity is not an easy task since we don't know all the factors mentioned above. The linear sum proposed in §3 is a strong simplification. Nevertheless belief formation by evaluation of the causal strengths and of propensities is a rational way of processing the available information since it is based on the very structure which underlies the objective features of chance.

EXAMPLE 2 (The tennis match). A tennis match is a process whose outcome is objectively indeterminate and from an epistemic point of view uncertain. The two mutually exclusive outcomes are: e_1 = player A wins and e_2 = player B wins. The performance of each player is due to some stable causes whose capacities say C_1 = technical level, C_2 = age, C_3 = health, C_4 = mood, support with different strengths the two players. The strengths, c_1, c_2, c_3, c_4 have various positive or negative values depending upon the outcome they support.

Player A is a well-known champion with an excellent technical level, but he is no longer young and he has a touch of arthritis. Consequently A's capacity for winning is strongly supported by c_1, but thwarted by c_2 and c_3 (negative values). c_4 is neutral. Player B is young and healthy (high c_2 and c_3) but he is a beginner (small c_1) and has a family problem (negative c_4).

Normally, player A has a greater propensity to win (we say that he has a greater chance) than player B as the sum of all his constant strengths is greater. Nevertheless, he is not winning in all the sets because of irregular events (with variable strengths) which can influence the outcome (a cramp in a leg, a gust of wind, a momentary lack of attention etc). These irregular events combine with the stable ones and lead to instantaneous propensities which can influence alternatively either player. Surely, in the long run the influence of the irregular events will vanish (null mean value) and the stable entities will prevail, so that the tournament will eventually be won by A.

Now if we want to appreciate numerically the chances of each player that is its constant propensity, according to the relation given in §3 we have to evaluate the algebraic sum of causal strengths which condition his winning. This means that we have to assign numbers to each of the causal strength depending on the information we have about the players and on our ability to encode qualities by numbers. By adding these numbers (with their respective signs) it results in a subjective measure of chance structured in the frame of propensity theory. This is an *absolute* value of chance.

The above measure of chance is not yet a subjective probability in the modern sense. When the concept of probability was invented, centuries ago, in order to treat is as a number, it was conceived of as a measure of our knowledge, a *degree* of certainty or as a part from a whole. If it was a fraction, then the whole was all possible outcomes.

In the same vein, we have to replace the absolute value of chance given above (the absolute value of the propensity) by its relative value that is, by the ratio of the absolute value of the propensity to the absolute value of the total propensity involved in the process:

Relative propensity (e_j) = absolute propensity (e_j) / total propensity involved in the process

$$= \sum_{i=1}^{n} c_{ij} / \sum_{i=1}^{n} \sum_{j=1}^{m} c_{ij}.$$

The relative propensity represents the *proportion* of causal strengths through which the various capacities determine the issue of the process, in other words, the *weight* of the determination of an outcome in the whole process. This represents a model of chance more adequate to human ability to evaluate proportions than absolute values.

By replacing the absolute propensity with its relative value, we move away from the objective (absolute) value of chance given in §5. The human measure of objective chance has a subjective flavour. Nature does not exhibit *proportions*. Nature works with absolute propensities, that is, with internal forces (strengths of capacities) which contend to produce their respective outcomes. In order to compare these internal forces and to find a measure of their relative efficacy without being bound to any specific unit of measure, we recast them as proportions, and thus obtain a universal measure of chance.

As Weatherford (1982, p.252) rightly said, the goal of probability theory is "to bring the uncertainty in our world view as closely as possible into congruence with the uncertainty in the world". By relative propensities (proportion of strengths), we brought the objective chance (absolute propensities) into our world view. Proportion is the key concept for probability (Belis and Snow, 2002).

7 The random singular case

In the evolution of the concept of probability, the "single case" was taken into consideration relatively late, after a standard theory had taken hold. Standard probability theory was born from practical necessities, first from games of chance and then extended to statistical data. In both domains, repetitive or mass phenomena were commonplace, so that "equipossible outcomes" and finite frequencies were useful tools to encompass this concept. Actuarial cases for insurance, mortality, or social sciences developed, reinforcing the status of this kind of probability.

Quantum mechanics drew the attention of scientific community to an *objective* evaluation of the indeterminacy of singular events. Born's probabilistic interpretation of the solution of Schrödinger's equation did not square with subjective or frequential interpretations in use at that time.

One of Popper's main objectives in introducing propensities was to justify asserting the existence of objective probabilities for solving this problem.

Popper focused on the generating or relevant conditions of the random singular event. He claimed that propensity ensued from these conditions. He did not explain how these generating conditions give rise to propensity nor why this propensity can be considered a probability. The lack of a rigorous definition of propensity inspired many doubts and questions. Moreover, propensity's untestable character imparted to it a metaphysical character, in the bad sense.

On the other hand, long-run objective probability is defined with respect to a "reference class" which can be viewed as a set of repeatable conditions. The problem is that the reference class is seldom a complete set of relevant conditions. This prevents the transfer of the objective value obtained in the long run to the single case.

A well-known example is the probability of whether a man now aged 60 lives to be 80. As this probability does not include any particular features of this man (country, lifestyle, heredity, etc.), it cannot be assigned to a particular person of this category. It is, in fact, the probability of survival for a virtual person devoid of real existence, since he lacks concrete features, a number which can serve only statistical purposes. Moreover, since the features of any particular reference class are established by the observer according to his knowledge, the resulting probability has a subjective character.

The evaluation of the propensity of the singular event from its relevant conditions is also a problem as the complete set is seldom known, nor is the relation which connects the conditions to propensity. Usually, the conditions are unrepeatable, apart from some experiments like the urn with balls or coin tossing, so that the long-run method for evaluating the propensity is rather virtual. However, to measure propensity is important not only from a conceptual but mainly from a practical point of view, since such singular indeterminate events occur frequently in everyday life.

The relation between propensity and its causal strengths provided in §3 is a way to make explicit the connection between propensities and the relevant conditions of the event. A more detailed relation can be provided taking into account not only the force but also the importance of the influence exerted by the strengths (Belis and Snow, 1998). It is by a relation of this kind that human beings evaluate the chance of the singular event (example: the tennis match in §6). It is a subjective evaluation, since it is based on incomplete or incorrect data (nevertheless objective), but it approaches the true objective value as information becomes more nearly complete.

Propensity theory justifies and makes intelligible the singular random case. This happens because propensity is causally based. That is, propensity goes to the heart of the singular process. Moreover, it helps us to explain some mysterious data provided by the frequency method. The limiting frequency of this method is postulated together with the value of the objective probability which is obtained in the long run. Why things happen that way nobody knows. Incredulous and patient people have tossed a coin tens of thousand of times to verify it.

Popper considered it the fundamental problem of the random theory. And he rightly thought that if the frequentist has to admit that the probability, which is equal to the relative frequency, depends on the generating conditions of the experiment, this means that these conditions are endowed with a tendency or disposition or propensity to produce sequences whose frequencies are equal to probabilities.

Indeed, each time the process is repeated in the same relevant conditions, the same set of propensities gives rise to the various outcomes according to their chances. As was argued in §5, in the long run each event will occur in proportion to its constant propensity. This intuitive explanation will be reinforced by the communication model given in §9.

Concerning propensity's measurement, this can be done subjectively in simple cases (as shown in §6) and objectively in some games of chance and scientific experiments where causal strengths can be objectively evaluated. In the urn paradigm, for example, the causal strengths of each ball to be drawn are the same, since the balls' sizes and weights are the same. The causal strength of each colour to be drawn is equal to the number of balls of the same colour. The propensity of each colour to be drawn, its chance, is equal, according to the relation provided in §6, to the proportion of balls of this colour to the total number of balls. The drawing mechanism represents the "noise events" enabling constant propensities to be realized. The long run verifies the values of the propensities.

To summarize, the single case is important for the following reasons:

- it provides an insight into the heart of the causal process (§4);

- it explains why stable frequencies occur in the long run through variable issues (§4 and §9);

- it provides a realistic explanation of the concept of chance (§5);

- it provides a model for the subjective evaluation of chance based on objective data (§6).

8 Are propensities probabilities?

In support of propensities being probabilities, Popper (1959, p.35) wrote that

> ... probabilities are conjectured or estimated statistical frequencies in long (actual or virtual) sequences. Yet by drawing attention to the fact that these sequences are defined by the manner in which their elements are generated—that is by the experimental conditions — we can show that we are bound to attribute our conjectured probabilities to these experimental conditions: we are bound to admit that they depend on these conditions, and that they may change with them. This modification of the frequency interpretation leads almost inevitably to the conjecture that probabilities are dispositional properties of these conditions — that is to say, propensities.

The reproach that has been raised against probabilities based upon propensities is that they seem to depend on the properties used to describe the event and not on the event itself. I think that this reproach is unfounded as probabilities of any kind always depend upon the relevant conditions, even if the dependence is implicit. Von Mises' "collective," renamed "reference class" by Reichenbach, is the frame in which the frequency theory provides an objective probability by relative frequencies. As I have argued in §7, the reference class is nothing but an incomplete set of relevant conditions, and the objective probability which is measured by the relative frequency is implicitly determined by them. An event cannot be conceived independently from its conditions. As Popper argued, the expression $P(x)$ is an abbreviated expression of $P(x/S)$ where S is a set of conditions.

The most striking objection against the probabilistic nature of propensities was formulated by Humphreys (1985, p.557) who argued that "the causal nature of propensities cannot be adequately represented by standard probability theory". But his contention is bolder: "... rather than this being construed as a problem for propensities, it is to be taken as a reason for rejecting the current theory of probability as the correct theory of chance".

I agree with both of these opinions. Causal probability based on propensities is different from standard probabilities, and the standard theory is not the only theory of probability.

Many of the problems created by this new concept are due to the trend initiated by Popper himself: to integrate his theory in the general frame of probability theory which already had a sound theoretical base. Being a former frequentist, he interpreted the probability of the singular event as "a property of the singular event itself, to be measured by a conjectured *poten-*

tial or virtual statistical frequency rather by an actual one," which tacitly supposes that this new probability has the same virtues as the standard one.

While the frequentist interpretation of Popper's probability is correct, this does not justify assigning to propensity, nor to the probability derived from it, *all* the properties prescribed by the standard probability theory. The causal origin of the concept of probability derived from propensity has specific features lacking in the standard theory, specifically, the following two:

- *causal productivity*, which differs from the classical *correlation*;

- *causal directedness*, which cannot satisfy *inverse conditional probability* issued from the standard axioms.

Any attempt to adapt propensity to standard theory is doomed because of these particularities. Instead of making desperate efforts to clothe new concepts in the straight jacket of old ones, we would do better to accept these new models of the world which deepen our knowledge and our control of it. We celebrate our inability to adapt relativity theory to Newtonian mechanics and non-Euclidian geometry to the Euclidian.

The above considerations imply that causal probability is a valuable concept which adds a new dimension to our way of modelling the world's indeterminacy. Causal probability has deep connections with reality which enable it to accommodate the objective probability of the singular event. The general axioms of causal probability must reflect its specific features. Valuable theoretical progress has already been made in this direction.

9 Causal transmission channels

Causal processes can be viewed as transmitting information between cause(s) and effect(s). In his 'at–at' theory of causal transmission, Wesley Salmon (1984, p.146) claimed that

> Processes that transmit their own structures are capable of transmitting marks, signals, information, energy, and causal influence. Such processes are the means by which causal influence is propagated in our world.

In the following, a causal process is modelled by a communication channel (Belis, 1973). The input signals to be transmitted represent the constant propensities, that is tendencies, dispositions, causal forces (a combination of causal influences) and the output signals represent the issues of the causal process. Hidden propensities become observable outcomes by means of the causal transmission.

Each of the propensities existing in a causal process is transmitted on its own channel in order to produce its outcome. During transmission, variable influences transform the constant propensities into variable, instantaneous ones (§4). These erratic influences, modelled as noise (with uniform distribution and zero mean), exist in every real transmission channel. The influence of the noise is decisive in producing the output. That is why at each transmission the output is indeterminate. In a real causal process the transformation of the propensity into an outcome happens when the propensity reaches the threshold of qualitative transformation (§4). This threshold corresponds, in our model, to a gate which enables the signals of the required level to pass through.

Because of the unavoidable influence of the noise, in technical transmission systems various methods have been initiated in order to recuperate the original signal from noise. One of the simplest consists in repeating a great number of times the transmission and averaging the outputs. The greater the number of transmissions, the more exact the value recuperated, as the average of the noise tends toward zero.

In order to adapt this communication model to the frequency method we have to focus not on the specific features of the output but on the number of times of its occurrence. That is why, each time the gate is open, a binary digit is sent to a counter which adds up the occurrences of the respective issue. It can be rigorously proved that after a great number of transmissions the values of the various propensities are recovered by their respective counters in numerical form. In other words, the input propensity, the "chance" (§5), is converted into a number at the output. This number is already an objective measure of the chance, but by taking its ratio to the total number of transmissions, we obtain the "relative frequency" that is the probability.

Probability theory postulates (and the theory is verified by experience) that the limiting value of the relative frequency in the long run tends toward the objective probability, but the connection with the reality of this result is quite mysterious. Why do things happen in this way? Why are probabilities recovered from these repetitions? What is the ontological justification of this result? Bernoulli's proof is purely algebraic.

The above model of the causal transmission provides a deeper insight into the process of transformation of the input propensities into output probabilities. This analogy explains why the input signals (propensities or chances) are recovered the output after a sufficient large number of transmissions. It puts in the fore the origin of the limiting value of the frequency method which is nothing other than a signal which has been extracted from noise by repetition. This signal provides useful information (the objective probability) but the same objective information could be obtained, *without repetition,* from the input propensities.

To evaluate the input propensities (*prior* probabilities) the observer has to know the causal structure of the process as well as the strengths of the respective capacities. As already discussed, this is a difficult task in the general case of complex causal structures, and even for simple ones. Laplace placed himself at the input, but in a very particular case: the urn with balls. As we have seen in §7 this amounts to a proportion of number of balls.

In contrast, von Mises chose the output end of the transmission channel and by repeating the transmission he succeeded in extracting the signal from noise, that is, to recover the true values of the *posterior* probabilities. The great advantage of this method is that it provides objective values of probability without any knowledge about the causal structure of the process. Its restrictive constraint is that it requires the same input signals (same causal strengths) which is not always possible. Moreover, there are cases when the transmission cannot be repeated.

In spite of its constraints and limitations, the frequency method represents the simplest way to recover the values of the propensities that is, the objective values of probabilities in the case of complex causal processes. Moreover, its merit is to provide a *proof* of the objective existence of probabilities. But the error consists in taking this proof as a definition of the random phenomenon.

The essential character of a random process lies in the special causal structure described above and not in the repetitive character of the process. Repetition only has the merit in putting these features to the fore, in *measuring* the probabilities, not to define them. Frequentists mistake the concept for one of its tests, the same error that Humeans make towards causality. In both cases the information brought by repetition of the process is due to the information comprised in a singular event which is transmitted many times.

10 Conclusions

The standard theory of probability was built with repetitive and mass phenomena in mind. One can wonder why the concept of chance and its theoretical grounds were not connected from the beginning to indeterminate singular events with which human beings often faced during their everyday lives.

The uncertain singular event was ignored for several reasons. Repetitive games of chance drew prompt attention because of the interest in profit.

Later, Laplacean determinism diverted effort away from understanding and finding models for objective indeterminacy, since this was supposed not to exist. Later still, long-run experiments put to the fore a mysterious entity which necessarily had to be objective, since it happened without human intervention, and objectivity was by then already associated with frequency probability.

A further reason for the lack of scientific interest in the objective chance of singular event was undoubtedly the lack of formal knowledge concerning the concept of causality. How to use such a hypothetical concept to explain and define objective probability? The distrust of the former would have rebound on the latter. The probability of the singular event was persistently seen as subjective in spite of the method used by human mind to evaluate it, which was, like other mental algorithms, a mirror image of the objective situation.

Advances occurred in focusing on the singular level in causality, together with an independent trend to recognize the existence of objective indeterminacy at the singular level. The scientific community began to look more closely at the objective chance of singular events, and to craft models of this new type of probability. The present work is part of that movement.

The objective probability of the singular event, in fact a "causal probability", differs in some respects from the standard probability built on mass and repetitive phenomena, which are not necessarily causal. Nevertheless, there is also an important body of similarities: when causal events are repetitive, they necessarily provide the same frequentist results as are seen with standard probabilities.

The bridge between the causal structure of random phenomena and the probabilistic measure of chance is the causal transmission model, which provides an ontological explanation of the frequency interpretation of probability. The model's principal merit is to illustrate the dependency of frequentist probability upon the causal structure of the single case via propensities.

Nature creates interactions and capacities endowed with causal strengths. The human mind organises what nature provides into propensities detected in the single case and proportions that represent probabilities. Proportions, the keystone of any probabilistic measure, arise within the causal structure of the singular indeterminate process, which reveals the causal origin of this new type of probability.

Acknowledgments

I am indebted to Paul Snow and Andreï Popescu-Belis for fruitful suggestions.

Marianne Belis

Supinfo, Paris, France.
marianne.belis@supinfo.com

BIBLIOGRAPHY

Belis, M. (1973). On the causal structure of random processes. In R. Bogdan and I. Niiniluoto, editors, *Logic Language and Probability*, pages 65–77. D.Reidel, Dordrecht, Holland.

Belis, M. (1995). Causalité, propension, probabilité. *Intellectica – Revue de l'Association pour la Recherche Cognitive*, pages 199–231.

Belis, M. and Snow, P. (1998). *An Intuitive Data Structure for the Representation and Explanation of Belief and Evidentiary Support.* IPMU, Paris.

Belis, M. and Snow, P. (2002). *Comment cerner le hazard.* Supinfo Press, Paris.

Bunge, M. (1963). *Causality. The Place of the Causal Principle in Modern Science.* Meridian Books, Cleveland and New York.

Cartwright, N. (1983). Causal laws and effective strategies. In *How the laws of physics lie.* Clarendon Press, Oxford.

Cartwright, N. (1989). *Nature's capacities and their measurement.* Clarendon Press, Oxford.

Earman, J. (1986). *A primer on determinism.* Kluwer Academic, The Netherlands.

Fetzer, J. (1988). Probabilistic metaphysics. In *Probability and causality*, pages 109–132. D. Reidel, Dordrecht.

Gillies, D. (2000). Varieties of propensity. *British Journal for Philsophy of Science*, 51:807–835.

Nakashima, H. M. and Osawa, I. (1997). Causality as a key to the frame problem. *Artificial Intelligence*, 91:33–50.

Hume, D. (1739). *A treatise of human nature.* Penguin Classics, reprinted 1948.

Humphreys, P. (1985). Why propensities cannot be probabilities. *Philosophical Review*, 94:557–570.

Popper, K. (1959). The propensity interpretation of probability. *Birtish Journal for Philosophy of Science*, 10:25–42.

Salmon, W. C. (1984). *Scientific explanation and the causal structure of the world.* Princeton University Press, Princeton.

Settle, T. (1973). Are some propensities probabilities? In R.Bogdan and I.Niiniluoto, editors, *Logic Language and Probability*, pages 115–120. D.Reidel, Dordrecht, Holland.

Weatherford, R. (1982). *Philosophical Foundations of Probability.* Routledge and Kegan Paul, London.

Causality and the axiomatic probability calculus

ANDREA L'EPISCOPO

ABSTRACT. The paper starts with a brief analysis of the axiomatic probability calculus and claims that every theory of probability and of causality ought to place constraints on its objects, thus dealing with a restricted version of the corresponding intuitive concept. It then proceeds drawing a distinction between the intuitive and the physical notions of causality on the one hand, and between the conceptual and the empirical analyses of the concept of causality on the other. Phil Dowe's CQ theory of causation and probabilistic theories of causation are regarded as issues of empirical and conceptual analysis, respectively. Relying on the distinction between the two aforementioned kinds of analysis, it is shown that some counterexamples to those theories are misdirected. The paper ends by arguing for causal pluralism.

1 Introduction

The aim of this paper is to show how a distinction between empirical and conceptual analyses of causality, and a distinction between physical and intuitive notions of causality, can be of great help in individuating the actual tasks of different theories of causation, and, consequently, in evaluating, case by case, whether or not counterexamples to such theories are really problematic. We will be focusing on Phil Dowe's CQ theory and on probabilistic theories of causation. The former is very interesting because of its empirical coherence, and because of the cases Dowe acknowledges as problematic for it; the latter because of the additional (with respect to other theories of causation) constraints they pose on their object, constraints that come directly from the axiomatic nature of probability calculus. Given that we want to argue that a theory must be analyzed and, eventually, challenged only within its empirical or conceptual framework, we believe that the choice of the CQ theory and of the probabilistic theories of causation is an eminently appropriate for the development of our arguments.

In the first section of the paper, we deal with some of the fundamental features of Kolmogorov's axiomatization of probability calculus and with their repercussions on the intuitive notion of probability. Kolmogorov's calculus represents a constrained version of the intuitive notion of probability: it could not be otherwise, given its abstract nature. This consideration will lead us to the claim that the fact that a theory can only deal with its own objects cannot be seen as a drawback to the theory itself; on the contrary, the constraints are the positive limits which make the theory work. In the second section, we briefly explore the intuitive and the physical notions of causality and, following Dowe, we pose a distinction between the conceptual and the empirical analyses of causality. Given this last distinction, we will identify probability calculus with a conceptual analysis of a probabilistic constrained version of the intuitive notion of causality. In the third section, we review Dowe's tenets, showing how his CQ theory is a perfectly coherent empirical analysis of (physical) causality; how causation* and misconnections are not really problematic cases for it, and, in much the same way, how Dowe's decay case is not a sound counterexample to probabilistic theories of causation: intuitions about causality cannot be the topics of empirical analyses such as the CQ theory, neither can physical contingencies be the topics of conceptual analyses, such as the probabilistic theories of causation. Finally, in our conclusions, we briefly summarize our main claims and we draw our general conclusions in favour of causal pluralism: different kinds of analysis analyze different aspects of causality.

2 Kolmogorov's axiomatization of probability

In *Foundations of probability*, Kolmogorov is explicit that he is proposing a mathematical theory (Kolmogorov, 1950, p.1):

> The theory of probability, as a mathematical discipline, can and should be developed from axioms in exactly the same way as Geometry and Algebra.

Probability calculus is a formal theory, and exhibits important features which belong to formal theories:

- its starting points are undefined primitive terms and the statements in which these terms are, i.e., the axioms;

- the symbols in the theory do not stand for objects, they are simply signs to be manipulated according to theory-specific rules;[1]

---

[1] "...after we have defined the elements to be studied and their basic relations, and have stated the axioms by which these relations are to be governed, all further exposition must be based exclusively on these axioms, independent of the usual concrete meaning of these elements and their relations". (Kolmogorov, 1950, p.1)

- more generally, "...in the formal approach there seems to be no appeal to intuition, because definitions, axioms and rules of transformation are clearly laid out from the beginning, and the proof produced appeals only to the meaning of the axioms, definitions and rules of transformation" (Oliveri, 2005, p.22).

The undefined primitive terms of a formal theory obviously have to be chosen, and various considerations can lead to this choice. The motivation for Kolmogorov's choice is twofold: on the one hand, there are "...the analogies between measure of a set and probability of an event..." (Kolmogorov, 1950, p.V); on the other hand, there is a pragmatic stance, i.e., "...to achieve the utmost simplicity both in the system of axioms and in the further development of the theory..." (Kolmogorov, 1950, pp. 1-2). This motivation has nothing to do with any principled connection between the abstract[2] theory of probability and the intuitive or the empirical notion of probability, as Kolmogorov himself implicitly claims, by a comparative reference to Von Mises' and Bernstein's approaches to probability:

> There are other postulational systems of the theory of probability, particularly those in which the concept of probability is not treated as one of the basic concepts, but is itself expressed by means of other concepts. However, in that case, the aim is different, namely, to tie up as closely as possible the mathematical theory with the empirical development of the theory of probability. (Kolmogorov, 1950, p.2)

The elements of probability calculus are sets and functions, because probability calculus is "...a special application of the general theory of additive set functions" (Kolmogorov, 1950, p.8). Sets of something, of course, but the fact that they are sets of elementary events "...is of no importance in the purely mathematical development of the theory of probability" (Kolmogorov, 1950, p.1).

So far we have quoted Kolmogorov himself, almost without any need for comments. From now on, we will consider how his choice concerning basic elements and axioms, while being able to give substance to the axiomatization of probability calculus, places a significant restriction on the concept of probability. We see this restriction at work when we come to the relationship of probability calculus to experimental data.

---

[2] It is worth noting that the identification between "axiomatic" and "abstract" is stated by Kolmogorov himself. See Kolmogorov (1950, p.1).

In posing this relationship, Kolmogorov relies on Von Mises' frequentism[3], and he often refers to the concept of "practical possibility". In order to define a field of probability, in fact, first of all, we have to form the set E, which includes "...all the variants which we regard a priori as possible"; these variants are the different ways in which events can occur, and the events we consider are those which might result from repeatable conditions. Thus the definition of a probability field, in empirical applications, presupposes qualitative judgments about the a priori possibility of events and their variants, given the repeatable conditions in question. In other words, when we come to experimental data, the formal theory of probability turns out to rely on an informal notion, such as that of "conditional" practical possibility.

The relationships between probability and possibility are not investigated by Kolmogorov, but the remarks he makes after the empirical deduction of his axioms tell us something interesting about this topic. Kolmogorov, following Von Mises, claims that the probability of an event, given a large number of repetitions of the conditions, will differ only slightly from the ratio between the number of times the event occurs and the total number of repetitions; at the same time, however, he clearly acknowledges that this is not always the case. The approximate equality between $P(A)$ and $\frac{m}{n}$[4] is in fact a *practical* certainty, and it might be possible that it is not the case "...that in a very large number of series of n tests each, in *each* the ratio $\frac{m}{n}$ will differ only slightly from $P(A)$" (See Kolmogorov (1950, p.5), italics in the text). In other words, it might be possible to encounter frequencies which do not approximate the numerical value of probability assignments. The tension is between probability theory and frequencies: probability theory deals with an abstract, logical notion of possibility, whereas frequencies are concerned with contingent possibilities. Logically, possibility is a redundant attribute of all which is not a contradiction; on the contrary, contingent possibilities are a seemingly irreducible characteristic of actual events, and they render even the task of identifying sets of possible events very hard, even given determined conditions.

The same tension arises in the distinction Kolmogorov makes between zero-probability events and impossible events:

> To an impossible event (an empty set) corresponds, in accordance with our axioms, the probability $P(0) = 0$, but the con-

---

[3] "In establishing the premises necessary for the applicability of the theory of probability to the world of actual events, the author has used, in large measure, the work of R. v. Mises..." (Kolmogorov, 1950, p.3, n.4).

[4] With $P(A)$ being the probability of the event A, m the number of times A has occurred, and n the number of times conditions have been repeated.

verse is not true: $P(A) = 0$ does not imply the impossibility of
A. (Kolmogorov, 1950, p.5)

Zero-probability events are then practically impossible only a posteriori, otherwise they would not enter our probability field. This consideration matters with respect to Kolmogorov's definition of conditional probability, which requires that the probability value of the event on which we condition the probability of another event has to be greater than zero. Such a requirement is tenable only with respect to the logical notion of impossibility; on the other hand, when we are dealing with experimental data, the probability of an event equals zero only a posteriori, as we have seen, and the fact that the repetitions of the conditions have shown such an event to be practically impossible is presumably relevant to other events in the field of probability. We should be able to condition probabilities of events in zero-probability events of this kind, but (by the definition of conditional probability) we cannot. However, it is worth noting that Kolmogorov himself does not link the definition of conditional probability to the empirical application of probability calculus, and this could be read as the sign that such a definition only works properly in the ideal domain defined by the axioms of the abstract theory of probability. The abstract theory, dealing with sets and their functions, and not with events and practical possibilities, can in fact exploit the fully logical relationships which hold good, by definition, between the basic elements of probability calculus.

The sharp distinction between practical and logical possibility, and therefore between the abstract theory of probability and its application to experimental data, in our opinion gives substance to the following claim: by the application of the probability calculus to experimental data, we can obtain degrees of practical certainty, which are the probability values, starting from judgments about the practical possibility of events. This is not a drawback of probability calculus; it is just a consequence of probability calculus being an axiomatized theory, and so of the constraints Kolmogorov had to place on the concept of probability to make its theory work. Probability calculus is an abstract theory, and it works properly only when it deals with the abstract basic elements on which it has been built. This claim can be generalised to every theory, including non-axiomatic theories, as we will show in what follows.

3 The intuitive notion of causality

The intuitive notion of causality is one of our most important concepts. Nevertheless, it is not immutable, and our view of it has gone through many changes as the sciences, particularly physics, have developed over the centuries, and as new phenomena have been discovered. Nowadays, there seem to be sound reasons for distinguishing between an intuitive notion of causality and a physical notion of causality. [5]

It is very hard to define the intuitive notion of causality; in its rough version, it might be defined by the conjunction of the counterfactual and the manipulability definitions of causation, and it can be characterized by some general claims:[6]

- causal statements express qualitative judgments about what links actual facts;
- a cause is supposed to be necessary for the effect to occur;
- a cause is not supposed to be sufficient for the effect to occur, because it is characteristic of the intuitive notion of causality to informally distinguish between causes and conditions;
- the intuitive notion of causality implicitly presupposes determinism;
- causality is an undefined primitive term;
- causality is supposed to be part of the actual world or, at least, to be an universal way of interpreting the actual world.

On the contrary, the physical notion of causality[7] is related to scientific work on physical phenomena. It is roughly characterized by a quantitative approach: when dealing with physical phenomena, we try to individuate a quantifiable or, at least, empirically meaningful feature, which could be the reason why we see some processes and interactions as causal. For this reason, the physical notion of causality has to be specified more precisely than the intuitive notion, and in so doing the characterization of causality often tends to diverge from intuition. The physical notion of causality does not treat causal relations as a semantical matter; causal relations are rather seen as a matter of fact. The relevant features of the physical notion of causality, specular to those listed above for the intuitive notion, are the following:

---

[5] The distinctions between abstract and practical possibility, on the one hand, and between the intuitive and the physical notions of causality, on the other, are not directly linked. The connection between the two is based on another distinction, the one between conceptual and empirical analysis, which will be discussed shortly (for ease of exposition).

[6] Some of these claims are so widely accepted that they have became trivial; others, while being in our view very reasonable, are more controversial. Our present goal is not to formulate arguments in favour or against others' theses, but to roughly characterize the intuitive notion of causality, and thus to affirm the impossibility of formalizing it exhaustively.

[7] By "physical notion of causality" we do not refer to any existing theory of causality, as will become clear in what follows.

- causal statements express empirical or quantitative information about (causal) physical processes and interactions;

- the cause is necessary for the effect to occur;

- a cause is not sufficient for the effect to occur, but the distinction between causes and conditions has to be drawn much more sharply, or it is partly obviated by keeping the conditions fixed;

- the physical notion of causality does not presuppose determinism; quite to the contrary, it must be able to deal with indeterministic phenomena;

- causality being an undefined primitive term is a problematic topic;

- causality has to be part of the physical world; it cannot be simply an interpretative tool, otherwise the physical notion of causality would turn out to be empirically meaningless.

Comparing the lists of features of intuitive and physical notions of causality, we see that the main differences between the two notions come from their epistemological status. The intuitive notion of causality can be seen as an universal interpretative tool by means of which we confront reality. It gives substance to our representation of the relationships between events, and it propels most of our actions within events. The physical notion of causality, even though it has arisen as a highly specialized version of the intuitive one, has followed the historical development of the sciences, particularly of physics, and is now a different notion, which deals with different "objects", and in a context which is all but intuitive, because the objects and the laws (in one word, the world) of contemporary physics are not part of our experience, at least not directly (is not our world).[8]

In his *Physical causation*, Phil Dowe does not claim that there are two notions[9] of causality, even if he clearly distinguishes between conceptual and empirical analysis of causality.

---

[8] The paradoxical nature of many quantum mechanics laws and experiments is underlined by many authors in many works such as, for example, Pearl (2000) or, with regard to the notion of explanation in quantum mechanics, Salmon (1992). Finally it is worth noting that, even among physicists, many authors have acknowledged the counterintuitive character of quantum mechanics: in Feynman (1996), for example, it is explicitly claimed that no one understands it.

[9] Here, we discuss only two notions of causality, but, in our view, for a notion of causality to gain autonomy it is only required that its level of specialization marks significative differences compared to the existing ones. So it is clear that we have in mind a many-notions approach to causality.

Before going on to see how conceptual and empirical analyses of causality are outlined by Dowe, we would like to underline the small but meaningful difference between Dowe's and our own ideas on intuitive and physical causality: in Dowe's thinking, the concept of causality is unitary, and the difference between its conceptual and empirical aspects only consists in a different analysis; in our view, on the other hand, the distinction between intuitive and physical causality arises from the different approaches that the two notions[10] have to reality, from the different realities with which the two notions deal, and from the role intuition plays in each of them. In an extreme synthesis, in our view, the physical notion of causality, while being obviously linked to the intuitive one, has reached such a level of specialization as to become autonomous.

Coming back to Dowe's distinction between conceptual and empirical analysis of causality, he claims:

> Conceptual analysis is not just dictionary writing. It is concerned to spell out the logical consequences and to propose a plausible and illuminating explication of the concept. Here, logical coherence and philosophical plausibility will also count. The analysis is a priori, and if true, will be necessarily true. (Dowe, 2000, p.2)

> ..empirical analysis seeks to establish what causality in fact is in the actual world. Empirical analysis aims to map the objective world, not our concepts. Such an analysis can only proceed a posteriori. (Dowe, 2000, p.3).[11]

It is worth noting that, a few pages later, Dowe proposes a sort of convergence between the conceptual and the empirical analysis of causality, to the extent that the empirical analysis is also a conceptual analysis:

> ...any empirical analysis will still be a kind of conceptual analysis, for example, of the concept implicit in scientific theories. [...] So I am happy to think of the task of empirical analysis as a conceptual analysis of a concept inherent in scientific theories. (Dowe, 2000, p.11)

---

[10] In this respect, it is worth noting that causality, as it is conceived in everyday life, very often turns out to be unusable in the world of physics. At the same time, the physical notion of causality is often too far from everyday life's intuitions, and so it is also unusable if applied to everyday life.

[11] A relevant remark is made by Dowe on the next page: "Wesley Salmon (1984) seeks merely to articulate what causation is as a contingent fact, while others such as Bigelow and Pargetter (1990) try to establish what causation is as an a posteriori necessity. We will use the term 'empirical analysis' to cover the first of these options".

We do not agree with this view because the empirical analysis of causality, as outlined by Dowe himself, seems to have very little to do with concepts, especially in virtue of the fact that it proceeds only a posteriori: "a conceptual analysis of a concept inherent in scientific theories" cannot be, in our view by definition, an empirical analysis; it is a conceptual analysis. The difference between the two kinds of analysis, as characterized by Dowe himself, does not hinge upon their objects, it is methodological and epistemic.

Given the distinction we have made between the intuitive and the physical notions of causality, we want to argue that conceptual analysis, on the one hand, applies to the former because, by intuition, we a priori pose causality, and then we represent events on the basis of it: a priori analysis, we think, is better suited to deal with a priori notions. Empirical analysis, on the other hand, applies to the physical notion of causality, because the latter depends on what data tells us about the single process in question, so it can be seen as an a posteriori notion, which is better analyzed by the sort of test that aims to isolate and, when possible, to quantify some common features of actual events (of which the empirical analysis consists). Returning to the axiomatic probability calculus, given that it proceeds a priori, exactly as as conceptual analysis does, we think it is a conceptual analysis of a probabilistic constrained version of the intuitive notion of causality. Probability calculus can enter the empirical analysis of the physical notion of causality, as it actually does, but in this case it is merely a tool for the analysis (just like many physical theories, which have nothing to do with causality), it is not the analysis itself. Probability calculus cannot be an empirical analysis of the physical notion of causality because, as previously stated, it can only proceed a priori.

Now that we have explicitly distinguished the two notions of causality and the two kinds of analysis, and characterized probability on the basis of these distinctions, we can go on to Dowe's theses on causality and to his criticisms of probabilistic theories of causation.

4 Dowe's theories of causality

The declared goal in Dowe (2000) is to formulate an empirical analysis of causality. The empirical analysis at which Dowe arrives is the well known Conserved Quantity (CQ) theory of causality:

> **CQ1.** A *causal process* is a world line of an object that possesses a conserved quantity.
>
> **CQ2.** A *causal interaction* is an intersection of world lines that involves exchange of a conserved quantity.[12]

---
[12]See Dowe (2000, p.90), italics in the text. A few lines below: "A world line is the

The CQ theory of causality is empirical, contingent with regard to the identity of causal processes,[13] and particularist.[14] It provides a clear and scientifically meaningful characterization of what it is, for a physical process to be causal, and this characterization is based essentially only upon scientific theories and their "objects". An example of the empirical coherence of the CQ theory of causality comes from Dowe's reply to an objection from Salmon: the objection regards the assumption of identity through time as primitive in Dowe's theory.[15] The reply consists in leaving the question of identity through time open, assuming identity only as a placeholder. We would further develop this reply, by claiming that if identity through time matters with regard to the notion of process, and if such a notion relies on current scientific theories, then the task of the analysis of identity through time has to be placed on the shoulders of scientific theories themselves. Otherwise, given that it relies on current scientific theories in the same way, why should we accept the definition of conserved quantities? If scientific theories can tell us what conserved quantities are, then they can, at least practically, give substance to the notion of identity through time, and an empirical analysis of causality can safely incorporate this notion.

Dowe's theory is also empirically coherent in not taking a position with regard to the direction of causality,[16] and in stressing that it is only plausible to suppose that conserved quantities might be the quantities which are typically associated with causality. Moreover, the CQ theory of causality is noncommittal with regard to probabilistic causality. But this theory, adhering to Dowe himself, runs into serious problems when we come to what he calls causation*. Causation* is causation by prevention and or by omission: A causes not-B, not-A causes B, respectively. Obviously, it is not possible to link by causal processes and interactions an event to a not-event, and this is the reason why Dowe develops a counterfactual theory of causation*, by means of which prevention and omission should be understood as counter-

---

collection of points on a spacetime (Minkowski) diagram that represents the history of an object". Next page: "A conserved quantity is any quantity that is governed by a conservation law, and current scientific theory is our best guide as to what these are".

[13] "In calling the analysis a contingent identity, we mean that it is contingent on the laws of nature and perhaps even on boundary conditions". (Dowe, 2000, p.95)

[14] "...a particular causal process is not analysed in terms of laws about that type of processes; rather, that a type of process is causal is a matter of generalisation over the particular instantiations of that process-type. The particular is basic." (Dowe, 2000, p.96)

[15] "...Salmon's objection to taking identity over time as primitive in a theory of causal processes appears to be that it violates the empiricist's stricture that one should not invoke empirically inaccessible elements as unanalysed or primitive in a philosophical theory." (Dowe, 2000, p.112)

[16] Apart from a brief suggestion: "...the cause is the incoming (earlier) processes, and the effect is the outgoing (later) processes". (Dowe, 2000, p.111)

factual truths about genuine causation. In our view, such a counterfactual theory of causation* cannot be seen as an extension of the CQ theory; it is in fact a conceptual analysis of the intuitive notion of causality, and indeed Dowe himself seems to be aware of this point in more than one passage. First of all, he presents his theory as "...a solution that is available not only for the Conserved Quantity theory, but for most theories of causation";[17] secondly, the distinction between genuine causation and causation* depends entirely on the so-called "intuition of difference";[18] thirdly, "...the similarity between causation and causation* is explained by the unity between the two concepts: one is a counterfactual truth about the other" (Dowe, 2000, p.145). If we accept that the counterfactual analysis of causation* is a conceptual analysis of the intuitive notion of causality, we cannot avoid the following question: can the counterfactual analysis of causation* be part of the CQ theory? As we have claimed before, and according to Dowe himself, we see conceptual and empirical analyses as different, directed at different objects, and with different aims; so the counterfactual analysis of causation* cannot be a constitutive part of the CQ theory. We also need to consider whether causation* is really problematic for the CQ theory of causality. Any empirical analysis of causality deals only with the physical notion of causality, and the CQ theory is not an exception. Are there, in the physical world, objects such as not-A? And, even if we could find such objects, is a physical notion of causation* at least conceivable? If no set of causal processes and interactions link causes* to their effects*, then simply there cannot be anything similar to physical causation*, and so we cannot see where the problem lies with regard to causation* in an empirical analysis of the physical notion of causality, such as the CQ theory. In the framework of the CQ theory, causation* simply should not be seen as causation, and we cannot see how this might represent a drawback to this theory; just as in the case of axiomatic probability calculus, this is merely a coherent consequence of the initial choices regarding the kind of objects, analysis, and aims proper of the theory. After all, if Dowe also wishes to extend the CQ theory to a conceptual analysis of the intuitive notion of causality, then prevention and omission will not be the only problems he has to face.

---

[17] Dowe (2000, p.123). See Schaffer (2001, p.811): "Dowe describes his account of causation* as a 'cross-platform solution' in that virtually any account of causation can be plugged in. But Dowe's can't. Since Dowe has only offered a contingent specification of how causation operates in the actual world, he has yet to say how causation operates in those nonactual worlds that his counterfactuals take us to (here a conceptual analysis is needed)".

[18] "I claim that we can recognise, on reflection, that certain cases of prevention or omission, ..., are not really cases of genuine causation. Call this the 'intuition of difference'. We also feel, however, that the 'mistake' of treating them as if they were causation doesn't matter for practical purposes." (Dowe, 2000, p.125)

Referring to Dowe's theory, Jon Williamson claims:

> The main limitation of this approach is its rather narrow applicability: most of our causal assertions are apparently unrelated to the physics of conserved quantities. While it may be possible that physical processes such as those along which quantities are conserved could suggest causal links to physicists, such processes are altogether too low-level to suggest causal relationships in economics, for instance. (Williamson, 2005, p.111)

This claim is a strong objection to the whole approach of the CQ theory, but in our view it is really cogent only if referring to the extended version of the CQ theory. A conceptual analysis of the intuitive notion of causality, or the hybrid one at which Dowe arrives by his counterfactual theory of causation*, must deal with objections such as the one made by Williamson. On the contrary, in our view, it is not surprising that an empirical analysis of the physical notion of causality does not justify intuitions for causality in economics: it is not the right kind of analysis, it cannot deal with these intuitions, but this does not mean that it is wrong or useless. An empirical analysis cannot be falsified by difficulties which arise from intuitive notions, because it does not apply to them.

The so-called problem of misconnections seems to be more serious for the CQ theory of causality. Misconnections arise when two events are linked by a set of causal processes and interactions, but "we would not call [them] cause and effect".[19] The CQ theory now seems to be too comprehensive, and is thus unable to exclude cases of "fake" causation, but again, if we look more closely at what matters, this is the case only with regard to the intuitive notion of causality. From a strictly empirical viewpoint, in fact, we only have two events linked by causal processes and interactions, and so, by definition, we have a cause and an effect. Mixed processes (the solution proposed by Dowe himself) involve both genuine causation and causation*, so causation by misconnections is still a kind of causation (as the CQ theory tells us). Obviously, it would be desirable to distinguish between genuine causation and causation by misconnections, but again this cannot be done by an empirical analysis of the physical notion of causality; this is the actual task of a conceptual analysis of the intuitive notion of causality. In Cartwright's example, a plant sprayed with a defoliant survives, and it is possible to link the spraying and the surviving by causal processes and interactions.

---

[19] "One such case is Cartwright's sprayed plant: a healthy plant is sprayed with defoliant that kills nine out ten plants, but this particular plant survives (Cartwright, 1983, chap.1). We can provide a set of causal processes and interactions linking the spraying and the surviving, yet spraying does not cause the plant's survival". (Dowe, 2000, p.148)

In this case, given Dowe's account, we have an event (the spraying) which initiates two processes, one causing and one preventing another event (the death of the plant), and improbably, in the case in question, the second one is successful. If we recall some of the more important features of the CQ theory, we cannot see why such a case is so problematic: the CQ theory is empirical, contingent, and, more to the point, particularist; so, in its framework, processes cannot be empirically probable or improbable,[20] they are merely processes; they may or cannot be causal, but only case by case. In our view then, a coherently particularist account of causation encounters no problems in dealing with causation by misconnections.

We think we have shown that causation* and misconnections cannot be the right counterexamples to the CQ theory of causality; we have shown this by relying on the distinctions made in the previous section. Now we will continue along the same path, to show how well known counterexamples to probabilistic theories of causation are also misdirected. In doing this, we will also rely on some peculiar features of the axiomatic probability calculus, features that we outlined in the first section.

Dowe's criticisms of probabilistic theories of causation are closely connected to the problem of misconnections. First of all, Dowe uses examples to argue that the existence of a probabilistic relationship between two events "is not a necessary condition for singular causation between those events". Secondly, he refers to well-known examples of the so-called chance-lowering causation, which can be seen as a special case of the problem of misconnections.

On the basis of what we have said about Kolmogorov's axiomatization of probability calculus, in our view probabilistic relationships can only be of two kinds, depending on what they link: they can be logical relations between abstract sets, or de facto quantitative relations between actual events, if probability calculus is applied to empirical data. In the first case, probabilistic relations are logical consequences of the definition of abstract sets, so they hold a priori; in the second case, they are the result of empirical tests over actual events, and so they obviously only hold a posteriori. It seems clear that, with regard to the problem of misconnections, we have to look at

---

[20] "I expect that the probabilistic element in the theory must enter as a propensity". (Dowe, 2000, p.113). "Perhaps we need to analyse chance in terms of causation. Although this theme is beyond the scope of this book, I think this is exactly how we should account for chance" (Dowe, 2000, p.168). Propensities are intrinsically causal; moreover, Dowe goes on, in the same passage, claiming that "...propensities should be regarded as referring to the operation of objective, indeterministic causal processes and interactions, such that the propensity takes values between 0 and 1 only where there is genuine indeterminism." Chances, more explicitly, have to be analysed in terms of causality. Dowe has no empirical notion of probability in mind, because probability has to be analyzed in terms of causality; for this reason, probability has no role in the formulation of the CQ theory.

the second kind of probabilistic relationships, and it is all but clear how an a posteriori contingency could provide a necessary condition for single-case causation. Moreover, even if we do not want to enter the long-debated question regarding the interpretations of probability calculus, it is worth noting that we still have no widely accepted objective interpretation of single-case probability.[21] Without such an interpretation, it is very hard to see how we could have a probabilistic condition necessary for single-case causation. It is true, then, that the existence of a probabilistic relationship between two events "is not a necessary condition for singular causation between those events", but we think that such an objection should be directed against any expectation of probabilistic theories of causality being well suited even for singular causation, and not against the actual framework of those theories.

With regard to Dowe's critical discussion of chance-lowering causation, we would like to develop two arguments. The first one deals with the decay example by means of which Dowe rejects Lewis' and Menzies' chains as successful proposals (Dowe, 2000, p.150). If we look at this example, and at the conclusions Dowe draws from it, in the light of the distinction between empirical and conceptual analyses of causality, both the example and the conclusions turn out to be problematic, or even completely misdirected. First of all, given that Dowe himself proposes the distinction between empirical and conceptual analyses, do probabilistic theories of causation belong to the former or to the latter kind of analysis? Dowe does not give us any answer, and, because of what we said earlier, we tend not to view them as empirical analysis. Considering probabilistic theories of causation as a particular kind of conceptual analysis of causality, we should have a clear and general understanding of what it means, from a conceptual point of view, to say that the production of the atom C is the cause of the production of the atom E. The process, in the given example, can be seen as causal only a posteriori, so no a priori notion of causality plays any role: we cannot see here what might be the object of a proper conceptual analysis, and we do not think we can consider probabilistic theories of causation as empirical analysis (nor does Dowe give us any reason to think that we should have to do so). For these reasons, we do not think that the decay example is a proper counterexample to the probabilistic theories of causation.

The second argument focuses on the Positive Statistical Relevance (PSR) theory of causation. According to this theory, causes, by definition, must increase the chances of their effects, so cases of chance-lowering causation

---

[21] Chances and propensities, which are nowadays the prominent candidates as objective interpretations for single-case probability, are in fact subject to a great deal of criticism. Moreover, many of these criticisms emphasize the causal character of such interpretations; if they are right, then these interpretations providing a necessary condition for single-case causation would turn out as circular.

appear to invalidate the theory itself. However, recalling what was said in the first section about axiomatic probability calculus, the fact that a theory rules out a particular feature of the intuitive concept it tries to formalize is not a good reason for rejecting it. The PSR theory of causation rules out chance-lowering causation by the actual definition of cause, just as the axiomatic probability calculus rules out conditional probabilities with zero-probability antecedents by the actual definition of conditional probability. We could say that the PSR theory, being a conceptual analysis of causation, should include chance-lowering causation in its definition of cause, because chance-lowering causation is at least conceivable, and that the axiomatic probability calculus should do the same with conditional probabilities with zero-probability antecedents, for the same reason. This is a reasonable desideratum, but it cannot be a good reason for disputing the validity of such theories: theories must exclude, by definition or otherwise, some features of their objects, or they would tend to coincide with the intuitive notions they try to formalize, losing a great part of their epistemic value.

5 Conclusion

We have seen how the axiomatization of probability calculus has placed severe constraints on what can be said to be probability and conditional probability, and how these constraints matter when we come to the application of calculus to empirical data. We made a distinction between the intuitive and the physical notion of causality, on the basis of the distinct features which they exhibit. Coming to Dowe's theories of causality, we recalled the distinction he makes between empirical and conceptual analysis of causality, and then we examined, on the basis of this distinction, not only his proposals, but also some of the most popular criticisms of probabilistic theories of causality. We believe we have collected evidence and arguments in favour of the following theses: any formalization of an intuitive concept is representative of merely a restricted version of it; there can be no essential convergence between the conceptual and the empirical analysis of causality. Bearing in mind the preceding two points, the cases which Dowe himself presents as problematic for his CQ theory turn out not to be real problems, and some of the main criticisms of probabilistic theories of causation turn out to be misdirected.

With regard to the formalization of causality, we think that whoever agrees with what has been claimed thus far should accept that we must choose between two alternatives: 1) to agree to deal with a restricted version of the intuitive notion of causality by a formal theory which performs well in restricted domains; 2) to take a pluralist stance, dictated by pragmatic considerations, and to make reference, case by case, to one of the theories which formalizes one of the several features of the intuitive notion of causality, constantly bearing in mind the whole intuitive notion itself as a heuristic one, for building models representing actual events. We prefer to the second alternative, because it preserves the heuristic value of the intuitive notion of causality, without losing so much in terms of the possibility of a formal representation of many of its features.

Acknowledgments

My gratitude goes to Jon Williamson, Federica Russo, Donald Gillies, Margherita Benzi and Gianluigi Oliveri for their support, their attention, and fruitful discussions. Thank to Giovanni Camardi, who first introduced me to these topics. Thank to two anonymous referees for their constructive comments and to Andris Ozols, who improved the English.

Andrea L'Episcopo
University of Catania, Italy.
andrea.lepiscopo@gmail.com

BIBLIOGRAPHY

Bigelow, J. and Pargetter, R. (1990). *Science and Necessity*. Cambridge University Press, Cambridge.
Cartwright, N. (1983). *How the Laws of Physics Lie*. Clarendon Press, Oxford.
Dowe, P. (2000). *Physical Causation*. Cambridge University Press, New York.
Feynman, R. (1996). *La legge fisica*. Boringhieri, Torino.
Kolmogorov, A. N. (1950). *Foundations of Probability*. Chelsea Publishing Company, New York.
Oliveri, G. (2005). Do we really need axioms in mathematics? In Cellucci, C. and Gillies, D., editors, *Mathematical Reasoning and Heuristics*, pages 119–135. King's College Publications, London.
Pearl, J. (2000). *Causality. Models, Reasoning, and Inference*. Cambridge University Press, Los Angeles.
Salmon, W. (1984). *Scientific Explanation and the Causal Structure of the World*. Princeton University Press, Princeton.
Salmon, W. (1992). *Quarant'anni di spiegazione scientifica*. Muzzio Editore, Padova.
Schaffer, J. (2001). Phil Dowe. Physical causation. *British Journal for the Philosophy of Science*, 52:809–813.
Williamson, J. (2005). *Bayesian Nets and Causality: Philosophical and Computational Foundations*. Oxford University Press, Oxford.

Two probabilities of dysfunction and two kinds of chance

FRANÇOISE LONGY

ABSTRACT. I shall investigate the nature of the probability of a manufactured item's being malfunctioning upon leaving its production plant. My first claim is that the sentence used to express the probability that an artefact will malfunction can be given two quite different readings in which the probability is supposed to refer to chances rooted in reality. In the first reading, the probability concerns the entity qua physical object, in the second qua instance of an artefact type. The probability corresponding to the first reading, the DYSF-PHYS probability, raises no major problem. The one of the second reading, the DYSF-ART probability, however, is a good deal more puzzling. This paper aims to analyse this last sort of probability. I intend to demonstrate that the DYSF-ART probability is often the one referred to when considering the possibility that artefacts might not work properly, and that this probability is as objective and as rooted in reality as is the DYSF-PHYS one. After presenting and explaining both probabilities, I elucidate the problem it raises to have two sorts of probability of dysfunction instead of one and explain why a frequentist interpretation of both does not solve it. I then investigate the causal basis of the chances measured by the DYSF-ART probability and explain how these chances may show no dependence on the physical make-ups of the artefacts. In the last part of the paper, I ask how we should interpret and qualify probabilities like the DYSF-ART one, and I make some connexions with biological cases to demonstrate how this investigation might prove helpful to the current discussion of probabilities in evolutionary biology.

1 The two probabilities of dysfunction

Consider an unused light bulb. What, we may ask, is the probability that the light bulb will burn out in, say, less than five minutes of continuous standard usage? There are two ways of understanding this question. The first is: what is the probability that this object, which has this particular physical structure will burn out in less than five minutes of standard usage? The second is: what is the probability that this object, which has been produced in such and such a factory as a bulb, will burn out in less than five minutes of standard usage? These readings, of course, could be made more precise if we took the time to stipulate conditions for 'continuous standard usage', but there is no need to do so explicitly. We need only suppose the conditions to be identical in each case. In fact, the two cases differ only in how one conceives of the light bulb under scrutiny. On the one hand, the light-bulb is conceived of as a physical entity without taking into consideration how it was made, and on the other hand, it is conceived of as an instance of an artefact type, made in some factory or other, without taking into consideration its physical make-up. In more abstract terms, it appears that there are two possible readings of the sentence "X has an objective probability p to malfunction" when X is a manufactured artefact; there is one reading if X is considered as a physical object (or as a member of a physical type), another one if X is considered as an artefact item (or as a member of an artefact type).[1]

If we cannot as yet give a precise interpretation of the probabilities appearing in these two readings, we can clear the way by making a little more explicit the terms of the question. We shall follow Mellor in distinguishing three basic kinds of probabilities: physical (or chances), epistemological, and subjective (or credences).[2] Credences are supposed to measure how strongly we believe in a proposition. Epistemic probabilities are supposed to measure to what extent evidence confirms or disconfirms hypotheses about the world. Chances are supposed to be objective features of the world. It is tempting to interpret the two probabilities under scrutiny as epistemological, that is as conditional probabilities relative to different pieces of evidence (e.g., knowing that the bulb has a deformed filament or knowing that it has been produced by Mazda). By doing so, one gets rid of every difficulty but misses the point. Interpreting probability in terms of credence takes us still farther away from our target. We should, at least to begin with, consider the probabilities appearing in the two readings as chances. The question addressed here is indeed whether or not there can be two different interpretations of the probability of dysfunction in terms of chances, as some ways of presenting or justifying such a probability suggest. We do not deny that

---

[1] By way of convention, we shall say that an item that malfunctions shows a dysfunction. 'Dysfunction' is the term currently in use in the literature on functions considered below, however the corresponding verb is missing.

[2] Mellor (2005, chap.1). Incidentally, let us precise that the term 'physical' is somewhat misleading. Mellor explains, in fact, that 'physical' designates not necessarily probabilities appearing in physics since chance probabilities can be found in many sciences, for example, in biology (the chances of mutation), in psychology, etc. The qualification of chances as "physical" is unfortunate because it suggests that chances are of one sort, the sort "physical", and it is also not very clear what "physical" means in this context: ascertainable by physics, rooted in some basic physical properties,...

there may be subjective and epistemic probabilities bearing on artefacts and even on their probability to malfunction, but we contend that there are also cases in which such probabilities are presented as an effect of a particular physical condition as well as cases where they are presented as an effect of a particular artefactual condition, as we shall see below. These cases will be the object of our investigation.

Let us consider the first reading, viz. the one a physicist normally makes. Later on the corresponding probability will be referred to as the 'physical probability of dysfunction', for short the 'DYSF-PHYS probability'. For the great majority of light bulbs, whether their filament will or will not burn out within five minutes of use is a determinate matter. However, there will probably be a small number of bulbs for which it is indeterminate, because say, their filament has some particular form of weakness. Here we don't mean an indeterminacy due to insufficient knowledge but rather an indeterminacy that persists even under the best possible human knowledge. We can suppose cases where the lit up bulb should be analysed as a non linear physical system, or as a system where quantum indeterminism matters (i.e., cases where, after five minutes of continuous usage under identical conditions, two perfectly identical bulbs might not end up in the same final state because the filament was highly sensitive to initial conditions, or because perhaps quantum effects would have 'percolated up'). I take it that there are such cases, though the matter is contingent. The point of the example is just to make visible with what each reading is concerned. In the first reading, the probability is understood to be what it is because of the physical properties of the object concerned. The basic question is whether or not these physical properties ground a disposition to burn out before five minutes in well-determined conditions. We have supposed that in most cases the physical properties will ground an objective sure-fire disposition to burn out or burn on, but that in some cases they will imply a probabilistic one.[3]

This first reading thus corresponds to what is usually understood when one speaks of an objective probability, rooted in some feature of the world. If there is something of the sort, one usually thinks, it must result from the physical make-up of the entity to which the probability is attributed.

---

[3]The distinction between sure-fire dispositions and probabilistic ones comes from Mackie. See for example Mackie (1977, 362): "a sure-fire disposition, one which the thing simply will manifest whenever the appropriate stimulus and/or conditions are present can be distinguished from a probabilistic or even merely possibilistic one, where even when the conditions are appropriate there is only some chance, or perhaps a mere possibility, that the thing will then manifest the disposition". Probabilistic dispositions are usually interpreted as propensities. However, propensities are often, but not always, understood as being probabilistic dispositions of entities or systems. Here, we shall be concerned only with probabilistic (non sure-fire) dispositions.

However, this inference, which often remains implicit, is not as evident as it may seem. The first reading is not the only possible one and it is not even the only one that matters in every day life and in science, as we shall now see.

Let us consider the production of light bulbs by some firm in a particular factory. A major concern for the managers and engineers of such a firm is the good functioning of the bulbs produced, since it is quite a decisive factor for people to continue to buy them. So the managers and engineers have certainly planned the assembly line and various mechanisms of control in order to keep the number of malfunctioning bulbs leaving the factory below some decided threshold. Since production costs and quality are two factors usually pushing in opposite directions, the different parts of the production process are, in general, quite accurately studied, planned and tested to determine how they may contribute to the risk of producing malfunctioning items.

Often, managers and engineers must calculate this risk and take such calculations into account in making choices. For example, according to the degree of refinement of the metal employed, the probability to obtain defective elements in an assembly line will usually vary. Suppose the possibility of buying some cost effective, but poorly refined, metal would appear in the market. The managers of our imagined bulb firm would certainly want to know whether they can take advantage of it, so they would ask the engineers whether or not the firm can maintain its decided quality level using the cheaper metal. Various scenarios would probably be envisaged. For instance, the engineers might reply to the managers with these two options: (1) installing a cheap and quickly operating control device, which will eliminate 85% of the defective metal sheets, at the front of the assembly line and then adding various new control mechanisms later on, or (2) installing a complex and costly device eliminating 99% of the defective metal sheets at the front of the assembly line, and reinforcing the control mechanisms only at the end. Thought experiments aside, we can regard the planning and controlling activities of the engineers as explaining why upon leaving the plant a manufactured object will have a quite precise probability of dysfunction. Later on, I shall argue that such a probability is, in fact, rooted in the mechanisms of production.[4]

---

[4] It must be stressed that this probability is not grounded in statistical data (we shall come back to this point later), even if statistics can be used to evaluate it. Moreover, this probability need not be obtained from statistical data. Granted, engineers may use statistics to justify or to test the assertion that out of every 100 bulbs, P will be malfunctioning, or that a determinate mechanism will produce proportion P of A-items with property B (e.g., of all metal sheets, P out of 100 will have such and such defect), but they are by no means restricted to the use of statistics. They may, for example, use

First of all, let us justify the claim that this probability of dysfunction, which we may call an "artefact-type probability of dysfunction" or for short "DYSF-ART probability", is of general interest and appears in every day life as well as in technical or scientific discourse. When we go in a shop and ask about the probability that a hairdryer or a bulb works well, we are not inquiring about the probability of a particular physical structure to work well, but about the probability that an instance of a determinate artefact type, or even more precisely an instance of an artefact type of a particular trade-mark, works well. Economists taking into account the role of quality in their studies on concurrence or on some other economical phenomena will be similarly interested in the ART probability of dysfunction, not in the PHYS one. Engineers or technicians using manufactured elements to build some complex machine will also use the DYSF-ART probability to calculate security margins or degrees of reliability; they will not usually conduct a physical inquiry about each manufactured element to determine more accurately their physical dispositions.

2 Identifying the problem

Before pursuing further our investigation aimed at determining whether there are indeed two sorts of objective probabilities of dysfunction, interpretable as chances rooted in some particular feature of the world, it is better to make sure that this investigation is worthwhile. Actually, it may seem that a frequentist interpretation of both probabilities would solve all the difficulties such a situation might raise. If this were the case, our investigation would lose most of its interest. However, contrary to a possible first impression, a frequentist interpretation does not solve all the problems or even the most important problems such a situation raises, it just makes them less apparent.

An analysis in terms of frequencies will indeed evade the most serious problems instead of solving them if it just explains the objectivity of these two probabilities in terms of the objectivity of the frequency relation. It is a trite remark that an event can be attributed many different objective probabilities when these are equated with frequencies. Frequencies, being relative to classes, are objective properties of a relational nature. If the probability of X to do A (our event) is interpreted as the frequency of A doers among the members of some reference class C, such probability refers indeed to something objective: the ratio of A doers in C.[5] Since different

---

mathematics and physical laws to deduce that when treated in a determinate way some quantity x of metal having a ratio i of impurity will yield 100 sheets of which a number between $p - j$ to $p + j$ will turn out to be defective.

[5] I speak of the probability X has to do A to keep up with the formulation used when

classes will usually mean different frequencies, X may have in this way many different objective probabilities of doing A: one for every reference class envisaged. However, the trouble lies not in the fact that the two probabilities under scrutiny may refer to some objective property, but in the fact that they appear to depend on no epistemic or pragmatic factors.

An interpretation of probabilities in terms of frequencies is not concerned with how the reference class has been determined or chosen. There is no a priori restriction on the choice of a reference class as long as it includes the event considered. One may attribute to John, let us suppose a bearded Bostonian liking guitar music who smoked heavily from 17 to 32 years of age, a probability of developing lung cancer after 45 considering as the referent class' determinant condition 'more than 10 years of heavy smoking', 'citizen of Boston between 10 and 40 years of age' or more oddly 'bearded man of 35 liking guitar music'. As a matter of fact, the choice of a reference class depends often on current knowledge—such and such statistics are available, such and such parameters have been measured, etc. Moreover, they are also often relative to specific interests—a town planner may want to know the probability of a woman to get pregnant in adult life insofar as she is a Boston citizen while a physician may want to know the probability of a woman to get pregnant in adult life insofar as she has suffered from some determinate illness in childhood.

Clearly, some perspectives will be much more informative than others about causal relationships and about 'causally grounded chance', that is chance grounded in a stable feature of the world, for instance in a determinate micro-structure or in a stable causal relation.[6] For instance, the number of years of heavy smoking is certainly a better parameter to obtain a probability delivering causal knowledge about the onset of lung cancers than musical preferences. However, insofar as epistemic and pragmatic factors are involved in the determination of the reference class and may affect through this channel the value of the probability attributed, the probability cannot be interpreted as measuring a chance rooted in reality. Nevertheless,

---

speaking of function and dysfunction. One can easily transform these phrases in order to obtain a more classical probabilistic formulation in terms of events: the probability of the event 'X does A'.

[6] According to some theories, for instance according to a frequentist theory of chance, chance can be relative to classifications of things – John will have a different chance of getting a lung cancer after 45 depending on whether he is classified by his number of years of heavy smoking, by his age or by his musical tastes (Mellor, 2005, 12 & 35-38). Since we want to distinguish between chances which can be analysed and explained as the effect of a stable feature of the world and chances which cannot because they depend, at least in part, on some arbitrary perspective, we have to introduce a new expression. Clearly, our concept of cause here is not a Humean one: causes may be micro structures, mechanisms producing stable dependencies, etc.

there are cases where probabilities are supposed to measure such chance, viz. chance depending only on how the world is made and structured and not on how it is viewed. A typical example is the probability of radioactive decay for radioactive elements. It is a paradigmatic case for a propensitionist interpretation of probability. However, a frequentist interpretation of such probabilities is also possible.

What we call here a causally grounded chance is similar to a propensity, understood as a non-sure fire disposition, in many aspects. It can be seen as a real property of the entity or the system under consideration. The probability that a radioactive atom in some physical conditions or other decays before, let us say 10 years, is a causally grounded probability when it is supposed to result simply from the chance behaviour of such a radioactive atom in such conditions. Similarly, the probability of an event that is the possible outcome of a non-linear dynamical system, as for instance the halting of a rolling die on 4 after a sufficient number of revolutions on a flat surface, measures a causally grounded chance if it is simply due to the non-linearity of the dynamical system so determined. In general, a probability will measure a causally grounded chance when an improvement of our knowledge has no reason to alter it. For instance, the probability 1/6 attributed by physicists to this well-equilibrated die to halt on 4 when thrown in such-and-such conditions is not supposed to depend on some statistical data about similar events measured with today insufficient precision, but on the very nature of the dynamical system such a throw will instantiate.[7] The two characteristic traits of causally grounded chance are in fact: 1) to be explainable as an effect of some general feature of our world, 2) to depend on no arbitrary condition from an ontological point of view like a particular interest or a particular state of knowledge or information. This idea of causally grounded chance is usually associated with the notion of propensity, however propensities are also associated with other ideas (singular case, tendency, underdetermination of physical effects by physical causes, ...) and furthermore they are strongly connected with some paradigmatic cases of physical indeterminism. By distinguishing causally grounded chances from propensities, some questions become easier to address and one is in a better position to investigate the case of two sorts of causally grounded chances, one of which does not perfectly fit the classical paradigm of physical propensities.

Now, we are in a position to identify where the difficulty lies with the two probabilities of dysfunction considered above: it lies not in the fact that

---

[7]One must distinguish the possibility of error (they thought erroneously that the die was well-equilibrated) from a dependence on some condition relative to our actual knowledge.

both probabilities could refer to an objective property simpliciter, but in the fact that they could both refer to causally grounded chances or, for short, to real chances. In fact, it is for this very reason that an interpretation of both probabilities as propensities may strike as odd (we shall come back to this point very soon). On the contrary, their interpretation as frequencies does not elicit the same alarm because frequencies are not associated to this sort of chances: frequencies may correspond to real chances but they are not supposed to do so since they may depend, in part or totally, on some perspectival factor reflecting our current knowledge or interests.[8] So, the frequentist interpretation does not resolve the problems that the existence of two probabilities corresponding to real chances may create, it just masks them by burying both probabilities in the vast number of conceivable frequencies.

Why does the attribution of both probabilities $P1$ and $P2$ (let us suppose $P1 \neq P2$) to the same event (X doing A) strike us as odd, if not indeed as contradictory, when $P1$ and $P2$ are supposed to measure chances rooted in reality? Real chances, unlike frequencies, are not intrinsically relational or not relational in the same manner. They are not supposed to depend on the *choice* of a more or less arbitrary reference class and by this means on whatever epistemic or pragmatic factor has determined such a choice. $P1$ and $P2$ are supposed to depend only on how the world is. Now, it is absurd to think that the way the world is could justify the attribution of two conflicting dispositions to exactly the same thing. It cannot be that Z (be it the world, some experimental setting or whatever other piece of reality) is such that, on the one hand, it makes it that X will do A in circumstances U with probability $P1$ and, on the other hand, it makes it that X will do A in circumstances U with probability $P2$. Nothing can possess incompatible properties, be they probabilistic dispositions or not. Actually, a finer analysis of our two probabilities of dysfunction renders a difference. So, we don't end up in an aporia.

As we have already made explicit, the difference in the two readings lies in the way the item X is considered, whether that be as a physical entity or as an exemplar of an artefact type. As a matter of fact, this difference implies a difference in the properties on which the probability bears. When

---

[8]The fact that a probability corresponding to a frequency may have nothing to do with a real chance resulting causally from some feature of the world is best seen in cases where our current limited knowledge is the reason for attributing a probability. One can use the knowledge that 20% of people demonstrating symptom S have been contaminated by virus V to attribute the probability 0,2 to the event "Mary has been contaminated by virus V". However, whether Mary has or has not been contaminated by virus V is a perfectly determinate matter, which could presumably be known with certainty, were we to have a better test (a fine blood test for example).

X is considered as a physical entity, what is at stake is the probability that the physical structure instantiated by X does F in circumstances C. When X is considered as instantiating an artefact type, what is at stake is the probability that X, as a member of some artefact kind, possesses a physical structure that does F in circumstances C. With so simple an artefact as a bulb the difference between the two probability attributions may not be readily visible. But if one takes as an example an engine, a plane, or a computer, it becomes evident. Belonging to an artefact type goes along with the possibility of possessing various physical make-ups. In fact, an artefact type is defined by a function, and functions, as is well known, are in general multiply-realizable. So the two probabilities of dysfunction are relative to the same event (X doing or not doing A in determinate conditions) but they are related to different features of X, one to the physical make-up of X, the other to its function. At present, we shall not further pursue the question by inquiring whether this distinction could be formulated in another way than by relating it to two features of X, for example by referring to different events or series of events. Such probabilities are for the time being analysed as being relative to objective probabilistic properties, namely non sure-fire dispositions, of the artefactual object X.

Before leaving the question, let us address a possible worry. Phrases like 'considered as a physical entity' or 'considered as instantiating an artefact type' suggest a choice in classifying the entity (or the event) in each case. Doesn't it mean that one chooses a determinate perspective? If so, doesn't it go against some of the previous suppositions and assertions: the supposition that the two probabilities of dysfunction measure real chances and the assertion that such chances, contrary to frequencies, are not relative to a particular perspective? As we have seen earlier, each probability of dysfunction is related to a particular feature of X. Being interested in one or the other probability is of course an arbitrary choice, but once this choice is made the world itself is supposed to lay down the law. No arbitrary parameter is supposed to determine, even partially, the value of the probability. So the contemplated choice does not involve any more relativity in the probabilities attributed than what is involved in the arbitrary choice to measure length rather than width in the case of measured distances.

Now we can resume our investigation on our two probabilities of dysfunction and focus on the puzzling one, the DYSF-ART one. Since the DYSF-ART probability is related to the feature of having a determinate function, we need to make clear what it means to have a function. It is by seeing whether and how function and physical make-up are related to each other that we can hope to resolve the question of whether or not DYSF-ART and DYSF-PHYS probabilities are really as independent of each other as they prima facie seem to be.

3   Functions, history and real kinds

What do we mean when we say of an artefact that it has some function or other? A simple and plausible response is the following: to have the function (of doing) F is to have the capacity to do F plus possibly some further condition like being currently used to do F. However, this plausible conception is false because it does not account for dysfunctions. A malfunctioning or non-working item is an item that has a function but lacks the capacity to perform it: it should do F in determinate circumstances but it is not capable of doing it.

In the last thirty years there has been intense reflection on the notion of function appearing in the biological sciences. A theory has been developed to account for the normative difference between items functioning correctly and the malfunctioning ones; that is the etiological theory of function, even more precisely, the selectionist etiological theory of function, abbreviated here as SEL.[9] Roughly, SEL says that a function is a historical property of items belonging to a special sort of class as, for instance, a species. An item X has function F if 1) some of the Xs that existed before F did, and 2) if their having done F is what explains the current presence of Xs. In biological cases, natural selection is the mechanism producing such a situation. Present hearts have the function of circulating blood and not that of making a rhythmic noise because previous hearts have contributed to the selection of their bearers thanks to their activity of blood circulator and not thanks to their noisiness. Since functions refer to historical properties and not to present capacities, one can easily account for cases of dysfunction. Something may have a function because of some historical fact concerning its ancestors but presently lack the capacity required to perform the function. In this way, one is able to explain why a diseased or malformed kidney remains a kidney, i.e., an organ whose function it is to filter blood, even if it is not any more able to perform it or even was never able to perform it. Various authors have claimed that SEL applies also to the functions of artefacts when these are numerous enough to form a conspicuous population, as is the case with many manufactured artefacts. In its main thrust, I think that such a theory of the functions of artefacts is correct.[10]

Now, if functions are historical properties, as SEL asserts, the function of an artefact does not depend in any direct way on its physical make-up or

---

[9]Some etiological theories of function don't suppose selection mechanisms, hence the precision.

[10]Obviously, I cannot present in this context all the reasons in favour of an etiological theory of function. Besides, I defend a more abstract version of the theory than SEL, but there is no need to go into such details here.

on some higher level physical property, contrary to what would happen if functions were some sort of capacities.[11] Dispositions to have some physical effect like dissolving salt, blowing hot air or flying in the sky (later labelled 'physical dispositions') are supervenient on physical make-ups. A function would be nothing else than a multi-realizable physical property, if it were nothing else than a physical disposition. In fact, to have function F would then mean to have one of the different physical structures that can do F. In such case, the idea that ART probabilities (probabilities relative to the function) might have a proper causal ground and be independent from PHYS probabilities (probabilities relative to the make-up) would become quite implausible. The probability that an item does Z, because it has function F, could not but be related to the probability that an item does Z, because it has such make-up, since to have function F would mean to have one of a determinate series of make-ups. Anyhow, we don't even need to go that far in the analysis to clear up the question. The very notion of DYSF-ART probability would collapse if to have function F would imply to have the capacity to do F since something having function F would necessarily (by definition, if you want) be capable of doing F. Indeed, dysfunctions would be impossible: all DYSF-ART probabilities would be equal to 0!

The importance of this point lies in the fact that it dissipates a confusion that may have helped ignore the case of causally grounded probabilities that are not relative to the physical properties of the entities concerned and the circumstances they are put in, that is to say, not relative to what Popper calls an experimental apparatus. As we have already noticed, for the advocates of propensities, the cases of physical indeterminacy are paradigmatic. This supports the tendency to think that every causally grounded probability should be rooted in some basic physical properties and relative to some sort of experimental apparatus. Probabilities related to functions, when acknowledged, had no reason to cast doubts on this preconception. The widespread idea that functions were some sort of higher-level physical properties (supervenient physical dispositions) supported the hypothesis that such probabilities could not but remand to real chances of a basic physical sort were they to remand to real chances. If ART probabilities are interesting to study, it is precisely because functions are not such physical properties, which is why ART probabilities raise the question of a different sort of real chance. Now, in order to understand better on what non-classical real chances may be grounded, we need to resume our positive investigation of functions.

What sort of entities are the classes in which functions can be rooted? In 1984, Ruth Millikan introduced the notion of 'reproductive family' in order

---

[11]'Capacity' and 'disposition' are considered here as perfect synonyms.

to answer this question. The notion was meant to apply to many classes of artefacts —from objects to linguistic items and cultural behaviours— as well as to biological species. However, the notion was complicated and probably also needlessly specific in supposing copying mechanisms.[12] By the end of the '80s, Richard Boyd and subsequently Ruth Millikan elaborated the more abstract notion of real kind.[13] Thanks to this notion, it then became possible to clearly distinguish two sorts of causally grounded kinds: on the one hand, those grounded on sameness of physical microstructures, and on the other hand, those grounded on external causes like, for instance, natural selection.

The notion of real kind can be seen as an extension of the notion of natural kind that originated in the semantic theories elaborated by Putnam and Kripke around the 70s. These theories explained how natural kind terms could refer directly to natural kinds in nature (e.g., to natural kinds from Nature's point of view). Contrary to what Locke claimed in his time, a direct semantic relation could exist that pointed towards real essences independently of our definitions (the so-called nominal essences). It did not matter what thoughts one might have about such real essences. The more conspicuous examples of natural kinds Putnam and Kripke gave were those of material substances like gold or water, and those of biological species like lions. In the case of material substances, we have a precise hypothesis about what these real essences may be: for water, H_2O; for gold, the atomic element number 79; etc. For biological species, by analogy, many have thought that insofar as they have a real essence it should be something similar, for example a genetic code structure. The tendency has been to think that real essences, if they exist, should be something like physical micro-structures. Boyd and Millikan have demonstrated that it was a supposition that needed justification; other sorts of essences were indeed conceivable.

At the end of the 80s, Boyd proposed to extend the theory of natural kinds to kinds defined by clusters of properties in the following sense: 'kinds such that the unity of the property-cluster which defines them is causal rather

---

[12] Millikan (see 1984, chap.1, 2).

[13] The notion of real kind has a long history, but under its modern form of natural kind it appears with John Stuart Mill who relates Aristotelian 'real kinds' (genders, species, ...) to 'Kinds in nature' (See Mill (1850, chap.7§4)). One can read Hacking (1991) for further information on the introduction in the modern philosophy of science of 'natural kinds' as a phrase and as a notion. In contemporary philosophy of science, 'natural kind' has become a current expression, but since the works of Putnam and Kripke, it has often been associated with the idea of a physical common essence or structure. So, in order to make it easier to distinguish between different sorts of 'Kinds in Nature', Millikan has reintroduced the Millian phrase 'real kinds'. Boyd himself rather uses the expression 'natural kind', but for clarity's sake we shall not systematically respect his own terminology.

than conceptual'.[14] It must be stressed that the extension concerns causal grounds and not at all conceptual means (necessary and sufficient conditions, family resemblances, etc). This is why it is in line with the ontological notion sketched above and not with the more traditional, conceptualist, understanding of 'natural kind'. The idea of property-cluster goes with those of imperfection and vagueness. The properties are supposed to be shared by most but not necessarily by all members.[15] Boyd called 'homeostatic' the mechanisms producing such a high similarity between the members of the kind relative to some property-cluster and spoke of 'homeostatic property cluster kinds' or of 'homeostatic cluster kinds'. For Boyd a perfect example of such kinds are the biological species:

> The paradigm cases of natural kinds —biological species— are homeostatic cluster kinds. The appropriateness of any particular biological species for induction and explanation in biology depends upon the imperfectly shared and homeostatically related morphological, physiological and behavioral features which characterize its members. The definitional role of mechanisms of homeostasis is reflected in the role of interbreeding in the modem species concept Boyd (1991, p.142).

Commenting on Boyd's theory, Millikan underlined rather the difference between eternal essences and historical essences. The emphasis should be put, she explained, not so much on shared properties as on historicity: 'Biological kinds are defined by reference to historical relations among the members, not, in the first instance, by reference to properties. Biological kinds are, as such, historical kinds' Millikan (1999, p.54). As for us, let us speak of 'non-physicalist real kinds', for short 'NP-real kinds', to highlight what will prove to be the most significant difference from our point of view. A NP-real kind is a kind whose unity is not rooted in identity (or similarity) of physical or chemical microstructure but in other, more external, causes. By being a member of a NP-real kind, an entity has a particular sort of essence, which has to do with how members of the kind are causally related to one another.[16] Such an essence can explain similarities that come from the existence of copying mechanisms, from the fact of sharing some common context, etc.[17]

---

[14] See Boyd (1989, p.15-18) and Boyd (1991, p.140-41)

[15] The idea that real essences or causally grounded unities could go with fuzzy borders and imperfect similarities is quite unsettling at first, especially for people trained in philosophy.

[16] An 'essence' is nothing else than the causal ground explaining why a member, as a member of some particular kind, possesses a whole series of features. A piece of gold has many features: colour, malleability, fusibility at such temperature, etc., because of its physical essence (being composed of atoms of atomic number 79). Similarly, an animal will have many features resulting from its belonging to a particular NP-real kind, a biological species, as we shall see more precisely below.

[17] That which causally explains the common features of the members of a same species

Millikan underlined also the fact that many artefact kinds are NP-real kinds:

> Besides biological taxa, there are many other historical kinds. In (Millikan, 1984) I spelled out why the 1969 Plymouth Valiant 100 was a real historical kind, there calling it a 'secondary substance':
> ... in 1969 every '69 Valiant shared with every other each of the properties described in the '69 Valiant's handbook and many other properties as well. And there was a good though complicated explanation for the fact that they shared these properties. They all originated with the selfsame plan – not just with identical plans but with the same plan token. They were made of the same materials gathered from the same places, and they were turned out by the same machines and the same workers ... or machines similar and workers similarly trained [on purpose] Millikan (1999, 56).

In order to establish in the clearest possible way, the independence of NP-real kinds from physical essences, let us explain in what sense an item can have a determinate feature because of its NP-real essence rather than because of its physical essence. In fact, it may be the case that a less proximate cause than the physical make-up is the most decisive one for explaining the possession of a particular feature. This is best shown by an example. Let us consider the phenotypic character of some organism, say a rooster with a red crest. It is possible to tell a physiological story explaining how the genetic code of this rooster is going to produce, in normal circumstances of development, the red crest. But when biologists ask why roosters have a red crest, often what they want to know is not how such crests may develop but why roosters have the capacity of developing red crests. As Mayr said and explained, the concern of biology is not only proximate causes but also ultimate ones.[18] If red crests in roosters result from various genetic make-ups and if more physiological stories are forthcoming (if this is not the case for red crests it is for many phenotypic characters),[19] then a rooster chosen at random has the capacity to develop a red crest not so much because it has a particular genetic code but because it has inherited a rooster genetic code whatever precisely this genetic code may be. Natural selection (or some other general mechanism acting at the level of the species) will explain why the gene pool of roosters includes only gene codes producing red crests in normal conditions of development. So at the end, what explains why this

---

is not only heritability (a copying mechanism) but also selection. Through selection, as we know, members of the same species get adapted to their environment (a common context).

[18] See Mayr (1961).

[19] Many biologists in the so called 'evo-devo' orientation stress the existence of different levels of integration: in a species a same phenotype may result from different developmental pathways and a same developmental pathway from different genotypes. Anyhow, the simple fact of heterozygosity is sufficient to show that the same phenotype may result from different genotypes.

particular individual will develop a red crest is the ultimate cause. It is because this individual is a rooster, which means a member of the species Gallus gallus and not a possessor of some genetic code or other, that it will develop a red crest.

The same thing can be said for many artefacts. The fact that your car engine power output is x is no doubt the consequence of the physical properties of this particular engine along with some correlative parts of your car. However if you want to know why, in general, the engine of a Renault 4L (or of a Plymouth Valiant 100) has this particular power output it will not be pertinent to consider the physical make-up of any particular Renault 4L engine. It will be more pertinent to consider the specifications of a Renault 4L engine. There may have been variations in the physical make-up of Renault 4L engines over the years, but what we are sure of is the following fact: if the specification of power output for Renault 4L engines has not changed over the years, then all the physical structures of Renault 4L engines will have the same power output property. Our confidence that it is so lies in our understanding of the mechanisms of firms' management and industrial production: we know that every change concerning the physical structure of the engines of Renault 4L has been planned, tested, and regulated in order to secure an identical power output for all Renault 4L engines. In fact, an engine will be malfunctioning, defective, if being a Renault 4L engines it fails to have the expected power output.[20]

4 Probabilistic properties and NP-real essences.

Boyd, Millikan or other advocates of non-classical real kinds have not addressed the question of probabilistic properties and of grounded probabilities. They contented themselves with the vague notion of largely shared properties. The question has been raised, but with a critical intent, by some philosophers of biology who remarked that Boyd's homeostatic cluster kind notion could not explain salient features of biological species. So it is that Ereshefsky and Matthen have recently criticized Boyd's notion because it is meant to account only for common features.[21] Such a notion, they explain, is not really useful in biology since it cannot apply to differences within

---

[20] For simplicity's sake and to avoid clumsy expressions, I have presented most of the examples of this section in a way so as to suggest that NP-real kind essence could cause sure-fire dispositions to develop a red crest or to manifest a determinate power output. However, NP real kind essences ground usually only probabilistic dispositions, hence the very notion of homeostatic cluster kind. To take the rooster example, what a NP real kind essence (or an ultimate cause) really explains is the fact that the individual has a determinate probability (a very high one in this case) to develop a red crest. The relation between probability and NP-real kind will anyway be the main point of next section.

[21] See Ereshefsky and Matthen (2005, §4).

taxa, some of which are of great biological importance. Such is, for example, the sexual dimorphism that exists in many species. The emphasis put on common features by Boyd and others is essentially due to circumstances.[22] A shift from the homeostatic cluster kind notion to the slightly more general one of NP-real kind makes it, indeed, possible to explain stable dimorphisms. In order to show it, let us try to characterise NP-real kinds a little more precisely:

NP-real kinds are classes such that:

1. There are mechanisms that make the members of the class depend causally on one another or on common causes and conditions.

2. These mechanisms make members at time $t+1$ quite similar to members at time t (relative to a pertinent time scale) and they make members at time t share to a large extent many features (this is the homeostatic cluster part Boyd emphasized).

3. These mechanisms can also be responsible for stable repartitions of properties between the members.

Many NP-real kind features will be largely shared all over the kind, but some can be shared only by a particular subgroup as is the case with males and females. The same general mechanisms which explain the features that all or almost all members of some species have in common (e.g., natural selection, heredity, and perhaps others) explain also why some will be male and others female. In our reflection on probabilities, we shall focus on a case where the repartition is usually a stable and causally significant feature, like with the male-female dimorphism.

Let us explain quickly what we mean here by a causally significant feature. Consider a species in which some of the members would have curly hair and others straight hair as the result of chance and drift. At no time would there have been any selective advantage in having one sort of hair rather than the other one. The repartition of curly and straight hairs at every generation would indeed be a consequence of the mechanisms responsible for the unity of the species (heredity), of the repartition of curly and straight hairs in the previous generation and also to some extent, of chance. It could even happen (for example, in a large population) that such a proportion would remain stable. However, the curly-straight hair dimorphism would be an insignificant feature having no causal role, it would just perpetuate itself.

---

[22]What explains Boyd's exclusive interest for common features is his particular epistemological interest: like many others, Boyd was wondering about the relationship between natural kinds and inductions.

It is quite different with the male-female dimorphism. In that case, any deviation from the expected ratio will have a series of effects which will turn out, in most circumstances, to restore this value in the next generation or in some later one. Moreover, the repartition of males and females in a species is a decisive causal factor affecting the future development of the species in many aspects.

This sort of causally significant ratio is what we want to focus on. If we suppose a species in which the favoured male-female ratio at birth in a determinate environment is 1:1, there will be a determinate probability for a member of the next generation to be male at birth, and this probability will result from the series of interdependent mechanisms that tend to produce this ratio and that possibly control it. It is at present a debated question whether there is, in fact, some specific mechanism that adjusts this ratio more or less finely to environmental conditions.

Let us now go back to artefact cases. They are simpler to analyse than biological ones because we have a better understanding of the mechanisms and conditions involved in artefact production than we have of natural mechanisms. The ratio of malfunctioning items to properly functioning ones for artefacts coming out of a factory appear to be similar to the male-female ratio, it results from mechanisms which are part of the whole causal net of mechanisms that are responsible for the unity of the NP-real kind under scrutiny (or to say it in another way, of the causal net which constitutes the NP-real kind essence). Our thesis is, in fact, that the DYSF-ART probability of an artefact item results from a series of mechanisms controlling the ratio of malfunctioning items. Let us now focus on these mechanisms and how they ground a determinate DYSF-ART probability.

There are in fact series of mechanisms that fit into each other. At the beginning of the article we discussed the mechanisms in a particular factory. These included mechanical devices as well as different human procedures to make different sorts of control and test. We can add to this the managing structure of technicians, engineers and foremen whose duty it is to control, maintain, repair, and restore all the elements involved in the mechanisms of production and control. A factory, understood as a whole of assembly lines, managing structures, and working procedures, grounds a determinate DYSF-ART probability in normal conditions of supplies, of employment, etc. We can widen our perspective to consider not just particular factories, but firms possessing more than one factory producing the same type of artefacts.[23] In this case we shall have to take into account, on the one hand, the research departments where engineers and managers plan the

---

[23] The type we are speaking of here is something like Renault 4L, not the type 'car' which will include sub-types meant to have different performances and quality levels.

assembly lines as well as the work procedures to be followed in the different factories, and on the other hand, the different systems of test and control that have been installed to assure a homogeneity of production among the different factories. From the existence of such firm level mechanisms, we know that the probability of producing a malfunctioning item in factory A of firm X must be the same as the one in factory B of firm X (within a determinate margin of error), even if the machines, the working procedures and the managing structures are quite different in A and B.

We can further widen the perspective and consider the mechanisms that drive different firms selling the same type of artefact within a determinate range of price in a shared economic context to adjust to each other and converge towards the same 'favourable' ratio of malfunctioning items. Here, the best examples are cases where the artefacts are widely used, have a relatively 'short life', and are neither a status symbol nor aesthetically valuable (think of bulbs, car batteries, etc.). It is no coincidence that the vocabulary one can use here is similar to the one used in evolutionary biology, since this convergence results from a selective pressure of an economical nature. The consumers act as a selective force: firms producing more malfunctioning items will have to sell their production at a lower price or improve quality if they don't want to be eliminated by the market.

The population at hand gets bigger every time we widen the perspective. First, it was the population of a type of artefact produced by a determinate factory while the structure of the assembly line, the work's procedures and the control mechanisms were constant. Second, it was the population of a type of artefact produced by a firm in any of its factory during the period in which the quality requirements and control at the firm level remained stable. Third, it was the population of a type of artefacts sold in the same range of price in the same economic context during the period in which some determinate economic conditions remained stable and no technical innovation or goods dumping occurred that could have created a new pressure for lower prices or better quality.

Every one of these three classes of items is a NP-real kind. The fact that such kinds are not necessarily mutually exclusive or in a clear genus-species relationship creates no real problem, even if this may appear puzzling at first.[24] Conversely, classes like the class of yellow hairdryers produced in the state of New-York between the 1st of May 2000 and the 1st of June

---

[24]We cannot develop this point here. Clearly, natural kinds and physical essences, because they go with clear-cut boundaries and simple genus-species relationships, are much more attractive than the more vague and more complex NP-real kinds with their external causal essences. However, this minor appeal should not be confused with low scientific value. It may be that a good causal theory of the world needs such kinds. Vagueness, like probability, does not necessarily indicate a lack of knowledge.

2000, which have arbitrary boundaries, will not be real kinds. There is no mechanism, responsible for the many similarities between the members of the kind, that may justify such boundaries. We shall take the simpler case for analysis, that of the NP-real kind corresponding to the production of one particular factory during the period in which the factory features are stable. We shall call the probability that an item X has as an artefact produced by a determinate factory, a Factory-ART probability.

5 How to interpret causally grounded ART-probabilities

Let us now come to a plain formulation of our thesis: the Factory-ART probability that an item will be malfunctioning is a causal result of the factory's general features as described earlier. Thus this Factory-ART probability is a property causally grounded in the factory's structure as the property of being a bulb or a hairdryer is a property causally grounded in being produced on a bulb or hair-dryer assembly-line. First and foremost, let us explain why this Factory-ART probability should not be understood as being relative to the real frequency of malfunctioning hairdryers or bulbs in the relevant actual population.

Suppose that during some period a group of saboteurs worked in the factory or suppose that a part of the production burned before it could be loaded for delivery, and by chance only items with no defects burned. This would not change the expected proportion of malfunctioning items such a factory would be letting out for sale under normal conditions, this being exactly what our Factory-ART probability refers to: the ratio of malfunctioning items that the factory, because of its structure and various control mechanisms is letting out for sale. Here the Factory-ART probability we are talking about has the modal meaning the notion of probability usually has, a meaning that prevents interpreting it in terms of real frequencies.[25]

The idea of interpreting such a modal meaning in terms of expectations about the real frequency on the basis of statistical data won't do, since statistical tests may indeed imply another probability than the Factory-ART probability. The hypothesis that contingent events may explain a discrepancy between the probability relative to the actual population (the ratio in the actual population that statistical tests make us expect) and the one to be expected in view of the pertinent features of the factory is conceivable. Moreover, it is sometimes the most rational one. After analysing

---

[25] As Hájek says in the general article on the interpretations of probability written for the Stanford encyclopedia of philosophy (Hájek, 2005, §2): "Probability is apparently, among other things, a *modal* concept, plausibly outrunning that which actually occurs, let alone that which is actually observed".

in detail the structure and functioning of a factory, experts may arrive at the conclusion that there is indeed such a discrepancy, even to the degree that after learning of the saboteurs, they would certainly conclude that it was one of the contingent events explaining the noted discrepancy. In the same manner, biologists may arrive at the conclusion that in some particular species the sex ratio at birth should be 1:1 although the population in the pertinent period has in fact shown another ratio. Different contingent events (predators, epidemics, etc.) must have by chance they will think, killed more mothers-to-be of males than of females.

The fact that the ART probability of dysfunction is causally grounded speaks in favour of interpreting it in terms of propensity, as does the possible discrepancy between this probability and the frequency of malfunctioning items in the actual population. Propensities allow for such discrepancies and give support to the methodology indicated above. As far as propensities are concerned, a discrepancy between the probability and the actual ratio is liable to two interpretations (1) the probability value is erroneous or (2) this value is correct but there have been disturbing factors or the population is too small to be significant (in the case of frequencies, only the first interpretation is possible).

However the interpretation in terms of propensities appears problematic. First, it would imply that an object could have two different sorts of propensities. Thus, we would need to use two different expressions, for example, 'physical propensity' and 'real kind propensity. This, in turn, would imply something that may appear quite odd: it would be possible to say that the physical propensity of X to do F is p, while its real kind propensity to do F is p', with p' being different from p. Here we touch a second difficulty. One spontaneously understands the attribution of a propensity as an attribution of a physically grounded chance because the idea of propensity is often associated with that of tendency. Intuitively, a tendency that something has is supposed to depend on what the thing is by itself. The idea of a real kind propensity attached to an entity is for this reason intuitively disturbing: it would be a tendency an entity would have not because of what it is by itself *hic et nunc* but because of its history, because of causal relations connecting it in its past with a particular group of other entities (the ones of some NP real kind).

The way to get out of these difficulties may be to make this relational aspect clearly visible, which means to present the DYSF-ART probability as a property existing at the population level rather than at the item level. However, as we have just seen this population property should not be identified with a real frequency, but with something like a propensity. In fact, if the Factory-DYSF-ART probability corresponds to a propensity, in a somewhat usual sense of the term, it is rather to a propensity attached to the factory than to an item produced by it. The factory instantiates a structure that can be seen as grounding a tendency to produce a type of event, for instance, the tendency to produce one defective bulb out of every N bulbs. In fact, an analysis of a factory (or of a Firm or of a determinate economic situation) can explain a tendency to produce a population of artefacts that will include a determinate ratio of malfunctioning items. However, it is to be noted that such a propensity will not fit any better into the model of physical propensities than what could be expected from having a propensity attached to the functional feature of a singular item. In fact, among the mechanisms that are supposed to determine causally the factory's tendency to produce a determinate ratio of malfunctioning items, there are such mechanisms as the market mechanisms, the management mechanisms or the factory mechanisms that depend directly on humans and on how the work is organized on the assembly line. In fact, as we have already seen, factory mechanisms include much more than just the physical mechanisms instantiated by machines.

If we choose to see in the DYSF-ART probability a property at the population level, then we need an expression making clear what sort of populational property is meant. The chosen phrase must make it explicit that the probability does not measure the frequency of defective items in an actual population more or less arbitrarily defined, but rather measures the tendency that some NP-real kind mechanism has to produce a determinate ratio of defective items (a series of connected mechanisms can be analysed as a mechanism). As a result, we end up with the idea of a general category of causally grounded probabilities containing two sorts: (1) propensities and (2) expected frequencies causally grounded in real kind mechanisms. Each sort of causally grounded probability seems to refer to a different type of chance: the propensities seem to refer to some basic physical indeterminacy while the 'real kind expected frequencies' seem to refer to the level of chance entailed or allowed by some NP-real kind mechanism.

6 Two sorts of causally grounded chance

Why should the study of a particular ART probability be of any interest in understanding probabilities outside technology? The answer is: because ART probabilities may be relatively simple. It is very likely that probabilities of the same sort in biology and possibly elsewhere will be much more difficult to analyse. The study of ART probabilities can thus help shed light on more complex cases. ART probabilities are easy to analyse, first, because we have a rather good understanding of the mechanisms involved, as mentioned earlier, and second, because this understanding makes it possible to leave aside a series of factors, whose involvement is a source of complexity, e.g., ignorance and physical indeterminism.

Let us first consider ignorance. The difference between a probability depending totally or in part on ignorance and a probability totally independent of it can be easily made in the artefact case because ignorance could presumably be eliminated, if ever we wanted to do so. As a matter of fact, it is almost certainly possible with our present means to determine with certainty, by physical analysis, for every particular bulb coming out of a factory whether or not it will malfunction. So, we could get rid of any probability about whether or not the bulb is defective, if we wanted to do so (we suppose for simplicity's sake that there are no indeterminate cases due to non-linearity or sensitivity to quantum effects). However this would not affect the value of the DYSF-ART probability of the bulbs. The DYSF-ART probability depends only on the proportion of malfunctioning items the factory tends to produce in normal circumstances. Our level of information about the physical constitution of bulbs or about any physical aspect of the production does not matter. In fact, even if we had perfect knowledge, the DYSF-ART probability could remain the same, for example, if for practical and economical reasons, the productive structure, i.e., the factory, would remain identical to what it is today.

Let us now come to the more difficult question of what relation ART probabilities may have to basic physical indeterminism. We need, first of all, to get a more precise idea of what the DYSF-ART probability measures. Why is it that the DYSF-ART probability measures real chance and not simply a statistical fact, the proportion of defective items in some arbitrarily defined population? It is because, as we shall see now, chance or indeterminacy intervenes in the production process in the form of authorized variations, and this results in a random distribution of the defective items among the non-defective ones. This point is better shown by an example. Suppose a situation in which instead of obtaining a proportion p of randomly distributed non-As (Xs lacking property A) among As, one obtains the same proportion p of non-As but these are produced systematically when and only when some switch is set on position O. Such could be the case, for example, with the property of being painted in yellow (the situation is harder to imagine with the property of being defective). There would then be no indeterminacy intervening in the production process and no randomness in the distribution of the non-As. The Xs would easily be separated in two populations produced in well distinguishable states of the world. When the switch is set on 1, As are produced; when it is on 0, non-As. There would thus be no more objective chance to be measured, and the probability of being A could not have the traits of the ART-probabilities considered till now. In fact, one could do without any probability at all by distinguishing two sub-populations, the population produced when the switch is on 1 and the population produced when on 0. It must be noticed that the feature of being A (or not) would then be analysable as a separable feature of the Xs, one could even say as a contingent feature of the Xs. In fact, the presence of the feature 'painted yellow' would be explainable by considering the autonomous mechanism producing it: the action of the painting machine when the switch is on 1. The causal story would not need to bring in any probability. It would also not need to involve any real kind since any object going through the painting machine with the switch on position 1 would come out yellow. As a consequence, the probability of being a yellow X would not be a causally grounded probability relative to the real kind of the Xs. In conclusion, if such were the situation, the probability of being a yellow X would be defined relative to an arbitrary class. It would just express a frequency, with no causal significance, nothing else and nothing more.

What does this chance, which results in a random distribution and which makes it impossible to get rid of probabilities, consist in? It consists in the variations and differences that the modes of production allow on the basis of a determinate cost/benefit calculation. This level of variations and differences could be reduced, and with it the part chance plays: if a more costly but more efficient method was used for quality controls, if the metal was purified in another way, if the machines were checked every day instead of every week, ... Now, less chance or indeterminacy in the production process means a minor variability in the produced items, hence a reduced proportion of defective items. Less chance would thus go with a lower Factory DYSF-ART probability.

Does such chance have anything to do with basic physical indeterminacy? The obvious answer is no. Of course, some variation or other may depend more or less directly on basic physical indeterminacies. Maybe, let us say, some small differences in metal sheets result from quantum mechanic effects. However, the variations may come from every sort of sources. Humans, for instance, may be less efficient on Mondays than on Wednesdays. Moreover, variations at source (every element entering the production process contributes to it: machines, workers, supplies...) are not really what determines the level of chance involved in the production process. This level seems to be settled rather by the level of random variations the whole production structure allows. To say it in a more colourful way, the disorder remaining at the end appears as dictated not so much by the level of original disorder as by how much order is being made (or by how much disorder is tolerated). Now, the level of order, or control, which is attained and aimed at in the production process, is the result, as we have seen, of a complicated series of mechanisms.

So, what sort of real chance does the factory DYSF-ART probability measure? Not some fundamental chance going back to primitive physical levels, but rather the level of variations that the production process allows. Actually, it is as if the series of mechanisms regulating the production process were acting as a force whose effect it is to attract constantly the NP-real kind population towards a definite ratio of defective items in a determinate series of conditions. If one takes a broader perspective, one can presumably also look at some economic mechanisms as a force, a sort of adaptive force, making the ratio of defective items depend on some determinate elements of the context, like, for instance, concurrence, technical means, consumers solvency and preferences, etc.

The question of the relationship between probability in evolutionary theory and indeterminism, especially physical indeterminism, is a hot topic in philosophy of biology, as Roberta Millstein says in her 2003 article on the subject. The study of the simple case of DYSF-ART probability shows, it seems to me, that some currently made inferences, with the alternatives they induce, are not as sound as they may seem prima facie. For instance, Bouchard and Rosenberg in their survey article on the current discussion on fitness describe roughly the following situation. First, one has to choose between a propensionist and a frequentist interpretation. Then, in view of the fact that usually the first option is preferred because the frequentist interpretation of fitness does not prove to be very satisfactory, one is faced with the following dilemma: supposing either 'quantum probabilities percolating up' or 'brute unexplainable probabilistic propensities at the level of fitness-differences'.[26] As for Roberta Millstein, she tries to free herself from these alternatives by suggesting that evolutionary probabilities would be best understood as 'propensities of population-level kinds'. The problem she raise as well as the sort of solution she is after (an interpretation of evolutionary probabilities as objective probabilities that are neither frequencies nor propensities resulting from basic physical indeterminism) is quite in line with the analysis presented here about DYSF-ART probabilities.[27]

In conclusion, the study of probabilities like the DYSF-ART one may help separate different issues and ideas that, according to me, are unduly put together. One reason of some unsatisfactory associations may be the major role physical indeterminism has as the paradigm of real chance. Now,

---

[26] Bouchard and Rosenberg (2004, p.696–97). Our doubts don't concern their exposition, for all we know it seems a rather good one, but definitely the soundness of these alternatives.

[27] See Millstein (2003, 1317 sq.)

physical indeterminism, taken alone, seems not a very good basis for thinking about the following issues: chance and levels of causality, chance and sorts of indeterminacy, chance and sorts of populations. The reflection on chance and physical indeterminism is, for instance, not much concerned with the nature of the population involved—real kinds (be they classical natural kinds or not) versus arbitrary classes—and with how causality may be related to types of population; its focus is elsewhere: singular case, irreducibility, etc. However, our present study as well as the importance of a particular sort of class in biology, the species, suggests that the relation between real chance and types of population might be a question worthy of serious investigation.

Acknowledgments

Let me thank, first, Joelle Proust who is at the origin of this work our discussions convinced me that probabilities of dysfunction raise a serious problem of interpretation. I am then much indebted to Philippe Huneman for his comments on the biological aspects and to John Vickers, Donald Gillies, Anouk Barberousse and Isabelle Drouet for very useful discussions and comments for all that concerns probabilities and chances. Finally, I am grateful to an anonymous referee whose pertinent comments have caused me, among other modifications, to rewrite the whole second section.

Françoise Longy
IHPST, Paris 1 University, France.
longy@ext.jussieu.fr

BIBLIOGRAPHY

Bouchard, R. and Rosenberg, A. (2004). Fitness, probability and the principle of natural selection. *Philosophy of Science*, 55:693–712.
Boyd, R. (1989). What realism implies and what it does not. *Dialectica*, 43:5–29.
Boyd, R. (1991). Realism, anti-foundationalism and the enthusiasm for natural kinds. *Philosophical studies*, 61:127–148.
Ereshefsky, M. and Matthen, M. (2005). Taxonomy, polymorphism and history: An introduction to population structure theory. *Philosophy of Science*, 72:1–21.
Hacking, I. (1991). A tradition of natural kinds. *Philosophical studies*, 61:109–28.
Hájek, A. (2005). Interpretations of probability. *Stanford encyclopedia of philosophy*, http://plato.stanford.edu/.
Mackie, J. (1977). Dispositions, grounds and causes. *Synthese*, 34:361–72.
Mayr, E. (1961). Cause and effect in biology. *Science*, 134:1501–06.
Mellor, D. (2005). *Probability, A philosophical introduction*. London: Routlegde.
Mill, J. (1850). *System of Logic, Ratiocinative and Inductive*. N.Y: Harper-Brothers.
Millikan, R. (1984). *Language, Thought, and Other Biological Categories*. Cambridge (Mass.): M.I.T. Press.
Millikan, R. (1999). Historical kinds and the 'special sciences'. *Philosophical Studies*, 95:45–65.

Millstein, R. (2003). Interpretations of probability in evolutionary theory. *Philosophy of Science*, 70:1317–28.

PART VI

CAUSAL PLURALISM

Causal dualism: which position? Which arguments?

MONIKA M. DULLSTEIN

ABSTRACT. In view of abundant philosophical accounts of causation, causal pluralism has become a fashionable idea. Causal pluralism may, however, imply different philosophical positions of varying strength and the question is which argument has to be given in order to support which position. In this paper, I take Hall's (2004) claim for causal dualism as an example and consider two ways of interpreting it. One option is to take it as an analysis of our way of thinking about causation. A second, philosophically more interesting option is to take it as a statement about two kinds of causal relationships in the world. I argue that the second interpretation is not supported by the argument that is given in favour of causal dualism.

1 Introduction

There seems to be a pacifying tendency at the moment in the philosophical literature on causation. Instead of continuing to contrast examples and counterexamples, 'causal pluralism' has become the new guiding idea. The claim for causal pluralism may, however, imply different philosophical positions of varying strength.[1] As neither the positions nor the arguments in favour of them are sufficiently separated, the fundamental question is what kind of argument has to be given in favour of which specific position.

I will concentrate on only one version of causal pluralism which is better called causal dualism. It consists of the claims that

1. The word 'cause' has two different senses:

    (a) A cause is something that is connected to the effect.

    (b) A cause is something that makes a difference to the effect.

2. Although these two senses cannot be accommodated within one unified, theoretical framework, both are equally legitimate explications of the concept of causation.

---
[1] See Hitchcock (2005).

Recently, this position has been defended by Hall (2004). Of course, the two senses as they stand are no definitions yet, but can be seen as the guiding ideas behind any attempt at further explication. The point of this version of causal pluralism is that there is no common characteristic that would make both ideas be senses of the word 'cause'. Therefore, it stands in contrast to weaker versions of causal pluralism such as, for example, Mackie's (1974, chap. 2). Mackie admits that the selection of 'the' cause may depend on pragmatic factors, but nevertheless identifies two core characteristics of our concept of cause.

There are two reasons for focussing especially on causal dualism. First, the distinction to which Hall points is an important one, because most of the recent accounts of causation can be grouped according to (a) or (b). While mechanistic and process theories might be counted as explications of the first sense, counterfactual, interventionist and probabilistic accounts all stress the idea that causes make a difference to effects. Second, causal dualism is not a new position. A number of other authors[2] have already been arguing for two senses of causation. Although they focused on slightly different aspects, one may wonder whether there is some deeper reason for defending two (not three or more) senses of causation.

The main concern of this paper is the interpretation of causal dualism. One option is to take it as an analysis of our way of thinking about causation. A second, philosophically more interesting option is to take it as a statement about two kinds of causal relationships in the world. While I have nothing to say against the first option, I will show that Hall's argument for causal dualism cannot be taken to support the second option.

After briefly summing up Hall's argument in section 2 and explaining the two possible interpretations in section 3, I will develop my argument in sections 4 and 5. It is centred around the two kinds of examples of causal relationships (overdetermination and negative causation) which Hall uses to establish his claim for causal dualism. A closer look at these intuitions reveals that they are based on different attitudes towards causal relationships in the world. While we do believe that there is something to be discovered 'out there' that makes the causal connection in cases of overdetermination, we do not care about this connection in cases of negative causation. In cases of negative causation we are satisfied by knowing how to control the effect. In our everyday life we can and do switch between the two attitudes and that is why a conceptual analysis that is referring to these intuitions results in causal dualism. A philosopher, however, being interested in the metaphysics of causation, has to decide which attitude he takes, i.e., whether he believes that there is something to be discovered about the causal con-

---

[2] For example Sober (1985) and Eells (1991).

nection or not, because this decision determines the philosophical method he uses and the options he has for arguing. In doing so, he weighs the two intuitions that are brought forward by Hall and accepts to dismiss one of them as not being relevant for the kind of analysis he aims at. To treat, by contrast, both senses of 'cause' as equally legitimate kinds of causal relationships in the world (as the second interpretation suggests), would mean to confuse incompatible epistemological and methodological points of view.

2 An argument for causal dualism

Hall (2004) founds his project on causation on a number of examples which we intuitively call causal relationships. In particular, he contrasts cases of overdetermination on the one hand with cases of omission and double prevention on the other hand. Alongside with preventions and preventions by omissions, omissions and double preventions are called negative causation because they involve a negative event being the cause, the effect or the intermediary event. Hall's main thesis is that the principles on which we rely to determine the cause in cases of overdetermination cannot be applied in cases of negative causation and vice versa.

Let us look at two examples to gain un understanding of the issue, first at the classic example for one form of overdetermination, late preemption:

> Suzy and Billy, expert rock-throwers, are engaged in a competition to see who can shatter a target bottle first. They both pick up rocks and throw them at the bottle, but Suzy throws hers a split second before Billy. Consequently Suzy's rock gets there first, shattering the bottle. Since both throws are perfectly accurate, Billy's would have shattered the bottle if Suzy's had not occurred, so the shattering is overdetermined.[3]

As the last remark shows, Suzy's throw did not make a decisive difference to the occurrence of the effect because, even without her rock, the bottle would have been broken by Billy's. Nevertheless, we tend to think that Suzy is the cause, not Billy or both of them. The idea of causes as making a difference to the effect seems not to fit this case. At this point, one might wonder, however, whether this is simply a question of correctly specifying the effect-event. Suzy does make a difference to the shattering as it happened (at its specific time and in its specific form), because the effect would have been slightly different if Billy had thrown. Hall rightly blocks this well-known solution by pointing out that examples might be construed in which the resulting effect is exactly the same, no matter which of the two possible causes has actually caused it. In addition, the strategy of event-specification

---
[3](Hall, 2004, pg. 235)

seems not to be the most obvious way we deal with overdeterminations. To argue that Suzy is the cause, we rely on a family of principles which point us to an actual connection between Suzy and the shattering, for example a spatiotemporal path along which Suzy's ball passes from her hand to the bottle.[4]

The reliance on a spatiotemporal connection, by contrast, does not help in cases of negative causation. Consider this example: Before going to bed, Suzy has secretly switched off Billy's alarm clock. Billy oversleeps the next morning and arrives late at work. What is the cause of his delay? A natural answer would be to cite either the failure of the alarm clock or Suzy's manipulation as causes. Both make a difference to the effect: if the alarm clock had rung, Billy would not have been late. Also, if Suzy had not turned off the alarm clock, it would have rung and Billy would not have been late. But as the failure of the alarm clock is a negative event there is no spatiotemporal connection between these causes and the effect.

As both kinds of examples have to be taken seriously, Hall argues, we are forced to admit that there are two notions of causation. The principle of a spatiotemporal connection is one way to explicate the idea that causes are connected to their effects. This is the sense of causation that Hall calls 'production' and for which he proposes an approach based on nomic sufficiency. In the case of negative causation, our leading question was whether the putative cause makes a difference to the effect. This is the sense of causation that Hall calls 'dependence' and for which he proposes a counterfactual approach. Nevertheless, being guided by his examples, one could also take interventionist and probabilistic approaches as accounts that are able to deal with negative causation, and process and mechanistic approaches as accounts that are able to explicate the production view.[5]

Let us concede for the moment that Hall has actually discussed and dismissed all the strategies that have been proposed to accommodate the possible counterexamples, i.e., that there is no way to explicate the dependence view that could solve cases of overdetermination and that there is no 'production-account' that is able to deal with negative causation. What I would like to focus on is the reasoning by which Hall defends his claim for causal dualism and, connectedly, the question of how to interpret it.

3 Two interpretations of causal dualism

The key to Hall's argument lies in the admission of both negative causation and overdetermination as intuitive examples of causal relationships. Long-

---

[4] Hall explicates two other principles, transitivity and intrinsicness, but, as the line of reasoning is basically the same, I will concentrate on the spatiotemporal connection.

[5] See Hitchcock (2005).

standing discussions on whether negative causation is 'real' causation show that this is not an evident point of departure for a philosophical analysis of causation. Why do all intuitions, shaky or not, have to be respected, as Hall claims? Hall justifies his procedure by referring to his method, conceptual analysis. His idea is to develop a coherent and logically consistent concept of causation which is able to accommodate as many intuitions, conceptual connotations and paradigmatic cases as possible. In most cases of intuitive causal relationships both of Hall's notions of causation can probably be applied. In the standard launching event between two billiard balls, for example, the cause is both spatiotemporally connected to the effect and makes a difference to the effect. One could therefore be tempted to formulate a concept of causation as the conjunction of both notions. It is exactly the contrast between negative causation and overdetermination that reveals, however, that one of the two notions seems to be enough for us to judge something as a cause. Therefore, Hall asks us to overcome the "methodological maxim"[6] of one, unified concept and to accept causal dualism as the fact that we apply different principles in different situations. So far, so good. It makes sense to take a variety of situations into account if the aim is to illuminate our causal judgements. Following this line of reasoning, Hall's claim for causal dualism is to be interpreted as a statement about our way of thinking about causation.

Although being conclusive, this first interpretation raises further questions. From a psychological point of view, one asks for the reason why we should have developed two ways of thinking about causation. Was it just a freak of nature? Or is there a ontogenetic or phylogenetic story to tell? From a philosophical point of view, the explication of our way of thinking about causation is only one half of the job that has to be done. A number of authors[7] have pointed out that there are two questions one might ask about causation, and that the second question concerns the objective structure of the world: What—if anything—is causation 'out there', 'in the objects'? Of course, the results of a conceptual analysis would be philosophically much more interesting if they had also bearing on this second question. Mackie explicitly links the two questions when he states that "[...] we may, or rather must, accept the use of causal language as a rough guide to what we are to take as causal relationships, as indicating—though not as authoritatively marking out—our field of study."[8] In Mackie's analysis, the linkage is obvious. After having identified core features of our concept of causation—causal priority and necessary-in-the-circumstances—

---

[6](Hall, 2004, pg. 254)
[7]For example Mackie (1974) and Dowe (2000).
[8](Mackie, 1974, pg. 1)

, Mackie looks for structures in the world that could serve as grounding for the corresponding conceptual features. If, however, our concept of causation is as dualistic as Hall claims, which of the two notions should then be taken as a guide towards 'causation in the objects'? And: if there was a common grounding, why then is it so difficult to come down to one concept of causation?

It is tempting to evade the last questions with a speculation. Maybe, there are, in fact, two kinds of causal relationships in the world. This reply is the second way of interpreting causal dualism, and it is the one that accounts for the 'revolutionary spirit' of dualistic and pluralistic positions. No doubt, it is an elegant solution for several reasons. First, this interpretation answers both philosophically relevant questions at once: causal dualism can be counted a statement about both our way of thinking about causation and about causation 'in the objects'. Second, it implies a simple theory of correspondence between our thinking and the world. Finally, it directs us to an obvious way of answering the psychological question: we have developed two ways of thinking about causation because there are two kinds of causal relationships in the world. While Hall remains vague as to whether he supports this second interpretation of causal dualism or not, this is the position Cartwright (2004) holds. In her eyes, there are not just two, but a great many different notions of causation corresponding to the richness of causative expressions in natural languages. These various "thick causal concepts"[9] have their counterparts in the world, i.e., every concept stands for a different kind of causal relationship, and there is no single common feature that would make them all causal relationships.

The question is whether this second interpretation—which might be called metaphysical dualism or pluralism—is just a speculation, or whether it can be supported by arguments. In particular, I will concentrate on the question whether Hall's argument for causal dualism can provide any reasons to believe in metaphysical dualism. Is his method and way of arguing appropriate? Can all intuitions about causation be respected if the aim is to find out about 'causation in the objects'? Or, are there external criteria that might force us to concentrate on only some intuitions while neglecting others?

4 Differences in our intuitions

Let us go back to the two kinds of examples Hall used to establish his claim for causal dualism. While both are intuitive causal judgments, they nevertheless reveal two completely different attitudes towards causal relationships. The difference is particularly clear if we think about the reasons

---

[9](Cartwright, 2004, pg. 814)

for not applying the notion of causes as being connected to their effects in the case of Billy and the alarm clock. If we understood causes in this sense, we would have to cite an event or fact as cause that actually occurred before his waking up, for example his sleep. Imagine a situation in which Billy is asked about the cause for his delay and he says: "Well, I slept, got up, took breakfast and came here." Even if this answer was correct, even if there was a spatiotemporal connection between the cause and the effect, we would protest and discard Billy's answer for not being informative. We are not interested in knowing how the effect came about, we are not interested in the mechanism,[10] but we are interested in knowing the difference between this morning and the ones in which Billy arrives in time. Only if we already knew that one of the events that Billy cited is unusual (for example that he normally does not sleep in the morning), we would accept his answer. This is the reason why the failure of the alarm clock seems to be a much more natural answer. If Billy explains that the failure was the cause for his delay, we interpret his statement on the basis of knowing that alarm clocks usually do ring. More generally, the dependence view of causation which is expressed in the notion of causes as making a difference to the effect has two characteristic features. First, the dependence view never just focuses on the actual situation, but always implies—implicitly or explicitly—a comparison between the actual situation and (at least) a second one which differs in regard to the effect. Second, it is irrelevant whether there is, in the actual situation, a connection of any sort between cause and effect or not. The fact that there is, for example, a spatiotemporal connection between the sleep and the delay did not prevent the causal statement from being uninformative; the fact that there is no such connection between the failure and the delay did not inhibit the statement from being informative.

I think that the dependence view is the way we normally deal with causal relationships. We are interested in facts and events that make a difference because they allow us to control and manipulate the effect. In cases of overdetermination, however, the dependence view reaches its limits, and it seems to me as if here we 'switched' to another way of seeing matter. We now are interested in the way the effect was brought about. By calling something a cause in the sense of the production view, we express our conviction that there is, in fact, a certain connection to be found between the cause and the effect. In the case of Suzy, Billy and the throw, there is an obvious reason to believe in the connection: if we were to observe the situation, we could see the ball passing from Suzy's hand to the bottle. In order to

---

[10]The term 'mechanism' is used and defined in various ways. I understand it in the sense of Machamer et al. (2000), whose main focus lies on the productive continuity that links cause and effect.

know what exactly it is that makes the cause being a cause for the effect, we are inclined to ask natural sciences to explain the mechanism. As we believe that there is, sooner or later, a story to tell about the way the effect came about, there is no relevance for contrasting the actual situation with a second one as in the dependence view of causation. Whether something causes the effect or not, will be revealed as soon as one knows enough about the actual mechanism that took place.

On comparing the production and the dependence view of causation, it is obvious that the production view implies a much stronger commitment in regard to what can be known about causation 'in the objects'. While the dependence view eschews the question by taking a second, contrasted situation into account, the production view is based on a firm realist intuition about causation. On calling something a cause in the sense of the production view, one has to believe that there is or will be something to be discovered in the actual situation that makes the relation between cause and effect a causal one. This realist intuition of the production view is, in my opinion, part of our everyday convictions and is called for when needed (as in the case of overdeterminations). Furthermore, the realist intuition might explain a somewhat surprising remark by Hall (2004). In comparing the two senses, he concedes that the production view seems to be more fundamental than the dependence view: "[...], in that production does seem, in some sense, to be the more 'central' causal notion. As evidence, consider that when presented with a paradigm case of production without dependence—as in, say, the story of Suzy, Billy, and the broken bottle - we unhesitatingly classify the producer as a cause; whereas when presented with a paradigm case of dependence without production [..] our intuitions (well, those of some of us, anyway) about whether a genuine causal relation is manifested are shakier."[11] Once we believed that there was more to be known about causal relationships than mere dependence, we would surely like to call a corresponding notion more central.

5 Intuitions and philosophical method

As Hall's conceptual analysis shows, we seem to use both notions of 'cause' depending on the situation we have to judge. Obviously, we do not have any problems in switching in between the two views and the corresponding commitments. But what about a philosopher being interested in an account of causation 'in the objects'? His aim is find out as much as possible about causation in the world. As we will see in this section, it does make a difference for him whether he trusts his realist intuition or not: this commitment determines the method he takes, the way he argues, and, finally, the options

---

[11](Hall, 2004, pg. 256)

he has for being a causal pluralist.

Let us start with the realist intuition of the production view. A philosopher who believes that there is something to be known about the causal connection aims at illuminating what it is. He will try to concentrate on the actual situation and to find out as much as possible about what there 'really' is and what 'really' happens. In a first attempt, the philosopher might analyse his perceptions and hope that the causal connection he looks for is directly perceivable. A second option would be to ask the sciences. The latter strategy has been labelled "empirical analysis"[12] of causation and is the one I want to focus on. Its point is not to pursue another conceptual analysis on the basis of scientists' usage of the term 'cause' (which might then be contrasted to the laymen's usage). Rather, its idea is to take scientific theories as the best current information source about the world and to try to extrapolate from them what can be known about the causal connection.

Two leading defendants of this kind of analysis are Salmon (2002) and Dowe (2000) whose common aim is nicely expressed in the following citation:

> My aim has been to examine causality at what might be characterized as the 'deepest metaphysical level'. The account that has emerged removes this concept from the field of metaphysics and transports it to physics. If this goal has actually been achieved, I count it as philosophical progress.[13]

It is from physics that Salmon and Dowe claim to have taken the distinction between causal and noncausal processes which is the grounding for their process theory of causation. The basic idea is that two events or facts are causally connected if and only if there is a suitable set of causal processes and interactions that could link them. Dowe defines causal processes as the world-lines of objects that possess conserved quantities. Conserved quantities are quantities for which a conservation law exists or will be found in physics, for example mass-energy, linear momentum or charge. A causal interaction is the intersection of two world-lines that involves an exchange of a conserved quantity. Applying these ideas to our example of the shattering bottle, we roughly get the following picture: what we call cause and effect (Suzy's throw and the broken bottle) supervenes on some more fundamental facts of objects possessing and exchanging conserved quantities (for example Suzy's body transmitting a certain amount of energy to the ball; the ball transmitting the energy to the bottle). These facts are connected via a chain of causal processes and interactions (for example the world-line

---

[12] (Dowe, 2000, pg. 2)
[13] (Salmon, 2002, pg. 131)

of the ball having this energy), such that every exchange is governed by the same conservation law. The emerging account of causation 'in the objects' is monistic and, as the example shows, it has no problems to accommodate overdeterminations. The price, however, is that cases of negative causation cannot be handled. Negative events like the failure of the alarm clock are not part of causal processes in the above sense because they do not involve objects possessing or exchanging conserved quantities. Negative causation has to be excluded and one kind of Hall's senses of 'cause' has to be dismissed as not being relevant for a metaphysical analysis of causation.

At this point, one might wonder where the two main features of the process theory, monism and the exclusion of negative events, come from. Are they a direct consequence of the method, empirical analysis? At least for monism, the answer is no, as a second approach shows: Machamer et al. (2000) are also engaged in an empirical analysis of causation on the basis of a strong realist background, but are—in contrast to Dowe—explicitly nonreductionist. They generalize Dowe's exchange of physical quantities to discipline-dependent "activities" in which the corresponding objects or "entities" of a certain field of study engage.[14] In doing so, the objects act as causes because they bring about new facts or events. For example, some of the currently known causal relations in neurobiology are molecules that collide, ions that attract each other or cells that transmit electrical impulses. According to Machamer et al. (2000), each time a plausible combination of entities and activities (i.e., a mechanism) is given to explain how a certain fact or event came about a causal connection has been identified. The integration of other sciences besides physics yields a number of differences between the process theory and the mechanistic approach, but the most obvious change is the resulting pluralism about causation. As the example of neurobiology shows, there might be various activities in one field of science that are taken to be 'real' (in the sense that they are not reduced to some more fundamental facts), and Machamer (2004) doubts that there is one common feature that would make all the activities within and between divergent fields be ways of causing. An empirical analysis of causation does therefore not necessarily yield a monistic theory of causation. Whether metaphysical pluralism is adopted or not, depends on the philosopher's view on sciences, i.e., on his arguments for or against reductionism.

It is striking, however, that the mechanistic approach also opts for an exclusion of negative causation. In a somewhat similar argument to Dowe (2000), Machamer (2004) suggests to understand negative causation (and the corresponding activities like 'preventing' or 'blocking') as counterfactual claims about genuine causation, i.e., about mechanisms that involve 'real'

---

[14](Machamer et al., 2000, pg. 3)

activities. For example, when we call the failure of the alarm clock a cause for Billy's delay, we implicitly talk about the following facts: Billy was late and there was a genuine cause for his delay, the sleep (or a 'real' activity, sleeping). The alarm clock did not ring, but if it had rung, it would have interrupted the actual mechanism with the consequence that Billy would not have been late. Of course, Machamer's point comes down to the question of what 'real' activities are and how a distinction between these and other 'not real' activities could be justified. In a similar way, it might be asked how Dowe justifies the exclusion of negative causation and the underlying distinction between negative and positive events or facts. Neither author has given a convincing scientific argument, but refers at the very end to the intuition of a difference. This intuition is, in my opinion, a direct consequence of their realist commitments. Both Dowe and Machamer aim at an analysis of causation that concentrates exclusively on the actual situation, on what is 'really' there and 'really' happens. Having this aim in mind and trusting that it can be achieved, made them engage in an empirical analysis—and it made them disregard causal judgements about negative causation as not being relevant for an account of causation 'in the objects'.

So far for the people who do trust their realist intuitions. Let us now take a look at a philosopher who is sceptical as to whether there is anything to be known about the causal connection in the actual situation. I will focus on Woodward because he still believes that the question about causation 'in the objects' is sensible. In contrast to subjectivist accounts of causation, he does not deny that causation is an objective feature of the world—but he doubts the prospects of an empirical analysis in the above sense.[15] What does he do instead? How does he ensure that his account yields as much information about causation 'in the objects' as possible? The point of departure for Woodward's analysis are, similarly to the ordinary conceptual analysis, causal judgements of laymen and scientists. In addition, he introduces two normative criteria by which an account of causation should be guided. Not only should it be able to account for paradigmatic cases and intuitions about causation, but it should also (1) "explain why causal knowledge is sometimes practically useful and what its practical utility consists in.",[16] and (2) "be accompanied by some epistemological story that makes it understandable how human beings can sometimes learn whether claims with that content are true or false from evidence that is actually available to them."[17] To respect these two criteria—to link our concepts

---

[15] See, for example, his statement that: "The account that I present is not reductive, and I am sceptical that any reductive account will turn out to be adequate." (Woodward, 2003, pg. 21)

[16] (Woodward, 2003, pg. 30)

[17] (Woodward, 2003, pg. 22)

of causation with the way causal knowledge is actually used and tested in the world—guarantees, in Woodward's eyes, that the results of the conceptual analysis can reveal objective features of causation 'in the objects'.[18] A theory of causation which gives some plausible (or maybe even empirically justified) answers has the legitimisation to dismiss or correct some non-fitting intuitions.

In order to accommodate the two criteria, Woodward's account is based on two fundamental assumptions, namely:

1. The main purpose of causal knowledge is to use it for controlling and manipulating.

2. The best methods that have been developed in science for discovering and testing causal relationships are experimental ones (for example randomised controlled experiments).

Once the assumptions are accepted, there follows a straightforward development of a notion of causation in which something can roughly be called a cause if and only if the effect changed in the case of some intervention on the cause. Woodward's account is primarily aimed at capturing the notion of causation in type-level causal claims. Cases of singular or actual causation as in Hall's examples have to be seen as instantiations of type-level relationships and are analysed in the following way: the failure of the alarm clock, for example, is an actual cause of Billy's delay because there is a way of intervening on the alarm clock (for example, restoring its proper functioning) that would change Billy's arrival time, given that all other possible influences on Billy's arrival time that are independent of the alarm clock (for example, whether his bicycle works or not) stay as they are. As this example makes clear, Woodward translates the claimed causal relationship between two facts or events in the actual case into a statement about the relation between two variables (status of the alarm clock, arrival time) that can take on at least two different values (functioning / not functioning, in time / delayed). The consequence he draws for causation 'in the objects' is that there must be objective structures in the world that make such statements about variables true.

---

[18]This line of reasoning is particularly clear in considering the fact that Woodward uses exactly these two criteria in order to argue against a subjectivist conception of causation: first, all the methods that have been developed to test causal relationships would be irrelevant if causal relationships were completely dependent on us. The mere fact that there are accepted methods for causal inferences in science reveals that there must be something objective to be discovered. Second, if there were no objective difference between causal and mere correlated relationships, it would have made no evolutionary sense that the concept of causation had ever emerged.

By taking variables as the causal relata, Woodward is clearly a proponent of the dependence view of causation whose main characteristic is to compare the actual situation with a second possible one. Consequently, he has no problems in accommodating cases of negative causation. A negative event like the failure of the alarm clock is simply one value of one of the variables between which a causal relationship is claimed. Connectedly, it might not be surprising that it is exactly the cases of overdetermination that cause most of the problems for Woodward. If one does not admit the strategy of event-specification (which would mean, in Woodward's terms, a redefinition of the values of the variable that are taken into account),[19] Woodward's original analysis of actual causation yields the counterintuitive judgment that neither Suzy nor Billy are causes of the broken bottle. In a second attempt, Woodward suggests a more refined analysis (actual causation*) in which, however, both Suzy and Billy would have to be cited as causes. Neither analysis yields the seemingly obvious result that only Suzy is the cause of the broken bottle. Woodward's account has no resources to capture the idea of causes as being connected to their effect which seems to underlie our intuitive causal judgment.

At this point, one can ask a similar question as in the case of the process theory: Is the resulting monistic theory of causation and the neglect of the production view a consequence of Woodward's philosophical method? In regard to monism, one might well imagine a theory of causation that is based on various equally weighted methods for discovering and testing causal relationship yielding a kind of methodological pluralism. Arguments for or against pluralism would thereby depend on what methods are taken to be the best. Similarly, one might think of an account of causation that is developed in correspondence to various ways of using causal knowledge. Arguments for or against pluralism would thereby depend on what functions of causal knowledge are taken to be central. Therefore, monism about causation seems to be a consequence of Woodward's assumptions rather than of his method.

The case is different, however, in regard to the neglect of the production view. At first glance, Woodward's justification for dismissing the production view also seems to rely heavily on the fact that he takes controlling to be the main use of causal knowledge and experiments to be the best method. The production view, he argues, can be neglected because (a) knowledge about the causal connection is neither necessary nor sufficient for knowing how to control the effect; and (b) the best methods that have been developed in sciences (i.e., experimental ones) are not based on the idea of tracking causal processes. A defender of the production view might, in contrast,

---

[19] See Hall (2004), and section 2.

argue that knowledge about causal connections is useful for understanding the true structure of the world, and that the methods that have been developed in order to track causal processes are actually the best methods we have. It is Woodward's reply to this objection that leads us back to his scepticism: Woodward does not believe that we are able to find out about the true structure of the world and about actual causal connections. Causal knowledge about actual causal connections is, in his eyes, not practically useful because we cannot know it. In contrast to monism, Woodward's neglect of the production view can therefore be seen as a direct consequence of his epistemological commitments.

6 Conclusion

Let me sum up what can be learnt from Woodward's and Dowe's approaches. Both go beyond a conceptual analysis of causation. The authors not only want to illuminate our way of thinking and judging about causation, but are interested in extracting the objective core of causation, in linking our way of talking about causation with causation 'in the objects'. Their divergent epistemological commitments, i.e., whether they believe that there is something to be known about the causal connection or not, leads them to two completely different methodological approaches. Trusting his realist intuitions, Dowe suggests an empirical analysis of causation and develops a monistic account of causation that cannot accommodate negative causation. Woodward, being in contrast more sceptical, sticks with the traditional conceptual analysis, but introduces two further normative criteria and develops a monistic account of causation that cannot accommodate the idea of causes as being connected to their effects. The resulting monism is in neither case a consequence of the method that is taken. The neglect of the dependence view in Dowe's theory and of the production view in Woodward's theory, however, seems to be directly linked to the chosen method and to refer back to their different epistemological commitments: Dowe excludes negative causation because he wants to focus on what is 'really' there. Woodward dismisses the production view because he does not believe that this kind of causal knowledge is achievable and doubts its practical utility.

In short, the philosopher's decision for a certain epistemological point of view not only determines the philosophical method that is taken, but also implies a weighting of the two kinds of causal intuitions that Hall has identified. Hall's equally intuitive ways of judging causal relationships cannot be treated equally in an account of causation that aims at more than just a description of our way of thinking about causation. The different epistemological commitments that are involved in both senses of 'cause' act as external constraints on developing an account of causation 'in the objects'. A direct translation of Hall's claim for dualism into a statement about two kinds of causal relationships in the world does not make sense because it would mean to mix up incompatible points of view. One cannot be sceptical and nonsceptical at the same time in regard to the question whether there is something to be known about the causal connection in the actual situation or not.

On the one hand, the reference to the underlying epistemological commitments therefore blocks the second metaphysical (and more interesting) interpretation of Hall's claim. On the other hand, it is exactly this difference in regard to epistemology that might explain where conceptual causal dualism comes from. As we switch between the two convictions, we switch between two ways of understanding causal relationships.

Finally, while Hall's argument cannot be taken as support for any metaphysical claim, other ways of arguing for metaphysical dualism (or pluralism) have emerged on the way. Following an empirical analysis in the sense of Dowe, pluralistic positions might arise on the basis of arguments against reductionism. Following Woodward's strategy, pluralistic positions might arise on the basis of arguments for different equally legitimate methods to discover and test causal relationships, or on the basis of different ways of using causal knowledge. In short, there are options for being a causal pluralist in a sense that goes beyond a mere psychological statement, but the options are restricted: one can be a causal pluralist based on the intuitions of the production view, and one can be a pluralist based on the intuitions of the dependence view, but one cannot have a little bit of both.

Acknowledgments

I thank Peter Machamer, Daniela Bailer-Jones and two referees for discussions on this topic and comments of earlier versions of the paper. It was partly written during my stay at the Department of History and Philosophy of Science in Pittsburgh which was financially supported by DAAD (German Academic Exchange Service).

Monika Dullstein
Department of Philosophy, University of Heidelberg, Germany.
monika.dullstein@uni-heidelberg.de

BIBLIOGRAPHY

Cartwright, N. (2004). Causation: One word, many things. *Philosophy of Science*, 71:S805–819.
Dowe, P. (2000). *Physical Causation*. Cambridge University Press, Cambridge.
Eells, E. (1991). *Probabilistic Causality*. Cambridge University Press, Cambridge.

Hall, N. (2004). Two concepts of causation. In Collins, J., Hall, N., and Paul, L. A., editors, *Causation and Counterfactuals*, pages 225–276. MIT Press, Cambridge, MA.

Hitchcock, C. R. (2005). How to be a causal pluralist. *Talk given at the 7th Pittsburgh-Konstanz Colloquium in Philosophy of Science, Konstanz, May 2005.*

Machamer, P. (2004). Activities and causation: The metaphysics and epistemology of mechanisms. *International Studies in the Philosophy of Science*, 18:27–39.

Machamer, P., Darden, L., and Craver, C. F. (2000). Thinking about mechanism. *Philosophy of Science*, 67:1–25.

Mackie, J. L. (1974). *The Cement of the Universe*. Oxford University Press, Oxford.

Salmon, W. C. (2002). A realistic account of causation. In Marsonet, M., editor, *The Problem of Realism*, pages 106–134. Ashgate, Hampshire.

Sober, E. (1985). Two concepts of cause. In Asquith, P. D. and Kitcher, P., editors, *PSA 1984*, pages 405–424. Philosophy of Science Association, East Lansing.

Woodward, J. (2003). *Making Things Happen*. Oxford University Press, Oxford.

Can one deny both causation by omission and causal pluralism? The case of legal causation

AMIT PUNDIK

ABSTRACT. Various distinguished philosophers (e.g., Dowe, Armstrong, Salmon and Beebee) deny causation by omission by holding that causing things involves a physical interaction and therefore omissions cannot be causes. With only rare exceptions (e.g., Menzies), it is also common to deny causal pluralism by assuming that even though the concept of causation appears in various disciplines, it has a core meaning that can be captured by a single unified concept. In this paper, I argue that one cannot coherently deny both causation by omission and causal pluralism. At least one of them has to be accepted. My argument stems from a discussion of the usage of causation by omission in the law. In various legal fields, lawyers commonly regard omissions as causes. I show that these cases of causation by omission have to be captured by the concept of causation because the action/omission distinction is much more fragile than is currently conceived. I distinguish between two types of omission, qualitative and quantitative: whilst the literature focuses on the first, I show why the second creates more serious problems, especially for those who seek to deny causation by omission. I then show why dismissing legal causation as a mistaken usage of the concept of causation requires an acceptance of causal pluralism.

1 Introduction

One could deny *causation by omission* by holding that causing things involves a physical interaction and therefore cases of causation by omission are not genuinely causal. One could deny *causal pluralism* by holding that even though the concept of causation appears in various disciplines, it has a core meaning that can be captured by a single unified concept. Can one coherently deny both causation by omission *and* causal pluralism? In this paper, I seek to show that one cannot: either one has to accept causation by omission or one has to accept causal pluralism.

My argument is based upon the way causation is used in the legal context. First, I argue that legal causation should include omissions. I do so by showing that there are some cases of omission for which one should be legally liable. I then explain why the law should require a connection between the omission and the harm. I then proceed to the main part of my argument and show why this connection has to be causal *per se* and why these omissions cannot be captured or explained away with an alternative concept. First, I show the insufficiency of some specific concepts that have been suggested by various deniers of causation by omission (Dowe (2001, 2004); Hall (2004); Beebee (2004) in particular). Then, I highlight some fundamental problems that any alternative concept would face. Mainly, I argue that the action/omission distinction is much more fragile than is currently conceived. I distinguish between two types of omission, qualitative and quantitative. The first is the omission to do something, whilst the second is the omission to do *enough* (for example, the failure to drive slowly enough). Whilst the literature focuses on the first, I show why the second creates more serious problems, especially for those who seek to suggest an alternative concept for capturing omissions. Legal causation should therefore include omissions.

Second, I argue that legal causation cannot be dismissed as a mistaken usage of the concept of causation. I support this claim by showing that such a dismissal requires an acceptance of causal pluralism. Thus, it is consistent with my thesis that one has to accept either causation by omission or causal pluralism. I also hope that this section will elucidate why legal causation is a problem for philosophers.

Last, by investigating the motivations to resist causation by omission, I highlight two issues that are central to the legal discussion of causation, but have not received enough attention from philosophers of causation. The first is the existence of a distinction between causes and background conditions (or between 'a cause' and 'the cause'). The second is the existence of a normative element in the concept of causation.

My argument, if successful, should contribute to both the philosophical and legal discussions of causation. In philosophy, it poses another challenge to the denier of causation by omission (henceforth 'the denier'). Recently, Schaffer provided a detailed and powerful articulation of the case for causation by omission (Schaffer's case is in favour of negative causation rather than causation by omission, but it is unnecessary to distinguish between them for the purpose of this paper). He argues that causation by omission "is supported by all the central conceptual connotations of causation... features in paradigm cases of causation... is required by the useful theoretical applications of causation... and is routinely recognised in scientific practice" (Schaffer, 2004, p.203). My argument augments Schaffer's case in the following way. I argue that even if the denier is correct in holding that scientific applications of causation do not require causation by omission,[1] legal applications of causation do. If the denier wishes to bypass this problem by regarding the legal applications to be a mistaken usage of the concept of causation, s/he will have to accept causal pluralism.

My argument can also contribute to the intensive and extensive legal discussion of causation.[2] The traditional legal analysis of causation tends to distinguish between 'factual causation' and 'proximate causation'.[3] The first is considered purely factual whilst the second excludes some of these factual causes due to various policy considerations, and picks out only those factual causes for which one should be legally liable. Sometimes, lawyers use other terms to distinguish between factual and proximate causation, such as 'cause-in-fact' and 'legal causation'. To avoid confusion, *legal causation* in this paper is used to refer only to the first concept ('factual causation' or 'cause-in-fact') as it is used in the legal context. Several legal scholars from both the continental tradition[4] and the Anglo-American tradition[5] suggest splitting this concept into two, one for actions and another for omissions. However, I seek to show that at least in the legal context, the concept of causation must be a single concept that captures cases of both actions and omissions.

The structure of my argument is as follows. In section 2, I articulate what exactly is being rejected when one denies causation by omission or when one denies causal pluralism. In section 3, I show why legal causation has to include both actions and omissions. In section 4, I show that legal causation cannot be dismissed as a mistaken usage of the concept of causation without accepting causal pluralism. Lastly, in section 5, by discussing

---

[1] Dowe regards his theory as a product of empirical analysis (Dowe, 2000, p.7). But see also Schaffer's objection to this claim, arguing that scientists commonly refer to causation by omission in their research (Schaffer, 2004, p.208).

[2] Hart and Honoré (1985) are probably the most familiar source to both lawyers and philosophers. However, one of the purposes of this paper is to bring other legal literature on causation to the attention of the philosophical community, see, in particular, section section 5 below.

[3] This traditional distinction appears in criminal and tort law in particular. For this distinction in criminal law, see Card (2006, sections 2.24-2.25, pp.69-70); in tort law, see Simon Deakin and Markesinis (2003, pp.185-189). For a theoretical analysis and criticism of this distinction, see Hart and Honoré (1985, pp.90-94).

[4] Hart and Honoré provide a good survey of the issue in German scholarship (Hart and Honoré, 1985, pp.447-449).

[5] Becht and Miller (1961), for example, distinguish between simple causation for actions and hypothetical causation for omissions. However, Hart and Honoré observe that "the difficulty about omissions... has been felt more acutely in Continental than in Anglo-American law" (Hart and Honoré, 1985, p.447).

the motivations to resist causation by omission, I identify two issues that are central to legal causation and deserve more attention from philosophers of causation.

2 What exactly is being denied?

2.1 Causation by Omission

To deny causation by omission, one needs to insist that only actions can be causes and that we are mistaken whenever we identify omissions as genuine causes. For example, Armstrong argues that "omissions and so forth are not part of the real driving force in nature. Every causal situation develops as it does as a result of the presence of positive factors alone" (Armstrong, 1999, p.177). Beebee (2004) argues that causation by omission is over-inclusive because it acknowledges an infinite number of events that are obviously causally irrelevant. She also argues that acknowledging causation by omission introduces a normative element into the concept of causation (I describe and address both of her arguments in section 5).

The denier also needs to overcome, or at least explain away, some obvious cases in which it seems that omissions are causes. For instance, Alice died from dehydration because she did not drink enough water. Proximity aside, *prima facie*, one of the causes of her death was her omission to drink enough water. Some philosophers make an effort to explain such cases away by providing another concept to capture cases of causation by omission. Dowe (2001, 2004), for instance, uses *quasi-causation* to account for omissions as a counterfactual dependence; similar suggestion appears in German legal scholarship (Hart and Honoré, 1985, pp.447-449); and Becht and Miller (1961) use *hypothetical causation*. Another interesting idea is Beebee's suggestion that purported cases of caustion by omission are really cases of *causal explanation* rather cases of *causation* (Beebee, 2004). What is shared by all of these various attempts is the insistence that *genuine* causation does not include omissions.

2.2 Causal Pluralism

The monist, who denies causal pluralism, expects all cases of causation to be captured under a single concept. Although causation appears in various disciplines and contexts, the monist tries to extract the common features that make these various usages 'causal' and identify a single concept of causation.

In contrast, the pluralist considers causation to be a bundle of concepts. There can be different ways to construct this bundle. One way is to divide causation into types. For instance, Hall (2004) distinguishes between *dependence* and *production*. Another way is to match different forms of linguistic usages of causation with different metaphysical accounts of causation. Bennett, for example, distinguishes between fact-causation and event-causation. Fact-causation is used to express the causal story more precisely, whilst event-causation is used when we know only little of the causal story: "the relative uninformativeness of event-causation statements is their *raison d'être*" (Bennett, 1998, p.137).[6]

The version of casual pluralism on which I focus in this paper holds the concept of causation to be *discipline-dependant*. Similar to Gillies (2000, chaps. 8-9) who suggests that different disciplines may require different interpretations of probability, the proponent of this version holds that different disciplines may also require different interpretations of causation.[7] This version of causal pluralism is *contextual* in nature, because the truth values of causal statements are dependant on the context in which the question occurs (whether the questioner belongs to one discipline or another).[8] What makes this view pluralistic is its willingness to classify the exact same case as causal in one discipline and non-causal in another discipline whilst maintaining that *neither of the classifications is mistaken.*

There are at least three ways to construct the contextual version of causal pluralism. The most radical (and sometimes slightly caricaturised) option is to argue that the various discipline-dependant concepts are fully distinct and have no connection to each other whatsoever. A proponent of such a position would be justifiably required to explain why so many cases are classified similarly amongst different disciplines. Hart and Honoré use a similar argument to criticise those who think that it is impossible to specify principles according to which common sense is used to determine causal judgments. They argue that if this is true, these sceptics have to explain how and why we reach the same causal judgments most of the time (Hart and Honoré, 1985, p.26).

A more plausible option is to consider the various discipline-dependant concepts as sharing the same causal judgments about many (if not most) of the cases, but differing about a non negligible number of other cases. The shared cases could be explained by the common origin of the various concepts (e.g., ordinary language or common sense). Without necessarily accepting causal pluralism, Dowe recognises that "the word 'cause' as sci-

---

[6] However, although Bennett acknowledges event-causation, he explicitly argues for its inferiority to fact-causation (Bennett, 1998, sec.54, pp.139-142).

[7] Note, however, that most interpretations of probability share a common compliance with the probability calculus (e.g., Kolmogorovs axioms). The Baconian interpretation of Cohen (1997) is a noticeable and intended exception. In contrast, there is no such a recognised common denominator for causation. I thank Donald Gillies for this important observation.

[8] For attributor dependence as a feature of contextualism, see DeRose (1992).

entists use it in those scientific situations must make some 'historical' or 'genealogical' connection to everyday language" (Dowe, 2000, p.9). Hart and Honoré identify the source of the concept of legal causation in the common sense of the ordinary person. They devote the entire first section of their book to exploring the meaning of this concept in the ordinary language (Hart and Honoré (1985, chaps. I-V) and see also in Hart and Honoré (1985, pp.1-2)). They also show that the traditional approach of legal causation even attempted to solve complicated cases by reference to common sense (Hart and Honoré, 1985, pp.91-93). The disputed cases, in contrast, could be explained by the different needs which led each discipline to refine the imprecise ordinary-language concept of causation in a different manner. As Hart and Honoré observe

> We must not impute to ordinary thought all the fine discriminations that could be made and in fact are to be found in a legal system... Where there is no precise system of punishment, compensation or reward to administer, ordinary men will not often have faced such questions... Such questions courts have to answer and in such cases common judgments provide only a general, though still an important indication of what are the relevant factors. (Hart and Honoré, 1985, p.75)

This option is thus able to explain both common and disputed cases. Hart and Honoré advocate this sort of causal pluralism explicitly:

> The common notion of causation may have features which vary from context to context... there may be different types of causal inquiry. (Hart and Honoré, 1985, p.19)

Later on, they distinguish between effect, consequence and result, and conclude that "there is not a single concept of causation but a group or family or concepts" (Hart and Honoré, 1985, p.28).

A third option for constructing a contextual version of causal pluralism would be to preserve causation as a single concept, but to introduce a contextual element into it. For example, Menzies suggests "a way that integrates a certain contextual parameter in the relevant truth conditions" (Menzies, 2004, p.139). What is shared by all these options is the acknowledgment that *some* cases can be classified as causal in one discipline but as non-causal in another, *when none of the classifications is necessarily mistaken.*

It is also important to note contextual versions of causation that are *not* pluralist. Common sense seems to distinguish between salient causes ('the cause', such as the spark that started the fire) and background conditions ('a cause', such as the presence of oxygen) (Hart and Honoré (1985, p.11); Menzies (2004, p.143) and see section 5 for elaboration on this issue). Lewis (1986a, pp.215-216) adopts a contextual strategy to explain that it is the context which makes a cause to be salient. However, this is *not* a pluralist version, because the truth conditions of the causal propositions (whether something is *a cause*) are determined before the context highlights which cause is more salient (Menzies, 2004, p.151). To borrow Menzies' distinction, Lewis' theory is an *add-on* rather than *integrated* context-sensitive theory of causation (Menzies, 2004, p.151).

Note that the contextual version of causal pluralism can be easily denied by proponents of other versions of causal pluralism or causal contextualism. Nevertheless, I will argue that it is this version of causal pluralism (contextual and discipline-dependant) that one has to accept if one denies causation by omission. Henceforth, 'casual pluralism' is used to refer to this version only.

3 Why legal causation should include omissions

In this section, I argue that the same concept of legal causation that captures causation by action has to include causation by omission too. My argument consists of several steps:

1. Some omissions should be a possible basis for legal liability.

2. In those cases, a connection between the omission and the harm should sometimes be required to establish legal liability.

3. The connection between the omission and the harm has to be causal.

Therefore, legal causation should include causation by omission.

In each step of this argument, I examine a different way to avoid the need to classify causation by omission as genuine causation. In the first step, I examine the suggestion to avoid attributing legal liability to omissions at all. In the second step, I investigate whether other types of legal liability that do not require causation can be used to capture the cases of causation by omission. In the last step, I question whether lawyers can use one concept for actions and another concept for omissions. I examine whether either Dowe's *counterfactual dependence* or Beebee's *causal explanation* can serve as the alternative concept and I identify general problems that any suggestion for two concepts would face. Each of the following sections establishes the corresponding step in the argument above by rejecting one of these three different ways to avoid classification of the cases of causation by omission as genuinely causal.

3.1 Step 1: Why omissions should be a possible basis for legal liability

In this step I seek to show that even if one denies that cases of causation by omission are genuinely causal, it is difficult for one to deny that *at least in some cases*, a person should nevertheless be legally liable for the consequences of his/her omissions. Note that although this section sometimes includes causal terms (consequences, results, etc.), I do not assume that these cases are genuinely causal. All I argue in this section is that some cases of omissions should be subject to moral responsibility and legal liability (the role of the concept of causation in determining these moral responsibility and legal liability is discussed in section 3.3).

I take two alternative routes to establish this normative claim, in order to accommodate for conflicting approaches about the nature of the connection between law and morality. There is an extensive debate in the legal literature about whether this connection is really necessary; to what extent legal rules should match moral responsibilities; how these moral responsibilities are to be identified; etc. For the purpose of my argument, and with an acknowledged simplification, I distinguish between positivist and non-positivist approaches. Positivist approaches put more emphasis on the validity of the rule-making procedure rather than on the content of the rules themselves (and, hence, they put less emphasis on the *morality* of the rules).[9] Other approaches insist upon *some* necessary connection between law and morality.[10]

My first route aims to satisfy positivists by identifying several legal liabilities existing in the law (I assume that their legislative procedures were complied with the standards required by the positivists). In various legal fields, the law explicitly recognises some cases in which a person can be held legally liable for his/her omission. In criminal law, legislation sometimes includes specific references to clarify that omissions can be subjected to legal liability.[11] Thus, parents can be held legally liable for their omis-

---

[9] Hart is also a leading figure in legal positivism, see Hart (1994).

[10] Probably the most passionate objection to Hart's positivism comes from his former student Dworkin (see Dworkin (1978, 1986)). Another notable theory which subordinates law to morality is Finnis' theory of natural law, see Finnis (1982).

[11] See, for example, Children and Young Persons Act 1933, sec.1(1) (the offence of wilful neglect of a child by a person who has a responsibility for the child in a manner that cause the child unnecessary suffering or injury to health); Theft Act 1968, sec.24A (the offence of dishonestly retaining a wrongful credit); Gas Act 1986, sec.38(2) (the offence of failing without reasonable excuse to provide information required under Gas Act 1986); Zebra, Pelican and Puffin Pedestrian Crossings Regulations and General Directions 1997 and Road Traffic Regulation Act 1984, sec.25(5) (the offence of failing to accord precedence to a pedestrian on a 'zebra' pedestrian crossing); Road Traffic Act 1988, sec.170 (the failure to report a motor accident); and Person Act 1861, sec.34 (doing or omitting

sion to care for their children. For example, if parents notice their child drowning in a swimming pool, fail to take any action, and the child dies, then they might be held legally liable for his death.[12] Even offences that "may ordinarily require an act on the part of the defendant... can also in principle be committed by an omission to act" (Card, 2006, sec.2.10, p.57). The denier's objection that omissions cannot initiate causal processes is acknowledged but is considered as insufficient to prevent attribution of legal liability (Card, 2006, sec.2.39, p.86). In tort law, a person can be held legally liable for his/her failure to warn or to take adequate precautions. For instance, in the American case of *Tarasoff*,[13] a psychiatrist was held liable for failing to warn about the threats of his patient to murder his girlfriend. In contract law, a party who fails to perform an action specified in the contract (such as paying a certain amount by a certain date), can be held legally liable for the losses of the other party (Treitel, 2003, pp.753).[14] Indeed, legal liability for omissions exists only where there is a recognised duty of case to act in a certain way (Card (2006, sec.2.10, p.57) for criminal law and Simon Deakin and Markesinis (2003, p.149) for tort law). It might be also that "liability for omissions is exceptional" (Palmer (1999, p.159); and see also Simon Deakin and Markesinis (2003, pp.78-79)). However, all that is required for the purpose of my argument is that exceptional or not, in *some* cases, liability for omissions exists.

My second route aims to satisfy the non-positivist positions (which generally connect law to morality) by identifying cases in which the moral responsibility for omissions is so clear that it would be difficult to argue against legal liability. Note that this is a very modest claim, as can be seen by looking into the arguments of those who object omissions. Some state that imposing a duty to act upon individuals requires more consideration, because such a duty limits the individual's autonomy more significantly than a restriction not to act in a certain way (Ashworth, 2006, p.110). Others require more conditions to be satisfied before such liability is imposed. For example, a limitation of the risks that one can be reasonably expected to take when fulfilling one's duty (Card, 2006, sec.2.14, p.62). However, whilst this objection might be sufficient to establish a difference in gravity between the moral responsibility that should be attached to actions and to omissions,

---

anything to endanger passengers by railway).

[12] Children and Young Persons Act 1933, sec.1(1); the cases of *R v Instan* [1893] 1 QB 450 and *Stone and Dobinson* [1977] QB 354, [1977] 2 All ER 341.

[13] *Tarasoff v University of California* 17 Cal. (3d) 425 [1976]. For this issue in general, see Simon Deakin and Markesinis (2003, pp.152-153).

[14] The liability for failure to perform is traditionally considered as strict liability, that is liability that exists even when the failure was faultless, Atiyah and Smith (2006, pp.167-172).

it is hard to rely on this argument to support a case for *no* moral responsibility at all in any possible case of omission. Hence, my argument does not require one to take any particular position in the debate about whether the gravity of the moral responsibility for omissions should be of similar gravity to the moral responsibility for actions.[15] It also does not require one to take a position about the way moral duties determine our legal liabilities. All I argue is that there are *some* cases in which a person should be morally responsible and legally liable for the consequences of his/her omissions.

Assume, for example, that before Alice died of dehydration, she was taken to the hospital and treated by a vicious nurse, who instead of administering her an infusion of fluids, sat in a chair near Alice and peacefully drank tea. Had the nurse administered the infusion to Alice, Alice would have been saved. For the sake of simplicity, assume also that the vicious nurse had a clear intention for Alice to die. Indeed, demanding the nurse to stop drinking her tea and to start treating Alice might limit her autonomy more than demanding the nurse to refrain from certain actions (such as to refrain from shooting Alice). Be that as it may, this limitation on the nurse's autonomy should be weighed against the interests of other individuals, and of Alice in particular (Ashworth, 2006, p.46). It is therefore difficult to argue that the nurse has *no* moral responsibility whatsoever for Alice's death. In this extreme example, it is also difficult to argue that even if there is moral responsibility, the law should refrain from ascribing legal liability for such conduct ('conduct' is used in the legal context for both omissions and actions, Palmer (1999, p.158)).

This section has sought to show what it takes to reject my thesis by arguing against imposing legal liability for omissions. If one insists that "there is no single case of omission for which a legal liability should exist", then admittedly one can reject my argument. However, this position is difficult to sustain.

3.2 Step 2: Why legal liability requires a connection between omission and harm

Lawyers who were convinced by the denier's arguments against causation by omission might suggest another strategy: attributing other types of legal liability that do not require harm and hence, do not require a connection between the offender's conduct and the victim's harm (I do not assume that this connection is genuinely causal. See section 3.3 and section 4 for the role of the concept of causation in legal causation).

---

[15]This question arises, for instance, in cases of euthanasia. For that debate, see Fletcher (1967), who argues for a distinction between actions and omissions; and Kennedy (1977), who argues that it is morally irrelevant whether the conduct was consisted of body movements or not.

In criminal law, there are two main types of legal liability that do not require this connection. The first type of liability is through *conduct offences*, in which the offence is constituted only from certain behaviour of the individual, regardless of whether this behaviour has had harmful consequences or not (Card, 2006, sec.2.16, p.63). For instance, driving too fast attracts a legal sanction even if no accident has occurred as a result (Road Traffic Regulation Act 1984, sec.89(1)). Similarly, if Alice from the example above is under sixteen, the vicious nurse could be tried for cruelty to persons under sixteen, which applies "notwithstanding that actual suffering or injury to health...was obviated by the action of another person" (Children and Young Persons Act 1933, sec.1(3)(a)). The second type of legal liability which does require harmful consequences is *inchoate offences*, which impose legal sanctions for the mere *attempt* to commit another offence.[16] For example, in the case of *White*,[17] the accused put poison into his mother's drink with the intention of killing her. However, the mother died of heart failure, not of poisoning. The accused was acquitted of murder (because his conduct was not the cause of death) but convicted of attempted murder. Similarly, in Alice's case, the vicious nurse could be tried for attempted murder. In both types of liability, the law considers the conduct of the accused to be so condemnable that it deserves a legal sanction even if no harmful result has occurred. Since no harmful result is required, no causal connection is required either. Therefore, either type of legal liability could be put forward by those who deny causation by omission but want to avoid the radical strategy discussed in the previous step (section 3.1), according to which no legal liability should be imposed for any omission.

However, this strategy is difficult to sustain for several reasons. First, even if this strategy is successful in the criminal context, it faces much more difficulty in the civil context, because it is doubtful whether similar solutions can be found in that context. With few exceptions, civil wrongs require actual harm before compensation will be ordered: "damage is also an essential requirement" (Simon Deakin and Markesinis, 2003, p.76). Although there are some exceptions such as libel, trespass and false imprisonment, "in negligence, the claimant must prove that he sustained a loss or injury as a result of the defendant's negligence" (Simon Deakin and Markesinis, 2003, p.76). Similarly, contract law usually requires actual harm before awarding compensation: "a claimant cannot recover substantial damages if the breach has not adversely affected his position" (Treitel, 2003, p.934).

---

[16] Primarily, the Criminal Attempts Act 1981. Interestingly, this legislation uses the word "act" rather than "conduct", that can be interpreted as an exclusion of omissions. See Palmer (1999) and Dennis (1982) for objections to such an interpretation. For more on attempts, see Card (2006, sec.17.54-17.86, pp.689-706).

[17] *R v White* [1910] 2 KB 124.

Hence, this strategy might not succeed in civil law and hence, this solution is only partial.

Second, *prima facie*, there is a moral significance to the question whether the victim suffered actual harm or was merely exposed to a risk that was not instantiated (everything else being equal).[18] Regardless of whether one accepts or rejects this moral significance, the strategy is problematic. If one accepts that there is a moral difference, then including cases of causation by omission under both types of legal liability mentioned above is unsustainable. The reason is that this strategy diminishes the moral difference between omissions which result in actual harm to the victim and omissions which merely subject the victim to a risk (but luckily nothing happens). For instance, this strategy diminishes the difference between the example above (in which Alice dies), and a variation in which another nurse interferes and saves Alice by administering her an infusion of fluids. According to this strategy, the legal liability of the vicious nurse would remain the same regardless of whether Alice died or not. This is inconsistent with the initial acceptance of the moral significance of actual harm.

Of course, one can reject this moral difference and argue that it is morally insignificant whether or not the offender achieves her desired result. Hart, for example, asks "why should the accidental fact that an intended harmful outcome has not occurred be a ground for punishing less a criminal who may be equally dangerous and equally wicked?" (Hart, 1968, p.129). To put it more technically, one can argue that resultant luck should be irrelevant to moral responsibility.[19]

Indeed, this is a possible response that can be used to reject my argument. However, it has a high price. First, a claim of no moral difference between successful and unsuccessful attempts should be accompanied with a call to change every part in our current criminal law and tort law that depends upon this difference.[20] These changes amount to very radical reform. In criminal law (to the extent that moral desert is the basis of punishment), not only would the sanctions for unsuccessful attempts and successful attempts have to be compared, but also numerous offences would have to be either added or removed from our current criminal law. In every case where a successful attempt is subject to a sanction but an unsuccessful attempt is not, either the existing offence that captures the successful attempt should be removed, or alternatively, a new offence should be added to capture the

---

[18] Ashworth (2006, sec.2.4, pp.30-33). Ashworth bases the requirement of harm on Mill's Harm Principle and advocates minimalism in extending criminal liability for no-harm situations (Ashworth, 2006, sec.2.4, pp.33-35).

[19] Resultant luck is only one of the four types of moral luck identified by Nagel (1979).

[20] For the debate on resultant luck in the criminal law, see Davis (1986); Feinberg (1995). For the debate on resultant luck in tort law, see Waldron (1995).

unsuccessful attempt. In tort law, the reform is even more radical because most torts currently require actual harm.[21] It is interesting to notice that even those who accept the problem of moral luck and call to revise our practices do not go that far to suggest such radical reforms (see, for instance, Browne (1992)).

Second, this position is faced with its own difficulties. Withdrawing the requirements of harm and causal connection leaves the agent's conduct and the agent's mental state (which lawyers call *mens rea*) as the only elements that are required to constitute the offence. Thus, two out of four elements of the offence would be dropped. From an evidential perspective, proving the mental state of the accused at the time the offence was committed is extremely difficult.[22] Empirical research shows that juries have difficulty in accurately determining the state of mind of a witness *even when s/he stands in front of them*. One interesting example is the popular myth that the self-confidence of a witness reflects reliability and truth. In contrast, empirical research conducted on mock juries (it is forbidden to conduct research on real juries) shows that if self-confidence reflects anything, it is probably the opposite.[23] These difficulties intensify dramatically if the state of mind in question is unobserved by the jury and occurred in the distant past.[24] Whilst these difficulties are already apparent in the current system, leaving the mental state and the conduct as the sole elements of the offence might lead to a substantial increase in both erroneous convictions and erroneous acquittals. This is because juries and judges might rely more heavily on mistaken generalisations and dangerous prejudices.

I accept, however, that if one goes the entire way to argue that actual results are irrelevant to moral responsibility and legal liability, then one can reject my argument. If one claims that harm and causal connection are not needed for determining legal liability, then obviously causation by omission is not a problem. However, this is not a specific solution to the cases of causation by omission, but rather an overall abolishment of the requirements of actual harm and causation in the legal context, which is a very radical position to maintain. As long as one wishes to avoid this radical position, the question surrounding causation by omission cannot be

---

[21] See text after *supra* n.17.

[22] Cane, for example, raises a series of difficulties to prove intention. Although he brings them to explain the relatively minor role intention has in tort law, the same difficulties of proof exists in other legal fields and regarding other mental states (Cane, 2000, pp.542-544).

[23] See Roberts and Zuckerman (2004, pp.218-221), and the empirical studies they bring. In particular, see Blumenthal (1993, pp.1194-1195).

[24] For a more optimistic view of this issue, see Card (2006, sec.4.17-4.18, pp.141-143). However, the empirical basis for this optimism is unclear.

resolved by switching to other types of legal liability.

3.3 Step 3: Why the connection between omission and harm is necessarily causal

The first two steps showed that it would be hard to avoid attribution of any legal liability for omissions or to attribute only legal liabilities that do not require harm (and, thus, do not require causal connection). However, the denier of causation by omission can still insist that unlike the connection between action and harm, the connection between omission and harm is not genuinely causal. As noted above (section 2.1), several philosophical and legal suggestions were raised to elucidate how omissions can be captured with an alternative concept (quasi-causation, causal explanation, etc.). Let the concept that the denier advocates for genuine causation, whatever it is, be Concept A (for actions). Let the alternative concept that the denier suggests for omissions, whatever it is, be Concept O (for omissions). In this section, I seek to show that the cases of legal causation cannot be divided between two concepts, Concept A and Concept O, but have to be captured by a single concept that includes both actions and omissions.

I separate my discussion into two parts. In the first part, I address some of the specific alternative concepts that were suggested in the literature. In the second part, I identify generic problems that every approach which seeks to split legal causation into two concepts would encounter, whether or not the legal consequences of these two concepts are the same.

Specifics: the redundancy of the existing accounts

I divide the various suggestions for Concept O that appear in the literature into two main types: *counterfactual dependence* and *causal explanation*. For each suggestion, I seek to show that Concept O is sufficient to capture the current connotations of legal causation, making Concept A redundant. Furthermore, I show that Concept A fails to capture several important connotations of legal causation. In addition to omissions, Concept A has difficulty capturing some types of *actions* for which legal liability is commonly imposed (see section 4 for why this is a problem for the denier). Therefore, the workability of Concept O and difficulty of Concept A make it harder for the denier to suggest that legal causation is made up of two concepts, one for actions and another for omissions (attempts to use only Concept O to capture *all* cases of legal causation are discussed in section 4).

Counterfactual dependence Many philosophers distinguish between two possibilities for the concept of causation: the first is based on an actual *physical connection* between cause and effect (whatever this means) and the second requires only a *counterfactual dependence* between them. They all share the view that omissions can only be captured by the latter. Hall, for example, draws a distinction between two concepts of causation: *production* and *dependence* (Hall, 2004, pp.225-226).[25] Production is consistent with three theses about causation (transitivity, locality and intrinsicness) but is inconsistent with omissions. Dependence, on the other hand, is consistent with omissions but inconsistent with these three theses (Hall, 2004, sec.5, pp.248-252). Lewis calls the physical connection *biff* and emphasises that omissions can only be captured by counterfactual dependence (Lewis, 2004, pp.282-283). Schaffer divides the various accounts of physical connection into three types: *transference, process,* and *tie* (Schaffer, 2004, p.203). The common ground shared by all accounts of physical connection is the denial that cases of causation by omission are genuinely causal. Thus, a possible alternative concept could rely on the distinction between physical connection (Concept A) and counterfactual dependence (Concept O).

The denier's strategy to explain away cases of causation by omission using this distinction is based on two steps. First, the denier establishes that cases of causation by omission are captured only by counterfactual dependence. Second, the denier argues that only physical connection is genuinely causal whilst counterfactual dependence is not. For instance, Dowe puts forward another concept, *quasi-causation*, which captures the cases of causation by omission. Quasi-causation consists of a counterfactual dependence: the non-action (omission) O quasi-caused the effect E iff had the action occurred, the effect would not have occurred (Dowe, 2004).[26] According to Dowe, this counterfactual dependence is the reason why we intuitively yet mistakenly consider some cases of causation by omission to be genuinely causal. However, only physical connection is genuinely causal and counterfactual dependence is not. Therefore, according to Dowe, cases of causation by omission only seem genuinely causal, when in fact, they are not.

Schaffer raises three arguments against Dowe's theory (Schaffer, 2004, pp.212-214). First, he argues that the theoretical difficulties involved in acknowledging omissions that the denier identified are, in fact, difficulties with the physical connection theories themselves. Second, the concept of quasi-causation does not really work, as cases of over-determination between omissions show. Last, the various connotations of causation show that quasi-causation, "that paddles, waddles and quacks like causation" (Schaffer, 2004, p.213), is a better candidate for genuine causation than physical connection.

---

[25] Hall recognises that these might be two kinds of causation rather than two concepts, but dismisses this objection by noting that "in the case at hand it doesn't matter in the slightest" (Hall, 2004, p.256). Thus, I take his suggestion to be for two concepts.

[26] Like Beebee (2004, p.294), I use a simpler version that the one actually used by Dowe, as I focus on the counterfactual dependence, which exists in both versions.

Schaffer's third argument can be augmented by examining the usage of causation in the legal context in detail. In the law, *counterfactual dependence* is actually very similar to the legal test of causation, which is used in various legal fields for both omissions and actions. The traditional test in law is the *sine qua non* test, also known as the *but-for* test: the claimant has to prove that *but for* the alleged conduct of the defendant, the harm would not have occurred.[27] Noticeably, this test is very similar, if not fully identical, to Dowe's test for quasi-causation.[28] This traditional test of causation is criticised by many distinguished scholars. Hart and Honoré, for example, devote an entire chapter to describe in detail the deficiencies of the *sine qua non* test (Hart and Honoré, 1985, chap.5). However, other suggested tests also include an element of counterfactual dependence. For example, Wright suggests the NESS test: a cause is a Necessary Element in a Sufficient Set of conditions (Wright (1985, p.1777) and Wright (1988)). This test is based on Mackie's INUS (Mackie, 1974). This test has an element of counterfactual dependence because the requirement for the element in a set to be necessary (Mackie, 1974, chap.2). Like *sine qua non*, NESS has no difficulty in identifying omission as a cause (Fischer, 1992, p.1340). Another legal test is the requirement that cause be a *substantial factor* or *material contribution*. This test was justifiably criticised for lack of definition as it effectively allows juries to reach whatever conclusion they like (Fischer, 1992, p.1347). Since it has no clear meaning, it is hard to know whether counterfactual dependence has any role in this test. In any case, this test is only used to address specific problems such as cases of overdetermination (Fischer (1992, p.1345), referring to these cases as 'multiple sufficient causes'). Therefore, counterfactual dependence is an integral part of legal causation in most cases.

Not only is the legal test of causation more similar to Dowe's quasi-causation, it is also questionable whether any account of physical connection can be successfully applied in the legal context. One of the main problems in applying physical connection accounts of causation in the legal context arises in cases when a person commits a certain offence having been persuaded, forced, induced, ordered, offered, or threatened to do so by another person (Hart and Honoré, 1985, p.51). Many of these conducts do not necessarily involve the physical transfer/process/tie that is endorsed by the physical connection accounts. Indeed, these situations involve complicated issues of mental causation, which are outside the scope of this paper. Nevertheless,

---

[27] In criminal law, see Ashworth (2006, p.124); Card (2006, sec.2.24, p.69); Ormerod (2005, p.53). In tort law, see Simon Deakin and Markesinis (2003, p.189) and McBride and Bagshaw (2005, p.530). See also Fischer (1992, p.1338) and Williams (1962, p.63).

[28] The counterfactual element of the but-for test is acknowledged in the legal literature too, see Grady (1984, p.392).

the question whether of the offender acted *because* of the other person's conduct can have significant legal consequences. In particular, the offender can try to avoid a conviction by raising the general defence of duress (Card, 2006, sec.19.15-19.27, pp.781-793).

For example, assume that the Queen of Hearts *implicitly* threatens that she will kill Humpty if he does not steal the Mad Hatter's teapot for her (both explicit and implicit threats can suffice to constitute the defence of duress, see Card (2006, sec.19.15-19.27, pp.781-793)). Humpty tries but is caught and put to trial. Humpty can argue that he acted under duress, and, thus, should not be convicted of theft. The key question before the court would be whether Humpty acted *because* of the Queen of Hearts' threat. The important point to note is that it is hard to apply the physical connection theories to law, because it might be difficult to identify a physical connection between Humpty's action and the Queen of Hearts' conduct (at least as long as 'physical connection' does not include cases of conversation, implicit threats, hints that can be understood as threats only in particular contexts, etc.). In contrast, counterfactual dependence provides the law with a meaningful test with which these cases can be analysed. For example, "would Humpty have acted as he did *but for* The Queen of Hearts' threat?" No wonder that the current legal test of duress includes a counterfactual element (Card, 2006, sec.19.17, p.782).

Therefore, physical connection accounts are more difficult to apply in the legal context not only because they cannot capture omissions, but also because they face difficulties in capturing some types of actions for which moral responsibility and legal liability should be attributed.[29]

Note that I do not argue that using counterfactual dependence in the legal context is problem-free. However, most of these difficulties are common to both counterfactual dependence and physical connection, because they result from a lack of sensitivity to the context and to the manner of description, which makes these problems equally applicable to both types of account.[30] Similarly to Schaffer (2004, p.213), in this section I argued that in the legal context, the concept of counterfactual dependence is more successful than the concept of physical connection. This is because counterfactual dependence can capture both actions and omissions and physical connection has serious difficulties in capturing either of them. As Hart and Honoré observed a long time ago with reference to cases in both English

---

[29] Some attempts were made by scholars and courts to use "forces" as an account of causation. These attempts are presented and justifiably criticised by Hart and Honoré (1985, pp.96-97).

[30] For elaboration on some of the legal difficulties of the *sine qua non* test (counterfactual dependence), see Hart and Honoré (1985, pp.68-70) and also Hart and Honoré (1985, chap.V).

Common Law and French Civil Law, "the analogies with a thing which is active or moves have darkened and continue to darken many a discussion of causation" (Hart and Honoré, 1985, p.30).

Causal explanation Beebee (2004) suggests using the concept of *causal explanation* to account for cases of causation by omission whilst preserving genuine causation for actions only. According to Beebee, causal explanation can contain information other than causes (Beebee, 2004, p.302). Cases of causation by omission are causal explanations (*e because c*) rather than genuine causes (*c causes e*), and describing them as causes is philosophically mistaken (Beebee, 2004, p.305). As Beebee admits (Beebee, 2004, p.306), her alternative concept does not rule out the possibility of accepting causation by omission, but rather gives the denier a convincing response to explain away intuitions about causation by omission. (However, Beebee does raise other arguments against accepting causation by omission, which I describe and criticise in section 5.)

Similar to Dowe's quasi-causation, Beebee's causal explanation fits the legal context in a way which renders genuine causation redundant, at least from a legal perspective. Causal explanation seems to fit the legal context better for two reasons. First, as Beebee emphasises (Beebee, 2004, p.307), causal explanation is context-sensitive. In particular, it takes into account the attribution of *blame* and *fault*, which are important features of the legal context, mainly in criminal law, but also in tort law.[31] As argued by Hart and Honoré, to state something as a legal 'cause' does not merely state what the cause is, but also serves as *an explanation* when one is sought after (Hart and Honoré, 1985, p.32).[32]

Second, causal explanation is better equipped to address a deficiency from which most accounts of genuine causation suffer. Most accounts do not provide any mean to distinguish between the offender stabbing a knife into the victim's heart and the great-great-grand mother of the offender gave birth to his great-grand father a hundred years ago. This problem is recognised by several philosophers and lawyers. Schaffer titles it as *selection* (Schaffer (2003, sec.2.3), as do Hart and Honoré (1985, p.22)), whilst Menzies calls it *the problem of profligate causation* (Menzies, 2004, p.139). In the legal liter-

---

[31] The debate about the role of moral blame in law, in relation to other utilitarian goals such as deterrence or efficiency, is extensive and appears in many legal fields. However, all I assume is that moral blame has some importance for the law, even if this importance is limited or can be accommodated within and even reduced to utilitarian goals.

[32] Hart and Honoré refer to the ordinary usage of the word 'cause' rather than to its usage in the legal context. However, they also argue that the legal concept of causation should be strongly connected to its ordinary meaning and to the common sense (Hart and Honoré, 1985, p.26). Therefore, I think it is safe to apply their discussion of causal explanation to the legal context.

ature, this problem appears as *proximity*.³³ Interestingly, whilst disagreeing on the nature of causation, Lewis (1986b, p.198) and Mackie (1974, p.34) both consider this distinction to be merely contextual.³⁴ As Beebee notes (Beebee, 2004, p.307), unlike the concept of causation, causal explanation can easily address this problem by distinguishing between *true* explanation and *appropriate* explanation. Similar to Dowe's quasi-causation, Beebee's causal explanation seems to suffice on its own, making the concept of genuine causation redundant, at least in the legal context.

Beebee might respond that this redundancy poses no problem for her, as the alternative concept should be used to capture *all* cases of legal causation (or, in her terms, all cases of legal 'causal explanation'). I address this response in section 4. However, before doing so, I turn to complete my argument against splitting legal causation into two concepts (one for actions and another for omissions) by highlighting some generic difficulties that every such account faces.

Generics: the difficulties every account has

If an event falls under Concept O, this classification will have precisely the same legal consequences as if it had fallen under Concept A (assuming everything else to be equal), or not. There is no third option. In the first part of this section, I identify the problems which arise when the legal consequences are the same. In the second part, I highlight the problems which arise when the legal consequences are different.

Same legal consequences. If Concept A and Concept O lead precisely to the same legal consequences, what is the point, from a legal perspective, in distinguishing between these two concepts? Would it not be better, for the sake of conceptual parsimony, to have one unified concept that captures both actions and omissions?

One can insist that conceptual parsimony is not so important, and that if these two concepts are really separated, then the law should reflect this. Indeed, this is a question in which both philosophers and legal scholars might find interest. Nevertheless, my argument in this section (2) is merely that causation has to include omissions *in the legal context*. Therefore, if classifying a given event under Concept A or Concept O makes no difference to the legal consequences, one has to explain why and for whom this is an interesting *legal* question. Legal practitioners would not and should not invest efforts in this question. Litigants can reasonably expect their legal representatives to invest their billable time only in legal questions that

---

³³ *Supra* n.3.
³⁴ However, Menzies argues that this admission is not enough to solve this problem, and the contextual element should be integrated into the truth conditions, Menzies (2004, sec.4, pp.145-151).

might have consequences in the actual case at hand. As for the courts, investing resources (the time and efforts of judges and administrators) in addressing this question may be justifiably criticised as a waste of the taxpayer's money. Moreover, one can argue that even if this question should be addressed by the courts at some stage, the priority should be given to questions that have legal consequences. Thus, instead of investing resources in this question, courts should use these resources to provide a timely resolution of other legal questions that do have legal consequences (reducing delays was one of the main motivations of Lord Woolf's recent reform of the Civil Procedure Rules 1998 (CPR)).[35] The legislature can also be reasonably expected to deal only with legal questions that have an impact upon our rights and duties, or at least can be expected to address those questions first before investing efforts in legal questions which have no impact upon us. Hence, if one insists that this is an important *legal* distinction, one has to explain why and for whom it is so important.

Of course, one can insist that whether an event falls under Concept A or under Concept O is still an interesting philosophical question and it does not really matter that it has no importance for legal practitioners, judges and legislators. Dowe, for instance, argues that "it does not matter that for practical purposes we don't bother to, or can't distinguish quasi-causation from causation. The distinction only becomes important theoretically, in metaphysics" (Dowe, 2004, p.194). Be that as it may, *in the legal context*, using two separated concepts which have precisely the same legal consequences is a waste of effort that should be better invested somewhere else. Thus, if the legal consequences are the same, causation *in the legal context* should capture both actions and omissions (and see section 4 why legal causation is a problem for philosophers of causation).

Different legal consequences. The denier can respond to the last argument by accepting that Concept A and Concept O should have different legal consequences. For instance, the ethical literature on euthanasia includes an interesting debate about whether it morally and legally matters if the doctor or nurse's conduct consists of an action or omission.[36]

Two strong arguments have been raised in the legal and ethical literature to show that Concept A and Concept O should *not* have different moral and legal consequences. First, it is morally insignificant whether the agent achieved his/her intended result with or without body movements

---

[35] For a detailed discussion of the problems of delays in general, the problem of the lengthy delays experienced in the pre-CPR period, and the attempts of the reform to solve this problem, see Zuckerman (2004, sec.1.21-1.35, pp.10-15) and also Zuckerman (2004, sec.1.59-1.78, pp.26-33).

[36] *Supra* n.15.

(Kennedy, 1977).[37] Second, any omission can be described as an action merely by changing the *scope* of the description. Hart and Honoré note that "(a)n event... may be identified or described by reference to features which need not be causally connected with the outcome, provided that there are other features of the event which are causally connected to it" (Hart and Honoré, 1985, p.lix). They illustrate this observation with the case of *Empire Jamaica*[38] (affirmed on appeal in *Empire Jamaica* (CA)[39] and *Empire Jamaica* (HL)[40]), in which shipowners did not certify their officers and an experienced yet uncertified officer was on duty when a collision occurred. An important point can be extracted from their observation: the event can be described either as an omission (the shipowners' failure to certify the officer) or as an action (the officer's skilful navigation).[41] Ashworth also brings some examples for omissions that can be described both as actions and omissions and highlights how "courts have sought to exploit this ambiguity" (Ashworth, 2006, p.111). No wonder that Fischer concludes that "any omission can be characterized as part of a larger encompassing act" (Fischer, 1992, p.1339).[42]

Both arguments can be used to block the denier's attempt to sustain two separate concepts in the legal context by suggesting that they should have different legal consequences. The first argument, the lack of moral significance of body movements, can be used to question the justification for having different consequences for each concept. If the denier accepts this objection, s/he has to withdraw to accept the same legal consequences, and in that case, s/he faces again the difficulties described above (section 3.3, under "Same legal consequences"). The second argument, the problem of two possible descriptions, can be used to question the workability of the denier's conceptual framework, as it is unclear how to divide the cases between these two concepts.[43]

---

[37] Note that this argument assumes that physical body movements account for the difference between actions and omissions.

[38] *NV Koninklijke Rotterdamsche Lloyd v Western Steamship Co Ltd (The Empire Jamaica)* [1955] 1 All ER 452.

[39] *NV Koninklijke Rotterdamsche Lloyd v Western Steamship Co Ltd (The Empire Jamaica)* (1955) 3 All ER 60 (Court of Appeals).

[40] *NV Koninklijke Rotterdamsche Lloyd v Western Steamship Co Ltd (The Empire Jamaica)* (1956) 3 All ER 144 (House of Lords).

[41] Hart and Honoré make this observation to point out that the proper way to describe an event focuses on the unlawful elements (Hart and Honoré, 1985, p.lix). However, the criterion to prefer one description over another is normative rather than causal.

[42] See also Becht and Miller (1961, pp.178-179); Hart and Honoré (1985, pp.138-139); and Green (1962, pp.546-547).

[43] "This demonstration of the fragility of the act-omission distinction... indicates that it may be simplistic to oppose omissions liability in principle", Ashworth (2006, p.112). But see Dowe (2004, p.194) for a dismissive and unestablished response to the lack of

In this section, I raise an additional argument to show that the problem of two possible descriptions is much more serious than has been hitherto articulated. The purpose of this argument is to emphasise the difficulties that the denier would have if s/he was suggesting that cases of causation by omission should be accommodated by a different concept with different legal consequences than the concept of causation.

The structure of my argument is as follows. First, I distinguish between two types of omission, qualitative and quantitative. Whilst the literature focuses on the first, I show why the second is even more problematic. Unfortunately, for the denier, the second type is very important in the legal context. I then argue that if the two concepts were to have different legal consequences, each litigant would seek to describe the event in the way that promotes her interests best. How should the court decide between the two descriptions? I argue that in the more difficult case of quantitative omission, there are no clear criteria, either factual or normative, for how to choose between the competing descriptions.[44]

Qualitative and quantitative omissions. Two types of omission can be distinguished: qualitative and quantitative. Qualitative omissions are situations in which the agent was required to take a particular action to achieve a certain effect, but instead took *a different action* (or took no action at all). For example, when the vicious nurse peacefully drinks her tea instead of administering an infusion of fluids to Alice. Drinking tea is a different type of action than administering an infusion. This is the type of omission that appears in the writings of Fischer and of Hart and Honoré (see above, in section 3.3, under "Different legal consequences"). In contrast, quantitative omissions are situations when some action of the required type was taken, but not to the required level. In other words, a quantitative omission is the omission *to do enough*. For instance, assume that before she was dehydrated, Alice drank only one glass of water. Had Alice drunk two litres of water instead of one glass, she would not have been become dehydrated. Alice's failure to drink *enough* water resulted in her dehydration.[45]

Action and omission as two possible descriptions. Both types of omission can be described either as omission or as action. A qualitative

---

practical means to distinguish between his two concepts (quasi-causation and causation).

[44] My argument may be useful in the ethical debate about whether actions and omissions have different moral and legal consequences, as it reveals the difficulty to distinguish between actions and omissions when making an argument for a moral difference between them. However, this ethical debate is outside the scope of this paper.

[45] I do not assume that this is a case of causal connection. This connection might well fall under Dowe's concept of quasi-causation which in turn based on counterfactual dependence, see section 3.3, under "Counterfactual dependence".

omission can be described as an action by emphasising what the agent *actually did* (e.g., the nurse drank tea), or as an omission by emphasising what the agent *failed to do* (e.g., the nurse did not administer the infusion). With quantitative omissions, changing the emphasis of the description is even easier. As an action, the event can be described as *doing too little* and the description can emphasise the insufficient part of the action that was done (e.g., Alice drank too little water). As an omission, the same event can be described as *not doing enough*, a failure to do enough to achieve a certain effect (e.g., Alice failed to drink enough water).

There are many varieties of quantitative omission because quantity can be expressed in various ways. The quantity can be in terms of intensity (e.g., "drinking too little" vs. "not drinking enough"), time (e.g., "continuing to press on the accelerator until it was too late" vs. "not pressing the break pedal soon enough"), or distance (e.g., "standing too close to the cliff" vs. "not standing far enough from the cliff"). Thus, any event of quantitative omission can be described either as an action (doing too little) or as an omission (not doing enough).

The Legal Importance of Quantitative Omissions. Quite surprisingly, most of the common examples in the philosophical literature are qualitative rather than quantitative omissions. All of the eight examples brought by Dowe are for qualitative omissions (Dowe, 2004, p.191). Although disagreeing with Dowe about whether these cases should be described as actions or omissions, Schaffer also confines his discussion to qualitative omissions (Schaffer, 2004, pp.199-200). All of the examples Beebee uses are also qualitative (Beebee, 2004, p.294).

Nevertheless, in the legal context, quantitative omissions have a very important role. In criminal law, some types of conduct are unlawful only when they are done excessively. For instance, if a person drives over the speed limit (Road Traffic Regulation Act 1984, sec.89(1)), his conduct can be described either as an action (driving too fast) or as an omission (failure to drive slowly enough). In tort law, rather than not supplying any information at all, negligent professionals are often accused of not supplying *enough* information to enable an informed decision. For instance, in the case of *Barkes*,[46] an insurance broker did not supply enough information about an insurance policy. His conduct can be described either as an action (supplying too little information) or as an omission (failing to supply enough information). In contract law, parties to contract sometimes perform their obligations in a partial way (Treitel, 2003, pp.816-819). For instance, if a party breaches the contract by paying only part of what she was obliged to pay, her conduct can be described either as an action (paying too little) or

---

[46] *George Barkes (London) Ltd v LFC (1988) Ltd* [2000] P.N.L.R 21.

as an omission (failure to pay enough).

Similar cases can be found in many other areas of law. In humanitarian law, "prisoners of war shall be evacuated, as soon as possible after their capture, to camps situated in an area far *enough* from the combat zone for them to be out of danger" (1949 Geneva Convention III Relative to the Treatment of Prisoners of War, article 19, my emphasis). In welfare law, a homeless might be considered ineligible for housing accommodation if "he, or a member of his household, has been guilty of unacceptable behaviour serious *enough* to make him unsuitable to be a tenant of the authority" (Housing Act 1996, sec.160A(7)(a), my emphasis). In administrative law, "(a) registration authority is under a duty to ensure that there are *enough* registrars for its area" (Civil Partnership Act 2004, s. 29(2), my emphasis). Another example is school planning, which should be done "with the aim of ensuring that there are sufficient places available on relevant courses so that there are *enough* teachers with the right skills in schools" (Education Act 2005, sec.114, article 236, my emphasis). In environmental law, when the Secretary of State considers granting an exclusion from the controls imposed by waste management licenses, s/he shall regard "any deposits which are small *enough*...that they may be so excluded" (Environment Protection Act 1990, s. 33(4)(a), my emphasis). Another example is the regulation of an access to open country: "the strip of adjacent land comprised in any access order shall be wide *enough* to allow passage on foot along the water and wide *enough* to allow the public to picnic at convenient places" (Country Side Act 1968, s. 16(3), my emphasis). In constitutional law, "the Lord Chancellor can ask for reconsideration if he feels there is not *enough* evidence that the person is suitable for office; if he feels there is not *enough* evidence that person is the best candidate on merit; or if there is not *enough* evidence that the judges of the Court will between them have *enough* knowledge of, and experience in the laws of each parts of the United Kingdom, following the new appointment" (Constitutional Reform Act 2005, sec.30, article 120, my emphasis). In intellectual property law, "provision may be made by rules prescribing the circumstances in which the specification of an application for a patent, or of a patent, for an invention which requires for its performance the use of a micro-organism is to be treated as disclosing the invention in a manner which is clear *enough* and complete *enough* for the invention to be performed by a person skilled in the art" (Patent Act 1977, sec.125A). These are merely some non-exhaustive examples to illustrate the intensive legal usage of quantitative omissions.

How to Choose between the Descriptions? Back to the denier. If the denier was to suggest that Concept A and Concept O have different legal consequences, surely each party would describe the event in the manner that fits her interests best. How should the court decide between the two possible descriptions?

Even qualitative omissions can be difficult to classify. For instance, in the case of *Fagan*,[47] the court deliberated whether driving a car on to a police officer's foot and then lingering before reversing it amounted to action or omission. In the context of euthanasia, Ormerod notes that "ending and not continuing look uncommonly like the same thing" (Ormerod, 2005, p.88). However, for qualitative omissions, there might be a relatively simple solution. Regardless of the causal status of these omissions, one can argue that from a moral or legal perspective, the agent should not be morally responsible or legally liable for the action itself but only for the omission (Hart and Honoré give a detailed account of this solution in Hart and Honoré (1985, pp.lix-lxi)). For instance, in the example of the vicious nurse, there is nothing morally wrong or unlawful about peacefully drinking tea. It is the nurse's failure to administer an infusion to Alice that is morally wrong and unlawful. Whether she drank tea, played golf, or went swimming instead, is morally and legally irrelevant to the determination her moral responsibly and legal liability respectively.

Note that this point blocks another possible attempt to explain away cases of causation by omission. One could suggest that the cause is the active aspect of the event (e.g., it was the drinking of tea that caused Alice's death). However, not only does the action that accompanies the omission (drinking tea) seem causally irrelevant (Lewis, 1986b, pp.192-193), it may also be perfectly lawful. Thus, legal liability cannot be attached to such causes (and see 3.1 and 3.2 for why a legal liability is required and why establishing it requires a connection between the conduct and the harm).

Such a convenient way out is unavailable when quantitative omissions are involved. Why should driving too fast more (or less) morally condemnable than failing to drive slowly enough? Why should providing too little information about an insurance policy be subject to more (or less) severe legal sanction than failing to provide enough information?

If actions and omissions differ in their legal consequences, how should the court choose the right description, when each party seeks to convince the court that the description that fits her interests best is the correct, right or fair way to describe the event? Assume, for instance, that instead of administering no infusion at all to Alice, the nurse administered a single unit to Alice (knowing that three units were required to save her life). Assume further that the criminal law states that death caused by action is subject to ten years imprisonment, whilst death caused by omission is subject to only

---

[47] *Fagan v Metropolitan Police Commissioner* [1969] 1 QB 439.

five years imprisonment.[48] The prosecution might seek to describe the event as the nurse administering *too little* infusion, whilst the defence might seek to describe the event as *not* administrating *enough* infusion. How should the court decide between these two descriptions?

The challenge that quantitative omissions pose to those who seek to distinguish between actions and omissions is much more serious than the challenge posed by qualitative omissions. *Inter alia*, my argument will pose a difficulty for the denier if s/he is to suggest dividing legal causation into two concepts that have different legal consequences. Even if the denier can provide two concepts, it is difficult for him/her to specify when each concept applies.

3.4 Conclusion: legal causation should include omissions.

There are some cases in which omissions should be subject to legal liability (section 3.1) and cannot be properly captured by any form of legal liability which does not require a connection between the conduct and the harm (section 3.2). However, if a connection between the harm and the omission is required, it has to be a causal connection, because it is difficult to separate legal causation into two concepts, one for actions (genuine causation) and another for omissions (quasi-causation or causal explanation). Therefore, legal causation should include both causation by action and causation by omission.

4 Is legal causation really causation?

One might dismiss my argument by responding that identifying the appropriate concept of causation for legal purposes and justifying the attribution of moral responsibility or legal liability to cases of omissions are problems for lawyers and ethicists rather than for philosophers of causation. Whatever makes an agent legally liable for omissions has nothing to do with causation. These may be important or interesting problems but an account of causation can be complete without addressing these problems.

When asked to explain the fact that legal causation includes omissions, one can simply respond that maybe lawyers get it wrong. Maybe lawyers just do not understand the true metaphysical nature of causation. Maybe whenever lawyers use the term 'causation', we are mistaken in diverting from its true metaphysical meaning (for a remark in this vein, see Beebee (2004, p.305)). Instead of accepting my argument above as a problem, one

---

[48]Obviously, this is an oversimplification of the legal issue, as there are many more factors that should be considered, such as *mens rea* (the mental attitude of the accused), attenuating circumstances, etc. However, in the hypothetical example it is assumed that the rest of the factors are similar between the two offences, and, thus, they are not required for the purpose of my argument.

can rely on this argument to show that the legal usage of the concept of causation is, in fact, misguided.

Before addressing this objection, the extremity of this accusation should be noticed. If the legal connotations are to be dismissed, then the intensive reliance of the legal discipline on the concept of causation should be ignored and considered irrelevant to the discussion of causation.[49] If the legal usage is indeed misguided, it means that a whole community systematically and continuously uses the concept of causation wrongly.

However, an extreme accusation is not necessarily a wrong accusation, and thus, I offer the following response. Assuming that the legal usage of the concept really is mistaken, how should lawyers correct their ways? Three options are available. The first option for lawyers is to stop attributing any legal liability that requires a proof of causation to any case which is not genuinely causal. This is very radical demand because it is difficult to stop attributing legal liability to omissions (section 3.1), and legal liabilities that do not require causation are insufficient to capture the gravity of guilt (section 3.2). The second option is that every time lawyers talk about cases of omission, we should stop using the term 'causation' and switch to another term (maybe quasi-causation or casual explanation). However, this demand is either wasteful (if consequences are similar, section 3.3, under "Same legal consequences") or very difficult to follow (if consequences are different, section 3.3, under "Different legal consequences").

As a last resort, the denier might conclude that lawyers should use quasi-causation or causal explanation for both actions and omissions and leave causation alone. According to the denier's terminological reform, lawyers should only use Concept O and never use Concept A.

But does this response really differ from causal pluralism? According to the pluralist, there are several concepts of causation, each fitting a different discipline. Let concept L to be the concept of causation as it is used in the law and let concept N to be the concept of causation as it is used in the natural sciences (of course, the pluralist could identify a separate concept for each of the natural sciences. However, for the sake of simplicity, I assume a single concept for all the natural sciences.). Compare the pluralist's position to the denier's terminological reform. First, according to the pluralist, lawyers should always use a different concept (Concept L) to natural scientists (Concept N). The same can be said about the denier's concepts (with Concept O and Concept A respectively). Second, the crux of the pluralist position is that when considering whether a certain case

---

[49] For the same conclusion, see Hart and Honoré (1985, pp.84-85). For the centrality of the concept of causation to the legal discipline, see the examples brought in section 3.1, after *supra* n.10.

falls within their scope, these two concepts can lead to different conclusions (section 2.2). The same can be said about the denier's concepts.

The similarities between the positions go even further, as it seems that the denier's Concept O and the pluralist's Concept L resemble one another tremendously in both their extension and their intension. Both concepts have the *same extension*: they both arrive to the same results regarding every case. When lawyers classify a case as causal, the pluralist would classify it under Concept L (legal causation) and the denier under Concept O (quasi-causation or causal explanation). If the natural scientists classify a case as causal, the pluralist would classify it under Concept N (natural sciences causation) and the denier under Concept A (genuine causation). Moreover, since these basic classifications are the same, so are all of their derivatives. For example, the denier would identify any case that the pluralist identifies as falling within Concept L (legal causation) but not within Concept N (natural sciences causation), as falling within Concept A (genuine causation) but not within Concept O (quasi-causation or causal explanation). For example, the denier would classify the connection between the vicious nurse's conduct and Alice's death as falling within Concept O (either as quasi-casual or as causal explanation) and not within Concept A (genuine causation). The pluralist would classify the same case as falling within Concept L (legal causation) and not within Concept N (natural science causation, assuming again that this concept does not include omissions). Therefore, under the denier's terminological reform, the extension of the pluralist's concept of legal causation is identical to the extension of the denier's alternative concept.

Furthermore, both concepts have the *same intension*: the concept of causation in the law, even if mistaken, is based on counterfactual dependence (section 3.3, under "Counterfactual dependence"). Take, for instance, Dowe's quasi-causation, which consists of counterfactual dependence. What then is the difference in meaning between Concept L (the pluralist's legal causation) and Concept O (Dowe's quasi-causation)?

The only difference between the pluralist and the denier is in their attitude towards the difference in classifications. In particular, if lawyers classify an omission as causal whilst natural scientists do not, the pluralist would consider it as merely a legitimate case of disagreement between the disciplines. The denier, in contrast, would condemn the lawyers' usage of the concept of causation as mistaken (for example, see Beebee (2004, p.305)).

But one might wonder what makes the usage of this concept (or any other) to be mistaken or irrelevant to the metaphysical discussion. If one's theory of concepts is that concepts should *describe* our usage of causation, then attributing moral responsibility and legal liability is one of the main usages of the concept of causation (Schaffer, 2004, p.201). Omissions should sometimes be a possible basis for legal liability (section 3.1) and legal liability sometimes require a proof of causation (section 3.2 and 3.3). The same can be said if one's theory of concepts is that concepts should be defined according to their extension. If one's theory of concepts is that concepts are determined by the community which uses them, all the more so.[50] As shown above (section 3.1, after *supra* n.10), the legal community persistently uses the concept of causation to capture cases of causation by omission. Indeed, one might respond that for matters of causation, only the scientific community counts. Whilst it is unclear whether the scientific usage of the concept of causation is really evidence in favour of the denier of causation by omission or against him/her (see the debate between Dowe (2000, p.7) and Schaffer (2004, p.208) on this point), it is even less clear what the basis and justification are for preferring one community over another. For instance, when responding to the objection that one of his two concepts of causation might not really be causation, Hall accepts that the concept of dependence, which captures omissions, is less central than the concept of production, which captures actions (Hall, 2004, p.259). I fail to understand the basis for this concession. As shown above (section 3.3, under "Counterfactual dependence"), in the legal context, if any of the concepts is more central, it is certainly the concept of dependence rather than production. Even if the production contender suits science better, what is the basis of dismissing so quickly the centrality of dependence in the legal context? Therefore, the criteria according to which one can classify the legal usage of the concept of causation as mistaken or irrelevant remains unclear.

To summarise, even if lawyers are mistaken in their usage of the concept of causation, it is unclear how lawyers should respond to this criticism. Stopping imposing legal liability for omissions (section 3.1), or giving up the requirement of causal connection for those cases (section 3.2), is very difficult. Splitting legal causation into two concepts (genuine causation and the alternative concept) and requiring lawyers to use them simultaneously is also very hard to sustain (section 3.3). If the denier argues that only one of the concepts should be used by lawyers, then if lawyers were to use only the denier's genuine causation, a radical reform in the current law would be required. And if they were to use only the alternative concept provided by the denier, the difference between such a position and causal pluralism remains unclear. In addition to the general similarities between the positions, the concept that is provided by the denier has the same extension and intension as the concept provided by the pluralist. The only

---

[50]This argument is an instantiation of the general argument which appears in Schaffer (2004, p.207).

difference between the pluralist and the denier is that the denier convicts the lawyers of a systematic and continuous mistaken usage of the concept of causation. However, the basis and justification for such a verdict are dubious.

5 What motivates the resistance to causation by omission

In order to convince philosophers that legal causation is an interesting issue for philosophers of causation, it might be useful to identify the motivations to resist causation by omission and then compare them to the legal discussion of causation. Beebee expresses two serious worries that require the attention of those who accept causation by omission. These two worries reveal interesting differences between the issues that legal and philosophical discussions try to address when approaching the issue of causation. It is hoped that articulating these differences will draw the attention of philosophers to legal causation as a question which is interesting on its own right, rather than as a residual example in the general debate (Schaffer, 2004, p.201).

The first worry Beebee raises is that causation by omission is overinclusive because it includes cases that no one sees as causal (similar worry appears in Hall (2004, p.249) and Dowe (2004, p.192)). Even if my failure to water my plants caused their death, so does the failure of Tony Blair to water them. If causation by omission is genuine causation, then, according to Beebee, one has to accept an infinite number of causes to any effect, most of them clearly having nothing to do with the causal history of that effect.

This is a serious worry, but I think that Beebee *underestimates* its gravity. The same problem of over inclusiveness exists not only for omissions but also for actions. If a person was shot to death by a run-away criminal, it is hard to accept that the fact that 300 years ago the criminal's ancestors met in a Viennese Ball, is also a cause of the victim's death. As noted above, this problem was acknowledged as the problem of *selection* (Schaffer (2003, sec.2.3) and Hart and Honoré (1985, p.22)) or *profligate causation* (Menzies, 2004, p.139). Whatever is one's solution to this problem (e.g., Lewis (1986b, p.198) and Mackie (1974, p.34) claim that it is merely a contextual distinction), it can and should be applied to both actions and omissions. It can also address Beebee's worry by explaining those cases of omissions that are intuitively not causal. Therefore, Beebee's worry is not specific to the issue of causation by omission.[51]

---

[51] Menzies also identifies this worry, but correctly positions it within a wider problem for both actions and omissions (Menzies, 2004, p.145).

The important aspect that Beebee's worry highlights is the lack of a good solution to the problem of selection, which is very important to the legal context (the same observation is made by Hart and Honoré (1985, p.13)). The attempts to draw a distinction between causes and background conditions using the concept of causation were ridiculed by several philosophers. Hall regards objecting to an account of causation using this distinction as "to do so on the cheap" (Hall, 2004, p.228). A long time ago, Mill noted:

> Nothing can better show the absence of any scientific ground for the distinction between the cause of a phenomena and its conditions, than the capricious manner in which we select from among the conditions that which we choose to denominate the cause. (Mill, 1950, p.244)

Capricious or not, such a distinction is made every day in courts around the world. Some causes are singled out (e.g., the accused's shooting of the victim) and other causes are rejected (e.g., the victim's loss of blood, the intercourse that the accused's parents had, etc.). Justifiably, Hart and Honoré complain that although "the distinction between cases and conditions... [is] the source of most of the difficulties in understanding the causal concepts with which the lawyer works, [it has received] very little attention in the traditional philosophical discussions of cause" (Hart and Honoré, 1985, p.13). Even if Lewis and Mackie are right, and this selection is merely contextual, it might be still interesting to elucidate the process of selection in detail. For lawyers, this is an important issue with significant implications. For instance, an interesting legal question is the issue of *novus actus interveniens* when the causal chain is broken by nature, an action of the defendant/victim, or an action of a third party. For example, in the case of *Jordan*,[52] a man was stabbed in his stomach during a fight, taken to the hospital where he was treated by a negligent doctor. The question aroused was whether or not the stabber caused the death. This issue appears in many legal areas.[53] Maybe it also deserves more philosophical attention (Menzies (2004) is an important exception).

Beebee's second worry is that such a distinction might require introducing a *normative* element into the concept of causation. Beebee quickly dismisses this option by claiming that "nobody in the tradition of the metaphysics of causation that I'm concerned with here thinks that causal facts depend on human dependant norms" (Beebee, 2004, p.297). Whilst this

---

[52] *R v Jordan* [1957] 40 Criminal Appeal Review 152.
[53] In criminal law, see Card (2006, ss. 2.29-2.39, pp.72-85). In tort law, see McBride and Bagshaw (2005, pp.537-545) and Simon Deakin and Markesinis (2003, pp.201-203). In contract law, see Treitel (2003, p.976).

may well reflect the literature in metaphysics, in the legal literature this issue is in the centre of the debate about the nature of causation. Various legal scholars have acknowledged the role of normative and moral elements in causal judgments. Hart and Honoré use such elements to solve cases of over-determination (Hart and Honoré, 1985, pp.lxv-lxvi)[54] and Fischer uses such elements to solve cases of double prevention by omission (Fischer (1992, p.1348) and Fischer (1992, part IV)). Several other important legal scholars, such as Cooter (1987, pp.528-531) and Williams (1962, p.75), have expressed similar views. Some sceptics even go as far as arguing that whatever lawyers call factual causation has no factual element at all and it is used mainly as a deceiving veil to hide normative considerations (Malone, 1957).[55] In fact, one of the main goals of Hart and Honoré's famous *Causation in the Law* is to try to refute this sceptical claim (Hart and Honoré, 1985, pp.xxxiv-xxxv). Successful or not, the sceptics' claim is not easy to refute, as is testified by the intensive intellectual efforts that Hart and Honoré invest in trying to do so.[56] I suspect that most philosophers of causation, including those who accept causation by omission, would resist introducing any normative element into the concept of causation. Maybe they are right. It is hoped, however, that this challenge in particular, and legal causation in general, will be taken more seriously by philosophers of causation.

Acknowledgments

I thank John Gardner, Rhonda Powell, Sebastian Sequoiah-Grayson, Tal Ofek and the anonymous referees for their helpful comments.

Amit Pundik
Balliol College, University of Oxford, UK.
Amit.Pundik@law.ox.ac.uk

BIBLIOGRAPHY

Armstrong, D. M. (1999). The open door. In Sankey, H., editor, *Causation and Laws of Nature*, volume 14 of *Australasian Studies in History and Philosophy of Science*, pages 85–175. Kluwer Academic Publishers, Dordrecht, Boston.

---

[54] According to Hart and Honoré, the normative element is introduced into the determination of liability rather than into the concept of factual causation. However, they do not provide an alternative solution to solve this issue purely on causal grounds.

[55] Hart and Honoré present a good summary of the various criticisms on the concept of 'factual causation' (Hart and Honoré, 1985, pp.96-102).

[56] "In both this and the first edition of our book much of our argument has been directed against those writers whom we have termed 'causal minimalists'. They are theorists who allot only a minor role to causal issues in determining questions of legal responsibility" Hart and Honoré (1985, p.lxvii) (footnotes removed). See also their attempts to refute this sceptical claim Hart and Honoré (1985, pp.102-108).

Ashworth, A. (2006). *Principles of Criminal Law*. Oxford University Press, Oxford, fifth(1991) edition.
Atiyah, P. and Smith, S. (2006). *Atiyahs Introduction to the Law of Contract*. Clarendon Press, Oxford, sixth(1995) edition.
Becht, A. C. and Miller, F. W. (1961). *The Test of Factual Causation in Negligence and Strict Liability Cases*. St. Louis: Committee on Publications, Washington University.
Beebee, H. (2004). Causing and nothingness. In J. Collins, N. H. and Paul, L., editors, *Causation and Counterfactuals*, pages 291–308. MIT Press, Cambridge Massachusetts.
Bennett, J. (1998). *Events and their Names*. Hackett Publishing Company, Indianapolis.
Blumenthal, J. A. (1993). A wipe of the hands, a lick of the lips: The validity of demeanor evidence in assessing witness credibility. *Nebraska Law Review*, 72:1157–1204.
Browne, B. (1992). A solution to the problem of moral luck. *The Philosophical Quarterly*, 42:345–356.
Cane, P. (2000). Mens rea in tort law. *Oxford Journal of Legal Studies*, 20:533–556.
Card, R. (2006). *Card Cross and Jones: Criminal Law*. Oxford University Press, Oxford, seventeenth(1948) edition.
Cohen, J. (1997). *The Probable and the Provable*. Clarendon Press, Oxford.
Cooter, R. (1987). Torts as the union of liberty and efficiency: An essay on causation. *Chicago-Kent Law Review*, 63:523–551.
Davis, M. (1986). Why attempts deserve less punishment than completed crimes. *Law and Philosophy*, 5:1–32.
Dennis, I. (1982). The criminal attempts act 1981. *Criminal Law Review*, 1982:5–16.
DeRose, K. (1992). Contextualism and knowledge attributions. *Philosophy and Phenomenological Research*, 52:913–929.
Dowe, P. (2000). *Physical Causation*. Cambridge University Press, Cambridge.
Dowe, P. (2001). A counterfactual theory of prevention and causation by omission. *Australasian Journal of Philosophy*, 79:216–226.
Dowe, P. (2004). Causes are physically connected to their effects: Why preventers and omissions are not causes. In Hitchcock, C., editor, *Contemporary Debates in Philosophy of Science*, pages 189–196. Blackwell, Oxford.
Dworkin, R. (1978). *Taking Rights Seriously*. Duckworth, London.
Dworkin, R. (1986). *Laws Empire*. Fontana, London.
Feinberg, J. (1995). Equal punishment for failed attempts: Some bad but instructive arguments against it. *The Arizona Law Review*, 37:117–134.
Finnis, J. (1982). *Natural Law and Natural Rights*. Clarendon Press, Oxford.
Fischer, D. A. (1992). Causation in fact in omission cases. *Utah Law Review*, 1992:1335–1384.
Fletcher, G. (1967). Prolonging life. *Washington Law Review*, 42:999–1016.
Gillies, D. (2000). *Philosophical Theories of Probability*. Routledge, London.
Grady, M. F. (1984). Proximate cause and the law of negligence. *Iowa Law Review*, 69:363–449.
Green, L. (1962). The causal relation issue in negligence. *Michigan Law Review*, 60:543–576.
Hall, N. (2004). Two concepts of causation. In J. Collins, N. H. and Paul, L., editors, *Causation and Counterfactuals*, pages 225–276. MIT Press, Cambridge Massachusetts.
Hart, H. (1968). *Punishment and Responsibility*. Clarendon Press, Oxford.
Hart, H. (1994). *The Concept of Law*. Clarendon Press, Oxford, second(1961) edition.
Hart, H. and Honoré, T. (1985). *Causation in the Law*. Clarendon Press, Oxford, second(1959) edition.
Kennedy, I. M. (1977). Switching-off life support machines: the legal implications. *Criminal Law Review*, 1977:443–452.

Lewis, D. (1986a). Causal explanation. In Lewis, D., editor, *Philosophical Papers II*, pages 214–241. Oxford University Press, Oxford.
Lewis, D. (1986b). Causation. In Lewis, D., editor, *Philosophical Papers II*, pages 159–213. Oxford University Press, Oxford.
Lewis, D. (2004). Void and object. In J. Collins, N. H. and Paul, L., editors, *Causation and Counterfactuals*, pages 277–290. MIT Press, Cambridge Massachusetts.
Mackie, J. J. (1974). *The Cement of the Universe*. Clarendon Press, Oxford.
Malone, W. S. (1957). Ruminations on cause-in-fact. *Stanford Law Review*, 9:60–99.
McBride, N. J. and Bagshaw, R. (2005). *Tort Law*. Longman, Harlow, second(2001) edition.
Menzies, P. (2004). Difference-making in context. In J. Collins, N. H. and Paul, L., editors, *Causation and Counterfactuals*, pages 139–180. MIT Press, Cambridge Massachusetts.
Mill, J. S. (1950). *A System of Logic*. MacMillan, New York.
Nagel, T. (1979). *Moral Questions*. Cambridge University Press, New York.
Ormerod, D. (2005). *Smith and Hogan: Criminal Law*. Oxford University Press, Oxford, eleventh(1965) edition.
Palmer, P. (1999). Attempt by act or omission: Causation and the problem of the hypothetical nurse. *Journal of Criminal Law*, 63:158–165.
Roberts, P. and Zuckerman, A. (2004). *Criminal Evidence*. Oxford University Press, Oxford, second(1989) edition.
Schaffer, J. (2003). The metaphysics of causation. In Zalta, E. N., editor, *Standford Encyclopedia of Philosophy*. The Metaphysics Research Lab, Stanford University, Stanford.
Schaffer, J. (2004). Causes need not to be physically connected to their effects: The case for negative causation. In Hitchcock, C., editor, *Contemporary Debates in Philosophy of Science*, pages 197–216. Blackwell, Oxford.
Simon Deakin, A. J. and Markesinis, B. (2003). *Tort Law*. Oxford University Press, Oxford, fifth(1984) edition.
Treitel, S. G. (2003). *The Law of Contract*. Sweet and Maxwell, London, eleventh(1962) edition.
Waldron, J. (1995). Moments of carelessness and massive loss. In Owen, D., editor, *Philosophical Foundations of Tort Law*, pages 387–408. Clarendon Press, Oxford.
Williams, G. (1962). Causation in the law. *Cambridge Law Journal*, 1961:62–85.
Wright, R. W. (1985). Causation in tort law. *California Law Review*, 73:1735–1828.
Wright, R. W. (1988). Causation, responsibility, risk, probability, naked statistics and proof: Pruning the bramble bush by clarifying the concepts. *Iowa Law Review*, 73:1001–1077.
Zuckerman, A. (2004). *Civil Procedure*. LexisNexis, London.

PART VII

GENERAL FRAMEWORKS FOR CAUSAL ANALYSIS

A conditional view of causality

FRIEDEL WEINERT

ABSTRACT. This paper argues for a conditional view of causality: a view which takes its starting point from actual causal conditions as they obtain in the natural and the social world. The actual obtaining causal conditions allow projections to counterfactual situations, which in turn depend on the availability of certain kinds of regularity. The paper defends a single model of causation for both the natural and the social sciences. This is illustrated with respect to the Franck-Hertz experiment in quantum mechanics and Weber's notion of adequate causation. This view is based on a naturalistic view of the social sciences, e.g., that there are only differences in degree between the natural and the social sciences.

1 Introduction

The goal of this paper is to investigate whether a general model of causality, covering such diverse fields as quantum mechanics and the social sciences, can be formulated. The advantage of such a general model would be its applicability to a wide variety of causal situations, whilst many established accounts of causality limit themselves to either everyday causal relationships (Mackie's *INUS* account) or mostly to the physical sciences (Salmon's and Dowe's mechanical models and Woodward's interventionist model). In view of the aim of this paper, Mackie's *INUS* account appears be the most likely candidate to investigate whether it can be extended from its usual context of everyday causal relationships into the area of quantum mechanics; and whether it can shed light on Weber's notion of adequate causation in the social sciences. Mackie's *INUS* account is however too closely associated with Humean-type regularity accounts of causation. It will therefore be better to interpret it more generally as a *conditional* model of causation. Such a model is concerned with the actual causal conditions, which obtain in particular situations. The primary motivation for a conditional model of causation is twofold: **A)** Causal interpretations of quantum mechanical and social events force us to abandon the link between the notions of conditionality and regularity. The guiding idea behind a conditional model is that causal relations exist even in the absence of lawful regularities, as they are expressed in the laws of science. In quantum mechanics, only probabilistic regularities occur. And in the social world, as this paper will argue, only trends obtain, which do not have the character of genuine social laws. **B)** A conditional model also differentiates itself from counterfactual analyses of causation, as they appear for instance in various interventionist models of causation. This paper will argue that counterfactual analyses of causation presuppose knowledge of lawful regularities in the physical world in order to impose constraints on the counterfactuals, which appear in hypothetical inferences. In the absence of such regularities, counterfactuals are much more difficult to evaluate.

The argument of this paper will proceed as follows: Section Two will start with an evaluation of counterfactual-interventionist approaches and show that they presuppose knowledge of actual obtaining conditions and lawful regularities (§2.1). This result leads to an analysis of a conditional approach to causation, which generalizes ideas introduced in Mackie's *INUS* account (§2.2). Section Three applies the conditional model to quantum mechanics, e.g., the Franck-Hertz experiment; Section Four does the same for the social sciences, e.g., it shows that Weber's notion of adequate causation can be shown to be a version of the conditional model. Section Five concludes the argument and compares the conditional model to other models of causation.

2 From actual to counterfactual conditions

This section will argue that counterfactual analyses of causation typically presuppose knowledge of actual causal conditions. We take Woodward's recent analysis of causation as a paradigm of a counterfactual-interventionist approach, because of its relevance to the sciences.

2.1 Counterfactual-interventionist evaluations

The need to control counterfactuals, which are employed in counterfactual analyses of causation, is emphasized in Woodward's recent interventionist model of causation. Woodward distinguishes between 'impractical' and 'practical' accounts of causation (Woodward, 2003, Ch.2.1) Practical accounts are to be preferred, because causal knowledge is useful for purposes of control and manipulation. Woodward hopes to handle counterfactuals in his manipulationist model of causation by exploiting the idea of hypothetical interventions. In order to appreciate the need to control the counterfactual situations, which are envisioned in the employment of hypothetical experiments, let's contrast two counterfactual statements:

C_1: If the Franck-Hertz experiment had been conducted in Copenhagen, the discreteness of quantum energy levels in atomic systems would have established.

C_2: If Hitler had been hit by the bullet, which killed his neighbour in a march on November 9 1924, the Second World War would not have happened. (Kershaw, 1998, Vol. I, p.211)

It is easy to evaluate C_1 because of the presence of invariant relationships, which are expressed in lawful regularities. Woodward also stresses that 'invariance is the key to explanatoriness' (Woodward, 2003, p.183 & Ch.6). Such lawful regularities are subject to symmetry principles, i.e., such that the conduct of the experiment is symmetric with respect to place and time. Neither the time *when* the experiment is conducted nor the place *where* the experiment takes place has an effect on the measured outcome. Such spatial and temporal symmetries are widespread in physics. Note that no additional information is required to evaluate this counterfactual statement; e.g., neither knowledge of the history of experimental practices in quantum mechanics, nor details about the experimenters or the experimental situation.

It is much more difficult to evaluate C_2—without any additional evidence—because of the huge number of contingent conditions and the lack of any known lawlike regularities. Historians state, of course, that in the event of Hitler's death on November 9 1924 the 'course of history' would have been different. Given additional evidence about the Weimar Republic, the treaty of Versailles and general knowledge about human conduct in general, historians will find it easier to say that the socio-economic and political conditions of the 1920s raised the probability of another world war. The difference lies in the precision with which the evaluation can be carried out. The availability of quantitative laws permits a numerically precise evaluation of the counterfactual scenario, for instance what the gravitational force would be between the moon and the earth *if* the distance between these two bodies were doubled. In the absence of such quantitative laws the evaluation remains broad and imprecise. Although this paper will argue that no 'social' laws exist to aid the evaluation of counterfactuals like C_2, there are nevertheless social trends, which establish certain regularity pattern. The rise of Hitlerism can, however, be regarded as a unique historical phenomenon, which does not fit into any patterns. The existence of trends in the social world will improve the evaluation of counterfactuals in this area. Nevertheless, the availability of lawful regularities (as opposed to mere trends, which can be broken) makes a significant difference to the evaluation of counterfactuals. The above examples suggest that it is the presence of lawlike regularities, which permits the relatively easy evaluation of the counterfactual statement C_1. Contrary to what Woodward claims, the evaluation of the counterfactual is not dependent on the construction of a hypothetical experiment; rather it is the presence of a law of nature, even if statistical in form, which supports the evaluation of the counterfactual.

The outcomes of hypothetical interventions, which Woodward regards as essential for the evaluation of the interventionist model of causation, significantly presuppose lawlike knowledge of actual interventions. Although Woodward is preoccupied with hypothetical experiments, science does not seem to be concerned with 'what-if-things-had-been-differen' questions. Causal explanations of scientific experiments, social events and accidents seem to be concerned with the actual causal situation. That is, they seem to be concerned with the actual causal conditions, which can be shown to have produced an effect. This is not to deny the importance of hypothetical constructions. But these hypothetical constructions are often made dependent on lawlike knowledge of the actual world. Pearl even suggests that the counterfactual is a 'conversational shorthand of the predictive form' of Ohm's law (Pearl, 2000, Sec. 7.2.2). Counterfactuals in hypothetical interventions can be evaluated and controlled because they are projections from the knowledge of actual causal situations. Knowledge of actual causal situations then constrains the interpretation of counterfactual situations, which can be imagined to occur. Generally, agency-related models of causation, like von Wright (1971), Woodward (2003), Gillies (2005), presuppose knowledge of actual causal situations in the physical world. (Cf. Pearl (2000, p.211)) It seems that the better our understanding of the lawful regularities of the natural world is, the better is our control over the evaluation of counterfactuals. This important, yet unrecognized presupposition in counterfactual models of causation is also demonstrated by considering the difference between C_1 and C_2. We regard C_1 as true because the lawful regularities underlying this counterfactual already carry a counterfactual commitment. As is well known, laws of nature support counterfactuals. In consideration of C_2 historians may assess C_2 as 'more or less probable in view of the evidence' from the historical circumstances but there is no guidance from any known underlying laws. Moreover, there is no closed set of conditions, which could impose some control on the evaluation of the counterfactual situation. It seems that the evaluation of counterfactuals in the social world is much more difficult to control and evaluate. A reason for this failure of projection suggests itself: while there are trends in the social world, they do not easily qualify as lawlike regularities, as we shall argue. This makes the evaluation of counterfactuals like C_2 so uncertain. Furthermore, as Woodward admits:

> We are often more interested in accounting for actual rather than merely hypothetical or counterfactual variation (especially when the latter is regarded as unlikely to occur or far-fetched). (Woodward, 2003, p.231)

How could such a conditional model of causation be developed?

2.2 A conditional model of causation

It will be assumed, without further discussion, that both general and singular causation exist and that singular causation is an instantiation of general causation under the employment of specific boundary conditions. A conditional model of causation is concerned with the actual conditions, which obtain in a causal situation. It is called a *conditional* model in order to emphasize the consideration of antecedent and consequent conditions in actual causal situations, without any prior commitment to a regularity theory of causation. In view of the existence of probabilistic relations in quantum mechanics and the lack of known lawlike regularities in the social sciences, the abandonment of regular succession seems to be unavoidable. In fact Mackie himself accepted that conditionality cannot be completely understood in terms of the regularity theory (Mackie (1980, pp.86 and 194); cf. (Spohn, 2006, p.98 Fn 8)). For this reason analyses of causation, as they can be found in Reichenbach and Mackie, will be generalized to a conditional model. The practice of causal explanations in the natural and social sciences tells us that we seek answers to 'why questions' concerning actual conditions, not to 'what-if-things-had-been-different' questions. It is customary in the literature to distinguish between deterministic and stochastic causality. But in view of quantum mechanical relations and social events a model of causation should be formulated in such a way that deterministic causation turns out to be a limiting case of probabilistic causation in the limit of a narrowing of the probability distribution of the posterior conditions, conditionalized on a set of prior conditions. A conditional model seems to be particularly apt for such an undertaking.[1]

Mackie's INUS Account

In *The Cement of the Universe* (1980) Mackie made no attempt to measure the adequacy of his *INUS* account against classical physics, let alone quantum mechanics. Rather, Mackie tried to develop a general model of physical causation. Causation as it works in the real world is the cement that holds the universe together. This model stands in the tradition of D. Hume and J.S. Mill. Any questions of the existence of a causal bond in the physical universe between correlated events, over and above their succession, are treated with caution. In the physical world, causation is only regular succession of events. In order to apply Mackie's account to quantum mechanical and social events, the idea of succession in the Humean sense must be abandoned.

---

[1] Several passages from sec. 2.2 and sec. 3.1 are reprinted with kind permission of Springer Science and Business Media from my book (Weinert, 2004, pp.261-2,263-6).

Causation is not something *between* events in a spatio-temporal sense, but is rather the way in which they follow one another. (Mackie, 1980, p.296, italics in original)[2]

For Mackie, a cause is an ***INUS*** condition, i.e., an ***I***nsufficient but ***N***on-redundant part of an ***U***nnecessary but ***S***ufficient condition for some effect E. (Mackie, 1980, p.62) Thus there is a cluster of factors, making up the cause C, which bring about the effect E. Unlike Mill, however, who took the cause to be the sum total of the conditions, Mackie makes a distinction between *necessary* and *sufficient* conditions.

If X is a *necessary* condition for Y, then in the absence of X, Y cannot occur. Oxygen (X) is a necessary condition for fire (Y). In Mackie's words, 'whenever an event of type Y occurs, an event of type X also occurs' (Mackie, 1980, p.62).

If X is a *sufficient* condition for Y, then in the presence of X, Y will occur. Rain (X) is a sufficient condition for the street to get wet (Y), since in its presence the streets get wet.

Given that Mackie analyses causation in terms of a cluster of conditions, an *INUS* condition is a *partial* cause. The whole cause may consist of a cluster of conditions ABC, which combined are sufficient but not necessary for E, since the presence of the clusters DGH or JKL may also bring about E. If the cause is the cluster of conditions ABC, none of these individual factors is redundant.

Let us illustrate how this availability of a cluster of causal conditions presents itself in such diverse fields as quantum mechanics and the social world.

(**a**) For instance, in the Franck-Hertz experiment electrons are made to collide with mercury atoms, under specific experimental conditions, to test the discreteness of the energy level of the mercury atoms. The cluster of conditions, say EQ (an appropriate *electronic* input and appropriate atomic *quantum* states), will be sufficient to produce energy level transitions in the mercury atoms. Each one of those E, Q will be non-redundant or necessary parts of the conditional state EQ. Different atoms can be chosen to produce the characteristic curves (see Fig.2), but in each case E and Q must be present. The electrons will induce state transitions in the atoms. How this happens depends physically on EQ. The advantage of causal relations in laboratory situations is that the cluster of conditions can be manipulated to a large extent. The cluster of conditions is closed, for all practical purposes. The manipulation of conditions covers both the inclusion and exclusion of

---

[2] For a more detailed discussion see Weinert (2004, Ch.5.3.5).

parameters. For instance, when Rutherford and his collaborators tried to establish the presence of a nucleus in atoms in 1909-1911, they used so-called scattering experiments. In these experiments, the target system consists of gold atoms, embedded in a gold foil, which are bombarded with ionized helium atoms (e.g., stripped of their electrons). Rutherford found that 1 in 8000 helium atoms were scattered to angles larger than 90 degrees. Rutherford argued that the helium atoms could not have been scattered by an encounter with the electrons in the gold atoms because of energy considerations, e.g., the trajectory of heavy helium atoms would not be disturbed by the light electrons. Hence a heavier object must be present in the gold atoms, which could explain the surprising scattering angles. Rutherford identified this heavier object as the nucleus (see Weinert (2004, pp.89-90)). We see that Rutherford clearly identified negligible parameters (e.g., the electrons in the gold atoms) and causal parameters (e.g., the helium atoms and the nuclei in the gold atoms) in the experiment.

(b) In the social sciences the closure of the set of antecedent conditions is not possible because of the ramifications of human agency. Nevertheless, as Weber argued, the social scientist is interested in establishing causal relations in the social world; this means that the social scientist will want to establish a set of causal relations, which have the ability to causally explain an effect in the social world. For instance, in 1999 the Home Office in Britain predicted that the number of burglaries and thefts would increase by almost a third in a short time span of two to three years. The expected rise in crime is a consequence of the rise in the young male population in Britain. In this instance we have a causal analysis and a prediction rolled into one. The increase in the number of burglaries and thefts is blamed on an increase in the number of young adults and a growth in the amount of stealable goods. The latter are the causal conditions, which are said to lead to the effect, if no other conditions interfere. This example shows that the Home Office prediction is predicated on a *ceteris paribus* assumption, e.g., it is assumed that (a) no other factors will interfere to either prevent or reverse the predicted events; (b) that no other causal factors need to be considered. Neither of these *ceteris paribus* conditions is known to hold so that the Home Office prediction indicates a possible trend, if no other measures are taken. In this sense the set of causal conditions is not closed.

One obvious disadvantage of the minimal conditional view of causation presented so far is that it makes no distinction between conditions, whether necessary or sufficient, which are physically operative in the production of the effect and non-operative conditions, which are merely in the vicinity of the cause-effect relationship. This is sometimes expressed by making the distinction between the 'cause' and 'causal conditions' or by speaking of the 'cause' and 'contributing conditions'. Mill regards all conditions as candidates for the antecedent cause. But laboratory experiments in physics, as just indicated, are such that many background conditions can be excluded on quantitative grounds. To capture this distinction, Mackie introduces the concept of a *causal field*.[3] A causal field comprises the background conditions, which make the normal running of things possible. A causal event disturbs the normal running of things. The antecedent conditions C, and consequent conditions E, must make a *difference* within a field to establish a cause-effect relationship. Conditions are statistically relevant if their inclusion in the causal account affects the probability distribution of the outcome, otherwise they are statistically irrelevant. In the Franck-Hertz experiment it is statistically relevant whether the electrons are made to interact with the atomic systems, but the *presence* of mercury atoms is not statistically significant. Franck-Hertz used mercury atoms, but the experiment can also be run with atomic hydrogen gas. Whether a condition belongs to the causal field (the statistically irrelevant conditions) can be measured in the laboratory setting. It turns out that mercury atoms result in clearer experimental results, because 'hydrogen occurs naturally in the molecular form H_2, rather than in atomic form'. (Krane, 1983, p.170) In certain cases statistically relevant and irrelevant conditions can also be distinguished in social relations. In the previous example of the Home Office prediction the age of the potential perpetrators is statistically very significant. It is however not specified whether the socio-economic background of the young male population is significant. But it seems to be statistically insignificant in which part of Britain the young males live. (All of these factors may still turn out to be statistically significant.)

Causal fields

The notion of a causal field introduces context-dependent aspects. The same conditions, regarded independently, may be statistically relevant or irrelevant. It depends on the causal situation whether conditions are causally significant or negligible, as the previous examples illustrate. Interactions *between* and *within* causal fields are subject to certain regularities: either laws of nature in the material world or trends, and other particular conditions, in the social world. When conditions make a difference to a causal field in the natural world, they do so in a lawful way. For instance, a photon moves with momentum $h\nu/c$ towards the stationary electron, which has no momentum. After the collision the scattered photon has a new momentum, with a component in the x-direction $(\cos\theta h\frac{\nu'}{c})$ and and a component in

---

[3]Mackie (1980, pp.35, 63) and Bohm (1957, Ch.I.4) similarly discuss significant causes in a given context.

the y-direction ($\sin\theta h\frac{\nu'}{c}$). The electron, too, has acquired a momentum, or before the mercury atoms collide with the electrons in the Franck-Hertz experiment only spontaneous emissions and absorptions occur. After collision with the electrons the atoms will have changed their energy states. When conditions make a difference to a causal situation in the social world, they do so as a result of human agency. An intervention with selected social factors may prevent or slow the predicted rise in the young male population in Britain. Deliberate interventions to break the causal relation between drink-driving and road fatalities have lowered the number of alcohol-related car crashes. Such interventions rely on the knowledge of behaviour patterns, social trends, and knowledge about human nature. Insofar as such patterns and trends are relatively stable, they allow certain counterfactual considerations, subject to certain limitations (see sec. 3.5).

Conditions belonging to a causal relation derive from the interactions between causal fields. There are many examples of this situation both in the social sciences and in quantum mechanics. We have already noted the expected rise in crime *and* the rise in the number of young offenders; social scientist also investigate the relation between the socio-economic status of a household *and* the differential educational performance of children; economists relate inflation to the level of interest rates in a particular economy; Weber attempted to establish a causal connection between the Protestant spirit *and* the rise of capitalism. There are also many examples of this situation in quantum mechanics: the non-uniform magnet field *and* the beam of silver atoms (Stern-Gerlach apparatus); the nucleus *and* the α particle (scattering experiments); the photon *and* the electron (Compton scattering); the electron beam *and* the nickel crystals (Davisson-Germer experiment); the electrons *and* the mercury vapour (Franck-Hertz experiment). A causal relation is brought about by the interaction of a) lawful behaviour in causal fields in the natural world, or b) intentional behaviour in causal fields in the social world. The causal field would not, if left undisturbed, lead to the implication: $C \rightarrow E$.

We have now spelt out a conditional view of causality. It treats causal relations as facts about conditional dependencies between antecedent and consequent conditions. It is now time to consider particular applications of this model.

3 Quantum mechanics and the conditional model

This section considers the conditional model of causation in relation to the quantum world. In particular, we choose a famous experiment in quantum mechanics—the Franck-Hertz experiment—and we interpret it according to the conditional view of causation. Such a view was already anticipated by

Reichenbach who applied it to quantum mechanics. Reichenbach added the important notion of probability to causal considerations.

3.1 Reichenbach

To the author's knowledge Reichenbach was one of the first philosophers who attempted a conceptual model of causation, which would be compatible with quantum mechanics Reichenbach (1920, 1931). Reichenbach wrote his first contribution (1920) on this question at a time when the old quantum theory was still accepted, and his second contribution (1931) when the new quantum mechanics had become available. Reichenbach sought to achieve this compatibility by associating the notion of causation with that of probability. The first step in any physical situation, in which a causal connection may be suspected between the antecedent and the consequent, is the recognition that the antecedent consists of a number of factors. The antecedent factors contain the causal conditions. Of these factors, some will be measured parameters and others will be unmeasured residual factors. In structural equations these are expressed as variables X and U, respectively. A closer analysis may turn some of these residual factors into measured parameters but the residual factors can never be exhausted. Causation is concerned with the relation between individual measured parameters. Probability has to do with the distribution of the residual factors (Cf. Pearl (2000, Ch.3.2.1)). According to Reichenbach, the principle of causation cannot be formulated without the principle of a statistical distribution. Causal claims take the form of an implication ('If C, then E'). The factor C consists of observable measured parameters and unknown rest factors, which may equally have an influence over E.

> What we know of C, can only be expressed in terms of a statistical statement: we know that subsequent situations, *with great probability*, differ little from C. (...) We predict E only with probability, not certainty. *Every causal statement, applied to the prediction of a natural event, has the form of a statistical statement. (...) If an event is described by a finite list of parameters, the future evolution of the event can be predicted with probability. This probability tends towards 1, the more parameters are taken into account.*[4]

---

[4] Reichenbach (1931, pp.715-6; italics in original). Author's own translation: Reichenbach's letters A, B have been exchanged for C, E. Zilsel (1927) makes a distinction between dependent (effect) and independent variables (cause) in his discussion of the asymmetry of causation. The independent variable is sometimes accessible to human manipulation. Zilsel's discussion is reminiscent of Mackie's much later account of causal priority in terms of fixity, which is related to the possibility of intervention. See von Wright (1971), Mackie (1980, p.180), Woodward (2003). A conditional view of causality

This notion of probabilistic causation, which is implied in Reichenbach's combination of probability and causation, can then be regarded as a generalisation of the classical notion of causation. Only this new notion is applicable to quantum mechanics. The Heisenberg indeterminacy relations forbid a convergence of the predictability of particular, singular quantum events towards 1. Reichenbach interprets this as an extension, not an abolition of the notion of causation. Note again the difference between experiments in physics and in the social sciences. For a judgment of causal relations, laboratory experiments can be regarded as closed sets of causal conditions. The probability of the residual factors' influence on the effect can often be calculated, such that the probability of a causal disturbance on the residual factors is negligible. Furthermore, in the case of quantum mechanics the cause is probabilistically related to its effect, although the effect E cannot be deterministically predicted. It is significant for the causal analysis of social events that this closure cannot be achieved. A laboratory experiment in the social sciences can mimic a laboratory experiment like the Franck-Hertz experiment, but it cannot achieve closure because human agents carry symbolic dimensions. Reichenbach anticipated that a *conditional* view of causation may well be the most adequate answer in the light of the results from quantum mechanics. Our analysis of the Franck-Hertz experiment in quantum mechanics will show that the notion of probabilistic causation is essential for a proper interpretation of this experiment. Let's consider the case of the Franck-Hertz experiment.

3.2   A causal interpretation of the Franck-Hertz experiment

In 1913 Bohr took Rutherford's nucleus model of the hydrogen atom as the basis for the first quantized atom model in the history of physics. A year later, Bohr's postulate of energy quantization—the idea that atoms only exist in discrete energy levels—was confirmed in the *Franck-Hertz* experiment (Franck and Hertz, 1914). This experiment is also remarkable from a causal point of view. In this experiment electrons are ejected from a cathode, C, into a tube filled with mercury gas (see Fig.1). The energy of the electrons can be increased in a controllable manner by accelerating them towards the positively charged grid, G, through the potential difference V_a. Electrons fly through the grid towards anode A. Between G and A, a small retarding voltage, V_r, decelerates the electrons. They will only reach the anode A, if their energies V exceed V_r.

Collisions between the atoms and the electrons will occur. Only electrons with sufficient energy will cause the mercury atoms to make transitions to higher states of energy. The electrons will lose their energy to the atoms.

---

was also defended by Bohm (1957).

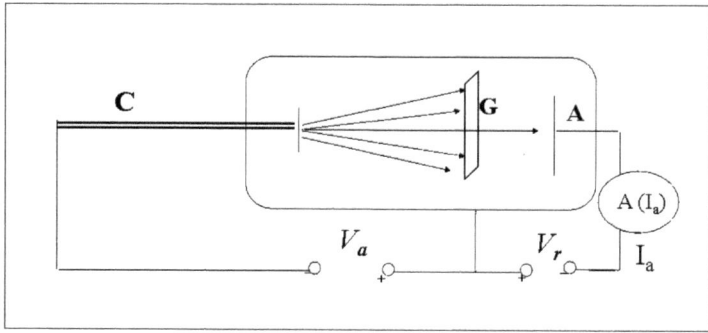

Figure 1. Franck-Hertz experiment (1914)

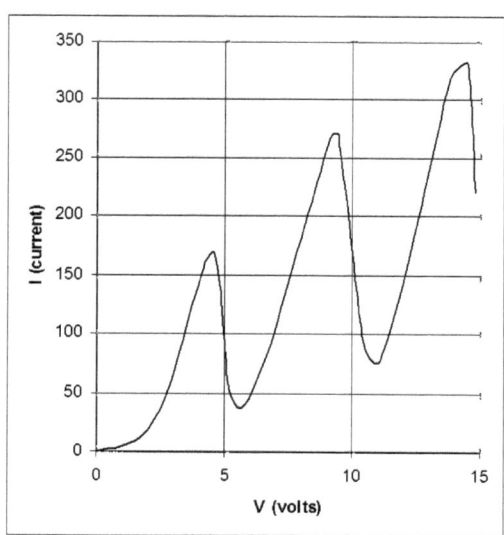

Figure 2. Franck-Hertz Experiment (1914)—Dependence of current on accelerating potential

When $V_a = 4.9$ V the curve drops very sharply. The causal explanations is that electrons near the grid lose all their energy to the mercury atoms and are unable to overcome the small retarding potential, V_r, to reach the anode. A drop in the current, I_a, is observed. When $V_a = 9.8$ V, another drop in the curve occurs. Electrons either excite the atoms to higher energy levels or lose 4.9V more than once. In turn, the excited mercury atoms will return to their ground energy state and emit photons with energies corresponding to the energy intake. The experiment confirmed the loss of the electronic energy at discrete levels and confirmed that the higher states of energy of the atoms corresponded to the Bohr postulate of discrete energy levels. The observable results are shown in Fig.2. In order to interpret this experiment causally we need a probabilistic notion of causality, since we are not in a position to predict which electron will collide with which mercury atom and how much energy it will transfer. There is no causal trace from the antecedent to the consequent conditions. The experiment does not give rise to predictive determinism in the manner of Laplace's demon. Despite the indeterminism of the situation, a causal explanation of the appearance of peaks and troughs in the graph can be derived from the Bohr model. In the experiment, a set of antecedent conditions is defined, with the rest factors having a negligible probability of occurrence. But the dependence of the current, I_a, on the accelerating potential, V_a, displays a probabilistic distribution of measured values.

If the conditional model of causation is to serve as a general model, which applies both to the indeterministic context of quantum mechanics and the exception-ridden, local causal relations between social events, it must give the right answer in the case of causation in the social sciences. The task of the next section is to show that Max Weber has developed a notion of 'adequate causation', which can be expressed in terms of the conditional model of causation.

3.3 The social sciences and the conditional model

An interesting feature in the physical sciences is that the set of antecedent conditions can be regarded as closed; for instance, the necessary and sufficient conditions can be clearly defined and any further conditions in the background information, B, can knowingly be excluded. In Mackie's terms, the normal running of things in the causal field can be held constant for the purpose of the laboratory experiment ($B = 0$). This closure is well illustrated in the Franck-Hertz experiment, where the relevant parameters enter the laboratory set-up, whilst irrelevant parameters are knowingly excluded. This epistemic situation contrasts sharply with the causal situation in the social sciences. If C_1, C_2, C_3 represent a cluster of antecedent causal conditions and E represents an effect in the social world, then generally $Pr_1(E|\,C_1,C_2,C_3;B \neq 0) > Pr_2(E|C_3,C_4,C_5;B \neq 0)$. The expression '$Pr_1 > Pr_2$' captures Weber's notion of adequate causation. As we have stressed repeatedly, the closure of a set of parameters is difficult to achieve in the social sciences ($B \neq 0$). The problem of closure leads to probability considerations, whereby the weight of the evidence (C_1, C_2, C_3) is taken to distribute the credibility of causal explanations in the social sciences differentially. According to the above formula, the probability of the effect, E, is considered to be greater on account of one set of causal factors (C_1, C_2, C_3) than on account of an alternative set (C_3, C_4, C_5). But the presence of some overlap (C_3) indicates that social scientists often disagree about the precise nature of the causal conditions, which are taken to have brought about the event, E. Event E may be a social or a historical event; such effects can range from the expected rise in crimes as a consequence of a rise in the young male population to the outbreak of the Second World War. This uncertainty in the stipulation of antecedent causal conditions, due to the failure of closure, is further compounded by the lack of lawful regularities in the social world. The social world, as this section will attempt to show, exhibits regular patterns or trends, which however do not acquire the status of genuine laws. As we argued at the beginning of this paper, this absence of 'social laws' has repercussions for the evaluation of counterfactual situations in the social world.

We will first specify Weber's notion of causation in terms of a cluster of necessary and sufficient conditions. Weber regards certain clusters of antecedent conditions as more adequate conditions to account for a social or historical effect. Secondly, we will defend the claim that there are no genuine laws in the social sciences.

3.4 Weber's notion of adequate causation

How does Weber characterize adequate causation? For Weber this notion is based on his ideal-type methodology. Ideal types can be understood as hypothetical (or *as-if*) models in the social sciences (Weber, 1949; Weinert, 1996). The use of ideal types emphasizes the need on the part of the social scientist to use abstractions and idealizations in a reconstruction of a social event. The aim is, of course, to throw light on the actual determinant factors, which produced some event in history or the social world. Social systems, however, are open-ended. The social scientist is faced with infinitely many potentially determining factors, which could be possible causal factors. Out of the complex of determining factors, the social scientist must distil a complex of possible causal relations, which "should culminate in a synthesis of the 'real' causal complex" (Weber, 1949, p.173). Guided by his/her interests, the social scientist must construct a causal model (an ideal type) of a sequence of social events. Weber speaks of 'adequate causation' when the social science model meets several conditions: the social scientist has isolated a number of conditions from the maze of social conditions; there is a high probability that the reconstruction of the social or historical event, on the part of the social scientist, isolates the 'likely cause of an actual historical event or events of that type'. The ideal type model of the causal sequence of social events therefore depicts an objective possibility. The model of the social scientist provides the most adequate causal conditions which are likely to have brought about the social event in question. By insisting that the social scientist is engaged in hypothetical model construction, Weber guards himself against a possible charge of nave realism. In fact Weber defends a 'primary of theory' view of the social sciences. They express the 'typical' behaviour of the 'average' social agent. They are constructed in order to measure degrees of deviation of real social agents from agent types. The ideal types are intellectual tools for the mastery of empirical data. How can a social scientist be relatively certain that a proposed model of, say, the outbreak of World War Two, captures the most adequate conditions, which are most likely to have brought about the outbreak of the Second World War? Weber insists that "it is possible to determine, even in the field of history, with a certain degree of certainty which conditions are more likely to bring about an effect than others" (Weber, 1949, p.183). The only way of achieving this degree of certainty seems to be to think of the actual obtaining conditions in social situations as *constraints* on the models. Weber considers counterfactually, what *would* have happened to a given effect *if* certain conditions had either been absent or modified in the antecedent causal conditions. These counterfactual situations are not used in the service of investigating hypothetical historical or social situations. For the reasons mentioned above, such hypothetical situations are difficult to evaluate in the social world. Weber hopes to throw light on the 'historical significance' of the actual determinant factors in the emergence of some historical event. Whilst historians may disagree about all the possible relevant factors, which, say, swept Hitler into power, their models at least overlap to a certain extent. Certain factors will be so improbable—the flapping of a butterfly's wings—that they can be excluded from the causal account. Furthermore, with the emergence of new empirical data, historians' models can be improved, i.e., the historical evidence will credit some factors as the relevant ones and discredit others as irrelevant. For instance, since the end of World War II, Hitler scholarship has developed a number of models: from psychological to structural models. Such models may not be completely refutable, like models in the natural sciences. Nevertheless, they are open to a certain degree of testability: some of these models will, in the light of the increasing historical data, be judged more adequate than earlier models, whose authors were not yet able to benefit from archive resources, which became available after the collapse of the Soviet Union.

In terms of a conditional model of causation, Weber clearly argues that some clusters of conditions are more adequate explanations of the event than others. It is to be expected from the nature of the social sciences and the concern for actual obtaining situations that causal analysis will mostly be concerned with sufficient conditions. Can the social sciences specify non-trivial necessary conditions? The historian will struggle to state how Hitler's death in the November march of 1924 would have prevented or enabled certain subsequent events in history (Cf. Kershaw (1998, pp.xxvii)). There is simply no sufficient regularity to evaluate this condition as a relevant necessary condition. But the social sciences do not necessarily have to forego the consideration of necessary conditions. Weber certainly felt that his ideal types could spell out necessary conditions for social phenomena like feudalism and authority. Sociologists regard the differentiation of social systems into various subsystems (like the economic and the cultural system) and the organization of the economic system as a liberal market system as necessary conditions for modern, complex market societies. Political scientists and philosophers treat the tripartite division of power structures into the legislative, the executive and the judicial arm as a necessary condition for democratic societies. These fundamental structures manifest themselves in different ways, depending on the sufficient conditions present in particular societies. There clearly exist patterns of regularity in the social world, which could improve the counterfactual evaluation of unrealized social events, especially in company of other types of evidence. The question arises, however, which epistemic status should be granted to such trends.

3.5 Background conditions and trends

Some writers have expressed a much more deterministic view of causal models in the social sciences:

> Only quantum-mechanical phenomena exhibit associations that cannot be attributed to latent variables, and it would be considered a scientific miracle if anyone were to discover such peculiar associations in the macroscopic world.[5]

---

[5] Pearl (2000, p.62). According to Pearl's definition 2.3.2 "A latent structure is a pair $L = \langle D, O \rangle$, where D is a causal structure over (variable set) V and where $O \subseteq V$ is a set of observed variables."(Pearl, 2000, p.45). It is worth noting that Pearl explicitly embraces a Laplacean 'quasi-deterministic' conception of causality (Pearl, 2000, pp.26 and 63).

The question that inevitably arises is whether this is true of an abstract model, rather than the social world. And if it is true of a model, how does the model represent social reality? Any model in the social sciences, to be representative of a section of social reality, must 'fit' the available data. Even if we define causality for a model world, the model must permit a projection of the model world—counterfactual or not—onto the actual world. Any such projection faces two issues because 'adequate' causation in the social sciences differs from causation in the physical sciences in at least two ways: **a)** background conditions, even in laboratory experiments with human subjects, cannot be set to zero ($B \neq 0$); the background effects are not calculable; they need to be added as an error term in structural equations but their effect is difficult to evaluate; **b)** it is questionable whether lawlike regularities exist in the social sciences, although trends clearly do exist in the social world.

Ad **a)** The failure to achieve closure in the social sciences can be partly attributed to the symbolic and social dimensions of human life and partly to the impossibility of precise predictability of human actions. As many social scientists have emphasized, the social sciences do not deal with brute facts. Rather, social facts and relations form part of a pre-interpreted social world, in which social agents entertain relations through the use of symbolic signs. That is, social actions are typically intended to have meaning. The intended meaning of a social act, like greeting, must be such that it is typically understood as the intended act by other social actors. This symbolic dimension is not restricted to relations between individual social actors. Social institutions, like political offices, educational or religious institutions and economic systems, are typically invested with symbolic meaning. A Western visitor to Muslim countries, for instance, will clearly recognize that certain buildings, e.g., mosques, are designated as places of worship, even if the details of the religious practices in that country escape her/him. Looking at the conventions, norms, values and rules, which govern social life in any country, it is clear that regularities exist in the social world. These are often riddled with exceptions. But it remains unclear in the social sciences whether the exceptions are *compatible* with the regularities, which is an indication that the regularity is lawlike, or whether they are *incompatible* with the regularities, which is an indication that the regularity is not lawlike. This situation occurs because the epistemic status of social regularities is unclear.

Ad **b)** Popper considered that the decisive aspect of trends, as opposed to genuine laws, was that trends are social regularities, which are inductively generated from initial conditions (Popper (1957, §27 and 29); cf. Weinert (1997)) The dependence of trends on initial conditions seriously compromises their ability to support counterfactuals. To see this weakness consider what will be called a *structural interpretation* of laws of nature. This structural view can explain why genuine laws support counterfactuals, while trends fail to do so. The structural view does not consider that laws of nature are to be understood as exceptionless generalities; rather they are seen as expressing structural aspects of physical systems (Weinert, 1993, 1995). For instance, in the Franck-Hertz experiment, Bohr's equation,

(1) $\quad E_m - E_n = h\nu$

is taken to convey the structural information about quantum systems. It says that the transition between energy levels of quantum systems is discrete and that the emitted or absorbed energy possesses a frequency ν (where h is the Planck constant). Laws of science are the symbolic expressions of structural properties of natural systems, which only require initial conditions for their instantiation. Once such structural information is available, the evaluation of counterfactual situations is facilitated. The structural information expressed in the laws of science tells us how components (relata) of physical systems are mathematically related to each other (relations). The structure of a physical system consists in a specification of its components or relata—these can range from galaxies to atoms—and the way the relata are quantitatively related to each other—the mathematical relations can vary from Kepler's laws to Bohr's equation. By way of illustration, consider the solar system. Once its heliocentric structure is known, it is easy to evaluate what would happen if, say, a satellite was put into orbit beyond the moon. It would obey Newton's laws because it would become a further component in a structure, which is governed by Newton's laws. What would happen if a lump of gold were expanded to a diameter of 100m? It would still be a lump of gold because there are no structural limitations to this particular expansion. Could an atom be extended to such a diameter? No, for such an expansion is not compatible with the structure of atoms. Laws of nature therefore constitute *structural* constraints on the behaviour of physical systems. But there exists no such physical structure or mechanism underlying the social world. The lack of an underlying physical structure is largely due to the symbolic dimension of human existence. Human agency does not follow mechanical laws. There are, however, social structures imbued with symbolic meaning. The symbolic dimensions of social structures give rise to trends, which act as *practical* constraints. Social trends are simply generalizations over initial conditions, which prevent them from constituting structural constraints.

It could be argued that there are low-level empirical laws in the natural sciences, which may equally be said to be inductive generalizations over an observed number of cases and dependent on initial conditions. There is, however, another decisive difference between empirical laws and mere trends. It resides in the fact that trends but not genuine laws can be modified or even reversed. To illustrate, consider the difference between Rutherford's decay law and a social trend. The radioactive decay law

(2) $N(t) = N(0)e^{-\lambda t}$

states a functional relationship, based on the natural logarithm e and a decay constant, λ, between the radioactive decay of an ensemble of particles, N, after time t, and the original ensemble N(0) at t_0. The decay law is statistical in nature since it refers to the behaviour of an ensemble, whilst the behaviour of individual members of the ensemble is unpredictable. Note, however, that the stochastic nature of this relationship informs us about a structural property of radioactive behaviour, which no physical interference could reverse. Similarly, the Franck-Hertz experiment provides structural information about the discreteness of energy levels in atomic systems. It can be exploited for the purpose of experimentation but neither changed nor reversed. Equations (1) and (2) therefore express structural constraints on the behaviour of the respective systems; they are laws of science, which these systems obey.

By contrast, a social trend like the relationship between drink-driving and road fatalities in a certain country can be reversed given, say, political interference or technical ingenuity. What is even more striking is that a social trend can be reversed using the very knowledge of the trend. If social research shows that there exists a causal correlation between the consumption of alcohol and car accidents, then human interference at the level of alcohol consumption on the part of drivers will change the very correlation which existed prior to the research. The existence of trends prevents us from speaking of probabilistic causation in the social sciences in a straightforward sense. We can say of an individual atom in an ensemble N that it possesses a 50% probability of decaying within the characteristic decay time t. For instance, the half-life of radium (^{226}Ra) is 1600 years. But this probability is identical with the probability of decay of the whole ensemble, which is given by equation (2). As a matter of fact we can also say of social trends that they are likely to produce certain effects. Recall that the Department of Social Affairs has predicted an expected rise in crimes as a consequence of a rise in the young male population in Britain. And it is possible, as a matter of fact, that this rise will be observed. Trends are therefore practical rather than structural constraints. But a trend or regularity pattern is only a *ceteris paribus statement*. Two kinds of interventions can be envisaged which could either stem the rise or prevent it. The first, controlled intervention consists in the employment of this very fact, which could in principle be used to prevent the rise from happening. A second, uncontrolled kind of intervention may arise from social or political situations elsewhere in the social system, which again may prevent the predicted occurrence of the rise—for instance an unpredictable external or internal political crisis. The social sciences can only rely on weak regularities to construct causal models. The social sciences need an adequate model of causation to express the causal realities of the social world.

The presence of mere trends has implications for the evaluation of counterfactuals in the social sciences. The more reliable the trend, the better the chances of a counterfactual assessment. The trend of high road fatalities in the absence of limits on alcohol consumption can be used to make counterfactual statements about the effects of a restriction of alcohol consumption for drivers in other countries, which have similar infrastructures and traffic densities. Note, however, that drink driving is a medical rather than a social condition, which makes the possible projection to other countries hardly surprising. By contrast, consider the introduction of some legislation, L, in a particular country, C. The effect of this legislation on the population of C will not be completely predictable (viz. the famous poll tax revolt under the Thatcher government in Britain in the 1980s). It will even be less predictable for a population in a different country D the more different the socio-economic landscape of D is.

The lack of lawful relationships or even trends makes a counterfactual scenario like C_2 difficult to evaluate—what would have happened if Hitler had been killed by a bullet in November 1924? It may be objected that socio-political evidence about the Weimar Republic will make counterfactual assessment relatively secure, even in the absence of practical regularities. Counterfactual evaluations based on socio-historical evidence must make *ceteris paribus* assumptions, e.g., that no interference will occur between the counterfactual scenario (Hitler's death in 1924) and the occurrence of a later event (the outbreak of World War Two). The problem is that even under such *ceteris paribus* assumptions, the evidence only acts as a practical constraint and therefore suffers from the same weakness as trends. The evidence from the Weimar Republic may suggest that World War Two would still have happened if Hitler had been killed in 1924. But social conditions are changeable and reversible, often due to human interference. The counterfactual evaluation based on evidence is affected by the same uncertainty as counterfactual evaluation based on mere trends. This analysis concurs with Weber's conclusion:

> We can (...) estimate the *degree* to which a certain effect is 'favoured' by certain 'conditions'—although we cannot do it in

a way which will be perfectly unambiguous or even in accordance
with the procedures of the calculus of probability. (Weber, 1949,
p.183; italics in original)

4 Conclusion

The openness of social systems and the lack of lawlike generalities in the social world make it unlikely that either causal-mechanical or counterfactual-interventionist models of causation can be successfully employed by the quantum physicist or the social scientist. W. C. Salmon believed that causal connections between events can be found in causal processes, which establish a physical link between C and E.

> Causal processes are capable of transmitting energy, information, and causal influence from one part of space-time to another. (Salmon, 1998, pp.71 and 16)

This may be true of classical physics. Quantum mechanics poses serious obstacles in the way of establishing such an account. All quantum mechanical experiments at one point fall short of delivering a continuous trace between C and E, most dramatically in the case of the non-local correlations of entangled quantum systems. Salmon's account must give the wrong answer because it insists on spatio-temporal continuity, which is violated in these experiments. In the Franck-Hertz experiment we observe a discontinuous change of energy, as permitted in the Dowe's conserved quantity account which does not require spatio-temporal continuity (Dowe, 2000). But the analysis of the experiment shows that insistence on conserved quantities is only a small subset of the whole set of causal conditions.

We agree with Woodward that a philosophical account of causation should account for scientific practices. Many of the causal accounts in circulation seem to be applications to small areas of scientific practice, as observed in classical physics and deterministic systems. Woodward's counterfactual-interventionist approach presupposes the availability of actual lawful regularities. If, however, there are no social laws, as argued in this paper, even counterfactual-interventionist models face difficulties in the evaluation of hypothetical (social) experiments.

By contrast, a conditional model of causation, which stresses the availability of actual antecedent and consequent conditions, seems to be applicable to such diverse fields as the Franck-Hertz experiment and the social sciences. Although there are significant differences in the sets of antecedent and consequent conditions, depending on whether the conditional model discusses a physical or social event, the model is sufficiently general to be applied to both the natural and the social sciences. The desire for a unified model of causation, rather than causal dualism, is motivated by a naturalistic view of the social sciences. Such a view of the social sciences claims a certain continuity between the natural and the social sciences, despite differences (lack of genuine laws, failure of closure). But the employment of ideal-typical models and the availability of patterns of regularity give rise to causal explanations of social events and limited predictions. The pervasiveness of symbolic dimensions in social life does not, on the present view, prevent the social sciences from attaining scientific credibility. A certain unity or at least continuity of method between the natural and the social sciences calls for a single model of causation.

Acknowledgments

The author would like to thank an anonymous referee for helpful comments on an earlier version of this paper.

Friedel Weinert
SSH, University of Bradford, UK.
f.weinert@brad.ac.uk

BIBLIOGRAPHY

Bohm, D. (1957). *Causality and Chance in Modern Physics.* Routledge and Kegan Paul, London.
Dowe, P. (2000). *Physical Causation.* Cambridge University Press, Cambridge.
Franck, J. and Hertz, J. (1914). Über Zusammenstöße zwischen Elektronen und den Molekülen des Quecksilberdampfes und die Ionisierungsspannung desselben. *Verband Deutscher Physikalischer Gesellschaften*, 16:457–67.
Gillies, D. (2005). An action related theory of causation. *The British Journal for the Philosophy of Science*, 56:823–42.
Kershaw, I. (1998). *Hitler.* Penguin.
Krane, K. (1983). *Modern Physics.* John Wiley and Sons, New York.
Mackie, J. L. (1980). *The Cement of the Universe.* Clarendon Press, Oxford.
Pearl, J. (2000). *Causality.* Cambridge University Press, Cambridge.
Popper, K. (1957). *The Poverty of Historicism.* Routledge and Kegan Paul, London.
Reichenbach, H. (1920). Philosophische Kritik der Wahrscheinlichkeitsrechnung. *Die Naturwissenschaften*, 8:146–53.
Reichenbach, H. (1931). Das Kausalproblem in der Physik. *Die Naturwissenschaften*, 19:713–22.
Salmon, W. C. (1998). *Causality and Explanation.* Oxford University Press, Oxford.
Spohn, W. (2006). Causation—an alternative. *The British Journal for the Philosophy of Science*, 57:93–119.
von Wright, H. (1971). *Explanation and Understanding.* Cornell University Press New York.
Weber, M. (1949). Objective possibility and adequate causation in historical explanation. In Shils, E. A. and Finch, H. A., editors, *The Methodology of the Social Sciences: Max Weber*, pages 164–88. The Free Press.
Weinert, F. (1993). Laws of nature: a structural approach. *Philosophia Naturalis*, 30:147–71.

Weinert, F. (1995). Laws of nature—laws of science. In Weinert, F., editor, *Laws of Nature*, pages 3–64. de Gruyter, Berlin.

Weinert, F. (1996). Weber's ideal types as models in the social sciences. In O'Hear, A., editor, *Verstehen and Humane Understanding*, volume 41 of *Royal Institute of Philosophy Supplement*, pages 73–93. Cambridge University Press, Cambridge.

Weinert, F. (1997). On the status of social laws. *Dialectica*, 51:225–42.

Weinert, F. (2004). *The Scientist as Philosopher*. Springer, Heidelberg, Berlin, New York.

Woodward, J. (2003). *Making Things Happen*. Oxford University Press, Oxford.

Zilsel, E. (1927). Über die Asymmetrie der Kausalität und die Einsinnigkeit der Zeit. *Die Naturwissenschaften*, 15:280–6.

The inference of common cause naturalized

AVIEZER TUCKER

ABSTRACT. Reichenbach failed in his attempt to deduce a principle of common cause from the second law of thermodynamics. Most philosophers attempted to explicate principles of inference of common cause. They confused the inference that some common cause existed without specifying its properties, with the inference of a concrete common cause with a unique set of properties. They also did not distinguish the inference of type common causes from token common causes. I present a new naturalized theory of the inference of common causes by examining how scientists actually infer common causes, for example: Darwin's inference that all the finches in the Galapagos Islands had a common cause, or the inference that all the Indo-European languages had a common cause. Inferences of common causes usually proceed in three stages, consecutive comparisons of the likelihoods of the evidence: first a comparison of likelihoods given some common cause, whose properties are unknown, and given separate causes; second, if the evidence is more likely given a common cause, five types of common causes and causal nets compete over which increases most the likelihood of the evidence. Finally, if it is possible to prove which of the five types is most probable, scientists attempt to infer the actual properties of common causes and the information transmission chains that connect them to the evidence by comparing the likelihoods of the evidence given competing concrete common causes.

1 Introduction

Inferences of common causes from their effects are ubiquitous in science and everyday life: evolutionary biologists infer the existence of extinct species from similarities between their descendants, and their properties from those of fossils, detectives and historians infer descriptions of past events from testimonies, textual critics infer properties of lost texts from present exemplars, geneticists and comparative linguists infer the migration routs of our pre-historic ancestors from the present geographic distributions of genes and languages, teachers infer plagiarism from similarities between exams. Many philosophers (Reichenbach, 1956; Sober, 1988; Cleland, 2002; Tucker, 2004) agree that much of our knowledge of the past is founded on inferences of common causes from their present effects.

The philosophical literature about the inference of common causes originated in Reichenbach's attempt (Reichenbach, 1956) to deduce a principle of common cause from the second law of thermodynamics. I argue first that this attempt failed. Most philosophers did not pursue Reichenbach's deductive project. Instead, they attempted to *explicate* principles of inference of common cause in the logical positivist *prescriptive* and *unempirical* senses of the term. It is difficult to debate philosophical explications because they can fall on their normative aspect when confronted with empirical counter-examples. Rather than refute, empirical evidence can demonstrate the *irrelevance* of these explications, as the history and sociology of science demonstrated the irrelevance of much of the logical positivist explication of science, spurring the introduction of new philosophies of science. I attempt to do so here. I show that explications of Reichenbach's common cause principle confused the inference that *some* common cause existed without specifying its properties, with the inference of a *concrete* common cause with a unique set of properties. They also did not distinguish the inference of common cause *types* from the inference of common cause *tokens*. I clear these confusions and argue that no exclusively *a priori* probabilistic silver bullet exists for the description of the inference of common cause. Additional assumptions and theories about the transmission of information in time must be added. The shortcomings of the existing philosophical approaches to the problem of the common cause invite the introduction of a new theory.

I present here such a new *naturalized* theory of the inference of common causes by examining how do scientists actually infer common causes, especially in cases that they recognize as successful such as Darwin's inference that all the species of finches in the Galapagos Islands had a common cause; Rasmus Rask's and Franz Bopp's inference that all the Indo-European languages had a common cause; and the inferences editors of scientific editions of classics make in attempting to infer the most authentic version of the texts they collate. I argue that scientific inferences of common causes usually proceed in three stages, i.e., three consecutive comparisons of the likelihoods of the evidence: first a comparison of likelihoods given some common cause, whose properties are unknown, and given separate causes; second, if the evidence is more likely given a common cause, five types of common causes and causal nets compete over which one most increases the likelihood of the evidence; third, if it is possible to prove which of the five types is most probable, scientists attempt to infer the actual properties of common causes and the information transmission chains that connect them to the evidence by comparing the likelihoods of the evidence given competing concrete common causes.

2 Reichenbach's God

Reichenbach (1956) believed that when $A\&B$ are probabilistic frequencies of a property or a set of properties in a population, *if* $P(A\&B) > P(A) \times P(B)$, *then* there is a common cause C for $A\&B$ that satisfies the condition $P(A\&B|C) = P(A|C) \times P(B|C)$.[1]

Reichenbach believed that he was engaged in applied theoretical physics rather than describing or explicating scientific practices. He sought to deduce this common cause principle from the Second Law of Thermodynamics with the "hypothesis of the branch structure", referring to the branching of a physical system from the general direction of entropy in the universe (Reichenbach, 1956, p. 157). Branch systems are isolated from the main system but are connected to it at their beginning and end; one end has low entropy and the other has high entropy. "In the vast majority of branch systems, the directions towards higher entropy are parallel to one another and to that of the main system" (Ibid, p. 136). Reichenbach considered these hypotheses as "empirical hypotheses which are convincingly verified" (Ibid). Since any physical system should display increasing entropy, cooling, disorder, however slowly, if and when we encounter increasing order, it must be caused by an intervention, a common cause, from outside the system in question. Causal explanation replaces then low single system probability with a high multi-system probability. Accordingly, causes are always in the past of their effects. Reichenbach even speculated that, historically, the concept of causation was introduced to explain otherwise improbable ordered states. Ordered states are defined by being constituted by simple rules. Arrangements that are rule indistinguishable have the same degree of order. *Therefore* when we perceive an increase of order in a system, it must have an external cause, as billiard balls neatly ordered in a row are proof of intervention. Wind and waves create high probabilities for "unordered" states of smooth or wavy sandy surfaces. "It is too improbable that such an ordered state should result by chance..." (Ibid, p. 149) "Ordered" footprints in the sand are highly improbable. The detection of order and disorder requires

---

[1] Reichenbach did not apply his own analysis consistently. His first example for a common cause is of a thunderstorm that produces lightening and wind. Lightening and wind are not distinct from the thunder storm, $A\&B$ are aspects of C. Analytically, understanding the meaning of thunder storm implies the defeasible presence of lightning and wind and does not require any statistical evidence, just understanding the meaning of 'thunder storm'. The lightning and wind do not represent higher level of order than the storm.

special knowledge, of geology, paleontology and archeology for example. Reichenbach acknowledged that the estimate of degrees of order/entropy is subjective. For example, why consider footprints on sand a more ordered state than waves of sand dunes?!

Reichenbach's research program was neither epistemic nor methodological, but metaphysical. He sought to demonstrate the direction of time from past to future by discovering asymmetries between past and future. "The cause produces the effect; but since the cause leaves traces in the effect, it can be inferred from its effect. Here is the origin of the distinction between past and future... although the past can be recorded, the future cannot, is translatable into the statistical statement: *isolated states of order are always postinteraction states, never preinteraction states,...* if positive time is defined as the direction of increasing entropy" (Ibid, p. 155). Reichenbach stressed the significance of the second law of thermodynamics for the exclusion of teleology. "The cause produces the effect, the effect records the cause" (Ibid, p. 156). Reichenbach did not consider that some causes neither leave traces in their effects nor generate records. The process of entropy is at play in the preservation of records too (Laplace, 1951, p. 123). The extent to which the "future" records the effects of the "past" has been the subject of a recent debate where both sides inferred from anecdotic cases conflicting conclusions about the possibility of knowledge of the past (Cleland (2002); Turner (2005); cf. the more cautious conclusions in Sober and Barrett (1992)).

Reichenbach sought to reduce the direction of time to asymmetric statistical relations between common causes and their effects, and common effects and their causes: arguably, the probability of the conjunction of causes is not affected by having a common effect (and so teleology is false), while the likelihood of effects is higher given a common cause. As time passes from past to future, entropy and disorder increase in separate systems and the common cause diminishingly increases the joint probabilities of its effects, until its effects fade completely (Reichenbach, 1956, pp. 164–165). The further effects are in time from their common cause, the narrower should be the probabilistic gap between $P(A\&B)$ and $P(A) \times P(B)$:

> A definition of time direction in terms of the principle of the common cause... thus defines time direction in terms of macrostatistics.... The registering of information reveals the unidirectional nature of time, because only a past cause can produce a correspondence between signs and physical data, and because an increasing amount of information represents the order of a space ensemble which calls for explanation in terms of a past cause. The time direction of information is the direction

imprinted upon the universe by its statistical isotropy. (Ibid, p. 186)

Eventually, Reichenbach failed to deduce his principle of the common cause from the Second law of Thermodynamics. On the macro level, entropy works very slowly and ordered states can emerge spontaneously in isolated systems. On the time scales of human history, comparative linguistics and textual criticism, entropy appears flat and can be ignored. Reichenbach considered any higher than expected correlation of frequencies, any positive probabilistic relevance, an ordered state in entropic terms. However, in the above human time-scales, descriptions of ordered correlations cannot be reduced to descriptions of rise of order in entropic terms and so do not imply a common cause.

From the extended time scale of the history of life, while entropy has increased in the universe, the system of life on earth has been sufficiently open to grow increasingly complex through the exchange of increase of order in the system for decrease in order outside of it. Therefore, the inference of common cause from states of increasing order in open systems, including correlations between frequencies in the history of life, let alone the history of humanity and languages, together with the Second Law of Thermodynamics, is false. This criticism can be stated conversely in a *reductio ad absurdum*. Reichenbach claimed that "in order to explain the coincidence of A and B, which has a probability exceeding that of a chance coincidence, we assume that there exists a common cause C" (Ibid, p. 159).[2]

This claim is assumed by the cosmic design argument for the existence of God that purports to infer, from the correlated complex intricacy of the universe, its ordered "probability exceeding a chance coincidence", correlations between the existence of life and the constants and laws of physics, or between the various parts of the eco-system, or between the various parts of the human organism, the existence of one and only common cause, that is God's design. The design argument compares the likelihoods of the highly complex correlations necessary for life, given an intelligent design common cause and given their chance coincidence, to conclude that these highly complex correlations are more likely given the first than given the second. Sober (2004) noted that the fallacy of this argument is in comparing the common cause/design hypothesis with a chance coincidence hypothesis rather than with alternative separate causes hypotheses, such as evolution by natural

---

[2]Earlier, Reichenbach claimed that "If an improbable coincidence has occurred, there must exist a common cause" (Ibid, p. 157). The two statements are inconsistent because improbable coincidence must have a low probability, whereas a chance coincidence can have a high probability. The latter is more loyal to the probabilistic form of the principle of the common cause.

selection.

Design arguments are based on two successive fallacies: first, that the relevant comparison of likelihoods of the ordered correlated complexity of life is between a common cause hypothesis and a chance occurrence hypothesis, rather than separate causes hypotheses, and second, that of all possible common causes, i.e., the best explanation of cosmic correlations is an intelligent design, though, even if there is a common cause, it does not have to be God. For example, the correlation between the physical constants and the existence of life has a common cause: the physical constants were partial causes of the emergence of life. Reichenbach's case for the common cause principle is assumed by the first fallacy of design arguments.

3 The explications

Only a few discussions of Reichenbach's principle of the common cause mention the Second Law of Thermodynamics (Sober and Barrett, 1992; Barrett and Sober, 1994). Most interpreters of Reichenbach's common cause principle have considered it an explication, an *a priori* prescription with few illustrations that is not subjected to systematic empirical testing. The prescriptive aspect of explications of science renders the examination of actual scientific practices redundant because they either conform to the *a priori* prescriptive recipe, or approximate it, or are just bad science. Empirical evidence cannot *refute* an explication that is based on an *a priori* formal model. Still, aspects of the explications of the common cause principle are vague and incoherent; virtually all of these explications, if not false, are *irrelevant* for actual scientific practice because they are either too narrow or too broad.

Reichenbach interpreted probabilities as frequencies. Properties of events may be more or less frequent according to the class against which they are measured. Choices of classes are often theory laden (Van Fraassen, 1980, pp. 141–157); (Tucker, 2004, p. 192). This is not a problem as long as it is clear that frequencies are properties of events or events under a description, relative to a specified class. In many successful cases of inference of common causes it is possible to make only a fairly rough estimate of frequencies for lack of evidence. For example, the frequencies of properties among classes of extinct species are difficult to estimate given the fossil record, yet paleontologists can and do infer from them cladistic trees.[3] The comparison class

---

[3] Hofer-Szabó et al. (2002) demonstrated in a *reduction ad absurdum* examination of Reichenbach's common cause principle that it is impossible to infer common common-causes, causes of common causes, like the genealogical tree models in Cladistics, using Reichenbach's principle, thereby proving the uselessness of Reichenbach's common cause principles for the actual construction of cladistic trees in evolutionary biology, textual criticism, and the study of testimonies in historiography and jurisprudence, the most com-

may be too small for supporting estimates of frequencies. Still, eye witness testimonies may be sufficient for conviction in legal contexts, without estimates of frequencies. Relevant correlations for the inference of common causes occur not just between *types* of events, frequencies, but also between singular *token* events (Berkovitz, 2000, p. 68). In many cases, it is easier to infer common causes of species or languages on the basis of evidence for causal chains that led to them, irrespective of whether the relevant frequencies are known or not. Therefore, *Reichenbach's principle is too narrow* to cover much of the relevant scope of scientific inferences of common causes. Narrow scope does not *falsify* the principle. But an alternative analysis of the inference of common cause that has a broader scope will be preferable.

Sober has noted that there are many correlations in the world that imply nothing about the causal processes that brought them about. For example, monotonically increasing quantities such as British bread prices and sea levels in Venice are positively correlated (Sober, 1988, 2001). Correlations between properties such as having wings or fins may be homoplasies, result from independent adaptations to similar environments at different times and places (Sober, 2001). The universe is replete with positive correlations between frequencies of events that do not result from any common cause. In popular culture, frequencies of winnings in sport, rising stocks and election victories are correlated with frequencies of properties of the weather, animal behavior patterns, somebody's uncle intensity of rheumatism pain, the voting patterns in some remote locations in Iowa, and the Zodiac. Some of these correlations may well hold. A frequent reply to such counterexamples is that false positives result from mixing populations with different causal nets. However, in order for us to know which populations are mixed and which are not, we must know many things about the world and how parts of it interact, that is, we have to add many assumptions to Reichenbach's two. Therefore, there can be no simple reduction of the inference of the common cause to probabilistic correlations, causal background conditions must be assumed (Berkovitz, 2000, p. 63). Berkovitz (2000, p. 55) noted that there are plenty of correlations that require no explanation at all. For example, perfect correlation between independent series of coin tossing is highly improbable, yet requires no explanation. Therefore, *Reichenbach's principle is too broad* to be relevant.

There have been attempts to salvage the common cause principle by adding criteria to limit the scope of relevant correlations for the inference of common cause, or by weakening the strength of the inference (cf. Sober (2001, pp. 341–342); Hoover (2003)). However, it has proved challenging

---

mon and universally acknowledged as successful cases of scientific inferences of common causes.

to find criteria that are neither too narrow nor too broad. Reichenbach retreated from his initial claim that the principle of the common cause implies that improbable correlation "must" result from a common cause to a probabilistic version, acknowledging that coincidences can and do happen. But to infer inductively any principle of common cause, it is necessary to generate a statistically meaningful sample of higher than expected correlations between frequencies of events and examine on independent grounds what is the frequency of common causes for these correlations. Nothing of the sort has been attempted. However, the most acute problem with explications of the common cause principle is the ambiguity of the concept of the common cause (Arntzenius, 1992).

4 Some common cause vs. a particular common cause

Reichenbach did not distinguish the inference of *the properties of a particular common cause*, from the inference that some *common cause existed* without characterizing it further. In the same vein, Sober claims that "it is one thing to infer the existence of a common cause, quite another to say what that common cause was like, on the assumption that it exists" (Sober, 1989, pp. 281–282). It is one thing to say that men and apes had some common ancestor; quite another to infer the properties of that ancestor. "In applying the principle of the common cause to examples, Reichenbach (1956) and Salmon (1984) often treat postulating a common cause and inferring the state of that cause interchangeably" (Sober, 1989, p. 282, note 3); see also Sober and Barrett (1992).

The histories of evolutionary biology and historical linguistics clearly display two stages: First, Rasmus Rask and Franz Bopp and Darwin established respectively that groups of languages and species had some common causes. A generation or more later, other linguists and biologists attempted to infer the properties of these common causes (Tucker, 2004, pp. 63–68, 85–91). "The methods used in phylogenetic inference assume that the species surveyed are genealogically related. The question is to determine which tree is best supported by the data" (Sober, 1999, p. 267).

Most of Reichenbach's examples, simultaneously sprouting geysers, light bulbs that blow simultaneously, and traveling thespians who are inflicted simultaneously with food poisoning, do not distinguish the inference of concrete common causes—for instance, an underground conduit between the two geysers, a burnt fuse, and a spoiled shared meal—from the inference that some common cause existed. One can easily imagine alternative concrete common causes to Reichenbach's. The concrete common cause for the food poisoning in Reichenbach's example is ambiguous as it can be either the meal, or the spoiled ingredient(s) in it (Sober and Barrett, 1992, pp. 2–3). Reichenbach went as far as saying that "If there is more than one possible kind of common cause, C may represent the disjunction of these causes" (Reichenbach, 1956, p. 159). But in many of the paradigmatic cases of inferences of a common cause, the alternative "kinds" of common cause hypotheses were unknown.

For example, Darwin was certain that the Galapagos finches had a common cause-ancestor. But he did not surmise which properties possible common ancestors could have had. Therefore, it may often be impossible to formulate the kind of disjunction that Reichenbach called for. Even if such a *disjunction* is possible, its members may form an alternative *conjunction* of *separate causes*. For instance, the common cause of food allergies after a meal of shrimps in cream sauce may be formulated as a *disjunction* of factors: allergy to milk *or* shellfish. But it can also be explained by a *conjunction of separate cause*, if some of those affected by food poisoning reacted to the cream whereas others reacted to the shellfish.

This ambiguity in Reichenbach's characterization of 'the common cause' is most devastating in undermining his "screening" condition. Reichenbach suggested that if for frequencies A and B, $P(A\&B) > P(A) \times P(B)$, it is likely that A and B share a common cause *given that A and B do not affect each other*. The common cause C "screens" A and B from each other, iff $P(A\&B|C) = P(A|C) \times P(B|C)$. But, *if C means that some common cause of A and B existed without specifying its properties, C could be identical with A, B, the conjunction of A and B, or several common causes that may include A and B*. In such a case, Reichenbach's *screening condition* may be satisfied, yet since C could be A or B or a conjunction of the two, they would still *not be screened* from each other. If $C = A$ then, $P(A\&B|A) = P(A|A) \times P(B|A)$.

For example, identical sentences in exams imply that there were some common causes; but perhaps one student copied another, or perhaps the students co-authored their exams, or perhaps the students copied a common source. Further, "where causes act probabilistically, screening off is not valid. Accordingly, methods for causal inference that rely on screening off must be applied with judgment and cannot be relied on universally" (Cartwright, 1999, p. 109). Markov conditions do not hold universally for all statements asserting causation, certainly not when all we infer is the existence of a common cause rather than its state, because we cannot distinguish sufficiently "parents" from each other or their "descendants." As Cartwright noted, this problem is particularly acute in medical cases in distinguishing causes from correlated effects (Ibid, p. 113). Reichenbach could then suggest a third necessary condition for the inference of some common cause: $A \neq B \neq C$. This third stipulation should restore the screening.

However, it requires sufficient knowledge of the properties of the common cause to distinguish C from A and B. Such knowledge requires more evidence than is typically available when the inference of the existence of *some* common cause is made.

Berkovitz (2000, pp. 57–58) argued further that screening conditions hold rarely in the world because when a correlation has a common cause, usually the correlated frequencies or events are affected also by separate causes. Common causes are usually partial causes and do not screen their effects from other causal factors. For example, when two species have a common cause-ancestor, their state is affected also by later mutations and natural selection.

Reichenbach could then have suggested to drop the inference of *some* common cause to deal exclusively with concrete common causes at the cost of narrowing the scope of his principle. But the properties of concrete common causes cannot be inferred from Reichenbach's conditions plus $A \neq B \neq C$. These conditions cannot discriminate between alternative hypothetical concrete common cause hypotheses, nor generate an exhaustive list of such possible hypotheses. Philosophers in Reichenbach's tradition have been making a fallacious transition in the argument from claiming to have proved that there must have been *some* common cause to discussing the likelihoods of correlations between the effects of *concrete* common causes.

5 Type vs. token common cause

"A token event is unique and unrepeatable; a type event may have zero, one or many instances" (Sober, 1988, p. 78).[4] For example, if we share the same type of biological causes, the same types of sperm and egg, we most likely belong to the same species; if we share tokens of these causes, we are twins. Reichenbach's examples for common causes are all tokens. But he did not characterize them so explicitly. Salmon (1984) and Hitchcock (1998) mixed common cause types with tokens in their interpretations of the common cause principle. Theoretical science study common types of causes, gravity, atoms, cells, the speed of light and so on. Theoretical science is not interested in particular tokens of a cell or an atom. The inference of common cause types is the standard result of successful replication of experiments with tokens of theoretical types. Types of causes that appear in several

---

[4]The type-token distinction has been useful for the analysis of myriad philosophical problems since Pierce introduced the terms, most notably for identity theories of body and mind. Nevertheless, the definitions of types and tokens and the clear statement of relations between them have been fraught with contention (Wetzel, 2006). For current purposes, it is unnecessary to enter these debates. It is sufficient to make the fairly uncontroversial claim that as particulars, tokens necessarily occupy a unique spatio-temporal location, whereas types do not.

theories allow scientific theories to attempt to intertwine to construct a net of theories that covers a domain of nature.

Salmon (1984, pp. 220–221) analyzed the determination of a universal constant, the Avogadro number, as the common cause *type* of different effects in different experiments. Salmon likened the various experiments that confirmed the Avogadro number to multiple witnesses to a murder whose testimonies agree on a core description, though each witness may present different additional information. Even if the witnesses were unreliable by themselves, "it would be too improbable a coincidence for all of them to have fabricated their stories independently, and for all these stories to exhibit such strong agreement in precise detail" (Ibid, p. 220). Salmon did not distinguish the inference of a *token* event, a murder, from that of a *type*, a universal constant. Van Fraassen (1980, p. 123) rightly criticized Salmon's presentation of the discovery of correlation between the number of molecules in one mole of gas, the Avogadro number, and the number representing electron charges, one Faraday, as an inference of a common cause. This correlation "... does not point to a relationship between events (in Brownian motion on specific occasions and in electrolysis on specific occasions) which is traced back via causal processes to forks connecting these processes. The explanation is rather that the number found in experiment A at time t is the same as that found in totally independent experiment B at *any* other time t', because of the *similarity* in the physically independent causal processes observed on those two different occasions" (Van Fraassen, 1980, p. 123). Arntzenius (1992, pp. 230–1) stressed the frequent confusion between types and tokens in discussions of the inference of the common cause too.

Hitchcock (1998, pp. 427–8) distinguished insignificant "statistical correlations" from "probabilistic correlations" that warrant in his opinion the inference of common causes. He interpreted statistical correlations as the inference of a common cause *token* from token effects, and probabilistic correlations as the inference of a common cause *type* from correlated types, frequencies of properties. Probabilistic correlations are of frequencies of a property or properties in a class. Statistical similarities between occurrences, without correlations between frequencies, do not indicate a common cause, claimed Hitchcock, they may be anecdotic. Hitchcock's example (Hitchcock, 1998, p. 436) for the inference of a common cause type from a probabilistic correlation between types is of hypothetical languages like proto-Indo-European that should explain the correlations between classes of languages. However, in historical comparative linguistics, descriptions of languages, whether hypothetical or historical, are of token processes that lasted for a definite period among people spread over space. Hitchcock may have confused the meaning of 'language' as a type studied by the philosophy of language or Chomsky's generative linguistics with its meaning as a token studied by comparative historical linguistics. Historical linguistics infers common cause tokens from correlations among the most slowly mutating subsets of token languages that refer to places, fauna, flora, immediate family members, body parts and the first few numbers.

There is no significant correlation between what Hitchcock called probabilistic correlations and the inference of common cause types, and statistical correlations and common cause tokens. Probabilistic correlations are insufficient for inferring the existence of a common cause since the world is full of insignificant correlations. McGrew (2003) concluded that common cause type hypotheses that satisfy the conditions of Reichenbach's common cause principle are less probable than competing hypotheses that do not satisfy these conditions. Statistical correlations are insufficient either because as Sober (1988, pp. 72–78) noted, overall similarity, e.g. pheneticism in biology, does not indicate common ancestry.

I do not think that the scientific inference of common cause types raises new issues in the philosophy of science. Probabilistic and statistical correlations may play a role in the context of discovery of common cause types, but the justification of the inference of common cause types or the choice between competing common cause type hypotheses is not distinct of the justification of scientific theories in general. The inconclusive debate about whether there are common causes of probabilistic correlations in quantum physics proves the absence of consensus in science or philosophy about the inference of common cause types from such correlations. There is simply no probabilistic silver bullet.

Sober (1988) and Hausman (1998, pp. 207–8) interpreted the inference of a common cause as that of a token. For instance, Sober claims: "A common cause explanation postulated a common *token* object;...That is why homoplasies induced by numerically distinct but qualitatively similar selection processes do not have common cause explanations" (Sober, 2001, p. 339). The inference of common cause tokens may well distinguish the *historical* from the *theoretical* sciences (Cleland, 2002; Tucker, 2004).

6 Naturalizing the inference of common cause

If it is accepted that all the attempts to explicate a principle of a common cause that I outlined above have failed, philosophers may either attempt yet another explication that would avoid the drawbacks of its predecessors, or they may attempt to approach the problem of inference of common cause from a different methodological perspective. I think that as in the philosophy of science, explication exhausted itself and the philosophers who advocated it. Yet, it has not been falsified, and some philosophers still pursue it. Perhaps they may even come with interesting results. But most philosophers of science, I believe, find explication irrelevant for understanding the actual practices of scientists and the history of science. Following Quine (1985), the ambiguous relations between the normative and descriptive aspects of explication render it too vague to be refutable.

It is theoretically possible to confirm principles of inference of common cause by *inductive inference*. But I know of no attempt to create a database of processes involving common causes and their effects against which principles of inference of common cause like Reichenbach's principle, its interpretations and alternatives can be tested by counting frequencies of correct inferences. The nearest approximation I found is a computer simulated model that refuted Reichenbach's principle (Barrett and Sober, 1994, note 7). Induction has been mentioned only by philosophers who argued against Reichenbach's common cause principle by demonstrating that it cannot be inferred inductively as there are too many exceptions to the rule (Sober, 1988, 1989, 1999, 2001). Developing this approach would require more resources, time and labour, to build such a database than is available to most academic philosophers like me. This explains why this method has not been applied.

In the rest of this paper, I use a *naturalized* approach to this problem. Following post-positivist philosophy of science, I attempt to analyze the success stories of inference of common cause in the history of science. I assume that the most successful cases have become so for a reason for making sound inferences of common causes. Analyzing such cases is likely to offer solid foundations for a philosophical analysis that is both descriptive and normative, descriptive of the best practices of scientists and normative in recommending to other scientists to use the very same tried and tested methods. It is likely that principles confirmed by the examination of the most successful cases of inferences of common causes would have also been confirmed by an inductive study. Scientific communities are likely to emulate the methods used in successful cases that would then become representative of scientific inferences of common cause in general.

Advocate of an explicative approach may argue back that Reichenbach's attempted deduction from theoretical physics, the explicative attempts at probabilistic inference, and the naturalized descriptions of scientific practice are separate rather than conflicting or competing projects. The connotations of each concept of the common cause are different: in Reichenbach, theoretical physics; in the explications, probability theory and formal logic; in naturalized and inductive approaches, the history and philosophy of science. Still, despite these different contexts, I believe there is a sufficient overlap between the references of these concepts, the scope of examples, and claims for puzzle solving to warrant a claim for a *replacement* of Reichenbach's deduction and its explications with a naturalized account. I envision the relation between the naturalized account of the inference of common causes and its predecessors to resemble that between succeeding scientific theories. Gravity does not mean quite the same in Einstein and Newton's theories. But the partly overlapping range of examples and puzzle solving, along with the problems of the earlier theories facing a range of anomalies, warrants theory replacement rather than an ad hoc revision. As in the case of naturalized philosophy of science, I do not expect to convince everybody. I do believe however that the fruitfulness, scope, and dynamism of the new research program I present here will make a decisive case for naturalization against explication.

7 Information preservation and the inference of the existence of some common causes

It is impossible to describe and compare the infinite number of properties of any set of events or objects. Therefore, the common cause and its effects are best considered as *properties of events* rather than events (Sober, 1999). Greg (1927) called sets (of texts) that share properties *variational groups*. I adopt this useful term. I ask then, how do scientists infer that a variational group had some token common cause? This question needs to be specified further because as Sober (1988, p. 100) noted, any collection of events shares trivially some common cause token, at the very least, the Big Bang. Most random collections of events would share many common causes; all events in the history of planet Earth share the birth of the planet and all events in human history share the mutation that gave rise to the human race, as common causes. Yet, the Big Bang and the birth of the solar system do not *explain* correlations between texts, languages or species. Sober suggested that scientists specify explicitly or implicitly contrast classes whose members were not affected by the requested common cause. Contrast classes specify which common cause is relevant for the explanation of a variational group. For example, if we ask for a common cause to explain the correlations between the properties of dolphins and whales, we also imply a contrast class of, say, land mammals. Biologists then examine whether there was a common ancestor of dolphins and whales that was not also a common ancestor of land mammals. If we formulate the contrast class differently, the requested common cause would differ as well. Likewise, when textual critics compare texts and infer their common ancestors, they group them according to certain variables that other texts do not share (Greg, 1927). The use of comparison classes is more flexible and can answer a broader range of questions than the common fallback quest for a *direct* common cause, the *last* common causes of a variational group.

The properties that scientists look for in variational groups in order to infer their common causes are those that tend *to preserve information*. For example, each throwing of dice erases the information about previous results.[5] By contrast, the exclusively male Y chromosome and the exclusively female mitochondrion are passed respectively from father to son and mother to her descendants. Thus, they tend to preserve information about their male and female ancestors respectively, the common causes of contemporary frequencies and correlations of Y-chromosomes and mitochondria, and are useful for the inference of genetic historiography. The selection of variational groups according to their information preserving qualities is theory laden. Information transmission theories may be as complex as the combination of genetics with evolutionary biology, or as simple as those of corresponding sounds in linguistics, such as the English sound 'W' and the German sound 'V' that allow the discovery of information about common ancestor words. Such information is *nested*, it can be inferred only with the aid of theories that link properties explicit in the information signal with information that is "nested" in it (Dretske, 1981, pp. 71–80). Theories about the preservation of information also solve Berkovitz's problem (Berkovitz, 2000, pp. 57-58) of partial common causes by separating the aspects of effects of common causes (the evidence) that preserve information about their origins from aspects that reflect separate causes. For example, which parts of genomes preserve information about common ancestry, and which reflect separate causes that influenced the genomes since their separation.

Some processes tend to preserve in their end states information from their initial state more than others. Processes have varying levels of *fidelity*. Fidelity measures the degree to which a unit of evidence preserves information about its cause. *Fides*, fidelity, is a term used by textual critics to evaluate the reliability of texts (Maas, 1958). Fidelity is used in this context as *reliability* is used in probability theory or *credibility* in jurisprudence (Friedman, 1987). Biblical criticism, classical philology and historiography leaped for-

---

[5]Reichenbach described information as negative entropy. "The deviation of the observed frequencies from a chance distribution $P(A\&B)$ presents us with information greater than 0... negative macroentropy of the process represented by the coincidence of A and B. A high amount of information means that the macroentropy is low. Therefore, unusual coincidences indicate low entropy. We know that a low-entropy state requires an interaction in the past for its explanation;... We conclude that the occurrence of information requires past interaction, that is, that information can be supplied only by postinteraction states" (Reichenbach, 1956, pp. 178–179). Reichenbach (1956, p. 163) acknowledged that some common causes violate the principle that $P(A|C) > P(A|-C)$. For example, the hand that throws two dice that coincide. However, he did not examine the transmission, preservation and destruction of information that distinguish which effects of common causes are indicative of their origins and which are not.

ward, following the recognition that oral transmission has a lower fidelity than written transmission, and that memorized prose has a lower fidelity than memorized verse. In Common Law, hearsay evidence is excluded because its fidelity is too low. Historical linguistics leaped forward when it discovered that names of places, proper names of fauna and flora, and words for the first few numbers, body parts, and immediate family members have higher fidelities than other parts of language. Since there is sufficient evidence to establish continuous historiographies of some languages over thousands of years, it is possible to infer an average rate of fidelity across all these languages. The fidelity of this high fidelity vocabulary is according to Maurice Swadesh 86% per thousand years.

When attempting to infer a common cause, scientists look first for high fidelity properties that are shared by members of a variational group. The evaluation of the fidelity of properties of events may also involve the examination of evidence for the causal chains that purportedly transmitted information from some common cause to members of the variational group, for example, the fossil record. *Ceteris paribus*, the more distant is a member of the variational group from the common cause on a causal-informational chain, the lower is its likely fidelity. Variations tend to multiply between source texts and their transcribed copies (Greg, 1927, p. 9) and between genetic and linguistic ancestors and descendants.

8 The meaning of the existence of some common cause

When scientists and ordinary people infer that a variational group had *some* common cause whose *properties are unknown or unspecified*, it may mean one of five following kinds:

1. *A single ancestral common cause.* For example, a variational group of exams may be explained by plagiarism from a common single common cause like a book that all the authors of the exams in the variational group copied.

2. *Several common causes.* For example, the exams in the variational group may reflect plagiarism of identical common sources.

3. *The common cause may be a member or several members of the variational group itself.* For example, one or more exams of better students were copied by less diligent students.

4. *All the members of the variational group may mutually cause each other.* For example, similar exams may be the result of the collective work of all who submitted them. Likewise, the "wave" theory

of language suggests that variational groups of languages may be the descendants of a group of unrelated languages that influenced each other in a historical period when they were spoken by geographically adjacent peoples who influenced each other's languages.

5. *Combinations of 1 or 2, with 3 or 4.* For example, the students consulted several encyclopedias (2) before composing together (4) the exam. Recent research suggests that humans and chimpanzees had a common ancestor (1) but also that after a period of separation between the species, there was again a period of hybridization (4) followed by final separation. The greater similarity between human and chimpanzee X chromosomes than between autosomes indicates later hybridization (Patterson et al., 2006).

9 Likelihoods of the variational group given common and separate causes

Sober has proposed in a number of publications that scientists infer common causes by comparing the likelihoods of a variational group given a common cause and given separate causes (Sober (1999, pp. 255–6) and Sober (2001, p. 242)):

$$\frac{P(\text{variational group} \mid \text{some common cause})}{P(\text{variational group} \mid \text{no common cause})} \frac{P(\text{common cause})}{P(\text{no common cause})}.$$

Sober (1988, p. 95) and Forster (1988) interpreted the comparison of likelihoods of a variational group given *some* common cause and separate causes as between the best *particular* common cause hypothesis that specifies the properties of the hypothetical common cause and the best separate causes hypothesis that likewise specifies the properties of the separate causes. There should then be two stages in the inference of *some* common cause:

1. Two "internal" comparisons among particular common cause hypotheses and among particular separate causes hypotheses over which hypotheses makes the variational group most likely.

2. A final match between the respective "champions" of the above "tournaments."

Sober (1988, pp. 105–110) acknowledged that such best-case analysis generates problems of nuisance variables. But a more significant problem is its lack of correspondence with the historical progression of successful inferences of common causes that have proceeded from the general recognition that there must have been *some* common cause of a variational group rather than separate causes to the later introduction of particular hypotheses that suggested the properties of these common causes. Darwin could not have compared the likelihoods of his variational group of men and apes given the best common cause hypothesis and the best separate causes hypothesis, because he did not possess them anymore. Nor Rask or Bopp could compare hypotheses about the properties of proto-Indo-European. It is indeed odd that often Reichenbach's principle has been examined in relation to quantum physics where there is no acknowledged successful story of inference of a common cause (cf. Berkovitz (2000, pp. 80–82)).

Later, Sober (1999, p. 259) proposed an alternative, the comparison of the likelihoods of the variational group given *all* the particular common cause hypotheses, multiplied by their priors, and given *all* the separate causes hypotheses multiplied by their priors. This resembles Reichenbach's original proposal above, and it is just as inconsistent with the historical development of successful scientific inferences of common causes.

Finally, Sober (1999) suggested that it is possible to compute the ratio of likelihoods of a variational group given some common cause and separate causes, if the rates of preservation of information, of fidelity, are uniform at the same time across separate causal chains: "If two branches are contemporaneous, then any conditional probability that describes the one also describes the other" (Sober, 1999, p. 260). Sober argued that given such uniform rates, a variational group is more likely given some common cause than given separate origins and a more recent ancestor is more probable than a more distant one. Sober has recognized, though, that the assumption of uniform rates is generally false. I should add that, even had the assumption of uniform rates been true (as it may be in particular local historical contexts), it would have been insufficient for distinguishing common cause *tokens* from separate causes of the same *type* because variational groups would be equally likely given a common cause token or contemporaneous separate tokens of the same type.

Forster (1988, p. 539) argued against the possibility of comparison of likelihoods of variational groups given *some* common cause and separate causes. He argued that such comparisons of likelihoods are incomplete because they do not mention priors and because the transition probabilities from some common cause or separate causes to their effects are not precisely known.

Sober (1999, p. 258) acknowledged that there is insufficient evidence to assign values to the prior probabilities of common or separate origins of all life on earth, a single start or several starts of life, concluding that

I doubt that much can be said *in general* about the circumstances in which common cause explanations are to be preferred beyond general remarks about priors and likelihoods. What more there is that needs to be said must come from specific empirical theories, not from general philosophic ones. The separate sciences provide the background theories that show how observations have evidential meaning. It is these that ultimately decide whether a common cause explanation is better supported than a separate cause explanation. (Sober, 1988, p. 111)

Pace Forster and Sober, in most cases it is actually possible to assess priors. It is often difficult to discover *precise* transition likelihoods, but usually it is unnecessary because *scientists do not prove the high likelihood of the variational group given the hypothesis they favor, as much as prove its negligible likelihood given the alternative hypothesis*. Scientists are able to breathe life into the Bayesian probabilities by utilizing auxiliary evidence and theories, especially theories about the transmission of information in time and evidence for causal chains that connect, or not, common causes with their effects. Suppose that a variational group is composed of two units of evidence, E_1 and E_2, which share certain properties that $E_3 \ldots E_n$ do not. C is the hypothesis that some common cause is the best explanation of the variational group. Before we can compute the likelihood of E_1 and E_2 given C, $P(E_1 \& E_2|C)$, it is necessary to consider the prior probability of C, of some common cause given everything else we know $P(C|B)$. Scientists estimate priors of common cause hypotheses by examining whether causal chains that extend backwards from $E_1 \& E_2$ could intersect and converge before they encounter the causal chains that extend backwards from the units of the contrast class. For example, if we wish to examine the hypothesis that the similarities between the English word 'tea', the Czech 'čaj', and the Hebrew 'te' that the Polish "Herbata" resulted from a common cause, relevant background information about the history of the words for tea and the drinking habit itself, would tell us that it was initially a Chinese beverage that was exported westwards. It is then probable that the causal-etymological chains that lead backwards from the words in the variational group converged in some common cause, some word in one of the Chinese dialects that was spoken during the European Middle Ages, though without further research one cannot know the properties of that word, what the original word for tea actually sounded like. By contrast, the hypothesis that ascribes the similarity between the Egyptian and Aztec pyramids to some common cause has very low prior probability because there is no evidence for intersection between causal chains that stretched back from Aztec and Egyptian architecture. Given pre-modern means of transportation and information technology, the prior probability of such an intersection is vanishing.

The assessment of the likelihood of a variational group given some common cause depends on background information, auxiliary evidence about links in the causal chains that could transmit information from the hypothetical common cause to the variational group and theories about the preservation of information through its transmission from one link to the next. For example, a fairly rudimentary probabilistic biological theory that holds that organic like begets like is at the foundation of cladistic inferences in biology because it increases the likelihood of similarities between related species, given some common cause (Sober, 1988, 1989, 1999). The discovery of evidence for links on the causal chains that connect some common cause with a variational group, such as transitional forms in evolutionary biology, increases the likelihood of the variational group given some common cause.

Formally, the likelihood of the evidence given some common cause is:

$$P(E_1 \& E_2 \& \ldots E_n | C) = P(E_1 | C) \times \ldots \times P(E_n | C)$$

This is also the equation Reichenbach (1956, pp. 157–167) presented. However, here C clearly stands for *some token common cause* without specifying its properties and without asserting screening conditions, since C could be one or even all of its effects.

Formally, assessing the likelihood of a variational group given separate causes $(S_1, \ldots, S_n)$ and their respective prior probabilities can be expressed by the following equation:

$$P(E_1 \& \ldots \& E_n | S_1 \& \ldots \& S_n) =$$
$$= [P(S_1 | B) \times P(E_1 | S_1)] \times \ldots \times [P(S_n | B) \times P(E_n | S_n)]$$

The likelihood of the variational group given separate causes has been assessed by considering the function, the adaptational advantage, or the rationality of the shared properties of the members of the variational group. Separate token causes of the same type may generate a high likelihood for variational groups. For example, prior to the discovery of columns and arches, the only likely shape of a big and tall building is that of a pyramid, a solid wide base that can support a lighter structure above it. This is the only solution that ancient Egyptian and Aztec architects could have devised for this type of problem. Tokens of this type of cause, the constraints on the shape of large buildings that do not use arches or columns, make the similarity between pyramids on both sides of the Atlantic highly likely. The prior probability of such causes depends on what we know of the historical context of the separate causes, whether indeed the builders did not know of columns and arches. Similarly, various cultures invented independently agriculture, the taming of domestic animals and the wheel because these are the best functional solutions for the universal human needs for food and transportation. *Homoplasy*, the similarity of biological traits that results from similar environments, different tokens of a type of cause, such as that of fish and sea mammals, is quite likely given separate causes. Natural selection favored wings and fins several times, though insects, birds and bats, fish, ichthyosaurs, and dolphins neither evolved out of each other, nor are closely related to each other, because they are the best functional forms for organic movement through air and water.

Variational groups that share properties that have no functional value, or are even dysfunctional, such as grammatical mistakes in texts or redundant properties of species such as the feathers of birds that cannot fly, which Darwin called *rudiments*, are highly unlikely given separate causes. Imperfections and errors in transcriptions are the best evidence of common causes because the odds against separate processes arriving at the same dysfunctional forms are vast (Dennett, 1996, pp. 135–143). Maas (1958) meant by textual *anomalies* what biologists denote by *homologies* or *rudiments*. Maas considered similar textual anomalies as evidence for some common source. If there were equally useful simpler ways to express what the anomaly conveys, the anomaly must be the result of some common cause, a mutation in copying. The likelihood of any single grammatical mistake or non-adaptive trait is low; the likelihood of several identical ones, given separate causes, is vanishing. The likelihood of non-adaptive traits given some common cause, low as it may be, is still significantly higher. Therefore, geneticists and textual critics alike cherish and assiduously search for them. This is the usual method by which scientists infer some common cause, *not by proving that it has high posterior probability, but by proving that the only alternative, the separate causes hypothesis, has a vanishingly low posterior probability.*

When the likelihood of each member of the variational group given separate causes is low, the effect of multiple members, such as several similar testimonies, is to decrease exponentially the likelihood of the variational group given separate causes. Therefore, evolutionary biologists, comparative linguists, as well as policemen and journalists, exert themselves to discover members of a variational group, such as testimonies. Multiple members of a variational group such as witnesses *do not increase the posterior probability of some common cause; rather they decrease exponentially the posterior probability of its only alternative, separate causes*. Still, beyond a small number of similar testimonies, additional testimonies become redundant because the marginal decrease in the likelihood of the variational group given separate causes is tiny. This "economics of witnessing" is apparent in the plea bargains prosecutors agree with criminals who turn state witnesses to receive reduced sentences for testifying against their accomplices. After obtaining two or three witnesses to the same events, the state ceases to offer plea bargains to other criminals whose testimonies have an identical content, because the likelihood of detailed testimonies given separate causes is miniscule and the interest of the prosecutor is to keep deals with criminals to the necessary minimum to achieve conviction beyond reasonable doubt.

Since it is usually impossible to determine *precisely* the valued parameters of priors and likelihoods, a significant gap between the *roughly estimated* likelihoods of the variational group given some common cause and separate causes is required:

$$\frac{\{[P(E_1|C) \times P(C|B)] \times \ldots \times [P(E_n|C) \times P(C|B)]\}}{\{[P(E_1|S_1) \times P(S_1|B)] \times \ldots \times [P(E_n|S_n) \times P(S_n|B)]\}}$$

Where $E_1, \ldots, E_n$ represent units of evidence—members of the variational group; C, some common cause; $S_1, \ldots, S_n$, separate causes, and B represents background knowledge. The upper part represents the likelihood of the evidence, given some common cause; the lower its likelihood, given separate causes. The ratio of the likelihoods determines the choice between some common cause and separate causes.

For example, in the historiography of ideas, Skinner (1988, p. 46) compared the likelihood that one thinker like Locke was influenced by another like Hobbes (common cause), against his developing those ideas by himself independently (separate causes). Relevant evidence is of one thinker having read the other (information transmission): "the necessary conditions ... for helping to explain the appearance in any given writer B of any given doctrine, by invoking the 'influence' of some earlier given writer, A ... at least have to include (a) that there should be a genuine similarity between the doctrines of A and B; (b) that B could not have found the relevant doctrine in any writer other than A; (c) that the probability of the similarity being random should be very low (i.e., even if there is a similarity, and it is shown that it could have been A that B was influenced by, it must still be shown that B did not ... articulate the relevant doctrine independently)". Sometimes the results of such comparison of likelihoods remain controversial. For example, Evolutionary geneticists debate the relative likelihoods of variational groups of genomes that share a single non-functional mutation in relation to a comparison group, given common or separate causes. But usually, when the variational group shares more than a single variable, the inference of a common cause enjoys a consensus.

10 Alternative common cause hypotheses

If the existence of some common cause increases the likelihood of the variational group more than separate causes, alternative common cause hypotheses compete, over which is the best explanation of the variational group. This comparison between alternative common cause hypotheses proceeds in two stages: first, the five possible causal nets mentioned above (a single or multiple ancestral common causes, mutual influences among some or all the members of the variational group, or a combination of ancestral causes and mutual influences) compete over conferring the highest likelihood on the variational group. Second, once one of the five possible causal nets is chosen, scientists may attempt to infer the actual properties of the common cause.

The comparison between the five possible causal net hypotheses requires the expansion of the scope of evidence beyond that of the effects of the common cause, the variational group. The distinction between hypotheses of common cause or causes (kinds 1 or 2) and those of mutual causal influences (kinds 3, 4 or 5) is between describing members of the variational group as *independent* of each other or not. Reichenbach and Salmon meant to isolate kinds 1 & 2 by a screening condition, though they rushed too quickly to *assume* it before proving that there was some common cause. Scientists, who are interested in inferring common causes that are not members of the variational group, find type 5 to be *a nuisance* as it interferes with the inference of distinct common causes. Textual critics call their influence *contaminatio* (Maas, 1958).

Independence of members of a variational group is the absence of intersection between the causal chains that connect them with their common cause. Ascertaining whether the causal-informational chains intersected usually requires evidence for links on the causal chains that connect common cause events with their effects. Evidence may also prove that causal chains could not have possibly interacted after their original common cause. Evidence may be about the history of a text that allows historians, biblical critics and classical philologists to reconstruct at least a part of the genealogy of the texts they study. When evidence is scarce, more than one of the five possible kinds of common cause hypotheses may make the evidence equally likely. For example, in historical linguistics, though it is certain that the Indo-European languages had some common cause, it is impossible to find whether it was a single language, proto-Indo-European, or whether several geographically proximate languages mutually influenced each other, or both, a single language had several descendants who later influenced each other.

If there is sufficient evidence to eliminate the probability of intersections between the causal chains, or if there is sufficient evidence to "peel off" layers of mutual influence in the fifth kind of causal net, scientists need to distinguish type one, a single common cause, from type two, multiple common causes causal nets. Historical evidence may mention its multiple sources, even if they are lost. Compilations of documents may preserve linguistic differences that indicate different times and places of composition. Historians and textual critics look for discontinuities in style, conceptual framework, and implicit values; internal contradictions, gaps in narratives, and parts that are inconsistent with the alleged identity of the author.

Often, there is sufficient evidence to determine which of the five possible common cause hypotheses is most probable. For example, textual critics proved that the Bible and Homer are most likely given the fifth kind of common cause: the various exemplars of the Bible and Homer's epics had initially multiple common causes, sources, and then they influenced each other in the process of editing. Such composite documents preserve linguistic and other differences that indicate different times and/or places of composition.

When it is possible to determine that there was a common cause or causes, scientists attempt to determine which of the common cause hypotheses that specify their properties increases the likelihood of the evidence more than its competitors. Scientists first evaluate the prior probabilities of competing common cause hypotheses, then they evaluate whether the hypotheses cohere with established beliefs, and finally they evaluate whether they are internally coherent. The prior probability of a specific common cause hypothesis C_1 can then be multiplied by the likelihood of the evidence, $E_1, \ldots, E_n$, given that common cause hypothesis:

$$[P(C_1|B) \times P(E_1|C_1)] \times \ldots \times [P(C_1|B) \times P(E_n|C_1)]$$

The comparison of competing common cause hypotheses is then simply

$$\frac{P(E_1 \& \ldots \& E_n|C_1) \times P(C_1|B)}{P(E_1 \& \ldots \& E_n|C_2) \times P(C_2|B)}$$

The likelihoods of variational groups given competing common cause hypotheses depend on the theories of information transmission and preservation that scientists use to distinguish which correlations and similarities are indicative of some common cause in the first place. In biology, when paired copies of a gene stop recombining, their sequences will diverge increasingly as time goes by. A relatively small number of differences imply recombination stopped fairly recently; a larger number means it halted long ago. By comparing DNA sequences across species, biologists can often calculate, roughly, when species separated. The assumption of fixed fidelities, of constant rates of mutation, which geneticists use in their inferences of common causes in the history of life, is a statistical rule like: *all transmission of information (like languages, genetic data, and oral traditions) generates mutations, including errors in transcription, additions and subtractions. Ceteris paribus, the average rate of mutation per medium per transmission is constant.* Constants of spontaneous variation would vary according to the type of information that is copied and the method of transmission/copying.

Beyond evolutionary biology, genetics and comparative linguistics, in historiography, textual criticism and detective work, fidelities are affected by myriad factors, most notably voluntary human agency, and so it is difficult or impossible to infer average constant rates of mutation of information. The evaluation of fidelities in these sciences requires the examination of the causal chain that transmitted information from the common cause to the units of the evidence. Historians search assiduously for *primary sources, evidence that possesses the highest degree of fidelity by preserving more information that is relevant for a particular historiographic hypothesis than other relevant evidence.* For example, historians exclude evidence written by authors that were separated by time from the events they describe, unless there is evidence for a credible causal chain that could have transmitted information from events to evidence. Historians and detectives may assess the general fidelity of some sources according to whether they tend to cohere with other independent sources of high fidelity consistently. Historians recognize the high fidelity of parish registries, while they consider the official statistics of totalitarian states unreliable. Alvin Goldman (1999, pp. 123–125) proposed that the fidelities of testimonies are evaluated according to the competence of the witnesses to detect the kind of information they offer and their record for honest reporting. Ranke assumed a theory of memory that implies low fidelity of memoirs and high fidelity of contemporary eyewitness accounts written immediately after the events.

The ultimate assessment of the fidelity of evidence has to consider all the above elements. Its complexity may account for the absence of algorithms for such computations in historiography or detective work. Historians, evolutionary biologists, and linguistics must examine the causal information chains that should connect hypothetical event with evidence. The more such evidence there is, the more certain is the assessment of the fidelity of the evidence. Therefore, paleontologists labor hard to discover missing links on the evolutionary tree, and value the discovery of new fossils of missing links.

If we use $F_1, \ldots, F_n$ to symbolize the fidelities of various information transmission processes, the likelihood of the evidence given a common cause hypothesis is the value of the fidelity multiplied by the prior probability of a particular common cause C_1 given background information B:

$$P(E|C_1) = [P(C_1|B)F_1] \times \ldots \times [P(C_1|B)F_n]$$

Note that, unlike in ordinary Bayesian computations, we do not add here the likelihood of the evidence occurring irrespective of the common cause $P(E|-C)$ to its likelihood given the common cause, because this comparison of alternative common cause hypotheses comes after the elimination of separate causes hypotheses in the first stage of the inference. At this third stage we assume that there was a common cause and attempt to compare the likelihoods of the evidence given alternative common causes C_1 and C_2:

$$\frac{[P(C_1|B)F_1] \times [P(C_1|B)F_2] \times \ldots \times [P(C_1|B)F_n]}{[P(C_2|B)F_3] \times [P(C_2|B)F_4] \times \ldots \times [P(C_2|B)F_n]}$$

The various fidelities $F_1, \ldots, F_n$ have the same value if we assume constant and uniform rates of mutation across species and languages. If we do not, or cannot, as in historiography, they may have different values.

11 Conclusion

Scientific knowledge of the past became possible once theories and background information demonstrated that common causes make variational groups more likely than separate causes. As Hull (1992, p. 72) recognized, one of the basic tasks of the historian is to distinguish patterns that result from natural regularities such as homoplasies, from those that have common causes, such as homologies. When the likelihood of the variational group given separate causes is vanishing, the common cause hypothesis wins by default.

I outlined three consecutive stages of inference of common cause:

1. *that there was a common cause rather than separate causes*,

2. *which of five causal nets of common cause is most likely*, and

3. *what the likely properties of the common cause were*.

This is a naturalized account of the scientific inference of common causes, the best practice that the most universally recognized successful cases of inference of common causes have followed. There is no probabilistic magic bullet for the reduction of causation to probability or of common cause to a correlation between frequencies. The inference of common causes requires additional evidence and auxiliary theories about the transmission of information in time.

Aviezer Tucker
School of Politics, International Affairs and Philosophy, Queens University

Belfast, UK.
avitucker@yahoo.com

BIBLIOGRAPHY

Arntzenius, F. (1992). The common cause principle. In *PSA: Proceedings of the Biennial Meeting of the Philosophy of Science Association*, volume II, pages 227–337.
Barrett, M. and Sober, E. (1994). The second law of probability dynamics. *British Journal for the Philosophy of Science*, 45:941–953.
Berkovitz, J. (2000). The many principles of the common cause. *Reports on Philosophy*, 20:53–83.
Cartwright, N. (1999). *The Dappled World: A Study of the Boundaries of Science*. Cambridge University Press, Cambridge.
Cleland, C. E. (2002). Methodological and epistemic differences between historical science and experimental science. *Philosophy of Science*, 69:474–496.
Dennett, D. C. (1996). *Darwin's Dangerous Idea: Evolution and the Meaning of Life*. Touchstone, New York.
Dretske, F. I. (1981). *Knowledge and the Flow of Information*. MIT Press, Cambridge MA.
Forster, M. R. (1988). Sober's principle of common cause and the problem of comparing incomplete hypotheses. *Philosophy of Science*, 55:538–559.
Friedman, R. (1987). Route analysis of credibility and hearsay. *Yale Law Journal*, 96:667–742.
Goldman, A. (1999). *Knowledge in a Social World*. Oxford University Press, Oxford.
Greg, W. W. (1927). *The Calculus of Variants: An Essay on Textual Criticism*. The Clarendon Press, Oxford.
Hausman, D. M. (1998). *Causal Asymmetries*. Cambridge University Press, Cambridge.
Hitchcock, C. (1998). The common cause principle in historical linguistics. *Philosophy of Science*, 65:425–447.
Hofer-Szabó, G., Rédei, M., and Szabó, L. E. (2002). Common-causes are not common common-causes. *Philosophy of Science*, 69:623–636.
Hoover, K. D. (2003). Nonstationary time series, cointegration, and the principle of the common cause. *British Journal for the Philosophy of Science*, 54:527–551.
Hull, D. (1992). The particular-circumstance model of scientific explanation. In Nitecki, M. H. and Nitecki, D. V., editors, *History and Evolution*, pages 69–80. State University of New York Press, Albany NY.
Laplace, P. S. (1951). *A Philosophical Essay on Probabilities*. Dover, New York. trans. Truscott, Frederick Wilson and Emory, Frederick Lincoln.
Maas, P. (1958). *Textual Criticism*. The Clarendon Press, Oxford. trans. Flower, Barbara.
McGrew, T. (2003). Confirmation, heuristics, and explanatory reasoning. *British Journal for the Philosophy of Science*, 54:553–567.
Patterson, N., Richter, D. J., Gnerre, S., Lander, E. S., and Reich, D. (2006). Genetic evidence for complex speciation of humans and chimpanzees. *Nature*. doi:10.1038/nature04789, accessed 25 May 2006.
Quine, W. V. O. (1985). Epistemology naturalized. In Kornblith, H., editor, *Naturalizing Epistemology*, pages 15–29. MIT Press, Cambridge MA.
Reichenbach, H. (1956). *The Direction of Time*. University of California Press, Berkeley.
Salmon, W. (1984). *Scientific Explanation and the Causal Structure of the World*. Princeton University Press, Princeton.

Skinner, Q. (1988). Meaning and understanding in the history of ideas. In Tully, J., editor, *Meaning and Context: Quentin Skinner and his Critics*, pages 29–67. Polity Press, Cambridge.

Sober, E. (1988). *Reconstructing the Past: Parsimony, Evolution, and Inference*. MIT Press, Cambridge MA.

Sober, E. (1989). Independent evidence about a common cause. *Philosophy of Science*, 56,:275–287.

Sober, E. (1999). Modus Darwin. *Biology and Philosophy*, 14:253–278.

Sober, E. (2001). Venetian sea levels, british bread prices, and the principle of the common cause. *British Journal for the Philosophy of Science*, 52:331–346.

Sober, E. (2004). The design argument. In Mann, W., editor, *The Blackwell Guide to the Philosophy of Religion*, pages 117–147.

Sober, E. and Barrett, M. (1992). Conjunctive forks and temporally asymmetric inference. *Australasian Journal of Philosophy*, 70:1–23.

Tucker, A. (2004). *Our Knowledge of the Past: A Philosophy of Historiography*. Cambridge University Press, Cambridge.

Turner, D. (2005). Local underdetermination in historical science. *Philosophy of Science*, 72:209–230.

Van Fraassen, B. (1980). *The Scientific Image*. Clarendon Press, Oxford.

Wetzel, L. (2006). Types and tokens. In *Stanford Encyclopedia of Philosophy*. www.plato.stanford.edu/entries/types-tokens.

Contexts for causal models
MARGHERITA BENZI

ABSTRACT. I distinguish between two kinds of philosophical approach to causal modeling: (i) approaches based on the idea of an underlying omnicomprehensive causal network, and (ii) approaches which relativise the construction of causal models to the context of inquiry. I analyze how the two approaches can be related to the choice of the interpretation of causality and probability.

1 Introduction

Asserting that causes are context-sensitive hardly raises any objections in contemporary philosophy of causality. However, there are several different kinds of contexts to which causal judgements can be referred. In this paper, I outline a taxonomy of contexts, as seen by some contemporary theories of causation.

As a first step, I distinguish between conceptions which view contexts as mere shortenings, introduced for pragmatic reasons, of richer complexes of circumstances, and conceptions which assign to contexts a fundamental role in defining causal relationships. This distinction, traced in Section 2, is connected with the problem of the distinction between causes and conditions. In Sections 3 and 4, I briefly outline sensitivity to context in both probabilistic causality and causal modeling. In Section 5, I consider some recent attempts to reduce the context-sensitivity of causation by hypothesising an all-embracive Bayesian net. Finally, in Section 6, I discuss this hypothesis in the light of different levels of sensitivity to context.

2 Causes, conditions and contexts

What factors should be considered in investigating the causes of a certain effect? The problem has a long history in the philosophy of causation and has been well summarised by Patrick Suppes:

> Perhaps one of the most puzzling and difficult aspects of the analysis of causality is the problem of how to handle the background that serves as a framework for the occurrence of the particular events under study. If, for example, we study the cause

of a match lighting, to what extent must we consider meteorological conditions, the composition of the atmosphere, the absence of meteorites, etc.?[1]

Suppes thinks that the choice of the factors that must be included in what he calls "background", or "field", is relative to the theoretical approach one is adopting, such that "in one theoretical approach the field will include only the consideration of macroscopic bodies and their characteristics, but in another, it will go deeper and consider as well atomic objects and their properties."[2]

In Suppes's theory of probabilistic causality, the assessment of a causal relation between two factors depends on an accurate inspection of the background in a substantive way, as the background could contain common ancestors which screen off the two factors, thereby turning a *prima facie* cause into a spurious one. But if backgrounds are relative to conceptual frameworks, then causes are as well:

> With respect to one field, framework or background, one event may be a cause of another, and yet, when the field is changed and the framework is extended by the consideration of additional variables, the cause may turn out to be spurious.[3]

Provided that the components of the causal background are clearly indicated, causal relativity is not a problem for Suppes, who finds it preferable to obscure references to "the infinite complexity of things".[4]

A different tradition in the philosophy of causality has rejected the relativity of causation. A well known example is given by J.S. Mill, who observed that in ordinary speech we single out one of the factors that antecede the effect as *the* cause, and call the other factors 'conditions'; this practice, however, is not based on a rigorous distinction:

> The real cause is the whole of these antecedents; and we have, philosophically speaking, no right to give the name of cause to one of them exclusively of the others.[5]

> Nothing can better show the absence of any scientific ground for the distinction between the cause of a phenomenon and its conditions, than the capricious manner in which we select from among the conditions that which we choose to denominate the cause.[6]

Mill sees causation as an *absolute* relation, where "real" causes are extremely rich complexes of factors and do not vary from context to context.

---

[1] (Suppes, 1970, p.74)
[2] (Suppes, 1970, pp.74–75)
[3] (Suppes, 1970, p.75)
[4] *Ibidem.*
[5] (Mill, 1961, p.214)
[6] (Mill, 1961, p.215)

We pick out from these enormous sets of antecedent factors smaller sets of conditions, which we choose to call "causes". The selection of these smaller sets depends on the kind of explanation we are looking for: agents with different aims and interests can focus on different subsets of conditions, and therefore the selection of what we consider as causes is, with respect to the "real" nature of causation, "capricious".

The metaphysical assumption that causation is an absolute relation, and not relative to context, is defended by several contemporary authors; David Lewis's (1973) theory of causation provides a particularly clear example.[7] In the remaining part of this work, I shall refer to it as to the Absoluteness Assumption.

Mackie (1965, 1974) provides a sophisticated account of the role played by pragmatic considerations in the evaluation of a causal relation. As is well known, he proposes an analysis of causation in terms of INUS conditions ("INUS" being an acronym of "Insufficient Non-Redundant[8] part of Unnecessary Sufficient"). Normally, an effect (like the burning down of a house) is brought about by a cluster of conditions (e.g., a short circuit combined with the presence of inflammable material and the absence of a sprinkler where the short circuit occurred, ...) which, taken together, constitute a complex condition that is sufficient for the effect; sufficient but not necessary, as the house could have been burned down in consequence of the occurring of another cluster of conditions (e.g., the act of an arsonist combined with the presence of a strong wind and the absence of a sprinkler, ...). The components of the complex condition actually occurring —which is the whole cause of the effect— can be called "causes" if they are non-redundant parts of it. Therefore, according to Mackie, we say that C is a cause of E (on a certain occasion) if C is an INUS condition of E, i.e., is an insufficient but necessary part of a condition which is itself unnecessary but exclusively sufficient for E (on that occasion). This definition presupposes

---

[7]On the absolute character of causal relation in Lewis' theories, see Menzies (2004). There are two aspects of the theory presented in Lewis (2000) that seem to suggest that Lewis see causes as relative to context. The first is given by the notion of not-too-distant alteration, which, as Menzies (2004, p.179, note 8) remarks, "introduces an important contextual element into the truth conditions of causal statements". The second aspect is that Lewis notes that in ordinary talk we recognise as causes only a small number of the factors that in his theory count as causes: we don't say that birth is a cause of death, or that the absence of nerve gas in my room is a cause of my writing this paper. The selection depends on pragmatic reasons, which would render inappropriate to mention all the causes. In the expanded version of his paper, Lewis (2004) appeals to Grice's (1975) theory of conversational implicature. However, these pragmatic considerations are not meant to eliminate the idea of the full causal history of an event, where causation is an absolute relation.

[8]Or "Necessary" (Mackie, 1993, p.34).

that in evaluating whether a certain factor C is a cause of E, we find a set of conditions which, conjoined with C, constitute a complex condition which is sufficient, but not necessary, for E.

Mackie calls "conditions" those factors which are explicitly mentioned in the statements which describe INUS conditions. However, besides these explicitly mentioned factors, there are other factors which play a role in causal analysis, though not, so to speak, an explicit one: these latter factors constitute a *causal field*.[9] According to Mackie, statements of the form "C caused E" are elliptical, and are expanded into "C caused E in relation to the field F". The causal field is what legitimates the choice, as potential causes, of factors of certain kinds, and not of others. Mackie clarifies the role of the causal field by means of a well-known example:

> 'What caused this man's skin cancer?'[...] may mean 'Why did this man develop skin cancer now when he did not develop it before'? Here the causal field is the career of this man: it is within this that we are seeking a difference between the time when skin cancer and times when did it not. But the same question may mean 'Why did this man develop skin cancer, whereas other men who were also exposed to radiation did not?'. Here the causal field is the class of men thus exposed to radiation. And what is the cause in relation to one field may not be the cause in relation to another.[10]

Even if not explicitly stated, or if only roughly defined, the causal field plays an important role in causal analysis, because, on the one hand, it defines the area under investigation, and, on the other, it allows us to omit the explicit mention of all those factors which belong to the area under investigation, but which we do not want to mention as causally relevant.

The distinction between factors that can be explicitly mentioned as possible causes, and factors that are relegated to the causal field depends on pragmatic considerations: according to Mackie (1993, p.40), it is "an arbitrary matter whether a particular feature is regarded as a condition (that is, a possible causal factor) or a part of the field, but it cannot be treated in both ways at once".

An interesting account of the different kinds of causal relevance has been proposed by Herbert Hart and Anthony Honoré (1959). These authors characterise causes primarily as those factors that make a certain situation different from what was to be expected. Deviance from normality is what allows us to distinguish causes from conditions: "Conditions are mere conditions as distinct from causes because they are normal".[11] The distinction between what is normal and what is abnormal is relative to context in two different ways. The first kind of context-relativity regards what Hart and

---

[9] Mackie credits Anderson (1938) for the notion of causal field.
[10] (Mackie, 1993, p.40)
[11] (Hart and Honoré, 1959, p.36)

Honoré call the *context of occurrence*. A classical example[12] illustrates this kind of context-dependence: consider the lighting of a match in a smoking area and the lighting of a match in a quality-control laboratory where matches are struck in supposedly evacuated chambers in order to test the hardness of their heads. In the first case the presence of oxygen is normal, and therefore should be regarded as a *condition* of the lighting, whereas in the second case it is abnormal, and therefore it can be sensibly considered as a *cause* of the match lighting. Thus the relativisation to the context of occurrence applies to events of the same kind but in different circumstances.

The second kind of relativity to context is what Hart and Honoré call the *relativity to context of inquiry*. The examples they employ to illustrate this kind of context are similar to the above example of skin cancer discussed by Mackie: the person who prepares meals for a man suffering from a gastric ulcer might establish that the cause of his indigestion was eating parsnips, because he does not have indigestion when eating *normal*—i.e., different from parsnips—food; whereas a doctor might say that the cause of the man's indigestion is his ulcer, because *normally* people don't get indigestion when eating parsnips.

3 Context-dependence in probabilistic causality

The inquiry of causal connections through the study of probabilistic relationships between variables has produced a sophisticated account of the notion of 'background'. Besides the ascertainment that no common ancestor screens off the putative cause from the putative effect, several other kinds of other possible causal influences must be checked in order to obtain the right causal conclusions. In general, it is required that the causal impact of C on E be evaluated with respect to *causally* homogenous situations, that is by taking into account the presence of other causal factors correlated with the effect.

Nancy Cartwright (1979) called conjunctions of relevant factors where each factor occurs negated or unnegated "test situations", and the "relevant factors" are all the causes of E, except C, plus the causal intermediaries between C and E (i.e., those other causes of E which are themselves caused by C).[13] An analogue definition is given by Eells (1991), who calls causally homogeneous situations "Causal Background Contexts":

> In assessing X's causal relevance to Y, we have to hold fixed all the other factors, $F_1,..., F_n$ that are causally relevant to Y, independently of X, and then observe the probabilistic impact of X on Y given each of these ways. With n such other factors, there are 2^n ways of holding each fixed, positively or negatively. That is, there are

---

[12]See Gorovitz (1965).
[13](Cartwright, 1979, p.423).

2^n conjunctions in which each of these n factors occurs exactly once, either positively (unnegated) or negatively (negated). Of these 2^n "maximal conjunctions", let $K_1,..., K_m$ be exactly those that have nonzero probability both in conjunction with X and in conjunction with $\neg X$. (Thus, for $i = 0, 1, ..., m$, $Pr(K_i \& X) > 0$ and $Pr(K_i \& \neg X) > 0$). These K_i are called "causal background contexts", relative to the assessment of X's causal role for Y.

Then we say that X is a *positive causal factor* for Y if and only if, for each i, $Pr(Y \mid K_i \& X) > Pr(Y \mid K_i \& \neg X)$.

Eells then defines the *negative causal factorhood* (X lowers the probability of Y in every causal background context) and the *causal neutrality* (X leaves the probability of Y unchanged in every causal background context). The requirement that positive, negative and neutral factorhood be defined with respect to every homogeneous situation is known as the "Context Unanimity Requirement",[14] and it is accepted by most authors in probabilistic causality. Non-unanimous causal relevance is called *mixed*.[15] Subsequent refinements concern the role of 'interacting factors',[16] and the role of causal intermediaries.[17]

Some authors rejected the Context Unanimity Requirement. Skyrms (1980), for example, proposed a weaker condition, called "Pareto-dominance Condition", according to which X is a positive causal factor for Y if X raises the probability of Y in at least one causal background context, and X lowers the probability of Y in no causal background context. Other authors have instead defended the definition which Charles Twardy and Kevin Korb (2004) call the "Objective Homogeneity Thesis", which relativizes causal claims to single causal background contexts.[18] This relativization is endorsed by recent research in causal Bayesian nets, which introduces a further kind of context-relativity, namely relativity to models, as we will see in the next sections.

4 Context-dependence in causal modeling

The area of research called "Causal Modelling" has brought about a clarification of the factors that must be taken into account in our causal judgments.[19] In this section I focus on a particular example offered by Judea

---

[14] See Dupré (1984).

[15] It should be remembered that the factors, $F_1,..., F_n$ that must be hold fixed in evaluating the causal influence of X on Y can positive, negative or mixed causes of Y, independently of X.

[16] See Eells (1991, p.161).

[17] See Cartwright (1989, p.96).

[18] Twardy and Korb (2004) also provide a detailed comparison of the 'Context Unanimity' vs. 'Objective Homogeneity' approach.

[19] For a detailed presentation of the different kinds of causal models, see Pearl (2000).

Pearl (1996) in order to show that Causal Background Contexts can sometimes be determined in a noncircular way in a Bayesian net. The probabilistic analyses of causation mentioned in Section 3 are circular, since if we want to determine the background contexts for the assessment of the causal relevance of X for Y, we must know in advance the other 'causally relevant factors'. Although the circularity of the definition of causal background contexts, along with the consequent problems for a reductive analysis of causation, is by no means a central problem in recent causal modeling research and in the philosophy of causal models, I think that this example makes it intuitively clear why some philosophers still consider the reduction of causes to probabilities as a possible option.

In causal modeling the relata of a causal link are *variables*, i.e., properties or magnitudes capable of taking more than one value. Bayesian nets represent the starting point of the theory of Causal models, and consist of two components: a directed acyclic graph (DAG) and a probability distribution. Each node of the DAG represents a variable, and the directed edges (arrows) represent direct dependence relations among the variables. Here the set of variables is denoted by "**V**". To each variable A is associated a probability distribution that specifies the probability of each value, conditional on each value of the parents of the variable.[20] The link between the graph and the probability distribution is provided by the following assumption:

MARKOV CONDITION: In a Bayesian net, every node is independent of its non-descendants, conditional on its parents.

The Bayesian net in Figure 1 is the standard example introduced by Judea Pearl (1988 and elsewhere). It shows the dependency relationships between the season of the year (A_1), the rainfall during the season (A_2), whether the sprinkler is on during the season (A_3), whether the pavement is wet (A_4) and whether the pavement gets slippery (A_5). The sets of values that each variable can take are specified in the following way:

Variable A_1: Season of the year, *values*: {spring ($= a_{11}$), summer ($= a_{12}$), fall ($=a_{13}$), winter ($= a_{14}$)}
Variable A_2: Rainfall during the season, *values*: {yes ($= a_{21}$), no ($= a_{22}$)}
Variable A_3: Sprinkler during the season, *values*: {on ($= a_{31}$), off ($= a_{32}$)}
Variable A_4: Wet pavement, *values*: {yes ($= a_{41}$), no ($= a_{42}$)}

---

[20] The set of nodes each directly reaching B with an arrow is called *the set of the parents* of B; and the set of the nodes that are reached by an arrow from A is *the set of the children* of A. The set formed by the children of A, the children of children of A and so on is called *the set of the descendants* of A.

Variable A_5: Slippery pavement, *values*: {yes ($= a_{51}$), no ($= a_{52}$)}.

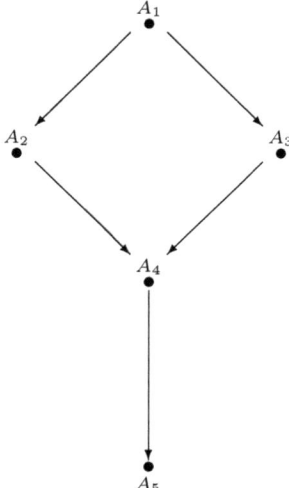

Figure 1. Example of Bayesian net.

As the main components of a Bayesian net are a DAG and a probability distribution, they do not represent causes. However, it is easy to view DAGs as representing *causal*, and not merely probabilistic structures. This requires further definitions. A *causal graph* is a directed graph with a causal interpretation, where an arrow from A pointing directly to B shows that A is a direct cause of B. A *causal Bayesian net*, or simply a *causal net* is a Bayesian net representing both probabilistic *and* causal dependence relationships among its variables. To give a Bayesian net a causal interpretation requires a further assumption, called the "Causal Markov Condition":[21]

CAUSAL MARKOV CONDITION. In a causal Bayesian net, every node is

---

[21] See Scheines (1997). Other important assumptions are *Causal Sufficiency* (the set **V** of the variables in the DAG includes all the common causes of the variables in **V**), *Minimality* (no proper subgraph on **V** of a graph on **V** satisfies the Causal Markov Condition) and *Faithfulness condition* (all the probabilistic independence relationships existing among the variables in **V** are a consequence of the Causal Markov Condition). This last condition assumes that there are no probabilistic dependencies that cancel each other by an exact balancing of their values. It is worth noting that Causal Sufficiency does not really function as a separate assumption, but it is an assumption under which Causal Markov Condition is supposed to hold.

independent of its non-descendants, conditional on all its direct causes.

One of the most important aspects in the area of causal modeling is the study of how causal knowledge can be obtained from statistical data. Due to statistical indistinguishability,[22] Bayesian nets cannot, in general, be directly 're-interpreted' as Causal nets. However, in some particular cases, they can. Consider again the Bayesian net in Fig 1, and suppose that we have a probability distribution P over the variables, that the ordering O of the variables coincides with their temporal order, and that the variables in the net are all the relevant variables for the phenomenon under study. Pearl (1996, p.7) remarks that in such a case we can use the Bayesian net to test whether a variable X is causally relevant (is a causal factor) for Y by means of the following procedure:

Consistency test: For each pair of variables labelled $R(X, Y)$, test whether
(i) X precedes Y in O, and
(ii) there exist x, x', y such that $P(y \mid x,z) > P(y \mid x',z)$ for some z in Z, where Z is a set of variables in the background context K, such that $I(X, Z)$ and $R(Z, Y)$,

where $R(A, B)$ stands for "A is relevant for B", $I(A, B)$ for "A is irrelevant for B", x and x' are values of the variable X, and y and z are values, respectively, of Y and Z. Like the definition of probabilistic causality, the Consistency test raises the question of what factors (here, variables) should be included in the background context. Pearl (1996, p.14) argues that the graph language allows a simplification and clarification of the notion of background context. Furthermore, it provides a "constructive, non-circular definition"(1996, p.14) of it:

Background context: In assessing the causal role of X relative to Y, the appropriate background context consists of all variables which are:
1. direct parents of Y or of any intermediate variable between X and Y and
2. non-descendants of X.[23]

---

[22] Here, 'statistical indistinguishability' designates the fact that the same set of statistical data can be represented by many distinct DAGs. For a detailed account, see Spirtes and Scheines (1993, Chapter 4).

[23] Pearl (1996, p.14). It should be noted that Pearl calls "Background Context" the set of variables that correspond to factors F_1, ..., F_n in Eells' definition, whereas in Eells's definition the denomination "Causal Background Contexts" designates conjunctions of (negated or unnegated) factors. This difference should not induce to think that Pearl's definition requires that the relevance of X on Y be ascertained against one single situation. It requires that the causal relation be evaluated against each possible vector of

Let us go back to our example in 1, and consider the graph before its causal interpretation. Given any two variables X and Y, we can select a set K of variables according to conditions 1. and 2. stated in the above definition of Background context. K will contain all and only the variables that must be 'held fixed' in evaluating the causal impact of X on Y. For example, if we want to assess the causal relevance of having the sprinkler on ($A_3 = a_{31}$) on having a slippery pavement ($A_5 = a_{51}$), the background context will be given by A_2. Consequently, the causal impact of the activity of the sprinkler on the state of the pavement ought to be evaluated both in situations where rain is present ($A_2 = a_{21}$) and in situations where rain is absent ($A_2 = a_{22}$).

While in probabilistic causality the definition of background context appealed to the other causes of the putative effect, thereby producing a circular definition of causality, here, as Pearl (1996, p.14) remarks, circularity is avoided:

> Since the notions of *parents*, *intermediate* and *descendants* are defined unambiguously in the graph, and the graph is defined constructively from the pair $< P, O >$, the background context, likewise, is well defined.

It should be stressed that Pearl is not defending the possibility, in general, of reducing causality to probability; on the contrary, he thinks that "due to circularity inherent in all definitions of causal relevance, probabilistic causality cannot be regarded as a program for extracting causal relations from temporal-probabilistic information but, rather, as a program for validating whether a proposed set of causal relationships is consistent with the temporal-probabilistic information" (Pearl, 1996, p.6); however, in some cases, under some particular assumptions (like minimality or stability), causal structures can be recovered from observed patterns of probabilistic dependencies[24].

In most cases, causal discovery is pursued by exploiting the graph formalism plus probabilistic knowledge and extra-probabilistic causal knowledge, the latter being provided, roughly, by our theories about the domain under study, or, in other words, by the 'context of inquiry'. However, the example suggests that *if* it were possible to find *the* uniquely determined causal net for a given situation, and *if* some assumptions about statistical correlations held, we could restrict the context-relativity of causation to background context-relativity. This raises the question if a similar restriction can be generalised, namely, if it is, in general, possible to *reduce* causal directionality "to undirected probabilistic laws between the cause, effects, *and* other

---

values of the variables picked out by conditions (i) and (ii) in the Consistency Test.
[24](Pearl, 1996, pp.17–18). See also Pearl (2000, Chapter 2).

events"(Papineau, 2001, p.16). If we assume that the set of probabilistic dependences holding among the totality of events (or factors) will display the unique 'true' causal structure of the world, we see how this particular form of reductionism can be combined with the "Absoluteness Assumption". In the next section I shall discuss two possible versions of this combination.

5 Absoluteness and reductionism: all-embracive Bayesian nets

Papineau (1993) argues for the reducibility of causes to objective probabilities, in the form of the hypothesis "[...] that every causal connection will be fixed, given sufficient information in the form of a *probabilistic* graph detailing conditional and unconditional correlations between the variables".[25] Given the fact that statistical data are often compatible with a multiplicity of Bayesian nets, reduction will not be immediate. But, as Papineau remarks, research in the area of causal modeling supports the thesis that statistical indistinguishability can be fixed by adding further variables from the background. In other words, ambiguous structures, i.e., graphs with edges whose direction has not been determined by the data, can be disambiguated by considering other variables which introduce further correlational asymmetries. Papineau concedes that it is a contingent matter whether these additional variables will actually be available; however, he offers several arguments supporting the reasonableness of such an assumption.

These arguments are refined in Papineau (2001), where the focus is shifted to the problem of violations of the Faithfulness Condition.[26] While the underdetermination of causal structures can be in principle eliminated by adding distal variables[27], failures of faithfulness cannot. However, Papineau argues, they could disappear in a more fine-grained description of the domain. This is, of course, a *metaphysical assumption*, resting, in turn, on the metaphysical assumption that at the *lowest* level, nature cannot be so conspiratorial as to host failures of faithfulness. Causal structures, therefore, are determined by an extremely rich underlying net of correlations, from which God can read off the causal facts:

Causal relationships at higher level are fixed by those at the lowest level. Patterns

---

[25](Papineau, 1993, p.241)

[26]A well known example of the failure of the Faithfulness Condition is given by the supposition that drinking cola stimulates people to exercise more, but also causes them to put on weight. If exercise has a negative influence on weight increase of exactly the same magnitude as the positive influence exerted by drinking cola, the two influences will be cancelled out and the overall correlation between drinking Cola and gaining weight will be zero. In this case we have a causal connection between drinking cola and gaining weight, but no corresponding correlation.

[27](Papineau, 2001)

of correlation can thus be misleading about causal structure at any higher level. But at the bottom level there is no metaphysical room for such failures of faithfulness, since there the causal order is simply constituted by the correlational order. What if there is no lowest level, if there is no limit to how fine we can cut up our mechanisms? Then reductionists can adopt a limiting procedure.[28]

Whereas Papineau's reductionist programme concerns the metaphysics of causation, Wolfgang Spohn (2001) argues for a reductionist *epistemology* of causation. Here it is worth noting that Spohn is worried about the relativity of causation apparently endorsed by causal modeling:

> For this purpose, let us look again at the proposed definition: the variable A directly causally depends, within the frame U [*the set of variables of the domain*], on all and only the members of the smallest set of variables in U preceding A conditional on which A is probabilistically independent from all the other variables. This definition hides two relativisations which deserve closer scrutiny.[29]

An initial relativisation depends on the granularity of the description of the domain, and regards the property of being a direct rather than an indirect cause: if the description is rough, with a small number of variables, we can say that C directly causes E; but if we refine the description of the domain by adding further variables, it is plausible that other factors are included in the causal chain from C to E. But the 'fragility' of the notion of *direct* cause is not a great problem, as the fact that C causally affects E is not cancelled. However, Spohn points out, things can get worse:

> The whole notion of causal dependence is frame-relative according to this definition: where there appears to be a direct or indirect causal dependence within a coarse-grained frame, there may be none within a more fine-grained frame, and vice-versa. This consequence seems hard to swallow.[30]

This second relativisation can be better examined, according to Spohn, by raising the question of how to conceive of probabilities. Spohn clearly prefers a mind-independent notion of causality and recognises that such a position should speak in favour of mind-independent, physical probabilities, such as chances.[31] But he also thinks that chances do not provide a suitable interpretation for the probabilities in Bayesian nets, as he does not see how chances could be accounted for in the social and medical sciences.

---

[28](Papineau, 2001, p.37)

[29](Spohn, 2001, p.162)

[30](Spohn, 2001, pp.162-163)

[31](Spohn, 2001, p.163): "The talk of conditional independence refers, of course, to an underlying probability measure. Where does it come from? It might come from reality, so to speak. This raises the question, of course, how to conceive of objective probabilities—a large question which I want to cut short by simply saying that they should best be understood as chances or propensities."

Consequently, the probabilities employed in causal models should be seen as mind-dependent probabilities, i.e., as subjective probabilities.

But this conclusion brings about a further relativisation: if causes are relative to Bayesian nets, and if probabilities are interpreted as degrees of belief, then the Bayesian nets themselves will be relative to the agent who holds the beliefs. How then could causes be connected to some external, mind-independent reality? Spohn's answer is that they aren't: it makes no sense to think of causes as a third, external component outside the patterns of dependence and independence displayed by the graphs. In other words, he posits no concept of causality independent of Bayesian nets. But the mental nature of Bayesian nets does not imply their arbitrariness, as we can imagine a perfectly refined Bayesian net where "the subjective probabilities are optimally informed and thus objective, at least in the sense proposed by Jeffrey" (Spohn, 2001, p.167). Under this hypothesis, the above mentioned relativisations of causal dependence will no longer hold:

> In the final analysis it is the all-embracive Bayesian net representing the whole of reality which decides about how the causal dependencies actually are. Of course, we are bound to have only a partial grasp of this all-embracive Bayesian net. Therefore it is important to have theorems telling under which conditions and to which extent our partial grasp is indicative of the final picture, that is, under which conditions the causal relations in a fine-grained Bayesian net are maintained in coarsening.[32]

With some simplification, the positions presented above imply the following theses:

1. A hypothetically perfect description of the world will dissolve all the apparent discrepancies between probabilistic structures and causal structures, delivering just one causal net;

2. The causal net will be associated to one Bayesian net;

3. Therefore, causes can be reduced to the probability, given the structure exhibited by the Bayesian net.

An obvious consequence of 1 is that in the ultimate description of the world provided by *the* underlying net, causes are not relative to *conversational* contexts (or 'contexts of inquiry' in Hart and Honoré's sense). In Section 1 above, we called the assumption that causes do not genuinely depend on conversational contexts "The Absoluteness Assumption". Therefore, it seems that accepting 1 implies an endorsement of the Absoluteness Assumption. Obviously, causal theories accepting 1-3 do not reject the thesis that causal relations are relative to contexts when the world 'contexts' designates

---

[32](Spohn, 2001, p.166)

the (causal) background contexts (or causally homogeneous populations) as illustrated above (Sections 3 and 4).

There is, at least, a third kind of context-dependence that we should consider with respect to Bayesian nets and causal Bayesian nets: namely, model-dependence. In the next section, I draw a list of the possible kinds of contexts which could be associated with the statement "Causes are context-dependent"; then I discuss some possible relations between these different kinds of context and criticise the idea of an all-embracive-Bayesian net.

6 Different levels of context-dependence

The accounts in Sections 1-4 show that the notion of context has received different interpretations in the history of philosophy of causation. In what follows I try to offer a 'taxonomy' of contexts based on different levels reflecting increasing specificity. The taxonomy is motivated by the question: "What other factors should be considered causally relevant in the assessment of the causal impact of C on E?".

L_0 (General background knowledge). This level is defined as the most general one. At L_0 one finds the most extensive possible description of the universe. Therefore, theories accepting the Absoluteness Assumption and stating that the other causes of a certain effect ought to be searched for in "the whole prior state of the universe", or in "the whole environment" should consider factors in L_0.

L_1 (Context of Inquiry). As previously mentioned, many accounts of causality stress that causal statements ought to be read as implicitly referring to a given set of circumstances, selected according to the needs of a specific causal inquiry. The aims and interests of the researchers determine what Hart and Honoré called the *Context of inquiry*. If we accept this thesis, we can say that context of inquiry selects a subset of the set of factors in L_0, and we can call these terms "L_1-factors". In Mackie's terms, the set of L_1-factors is given by the factors belonging to 'Causal field' and the factors chosen as 'conditions' in a given causal inquiry. In probabilistic causality and causal modeling, L_1-factors seem identifiable as the variables of the domain of interest in a given causal inquiry. Alternatively, we could view L_1-factors as those factors which are potentially statistically associated.

L_2 (Causal models). The causal modeling approach has introduced a further level of refinement. The construction of a Bayesian net, or of a causal net, requires the selection of a subset of the set of L_1-factors because all the variables of the context of inquiry will not normally be included in a single DAG. We call the elements of these subsets "L_2-factors". With respect to a particular DAG, the set of L_2-factors is equivalent to the set V of the variables in the DAG. A clear example of the inductive procedure of selection of L_2-factors is described by Jon Williamson:

> A good strategy here seems to be simply to observe values of as many variables in the domain of interest as possible and rule out as irrelevant those that are uncorrelated with the key variables. For example in a study to determine whether a mother's vegetarianism causes smaller babies, 105 variables related to the woman's nutritional intake, health and pregnancy were measured and then the small subset of variables relevant to the key variables (vegetarianism and baby size) were determined statistically [...].[33]

L_3 (Set of the Explicitly Relevant Factors). Finally, we have the set of factors which are more directly causally relevant for the assessment of a causal relation. At this level we find the components of Causal Background Contexts in probabilistic causality. In the causal Bayesian net the set of L_3-factors is determined by the procedure dictated by the definition of 'Background Context' given above (Section 4).

The reason I have proposed a hierarchy of levels of contextuality is that it shows how the distinction between causes and conditions seems to be a matter of degrees: at the most general level an enormous number of factors, perhaps all, can be regarded as relevant in the ascertainment of the causal influence of C on E, but as one steps to the next level, the set of causally relevant factors shrinks. Let us consider now the hypothesis of an all-embracive Bayesian net. Papineau argues that his theses are exclusively *metaphysical*, and do not claim to have any epistemological or methodological import, while Spohn defends the idea of an all-embracive Bayesian net as an epistemological hypothesis[34]. Leaving the metaphysical problem aside, I shall concentrate on the epistemological implications of the Absoluteness Assumption. Spohn accepts the thesis, advanced by Spirtes and Scheines (1993) that whenever there is a causal graph which is not a Bayesian net, there exists a suitable refinement of the original graph which is a Bayesian net (Spohn, 2001, p.165); this fact, if true, can be explained by admitting that the structure of a suitably refined Bayesian net exhausts our understanding of causation: we don't need any independent notion of causation. However, as probabilities are seen as subjective, and the graphs depends on the choice of variables, Spohn requires that probabilities are maximally informed and that the set of variables be "the all-embracive frame containing

---

[33](Williamson, 2005, p.148)

[34]It is worth remembering that Spohn (2006) qualifies as "doubtlessly obscure" the appeal to "the universal frame consisting of all variables whatsoever" and points to a possible alternative.

all variables needed for a complete description of empirical reality" (Spohn, 2001, p.167). These variables, which correspond to all the variables in $\mathbf{L}_0$ in our taxonomy above, would be represented by the nodes of an universal DAG where the three levels $\mathbf{L}_0$, $\mathbf{L}_1$ and $\mathbf{L}_2$ would collapse: there would be no proper context of inquiry, because contexts of inquiry, by definition, are incomplete representations of reality, or any proper model-dependence, as there would be just one model.

The hypothesis of an all-embracive Bayesian net receives some support by the consideration that the more we refine our models, the more they will deliver the right sets of dependencies and independencies among variables. So the all-embracive net provides the norm that the limited real models should approximate. However, it is not easy to see how such a net should be constructed, considering that $\mathbf{L}_0$ does not host any inner subdivision into contexts of inquiry. A possible solution would be to imagine a reduction of the ontologies a some microscopic level, such that he variables would be of the same kind.

But if we locate this ideal net at $\mathbf{L}_0$, we are forced to choose between two alternatives. On the first alternative, we should admit the duplication of causal explanations: for example, we could say that "Smoking causes lung cancer" is true with respect to a $\mathbf{L}_1$ chosen for pragmatic reasons, but is not true in at $\mathbf{L}_0$, where the set of factors is different.[35] Otherwise we should admit that our models are false, although perfectible, and that only the all-embracive Bayesian net is true; in this case, however, we would be forced to admit that a progressive improvement of our causal models would imply a progressive reduction of the ontologies of different disciplines to one: for example, the completion of causal models of sociological facts would require their progressive transformation into causal models of biological facts, then of physical facts, and so on and so forth.

Both alternatives hardly justify the use of causal models in, say, social sciences. Consequently, we should explore the possibility of accepting relativity to contexts of inquiry and model-relativity without paying the price of arbitrariness. One of the most accurate reflections on how causal models should be linked to the background knowledge and on how they could be justified has recently been provided by Jon Williamson (2001, 2004, 2005). Williamson's account of causation is based on the two notions of *objective Bayesianism* and *epistemic causality*. Objective Bayesianism, like subjective Bayesianism, interprets probabilities as degrees of belief. But while subjective Bayesianism views probabilistic coherence as the only condition for an agent's belief function being rational, objective Bayesianism imposes further constraints on degrees of belief, with the purpose of obtaining a

---

[35] For analogous remarks, see Menzies (2004, pp.150-151).

unique probability value as the rational degree of belief that an agent can assign to a hypothesis on the basis of a specific background knowledge. Moreover, if two rational agents have the same background knowledge, they "must adopt the same probabilities as their rational degrees of belief".[36] According to Williamson, the main constraints that ought to be imposed upon rational beliefs are of two kinds: 1) empirical information about the world (for instance, knowledge of relative frequencies), and 2) logical considerations (knowing only that the experiment I am performing has 6 outcomes, I should assign a degree of belief in each outcome of $1/6$).[37]

Williamson parallels the notion of objective mental probability with a notion of objective mental causality, which he denominates "epistemic causality". In order to define this notion, Williamson (forthcoming) requires that the causal beliefs that an agent ought to adopt are represented by a causal graph C_β, and that C_β is compatible with the agent's background knowledge β. As causal relationships are here viewed as sets of causal beliefs, causality is a mind-dependent, not a physical notion. But again, this does not imply that causality is arbitrary: if two agents with the same background knowledge disagree in regards to causal relationships then at least one of them must be wrong:

> Causality does not depend on the mind of any particular agent – it is a normative notion and causal relations are as mind-independent as the laws of logic. Causality is not subject to the whim of an agent: a rational agent can exercise little or no choice when she forms her causal beliefs; there is a little or no arbitrariness as what the correct causal relationships are. Causality is objective.[38]

The normative standard for causal graphs is given by the notion of a fully objective causal graph C^*, seen as the causal graph that an agent should adopt if he possessed some *ultimate background knowledge* β^*.

How should the ultimate background be constituted? Williamson gives slightly different characterisations. Williamson (forthcoming) rejects the suggestion that β^* should be seen as the limit of scientific enquiry, because science is not unanimous, and because "it is by no means clear that scientific knowledge will tend to a fixed limit".[39] Instead, he proposes a conception of ultimate background knowledge as that which optimises the convenience of the associated causal graph:

> Now causal beliefs will provide the most convenient representation of the world if they are based on the fullest knowledge of the world, i.e., if β^* contains knowledge of all probabilities, physical mechanisms, temporal relations, non-causal inducers of

---

[36](Williamson, 2005, p.12)
[37]See Williamson (2005, pp.65-66).
[38](Williamson, forthcoming, p.15)
[39]*Ibidem.*

probabilistic dependencies (semantic, logical and mathematical relationships, non-causal physical laws and boundary conditions) and so on. This strategy has the advantages that β^* is well defined (as long as the indicators of causality can be delimited) and that causality is not tied to a particular agent – indeed causality is not tied even to there *being any agents*.[40]

The requirement of a background knowledge containing "the fullest knowledge" of the world seems to imply a completeness requirement similar to the one required by the defenders of an all-embracive Bayesian net. If "fullest knowledge" is to be intended as "most complete", we probably should obliterate the distinction between causes and conditions and take into account lots of irrelevant variables (in my scheme, we would be operating at level L_0). This seems to contrast with the idea of convenience: an enormously rich knowledge base would probably be convenient for an epistemically ideal agent, who could easily consider an enormous quantity of variables, but not for real agents, who risk to lose the inferential advantages linked to causal modeling itself. Moreover, the construction of a graph dictated by this ultimate background knowledge seems to force us to some form of physicalism.

But perhaps this is not what Williamson means. A possible interpretation of the phrase "fullest knowledge of the world" is that it refers to fullest knowledge of the world as seen from a specific context of inquiry. A passage from Williamson (2001) can be read as supporting this last interpretation:

> Indeed there is one important sense in which causality is not a unique notion. Different people have different causal pictures if they individuate causes and effects differently. Health service economists, for example, are interested in causal relations among economic events such as government funding announcements and private insurance levels, while doctors are concerned with symptoms and diseases and medical researchers are interested in events on the molecular level. The point is that these people view causality in medicine in different but compatible ways. Thus it is not essential for the definition to yield a unique causal relation, although it may well do it if the causal relata are delimited and fixed in advance.[41]

Further evidence is provided in Williamson (2005), where "knowledge-independent objective causality" is interpreted as "those causal beliefs that an agent ought to adopt were she to have all the *relevant* information as her background knowledge"(italics added); here the adjective "relevant" could be interpreted as implying that the required "full knowledge" is the full knowledge of a specific domain. If this is correct, the pursuit of full objectivity advocated by Williamson appears to be compatible with a relativisation of causes to contexts of inquiry.

---

[40] *Ibidem.*
[41] (Williamson, 2001, p.25)

7 Concluging remarks

From an epistemological, or a methodological point of view, the acceptance of the Absoluteness Assumption implies a duplication of causal judgments (w.r.t the true all-embracive net, and w.r.t the local nets), and a commitment to some form of very strict physicalism. These consequences seem to offer a good reason to reject it. Moreover, the assumption is not the only option to avoid arbitrariness. If we consider relativity to context as relativity to Context of Inquiry, we can obtain objectivity without absoluteness: causal statements will be as objective as any other scientific statement. The price to pay is renouncing to a complete reduction of causes to probabilities, but it is highly dubious that, from an epistemological or a methodological point of view, this reduction offers any real advantage.

Acknowledgments

The author would like to thank the anonymous referees for their careful and helpful comments.

Margherita Benzi
Dipartimento di Filosofia, Università di Genova, Italy.
benzi@nous.unige.it

BIBLIOGRAPHY

Anderson, J. (1938). The problem of causality. *Australasian Journal of Psychology and Philosophy*, 16:127–142.
Cartwright, N. (1979). Causal laws and effective strategies. *Nous*, 13:419–437. Also in Cartwright (1983), pages 21–43.
Cartwright, N. (1983). *How the Laws of Physics Lie*. Clarendon Press, Oxford.
Cartwright, N. (1989). *Nature's Capacities and their Measurement*. Clarendon Press, Oxford.
Collins, J., Hall, N., and Paul, L., editors (2004). *Causation and Counterfactuals*. MIT Press, Cambridge [MA].
Dupré, J. (1984). Probabilistic causality emancipated. In French, P. A., Uehling, T. E., and Wettstein, H. K., editors, *Midwest Studies in Philosophy IX: Causation and Causal Theories*. University of Minnesota Press, Minneapolis.
Eells, E. (1991). *Probabilistic Causality*. Cambridge University Press, Cambridge.
Galavotti, M. C., Suppes, P., and Costantini, D., editors (2001). *Stochastic Causality*. CSLI Publications, Stanford.
Gorovitz, S. (1965). Causal judgements and causal explanations. *Journal of Philosophy*, 62:695–711.
Grice, H. (1975). Logic and conversation. In Cole, P. and Morgan, J. L., editors, *Syntax and Semantics*, volume 3. Academic Press, New York.
Hart, H. L. A. and Honoré, A. M. (1959). *Causation in the Law*. Clarendon Press, Oxford.
Lewis, D. K. (1973). Causation. *Journal of Philosophy*, 70:556–567. Also in Sosa and Tooley (1993), pages 193–204.

Lewis, D. K. (2000). Causation as influence. *Journal of Philosophy*, 97:182–197. An expanded version of the paper is in Collins et al. (2004), pages 75–106.
Mackie, J. L. (1965). Causes and conditions. *American Philosophical Quarterly*, 2:245–255 and 261–264. Also in Sosa and Tooley (1993), pages 32–55.
Mackie, J. L. (1974). *The Cement of the Universe*. Oxford University Press, Oxford.
Mackie, J. L. (1993). Causes and conditions. In Sosa and Tooley (1993), pages 32–65.
Menzies, P. (2004). Difference-making in context. In Collins et al. (2004), pages 139–180.
Mill, J. S. (1961). *A System of Logic Ratiocinative and Inductive: Being a Connected View of the Principles of Evidence and the Methods of Scientific Investigation*. Longmans, London.
Papineau, D. (1993). Can we reduce causal directions to probabilities? In Hull, D., Forbes, M., and Okruhlik, K., editors, *PSA 1992*, pages 238–252, East Lansing. Philosophy of Science Association.
Papineau, D. (2001). Metaphysics over methodology – or, why infidelity provides no grounds to divorce causes from probabilities. In Galavotti et al. (2001), pages 15–38.
Pearl, J. (1988). *Probabilistic Reasoning in Intelligent Systems: Networks of Plausible Inference*. Kaufmann, San Mateo [CA].
Pearl, J. (1996). Structural and probabilistic causality. In Shanks, D. R., Holyoak, K. J., and Medin, D. L., editors, *The Psychology of Learning and Motivation. Vol. 34: Causal Learning*. Academic Press, San Diego. Also in UCLA Cognitive Systems Laboratory Technical Report (R-237), ftp://ftp.cs.ucla.edu/pub/stat_ser/R237.pdf.
Pearl, J. (2000). *Causality: Models, Reasoning, and Inference*. Cambridge University Press, Cambridge.
Scheines, R. (1997). An introduction to causal inference. In McKim, V. R. and Turner, S. P., editors, *Causality in Crisis? Statistical Methods and the Search for Causal Knowledge in the Social Sciences*, pages 185–200. University of Notre Dame Press, Notre Dame [IN].
Skyrms, B. (1980). *Causal Necessity*. Yale University Press, New Haven and London.
Sosa, E. and Tooley, M., editors (1993). *Causation*. Oxford University Press, Oxford.
Spirtes, P. Glymour, C. and Scheines, R. (1993). *Causation, Prediction, and Search*. Springer, New York.
Spohn, W. (2001). Bayesian nets are all there is to causal dependence. In Galavotti et al. (2001), pages 157–172.
Spohn, W. (2006). Causation: An alternative. *The British Journal for the Philosophy of Science*, 57(1):93–119.
Suppes, P. (1970). *A Probabilistic Theory of Causality*. North-Holland, Amsterdam.
Twardy, C. and Korb, K. (2004). A criterion of probabilistic causality. *Philosophy of Science*, 71(3):241–262.
Williamson, J. (2001). On epistemic causality. Technical report, In philosophy.ai report pai_jw_01_b, Department of Philosophy, King's College, London.
Williamson, J. (2005). *Bayesian Nets and Causality: Philosophical and Computational Foundations*. Oxford U.P., Oxford.
Williamson, J. (forthcoming). Causality. In Gabbay, D. and Guenthner, F., editors, *Handbook of Philosophical Logic*, volume 14. Springer. Forthcoming.

Causal inference. How can Bayesian networks contribute?

Isabelle Drouet

ABSTRACT. Inference of causal knowledge from statistics is an old problem. Yet, for twenty years now, Bayes nets algorithms seem to have shed new light on it. Bayes nets algorithms have been extensively and heatedly debated. A fundamental criticism deals with the relationship they assume between causality and probability. What emerges from these debates is that some systems satisfy this relationship while others do not. On this basis, the paper aims at assessing the contribution of Bayes algorithms to causal inference independently from the problem raised by Bayes nets assumptions. The paper begins with a comparison between Bayes nets causal inference and traditional causal inference for social sciences systems satisfying Bayes nets assumptions. The conclusion is very positive: for these systems, Bayes nets causal inference well and truly has many advantages over the traditional methodology it competes with. In the second part of the paper, I go over the restriction to systems satisfying Bayes nets assumptions and show that those systems cannot be identified before causes are known. This leads me to propose a mixed methodology for causal inference which enables us to make use of Bayes nets algorithms in spite of the previously highlighted difficulty.

1 Introduction

Whether it is possible to infer causal knowledge from observational data and how to do it are very old questions for both philosophers and scientists. Since the beginning of the 1990s, it has been thought that Bayesian networks might provide new answers for those old questions. More precisely, various algorithms based on Bayesian networks have been introduced that are said to output causal knowledge from observational probabilistic information.[1] These algorithms have given rise to a heated debate, fed by papers ranging from the most enthusiastic support to the harshest criticism.[2] This de-

---

[1] A good and still recent review on these algorithms is Spirtes et al. (1993, chap. 5 and 6).

[2] McKim and Turner (1997) gives a significant sample.

bate has very quickly focused (and still focuses) on the assumptions about the relationship between causality and probability underlying Bayes nets algorithms. The discussion concerning those assumptions is obviously not over; yet it has already made it clear that some systems satisfy Bayes nets assumptions whereas other ones do not.

The present paper aims at assessing how Bayes nets algorithms can actually contribute to causal inference. This question is important from a practical point of view: one would like to know whether Bayes nets causal inference algorithms can be used, when and how they can be used, and what can be expected from this use. Yet this question is hardly treated independently from the heated discussion about the validity of Bayes nets assumptions. On the one hand, that Bayes nets assumptions hold in most interesting cases is a key claim of the proponents of Bayes nets causal inference (see Korb and Wallace (1997, p. 547), Pearl (2000, pp. 61-63), Scheines (1997, pp. 190-194), Spirtes et al. (1993, pp. 32-42)). On the other hand, that Bayes nets assumptions may be violated is a key argument of those who are skeptical about Bayes nets causal inference (see Cartwright (1999), Cartwright (2001), Freedman and Humphreys (1999, pp. 31-34)). I would like to slightly shift the emphasis of the debate, and to provide an assessment of Bayes nets algorithms that does not depend on how the possible violations of Bayes nets assumptions are interpreted—but just takes them for granted. The paper relies on the claim that taking those possible violations for granted gives means to broach the question of the contribution of Bayes nets algorithms to causal inference. Indeed, knowing that there exist systems satisfying Bayes nets assumptions allows us to focus on those specific cases, and knowing that Bayes nets assumptions do not hold universally enables us to identify as essential the question of what the consequences of this non-universality are. The outline of the paper accords with those two remarks. First, I restrict the analysis to systems satisfying Bayes nets assumptions; for these systems, the assessment is very favourable to Bayes nets algorithms. Second, I go back to the general case *via* an analysis of the significance of this restriction. The analysis shows that the existence of systems failing to satisfy Bayes nets assumptions precludes rough Bayes nets causal inference being performed in any real case. This leads me to propose the integration of Bayes nets algorithms to a wider mixed methodology for causal inference. The whole analysis is restricted to the social sciences—the domain that Bayes nets algorithms most directly address[3]—and to cases without latent variables. These restrictions enable us to keep the issue very

---

[3]The reason for this is that Bayes nets algorithms deal with causal inference in non-experimental contexts and that the impossibility of experiments is characteristic of the social sciences.

tractable from a technical point of view while already broaching essential philosophical issues.

2 Systems satisfying Bayesian network assumptions

As already stated, this section focuses on systems satisfying "Bayes nets assumptions". Those assumptions do not deal directly with systems (be they physical, economical, biological...), but rather with the sets of variables representing them. For such a set V, Bayes nets assumptions are the following ones:

- **Acyclicity**: the directed graph whose arrows represent direct causal relations among V variables is acyclic;

- **Causal Markov Condition**: any variable in V is probabilistically independent from all variables in V but its effects when one conditionalizes on its direct causes;

- **Faithfulness**: there are no probabilistic independencies among the variables in V but the ones that are entailed by the Causal Markov Condition.

I consider those Bayesian network algorithms that are most discussed by philosophers, that is algorithms that infer causal knowledge from probabilistic independences.[4] For cases without latent variables, the best-known amongst those algorithms are the IC algorithm introduced by Pearl and Verma (for a presentation, see Pearl (2000, pp. 50-51)) and the PC algorithm introduced by Spirtes, Glymour and Scheines (see Spirtes et al. (1993, pp. 84-88)). Once a set V of variables has been selected in order to represent the system under consideration,[5] those algorithms perform causal inference in three steps:

1. measure the values of the variables in V for the members of a representative and large enough sample of the population you are interested in;

2. compute the standardized correlations, absolute and partial, among V variables;

3. test each of the computed correlations against the null hypothesis that it is zero and identify the set I of correlations for which the null hypothesis cannot be rejected;

---

[4] In particular, I will not discuss works dealing with learning Bayesian networks using Bayesian techniques.
[5] As already stated, V is assumed to be causally sufficient.

4. construct a partially directed graph over V that represents all the causal knowledge that can be inferred from I owing to Bayes nets assumptions concerning the relationship between null partial correlations and direct causal relations. This graph is the output of the procedure. It represents (all and only) the binary causal relations shared by the causal structures which are compatible with the data plus Bayes nets assumptions.

The methodology I take up in order to assess Bayes nets causal inference when Bayes nets assumptions are satisfied is comparative. Therefore, the question is now: which methods for causal inference in the social sciences should Bayes nets algorithms be compared with? The most obvious answer is that they should be compared with the automated structure learning procedures they directly compete with, in particular those operationalized by the LISREL and EQS packages. There are two reasons why the present comparison will not be with procedures of that kind. First, Spirtes, Glymour and Scheines have given several convincing reasons why Bayes nets algorithms give more satisfactory results than those procedures.[6] Second, automated search procedures are scarcely used by social scientists, who still massively resort to more traditional techniques. Therefore, if one is to determine whether and how Bayes nets algorithms can contribute to actual causal inference, these algorithms should be compared with those more traditional techniques—which has not been done yet. It is true that this comparison raises methodological difficulties. I will discuss them as they emerge, and try to show that they can be overcome.

The "more traditional techniques" I will compare Bayes nets algorithms with stem from path analysis. The way social scientists actually use those techniques can and must be criticized. Yet this has already been extensively done by both opponents (see for instance Freedman (1991)) and proponents (see Kline (1998, chap. 12)), who widely agree on the main misuses of path analytic causal inference techniques, on some of the reasons for those misuses (on this point, see in particular Blalock (1991)) and on how they can be avoided. As a consequence, the present paper focuses on path analytic causal inference as it should be performed in order to give its best results. Accordingly, I consider a procedure which is not standard, but rather is a construct, which I think is the most rigorous and complete methodological proposal that can be extracted from usual social science practice. Once a set V of variables representing the system under consideration has been identified, the procedure runs as follows:

1' Specify a model M that you think might adequately represent the

---

[6](Spirtes et al., 1993, pp. 74-80)

causal relationship among V variables. As already stated, it is assumed that M is a directed acyclic graph over V, whose arrows (or 'paths') represent hypothesized direct causal relations among the variables in V. As far as possible, M should be overidentified by the data that will be collected;

2' measure the values of the variables in V for the members of a representative and large enough sample of the population you are interested in, and compute the standardized correlations amongst V variables;

3' use the data collected at step 2. in order to estimate the structural coefficients in M. Each of these coefficients evaluates the strength of exactly one of the arrows in M;

4' test M, that is:

- determine whether the causal relationship and the signs and absolute values of structural coefficients look plausible;
- if the model is overidentified, compute the correlation residuals, that is the differences between model-implied and observed correlations and check that none has absolute value greater than 1;[7]
- if the model is overidentified, identify the independent overidentifying restrictions, state them as null hypotheses, test these null hypotheses and check that they cannot be rejected;

If M does not pass these tests, reject it, come back to 1 and specify another model. If it does pass them, perform the following steps;

6' if M is overidentified, assess its fit to the data. Many model fit indexes are available and several of them must be performed;

7' reiterate steps 1' to 6' for another model;

8' compare the fit scores of the models that have been considered and were not rejected at step 5'. Identify the model $M*$ that seems to be the overall better data fitting;

9' identify models equivalent to $M*$ that look plausible representations of the causal structure over V, and provide a theoretical justification of your preference for one of them. This model is the output of the procedure.

---

[7] 1 is a conventional but generally accepted maximum for the absolute values of correlation residuals in an acceptable model.

The main difference between the two procedures that I have just detailed is one of Methodology. More precisely, Bayes nets causal inference is deductive whereas path analytic one is not. The case of Bayes nets is the simplest: Bayes nets algorithms output a "pattern" representing all the binary causal relations that deductively stem from the analysis of the data (2 and 3) under Bayes nets assumptions concerning the relationship between causality and probability. The case of path analytic causal inference as I have presented it is a little more intricate. From a methodological point of view, it is clearly composed of two parts—each of which is autonomous enough to provide the methodology for many social sciences papers. The first part, from $1'$ to $5'$, can be described as hypothetico-deductive. A hypothesis is formulated as a model is specified ($1'$), then consequences are deduced from it in the form of the estimation of structural coefficients ($3'$), and finally those consequences are tested against the data ($4'$ and $5'$). If the hypothesis does not pass the test, it is rejected; if it does, it is corroborated (end of $5'$). One could claim that path analytic causal inference has only the appearance of being hypothetico-deductive, and that this appearance badly hides crucial differences between the methodology of path analytic causal inference and standard hypothetico-deduction. In particular, the derivation of consequences from the initial hypothesis does not require auxiliary theoretical hypotheses but data. Moreover, the conclusion is not deductively drawn but only estimated from the premises. Similarly, the rejection of a hypothesis does not rely on a formal contradiction with the data, but on a probabilistic incompatibility. I would answer that the last two differences derive from the statistical nature of the data, but do not matter much at the level of methodological abstraction from which I am currently looking at things. The second part of path analytic causal inference is constituted by $8'$ and $9'$ It consists of identifying first the models that best fit the data amongst the ones that have not been rejected, and second the most plausible amongst the non rejected models that best explain the data. As fit to data corresponds to the capacity of explaining it, $8'$ can be considered as a form of inference to the best explanation. More generically, both $8'$ and $9'$ aim at selecting a hypothesis amongst competing ones. That is a matter of induction, though not of traditional enumerative induction. This kind of induction is described as "hypothetical induction" by Harman.

It is my contention that this methodological divergence results in (at least) three differences which all contribute to the superiority of Bayes nets algorithms over the path analytic procedure that has been described:[8]

---

[8]Note that claiming that Bayes nets causal inference is "superior" to path analytic one does not amount to renouncing to the approach that was assumed at the outset of the paper. Indeed, this approach does not consist of eschewing any kind of evaluation of

First, the procedure resorting to Bayesian networks is data-based, whereas the path analytic one is model-based. Data-baseness is an immediate consequence of deductiveness, together with the fact that analyzed data constitute the premises of the deduction. On the other hand, model-baseness is a feature common to hypothetico-deduction and hypothetical induction: one has to formulate a hypothesis before its consequences can be tested against the data, as one has to identify competing models before the best of them can be looked for. Model-baseness leads to the dependence of the final output on the causal order the scientist has been able to envisage, and that is clearly a drawback to path analytic causal inference over Bayes nets one. It could be argued that the difference between data and model-baseness is so essential from a methodological point of view that it makes the comparison unfair from the beginning. From that point of view, if Bayes nets algorithms are to be compared with a path analytic procedure, it must be with a procedure that runs through all possible models. Yet, such a procedure would clearly be computationally more demanding than Bayes nets algorithms, which are already criticized for requiring the computation of all correlations among V variables. More important, that would deprive path analytic methods of one of their essential features, and obscure the difference between them and the automated search procedures that I decided not to consider. Therefore, I will persist in considering a model-based path analytic procedure, and this model-baseness will count as a remarkable difference between path analytic and Bayes nets causal inferences. Obviously, this difference is in favour of Bayes nets causal inference. Relying on the specification of particular models (step $1'$) is a feature of path analytic causal inference that has been often criticized.[9]

Second, the Bayes nets procedure outputs a pattern, whereas the path analytic one outputs a single structure. As already stated, the pattern output by Bayes nets algorithms represents the binary causal relations that deductively stem from analyzed data under Bayes nets assumptions concerning the relationship between causality and probability. Therefore the fact that Bayes nets algorithms output a pattern well and truly derives from their deductiveness. Now this pattern stands for the whole class of acyclic causal structures sharing the binary causal relations it depicts; those structures that cannot be distinguished by data under the assumed relationship between causality and probability. By contrast, the output of the path analytic procedure is a causal structure, for the simple reasons that the hypotheses formulated at $2'$ are causal structures too and that hypothetical induc-

---

Bayes nets algorithms, but of evaluating them *independently from the debate on Bayes nets assumptions*. This is clearly what is going on at the present point of the paper.

[9] Freedman (1991, pp. 303-304 and p. 309).

tion aims at isolating *the* hypothesis to be preferred. It is true that step 9' requires the researcher to identify models that are equivalent[10] to the better-fitting considered model, to decide whether some of them are plausible and to provide a justification for preferring one of them to the others. Yet performance of the latter two tasks depends on both the state of scientific knowledge at the time causal inference is performed on and some of the scientist's convictions. Therefore, the nature of its output must be considered as an advantage of Bayes nets causal inference over path analytic one.

Third and lastly, and as another immediate consequence of deductiveness, the binary causal relations output by Bayes nets algorithms must be accepted as true as soon as zero partial correlation identified in 3 are. On the contrary, even one those accept estimations of structural parameters in 3' and identifications of the vanishing ones in 4', have no reason to believe that the structure output by the path analytic procedure adequately represents reality. On the one hand, some structures that have not been considered may better fit data than the output one; on the other hand, there is no guarantee that best-fitting data structures are causally significant. More generally, non demonstrativeness is a recognized feature of both hypothetico-deduction and hypothetical induction.

In this section, I have detailed how causal inference is performed by Bayes nets algorithms on the one hand, and by usual path analytic methods on the other hand. I have pin-pointed three features that make Bayes nets causal inference superior to path analytic one and shown that those features are correlates of deductiveness, as opposed to a combination of hypothetico-deduction and hypothetical induction in the path analytic case. But the whole analysis concerns only those systems that satisfy Bayes nets assumptions. In the next section, I analyze the impact of this restriction on the conclusions I have drawn and I come to the question of the contribution of Bayes nets algorithms to causal inference in the general case.

3 The general case

At first sight, the restriction of the preceding analysis to cases satisfying Bayes nets assumptions does not raise particular difficulties. First, it looks fair to focus on systems liable to Bayes nets causal inference when examining the contribution of Bayes nets to causal inference. Second, this focus does not seem to have consequences other than the following ones: the conclu-

---

[10]Equivalence of models in the path analytic context concerns the ability to explain observed correlations. It must not be mixed up with indistinguishability in the Bayes nets context, which corresponds to the depiction of the same probabilistic independencies.

sions are valid only for the systems that have been taken into account, and only these systems can benefit from the nice features of Bayes nets causal inference that have been highlighted.

The matter is that this first-sight analysis is misleading. What is at stake is not only that Bayes nets assumptions fail to universally hold, but also that we do not know when they do. Let me explain this idea focusing on the Causal Markov Condition ("CMC" from now onwards). There are two justifications for this focus. On the one hand, it is probably more central to inference of causality from probability than the other two Bayes nets assumptions. Correspondingly, it has been much more discussed and is better-known. On the other hand, this focus does not make any difference as far as the conclusions of the paper are concerned. Now it is true that we have a sufficient condition for the satisfaction of the CMC. More precisely, it is now well-known that the CMC holds for deterministic systems represented by causal functional models with independent exogenous variables.[11] In other words, given a system and a set of variables V causally sufficient for this system, V satisfies the CMC if:

a. the value of any V variable is functionally determined by the values of its direct causes in V;

b. any two disjoint non-empty subsets of the set of V variables that do not have any causes in V are probabilistically independent.

Although a sufficient condition for the satisfaction of the CMC, this cannot work as a criterion in the context of causal inference. Indeed, the very definitions of causal functional models and of exogenous variables imply that these notions are meaningful only for one who already knows of the causal structure of the system being considered. But that—knowledge of the causal structure—is exactly what we are looking for when performing causal inference. Furthermore, I do not know of any non causal criterion for the satisfaction of the CMC.[12] As a consequence, it is impossible to know whether Bayes nets algorithms can be rightfully used before causal inference is actually performed.

---

[11] See Pearl (2000, p. 30) and Spirtes et al. (1993, p. 32). The result also holds for pseudo-deterministic (Spirtes et al., 1993, pp. 27-28) and indeterministic systems. These extensions involve latent variables, hence they will not be taken into account in the present paper. Nevertheless, their existence constitutes an argument in favour of the possibility extending the results of the paper beyond the scope of the restrictions which were imposed at the outset.

[12] The current discussion concerning the relationship between the CMC, manipulability and modularity (Cartwright, 2002, 2006; Hausman and Woodward, 1999, 2004; Steel, 2006) may contribute to the formulation of such a criterion. But it seems clear to me that it has not yet, and I cannot see along which lines such a solution could emerge.

There exist arguments to the effect that Bayes nets algorithms should nevertheless be used each time no violation of Bayes nets assumptions has been detected beforehand.[13] Yet I think that no argument of this kind may be accepted. Let me indicate why, by first stating what the consequences of applying Bayes nets algorithms to a system that does not satisfy Bayes nets assumptions are. From a theoretical point of view, it must be clear that Bayes nets algorithms used under those circumstances will not output an adequate representation of the real causal structure. Besides, the absence of beforehand information concerning violations of Bayes nets assumptions makes it impossible to draw an upper limit to the deviation of the output from the true causal structure (under any plausible measure of this deviation). Important practical consequences follow. Indeed, if A causes B, acting on A is a good way to affect B.[14] Then, to wrongly believe that A causes B can lead to wasting time, energy and money in trying to affect B by modifying A. The losses are all the more dramatic and all the less affordable since the subject matter comes under the social sciences—which were assumed from the beginning in the present paper. To make it short, consequences of undue application of Bayes nets algorithms must not be overlooked. Now, whatever good reasons you may have to do it, resorting to Bayes nets algorithms each time no violation of Bayes nets assumptions is detected beforehand, will lead you to apply those algorithms to systems failing to satisfy Bayes nets assumptions. What is more, in the absence of any non causal criterion for the satisfaction of those assumptions, any system to which Bayes nets algorithms are actually applied becomes suspect, as being one of those problematic cases. With regard to the just discussed consequences of undue application of Bayes nets algorithms, this suspicion cannot be tolerated. The conclusion is that one should renounce Bayes nets causal inference.

This conclusion holds for Bayes nets causal inference as it was described in the preceding section of the paper. Yet it may not extend to all possible uses of Bayes nets algorithms. Indeed, several authors have envisaged integrating Bayes nets algorithms with a wider methodology for causal inference.[15] The proposition is most carefully defended by Williamson (2002, §3), under

---

[13] The principal two arguments are as follows: 1) according to Bayes nets proponents, Bayes nets assumptions are satisfied by most systems (references were already given in the introduction of the paper) and, they would probably argue, by nearly all causally sufficient ones; 2) following Williamson (2002, §4), it can be argued that the CMC must be accepted as a "default assumption" under objective Bayesianism. Both of these arguments call for specific comments, which will not be exposed here.

[14] This is a consequence of the commonly accepted relationship of causality with agency.

[15] This answer is developed in particular by Glymour et al. (1988, pp. 428-429) and in Williamson (2002).

the following form: use Bayes nets algorithms as a first exploratory step of causal inference, then deduce predictions from their output, test those predictions against evidence, consequently amend the hypothesized structure, and finally return to the deduction step for the thus obtained structure. My discussion of this proposition will focus on those consequences that are drawn from the hypothesized causal model. Williamson identifies three kinds of such consequences, respectively deriving from: "supposed connections between causality and probability",[16] the usual correspondence of causal claims with physical processes linking causes to effects and the "close relationship [causality has] with agency".[17] Now, the latter two facts are clearly irrelevant when one is interested in testing hypothesized causal relations in the social sciences. On the other hand, "supposed connections between causality and probability" are encapsulated in Bayes nets assumptions. As a consequence they will not be violated by structures just output by Bayes nets algorithms, and testing for them is of no use. What remains, then, is Williamson's later suggestion to remove from the amended structure those arrows representing dependencies that are found to admit of a non-causal explanation. I consider that this will not be enough to deal with all errors resulting from undetected violations of Bayes nets assumptions—and to reach a correct answer. As a consequence, I contend that the difficulty raised by the absence of a non causal criterion for the satisfaction of Bayes nets assumptions is not overcome by propositions of the same kind as Williamson's one.

Does it follow that Bayes nets algorithms must be altogether abandoned by anyone who takes seriously our current incapacity to identify systems satisfying Bayes nets assumptions? I would like to show that it does not. More precisely, I will indicate a way in which Bayes nets algorithms can contribute to inference of causal knowledge from observational statistical data in spite of the difficulty with Bayes nets assumptions. It is clear from what precedes that this implies that Bayes nets algorithms are integrated to a wider causal search procedure—like Williamson's—and that this procedure is such that Bayes nets algorithms are run only when Bayes nets assumptions are satisfied—unlike Williamson's. In other words, one has to envisage a mixed methodology such that a causal model is produced first and Bayes nets algorithms are run, if possible, subsequently. The most natural idea concerning the production of this initial causal model is to resort to the path analytic methods that have been discussed in section 2 of the present paper. This leads to propose the following methodology: given a system S represented by the causally sufficient set of variables V,

---

[16](Williamson, 2002, p. 7).
[17](Williamson, 2002, p. 8).

1″ perform steps 1′ to 9′ of path analytical causal inference methodology. Let M be the model that is output by this procedure;

2″ test whether Bayes nets assumptions would hold for S in case it would be correctly represented by M. If the test is not passed, then accept M as your causal model and go to 3″;

3″ perform steps 2 to 4 of Bayes nets causal inference for V.

- If the output of this procedure is not compatible with M, that is if M does not belong to the set of directed acyclic graphs represented by the output pattern, then consider M as refuted. Therefore, come back to 9′ in order to produce an alternative causal model. If a model equivalent to M seems satisfactory, then come back to 1″ with this model; if not, re-iterate steps 2′ to 9′ with new specifications in 1′ and come back to 1″ with the new path analytic output;

- If the output of the procedure is compatible with M, then accept M as your causal model.

The proposed procedure is clearly hypothetico-deductive. Path analytic methods enable to hypothesize a causal model M. If it cannot refuted that Bayes nets assumptions hold in case M is correct, the hypothesis that M is indeed correct leads to predict that it is compatible with the output P of Bayes nets algorithms is.[18] Then Bayes nets algorithms serve in the testing part of the procedure, with recommendations in 3″ following: M should be rejected when incompatible with P and is further corroborated when compatible with P. Now the hypothetico-deductiveness of the procedure sounds problematic in as much as all the nice features of Bayes nets causal inference that were highlighted in section 2 precisely stemmed from the deductiveness of the procedure. Actually, those features are lost in the context of the present proposal:

- the procedure is not data-based. Rather, it depends on the models the scientist has been able to envisage in the same way as path analytic causal inference does;

- although Bayes nets algorithms still output a pattern, the whole methodology outputs a single model. This point must be explained. Imagine that the output M of path analytic causal inference is such that Bayes

---

[18]The consequence is not a logical one since the test for the validity of Bayes nets assumptions is open to statistical errors.

nets assumptions would be satisfied if M were correct. Suppose further that M is compatible with the pattern output by Bayes nets algorithms. Then one may wonder why all the models compatible with this pattern cannot be taken as serious candidate models. I see two reasons why they cannot be. First, if plausible, those models have good chance to have been taken into account in 1″ If it is actually the case, I cannot see why they should be discussed once again after Bayes nets algorithms have been run. Second, even in case one of those models is correct, nothing guarantees that Bayes nets assumptions are satisfied—contrary to what happens if M is correct. As a consequence, compatibility with the output of Bayes nets algorithms does not provide any of those models with any kind of inductive support;

- as a direct consequence of its being hypothetico-deductive, the proposed methodology does not output a model which must be taken as true as soon as the identified zero correlations are. The output model is only well-corroborated by the statistical analysis of data.

The scene, then, is not very heartening. For one thing, the use I envisage for Bayes nets algorithms fails to retain the interesting features of Bayes nets causal inference as applied to systems satisfying Bayes nets assumptions. For another thing, they do not perform causal inference properly speaking anymore. What they do is provide an additional test for some (but not all) of the models already output by model-based usual path analytic methods—a contribution which obviously is not up to the initial ambitions. Yet the proposed methodology also has non negligible assets. Primarily, it settles the problem of identifying systems that satisfy Bayes nets assumptions—a problem whose important negative consequences we saw to be commonly overlooked. Indeed, Bayes nets algorithms are run only when one has good reasons to believe that Bayes nets assumptions are satisfied, more precisely when it cannot be refuted that the assumptions hold in case the model output by path analytic causal inference is correct. Positively, Bayes nets algorithms actually contribute to causal inference when used in the proposed way: path analytic causal inference being a non-deductive procedure, its results always call for more testing and any kind of further evidence in their favour is welcome and useful. Furthermore, there is no reason to believe that the test as such should be a trivial one. Quite the opposite, chancy compatibility of path analytic and Bayes nets outputs[19] is quite improbable since 1) path analytic causal inference relies on principles differing from

---

[19] By "chancy compatibility", I refer to a compatibility that would not stem from those outputs being correct.

Bayes nets ones and 2) Bayes nets algorithms enable to rule out a significant number of candidate causal models. Accordingly, the test relying on Bayes nets algorithms should enable to reject a certain number of models while providing others with substantial inductive support. This contribution is available each time Bayes nets assumptions are satisfied—in most cases according to one of the most common arguments of Bayes nets proponents.

4 Conclusion

In the present paper I have tried to assess how Bayes nets algorithms can contribute to causal inference. Taking for granted that Bayes nets assumptions sometimes hold and sometimes do not, led to a two-pronged approach of the question. In section 2, I focused on those systems that satisfy Bayes nets assumptions. This enabled us to pinpoint the exact ways in which Bayes nets causal inference is superior to current social sciences procedures. A methodological rationale was given for this superiority. Then, in section 3, I came to the general case through the analysis of the impact of the previous restriction. The difficulty quickly appeared to consist not only of the non universal validity of Bayes nets assumptions, but also of our present incapacity to determine beforehand whether a given system satisfies them or not. The difficulty and existing propositions aiming at overcoming it were analyzed. This led me to introduce a mixed methodology for causal inference in which Bayes nets algorithms are run only after good reasons have been provided for the validity of Bayes nets assumptions. Finally, it was explained how Bayes nets algorithms contribute to causal inference in this context.

The present analysis obviously suffers from its restriction to the social sciences, and to cases without any latent variable. These restrictions were formulated as a consequence of the impossibility to presently give an extensive treatment of the problem under consideration. Lifting them, as well as extending the analysis to algorithms learning Bayesian networks through Bayesian techniques, should form the subject matter of subsequent work.

Acknowledgments

I thank Anouk Barberousse and Jacques Dubucs for support, as well as two referees for very helpful comments.

Isabelle Drouet
Philosophy, IHPST, Paris 1 University, France.
Isabelle.Drouet@malix.univ-paris1.fr

BIBLIOGRAPHY

Blalock, H. (1991). Are there really any constructive alternatives to causal modeling? *Sociological Methodology*, 21:325–335.
Cartwright, N. (1999). Causal diversity and the causal Markov condition. *Synthese*, 121:3–27.
Cartwright, N. (2001). What is wrong with Bayes nets? *The Monist*, 84:242–264.
Cartwright, N. (2002). Against modularity, the causal Markov condition, and any link between the two: Comments on Hausman and Woodward. *British Journal for the Philosophy of Science*, 53:411–453.
Cartwright, N. (2006). From metaphysics to method: Comments on manipulability and the causal Markov condition. *British Journal for the Philosophy of Science*, 57:197–218.
Freedman, D. (1991). Statistical models and leather shoe. *Sociological Methodology*, 21:291–313.
Freedman, D. and Humphreys, P. (1999). Are there algorithms that discover causal structure? *Synthese*, 121:29–54.
Glymour, C., Scheines, R., and Spirtes, P. (1988). Exploring causal structure with the TETRAD program. *Sociological Methodology*, 18:411–448.
Hausman, D. and Woodward, J. (1999). Independence, invariance and the causal Markov condition. *British Journal for the Philosophy of Science*, 50:521–583.
Hausman, D. and Woodward, J. (2004). Modularity and the causal Markov condition: a restatement. *British Journal for the Philosophy of Science*, 55:147–161.
Kline, R. B. (1998). *Principle and Practice of Structural Equation Modeling*. Guilford Press, New-York, second (2005) edition.
Korb, K. and Wallace, C. (1997). In search of the philosopher's stone: Remarks on Humphreys and Freedman's critique of causal discovery. *British Journal for the Philosophy of Science*, 48:543–553.
McKim, V. and Turner, S. (1997). *Causality in Crisis? Statistical Methods and the Search for Causal Knowledge in the Social Sciences*. University of Notre Dame Press, Notre Dame.
Pearl, J. (2000). *Causality: Models, Reasoning, and Inference*. Cambridge University Press, Cambridge (UK).
Scheines, R. (1997). An introduction to causal inference. In McKim, V. and Turner, S., editors, *Causality in Crisis? Statistical Methods and the Search for Causal Knowledge in the Social Sciences*, pages 185–199. University of Notre Dame Press.
Spirtes, P., Glymour, C., and Scheines, R. (1993). *Causation, Prediction, and Search*. MIT Press, Cambridge (US), second (2000) edition.
Steel, D. (2006). Comment on Hausman and Woodward on the causal Markov condition. *British Journal for the Philosophy of Science*, 57:219–231.
Williamson, J. (2002). Learning causal relationships. Technical Report 02/02, London School of Economics, Center for the Philosophy of Natural and Social Sciences.

Counterfactuals, hypotheticals and potential responses: a philosophical examination of statistical causality

A. Philip Dawid

ABSTRACT. Statisticians have developed a variety of conceptions, frameworks and tools for causal inference. We study some of these from a philosophical angle, focusing in particular on two formal frameworks, "Potential responses" (PR) and "Decision Theory" (DT), and their use in problems of inferring the "effects of causes". Although PR is currently the predominant methodology, it is argued that DT is preferable both philosophically and pragmatically.

1 Introduction

Over more than three decades statisticians have developed, and applied with considerable practical success, a variety of formal frameworks for manipulating causal concepts and conducting causal inference. However, unlike the distantly related topic of "probabilistic causality" (Suppes, 1970), this extensive enterprise of "statistical causality" appears (with rare exceptions such as Glymour (1986); Spirtes et al. (1993)) to be essentially unknown territory to most philosophers. This article has been written in an attempt to redress that state of affairs: it describes, relates and compares a number of formal frameworks in common statistical use, paying particular attention to their philosophical underpinnings. Recognising that I am a rank amateur at this, I nevertheless hope that it will arouse enough interest among professional philosophers to stimulate them to improve on and extend my own philosophical fumblings.

In § 2 I introduce and emphasise the important distinction between *hypothetical* and *counterfactual* conditional queries, and in § 3 show how these relate, respectively, to the distinct tasks of inferences about *effects of causes* (EoC) and *causes of effects* (CoE).[1]

Section 4 shows how the familiar language of probability applies readily to EoC, but is unable to handle CoE. Section 5 then describes an alternative

---
[1] EoC corresponds loosely to what philosophers term "type causation", and CoE to "token causation".

language, that of *potential response* (PR) modelling, that can express CoE queries. Section 6 relates PR to some other popular frameworks, based on structural equations, functional models and Pearl's causal DAG models.

The rest of the paper concentrates on the use of these frameworks in addressing specifically EoC queries. Here both PR and DT can be applied, with PR very much predominating in current use. Extending views and arguments first laid out in Dawid (2000), I argue that this is a mistake: DT is preferable on both philosophical and pragmatic grounds. To this end, §§ 9 and 10 examine and compare the usefulness of PR and DT in problems of making inference from observational data, and § 11 extends this to a problem of dynamic decision-making. In § 12 we contrast the two approaches on a variety of philosophical dimensions.

If philosophers have paid little attention to statistical causality, it is equally true that statisticians have paid little attention to the philosophical issues underlying their causal methodologies. I believe that to do so can bring real practical improvements, and I hope this paper may also act as a stimulus towards this.

2 Several modalities

Consider the following queries:

1. I have just taken an aspirin for my headache. Will it have gone within 30 minutes?

2. If I were to take an aspirin now, would my headache be gone within 30 minutes?

3. I took an aspirin 30 minutes ago and my headache has now gone. If I had not taken the aspirin, would it have gone anyway?

These are all *conditional queries*, but within that common category all have distinct logical status.

Query 1 conditions on a state of affairs—my having taken the aspirin— that is known to be true at the time it is formulated. Queries 2 and 3, on the other hand, condition on a *conjectural*, rather than known, state of affairs: what a logician would call *modal* usage, and a grammarian might term *subjunctive* mood (Pollock, 1976), as against *indicative* mood in the case of query 1.

However, queries 2 and 3 differ in the following important respect. For query 3, it is already known that I did in fact take the aspirin. The conjectured condition is thus logically incompatible with known[2] facts—it is

---

[2] For our purposes the important issue is epistemological, in terms of what is *known* to be (or accepted as) true, rather than ontological, in terms of what is *in fact* true.

consequently termed a *counterfactual* conditional. For query 2, by contrast, I do still have the power, by taking an appropriate action, to bring the conjectured condition into actuality, and subsequently to observe the outcome I am now contemplating; I could similarly (and simultaneously) consider the consequences of not taking the aspirin. Such conditionals are not counterfactual, since their premises do not contradict any known facts: they are more appropriately described as *hypothetical*(a term which itself has more than one meaning, but will here be restricted to this one).

In much philosophical and statistical parlance these last two modalities are not distinguished, the description "counterfactual" being used indiscriminately for both.[3] However, it will be argued here that this distinction between the modalities involved in queries 2 and 3 is indeed real, and of fundamental importance to any sensible understanding of causality. Henceforth in this article the terms "hypothetical" and "counterfactual" will be carefully reserved to reflect the above distinction. Typically, though not universally, the former has a *prospective*, and the latter a *retrospective*, focus.

I have met considerable resistance to the thesis that counterfactuals and hypotheticals are truly distinct concepts. I shall respond to this by demonstrating that different formal languages are needed to express them. There are many other forms of modal conditional usage that might be distinguished, including: potential; conjectural; speculative; fictional; counterfictional; *reductio ad absurdum*; null hypothesis; Aunt Sally hypothesis; etc.

Some nice distinctions are made by Pollock (1976), although even he does not highlight that which I wish to emphasise here. Kevin Korb (2004) has made the obvious but important point: "It is a mistake to think that all modal concepts are counterfactual". It is however a mistake that seems particularly prevalent.

3 Contrastive causality

From a statistician's standpoint it is natural to take a contrastive view of causality, whereby we understand a "cause" to be a variable, X say, whose value might "make a difference", in some sense to be made precise, to that of some "response variable" Y of interest. In pragmatic terms, this means that we focus on measuring the *causal effect* of X on Y, loosely interpreted as an appropriate comparison between the consequences, for Y, of each of two or more values or settings for the variable X.

Consider now the following *causal queries*, relating to a single headache episode:

---

[3]The distinction between them has occasionally been remarked, as e.g., by Anderson (1951); Chisholm (1955)—but I am not aware of any serious analysis of its implications.

1. **Effects of causes (EoC)**   If I were to take an aspirin now, would that cause my headache to go away within half an hour?

2. **Causes of effects (CoE)**   My headache has gone away. Is it because I took an aspirin half an hour ago?

Given the knowledge that, having taken the aspirin, my headache has gone, the CoE causal query 2 might be addressed by constrasting this known outcome with the (unknown) answer to the counterfactual conditional query 3 of §2: if that answer were "no", we could say that the conjectured causal explanation is correct; if "yes", then taking the aspirin made no difference, and thus was not a cause of the observed response. More incisively, we might compare the durations of the headache under the actual and the counterfactual conditions, and take their difference as a measure of the causal effect.

Similarly, the EoC causal query 1 might be answered "yes" if the hypothetical conditional query 2 of §2 is answered "yes" at the same time as the similar query under the alternative hypothesis of not taking the aspirin is answered "no" (or again we might compare the durations of the headache under the two hypotheses).

There are still other varieties of causal query. For example: "I have just taken an aspirin. Has that caused/will that cause my headache to depart within the next 30 minutes?". However it will suffice here to restrict attention to the above two forms.

4   Probability

Causal queries are thus related to subjunctive conditional queries, as introduced in §2—questions of CoE requiring the counterfactual, and those of EoC the hypothetical, mode of reasoning. However, we will not usually be in a position to supply a firm answer to any such query. The best we can do is to assess a suitable measure of the uncertainty about its answer, in the light of all known information, and for this we might turn to the language of probability.

To focus further discussion, we shall use the following simple example of a comparative experiment.

EXAMPLE 1. The variables in the problem are X, the number of aspirins taken (with possible value 0 or 1), and Y, the logarithm of the time in minutes it takes for the headache to go away. (We use the "log-duration" merely to allow Y to take negative values; where necessary we use $Z := e^Y$ to denote that corresponding duration). The values of both X and Y for any headache episode can be observed.

We specify the conditional probability distribution of Y, given $X = x$, as normal, with a mean $m(x)$ that depends on x, and (for simplicity) variance 1 in all cases. This can be described by the *conditional probability*:

(1) $\quad p(y \mid x) = (2\pi)^{-\frac{1}{2}} \exp -\dfrac{1}{2} \{y - m(x)\}^2 .$

(However the specific form of this is of no substantive importance.)

We could interpret a distribution such as (1) as describing variation of the response Y over the population of headache episodes for which $X = x$ (this probability distribution typically being initially unknown); or alternatively, using fully specified probabilities, as expressing epistemic uncertainty about the response Y for a single episode for which $X = x$.

Although for many purposes these different interpretations call for different treatment, much of the following discussion will apply equally to either interpretation.

4.1 Conditional probability

From (1) we can calculate, by standard probability theory, $P(A \mid X = x)$ for any event A defined in terms of Y—for example the event A: "$Z \leq 30$" that the headache goes away within 30 minutes. Here the notation $P(A \mid B)$ represents the *conditional probability of event A, given event B*.[4] However, this expression has two quite distinct (though related) interpretations and corresponding uses, *indicative* and *subjunctive*:

Indicative To describe *actual* uncertainty in the light of available information. Thus if we know or learn that the variable X takes value x, the expression $P(A \mid X = x)$ describes the uncertainty for a (variable) event A in the light of the *known* information $X = x$.

Subjunctive To describe the uncertainty that *would* result from acquiring, or positing, information.

---

[4] We take the standard interpretation of conditioning, where the elaboration of $P(A)$ by means of the notational addition "$\mid B$" is understood as qualifying the probability P, not the event A. This was perhaps less ambiguously expressed by the older notation $P_B(A)$. There are however alternative logical approaches (see e.g., Milne (1997) and references therein) that attempt to interpret $P(A \mid B)$ as the *unconditional probability* of the *conditional event $A \mid B$*. But it turns out that we can not build a consistent theory of conditional events with ordinary 2-valued logic: "conditionals" can not be treated as being either TRUE or FALSE in the same say as regular propositions. The simplest viable assignment of truth value to the conditional $B \mid A$ has it TRUE if A and B are both TRUE, FALSE if A is TRUE and B FALSE, and VOID if A is false.

To pursue this approach we then need to determine appropriate rules to define "negation", "conjunction", etc. in this 3-valued logic, as well as interpreting probability suitably. However no such non-standard logic is required for our approach, involving conditional probabilities of (ordinary) events.

Before knowing X, we can contemplate various possible values x that X might take. For (now fixed) event A and (variable) hypothesised value x, $P(A \mid X = x)$ may be interpreted as describing the relevant uncertainty about A we should expect to hold in the light of a *hypothesised* value x for X.

Decision theory

We shall want to be able to interpret conditional probability in a slightly more general way than is envisaged in standard formal probability theory, which requires the conditioning variable X to be modelled as random. We might wish to allow X to be a "decision variable", under the control of an external agent. Or we might want to regard X as a parameter, labelling possible distributions we are considering (for example, in different geographical settings). The intuitive interpretation (be it indicative or subjunctive) of conditional probability in such cases is no different from the case of random X. It might be argued that these different variations require distinct formal treatments, but I am not aware of any case for this ever having been made. We shall assume here that the same notion of conditional probability is applicable to all these variations, and that standard formal manipulations are valid whenever they are defined (for example, when X is non-random we must avoid any mention or use of its probability distribution). In particular, and vital for our applications below, it is possible to extend the notions of independence and conditional independence consistently to cases where some of the variables involved are non-random (Dawid, 1979, 1980, 2002). This extended conception of conditional probability lies at the basis of statistical decision theory (Raiffa, 1968) where it has been fruitfully applied for decades. We shall therefore term this extended probabilistic framework the *decision-theoretic approach*, or DT.[5]

Hypotheticals

If I am currently considering whether or not to take an aspirin, I can interpret the conditional densities for Y, $p(y \mid X = 1)$ and $p(y \mid X = 0)$, in hypothetical subjunctive mood, as answers to the question: What is likely to happen to my headache if I take one (or no) aspirin? On the other hand, if I have just taken an aspirin, I could now interpret $p(y \mid X = 1)$ in indicative mood: What is likely to happen to my headache, now that I have just taken an aspirin? (At this point $p(y \mid X = 0)$ is of no further interest—except perhaps for the historian, as a record of my earlier hypothetical pondering,

---

[5]This is to be distinguished from the decision-theoretic framework for causality introduced by Heckerman and Shachter (1995), which has more in common with the PR framework introduced below.

before I decided what to do, about what might happen were I not to take the aspirin.)

DT is perfectly well suited to handle such hypothetical conditional reasoning, which only requires simple probabilistic conditioning—and consequently DT is perfectly well suited to handle causal queries of the EoCtype. We expand on this point in the context of some particular EoC problems in §§ 10 and 11 below.

Counterfactuals

But what about the other subjunctive form, counterfactualreasoning, as is apparently required for addressing causal queries of the CoE type? Here DT appears to face a problem.

Suppose that, after having taken an aspirin, I fall to speculating about the consequences I might have experienced had I in fact taken none. Since, within DT, both indicative and subjunctive conditioning are effected by the same formal conditioning rule, this would require conditioning my initial uncertainty both (indicatively on $X = 1$ and (subjunctively) and counterfactually) on $X = 0$. But the conjunction of these two conditions is the impossible event $\emptyset$—and conditioning on $\emptyset$ is not meaningful within DT. So if we are to express counterfactual uncertainty, it appears we shall need a different formal language.

5 Potential response

Within philosophy there have been various attempts to express counterfactual reasoning formally, either by constructing special-purpose modal logics or through e.g., "possible worlds" interpretations (Stalnaker, 1968; Lewis, 1973). However statisticians have approached the problem from a somewhat different standpoint.

Within statistics and cognate areas such as econometrics and epidemiology, a formal methodology known as *potential response modelling* (henceforth PR) has become the cornerstone of the enterprise of "statistical causality" ever since the pioneering work of Rubin (Rubin, 1974, 1978), itself foreshadowed by Neyman (1935). It has been successfully applied to a very wide range of problems: a particularly fruitful sub-enterprise, building on Robins (1986), uses PR to tackle dynamic situations, where, as time passes, a sequence of interventions can be made in response to continually accruing information.

The PR approach has now become the "industry standard" for statistical inference about causal effects. The following quotes illustrate the near total acceptance that this PR conception of causal inference (often loosely and somewhat misleadingly described as "counterfactual modelling") has achieved within these communities:

> "Today, the counterfactual, or potential outcome, model of causality has become more or less standard in epidemiology"
> (Höfler, 2005a)

> "In the past two decades, statisticians and econometricians have adopted a common conceptual framework for thinking about the estimation of causal effects—the counterfactual account of causality"
> (Winship and Morgan, 1999)

5.1 The PR framework

The novel special feature of the PR approach is that it represents a response ("effect") variable Y by two or more random variables, one for each of the possible values of the "cause variable" X under consideration. In general we can consider a variety of values that X might take, but for concreteness we will think here in terms of Example 1, with only the two possible values 0 and 1 for X.

In that example, instead of the single response variable Y, "the log-duration of my headache", the PR approach would introduce two *potential response* variables, Y_0 and Y_1, with Y_x interpreted as "the log-duration of my headache if I take x aspirins". (The corresponding potential durations are $Z_0 = e^{Y_0}$ and $Z_1 = e^{Y_1}$).

Both versions Y_0 and Y_1 of Y are conceived as existing simultaneously.[6] In particular (as is also implicit in Lewis (1973)), it is assumed[7] that "potential events", such as "$Z_0 > Z_1$", have determinate truth-values, obeying ordinary 2-valued logic,[8] and thus that "potential variables", such as Y_1/Y_0, have determinate numerical values.

There is then, formally at any rate, no impediment to assigning a probability distribution to represent joint uncertainty over all potential variables and events.

We might picture the two potential responses Y_0 and Y_1 as engraved on either side of a "diptych", i.e., a conjoined pair of "tablets of stone"

---

[6]This even though, in the light of their very definition, there is no world, actual or conceivable, in which both variables could be observed together. Their simultaneous existence must therefore be confined to some "Platonic heaven" of ideal forms, not fully accessible to real-world observation.

[7]To make such an assumption is easy; to explain how the truth-value is to be assigned or determined is quite another matter. This metaphysical issue is the focus of much philosophical debate about counterfactuals.

[8]This approach may be contrasted with the treatment of conditional events described in footnote 4.

inhabiting Platonic heaven. We further picture each side of the diptych as covered by its own "divine curtain". When, in this world, X takes value x, just one of these curtains—that associated with the actual value x of X—is lifted, so uncovering, and thereby rendering "actual" and measurable in this world, the associated potential response Y_x. However, because we can only apply one treatment on any occasion, we are never allowed to lift more than one curtain of the diptych.

There will be one diptych for every *unit*, or individual instance of our general story: in our example the unit is a headache episode. We will initially be uncertain about both the engraved values, on every diptych. Again we could use probability to express unit-to-unit variation, over the population of diptychs, of the pair of engraved values (such a probability distribution itself typically being initially unknown); or, using fully specified probabilities, to express epistemic uncertainty about the values engraved on a single diptych under consideration.

In either case, and in contrast to the univariate distributions assumed for Y of Example 1, we would now need a *bivariate* probability distribution, to describe the joint uncertainty about the pair of values (Y_0, Y_1). For consistency with the assumptions of Example 1, the marginal distribution of Y_x should be given by the density $p(y \mid x)$ of (1) (in which X is to be interpreted as a decision variable); but those assumptions impose no constraint on the dependence between Y_0 and Y_1. In particular, we could assign a bivariate normal distribution to (Y_0, Y_1), with the above margins, and a correlation coefficient ρ that is simply not determined by the description in Example 1. Some worrying implications of this indeterminacy were exposed in Dawid (2000).

The PR framework supplies a formal language in which it is possible to formulate counterfactual speculations: as soon as I have assigned a full joint distribution to all the variables in Platonic heaven, I can calculate counterfactual probabilities. Thus suppose I took an aspirin ($X = 1$) half an hour ago and my headache has just cleared up. The probability (in the light of what is known) that taking the aspirin *caused* my headache to go within 30 minutes can be expressed by the counterfactual conditional probability $P(Z_0 > 30 \mid X = 1, Z_1 = 30)$. The indicative condition ($X = 1, Z_1 = 30$) is introduced by straightforward probabilistic conditioning; while the subjunctive attention to the counterfactual state of affairs $X = 0$ is now expressed by focusing on the unobserved potential response Z_0. When we use the PR language, there is no formal impediment to making both these moves at once.

6 Variations

6.1 Structural equation

Another common approach to expressing and manipulating causal relationships—particularly popular in econometrics—involves representing response ("endogenous") variables as functions of other (both endogenous and "exogenous") variables, as well as of external "error" variables. Systems of this sort may consist of hundreds of such relationships, with associated assumptions about the joint distribution of the error terms. However the main philosophical issues are displayed by the simplest such system, consisting of just one equation, involving one exogenous variable X, one endogenous variable Y, and one error variable E. In particular, we consider the following 1-equation system:

(2) $\quad Y = m(X) + E$

for some (known or unknown) function m of X. We also need to make distributional assumptions about the errors in the problem. Here we simply assume a standard normal distribution for E:

(3) $\quad p(e) \sim (2\pi)^{-\frac{1}{2}} \exp -\frac{1}{2} e^2$

(again the specific form assumed is of no real consequence). It is also assumed that E is independent of X.[9]

How are we to interpret (2) (with (3))? One easy deduction is that, when it holds, the conditional distribution of Y given $X = x$ is normal with mean $m(x)$ and variance 1. That is, the distributional properties of (1) are valid. In particular, the structural model (2) allows us to formulate all the *hypothetical* queries that can be addressed by means of (1).

It is very common in statistical practice (so much so that it typically goes entirely unremarked) to write down an equation such as (2) when what is really intended is its distributional consequences such as (1)—as if it were simply an alternative way of saying the same thing. But to do this is to ignore the additional algebraic structure of (2), whereby Y is represented as a deterministic mathematical function of the two other variables X and E. Unlike the distributional formulation of (1), in (2) all the uncertainty is compressed into the single variable E, via (3). If we take (2) and its ingredients seriously, we can get more out of it.

In particular, it is implicit in (2) that the values of E and X are assigned separately, that of Y then being determined by the equation. Given that

---

[9] In more complex applications it is often unclear just what is being assumed about the dependence between the errors and other variables in the problem, which can be a point of confusion.

E takes value e, Y will take value $m(x) + e$ if we set X to x. Thus we can define potential response variables $Y_0 := m(0) + E$, $Y_1 := m(1) + E$. Being two well-defined functions of the same variable E, they do indeed have simultaneous existence—indeed, they are closely related, since it is known ahead of time that the difference $Y_1 - Y_0$ will take the non-random value $m(1) - m(0)$.[10] In particular the correlation between Y_0 and Y_1 is 1.

Since we thus do have potential responses available in this structural modelling framework, we can formulate and address counterfactual subjunctive queries in it.

An extension

An extension of the structural model (2) is given by:

(4) $\quad Y = m(X) + E_X$

where now we have two error variables, E_0 and E_1, with some bivariate distribution—again assumed independent of the value of X. In particular, when $X = 0$ we have $Y = m(0) + E_0$, with E_0 still having its initially assigned distribution; and similarly $Y = m(1) + E_1$ when $X = 1$. If the marginal distribution of each E_x is standard normal, with density as in (3), then (no matter what the correlation ρ between E_0 and E_1 may be) the same distribution model (1) for Y given X will be obtained. But again we can go further and define potential responses $Y_x := m(x) + E_x$, so allowing counterfactual analysis.

In this case the correlation between Y_0 and Y_1 will be ρ, and correspondingly their difference $Y_1 - Y_0 = m(1) - m(0) + E_1 - E_0$ will be a non-degenerate random variable. This quantity will be independent of the applied treatment $X = x$, but not, in general, of the actual outcome Y_x.

6.2 Functional models

Mathematically, all the models introduced in §5 and §6 have the following common functional form:

(5) $\quad Y = f(X, U)$

where X represents a cause, Y an effect, and U some further extraneous random variable; f is a (known or unknown) deterministic function. In particular, when X is regarded as a decision or parameter variable, uncertainty enters only through the distribution of U.

In model (4), we can take $U = (E_0, E_1)$ and $f(x, (e_0, e_1)) = m(x) + e_x$; the structural model of (2) is the degenerate case of this having $U = E$ and $f(x, e) = m(x) + e$.

---

[10]This property is termed "treatment-unit additivity" in the statistical literature.

In the case of a potential response model, we can formally take U to be the pair (Y_0, Y_1), and the function f to be given by:

(6) $\quad f(x, (y_0, y_1)) = y_x.$

Conversely, any functional model of the general form (5) is equivalent to a PR model, if we define $Y_0 = f(0, U)$, $Y_1 = f(1, U)$. (Any variation in U which is not reflected in variation in the pair (Y_0, Y_1) is entirely irrelevant to the relationship between X and Y.)

We thus see that (mathematically if not necessarily in terms of their interpretation) PR models and general functional models need not be distinguished; a structural model is simply a special case of such a model.

6.3 "Probabilistic causal models"

Pearl's influential book (Pearl, 2000) develops a methodology for causal modelling and inference largely based on representations by means of Directed Acyclic Graphs (DAGs). More than one interpretation of such a DAG model is either explicit or implicit within its pages, but that which is most favoured can be regarded as an extension of structural equation modelling.[11] We postulate a DAG over some set of variables (X_i), and for each i an associated "error variable" U_i. Each X_i is represented by means of a specified functional relationship:

(7) $\quad X_i = f(\text{pa}(X_i), U_i),$

where $\text{pa}(X_i)$ denotes the set of "parents" of X_i in the DAG. Pearl (2000), Defn. 7.1.1, essentially restricts the term "causal model" to a system of this kind. Uncertainty is introduced by assigning a joint distribution to the (U_i), when (7) becomes, for Pearl, a "probabilistic causal model"—a terminology that does not, however, clearly signify the specific feature of functional dependence that lies at the heart of this framework.

Such models can be used hypothetically: Pearl presents a semantics whereby we can read off the probabilistic consequences, for other variables, either of observing $X_i = x$ or of intervening to set X_i to x (these being different in general).

However we can get thus far without requiring functional relationships, as is evidenced by much of the development in the first 6 chapters of Pearl (2000), which only requires specification of a conditional probability distribution for each X_i given its parents.

However when we do posit functional relationships of the form (7), such models can also be used to effect counterfactual reasoning, where, in the

---

[11] Pearl presents this construction in a somewhat more general context than DAG models, but these will suffice for present purposes.

light of results obtained in this world, we consider what the result might have been in an alternative, unrealised world, differing in terms of what interventions are made to the system. To do this we have first to assume that the collection $\mathbf{U} = (U_i)$ of all error variables takes the same values in all the possible worlds, and that, for each i, the functional relationship (7) holds in all those worlds, except where overridden by an intervention to set the value of X_i. The value of any variable in any possible world is then determined as a function of $\mathbf{U}$: the solution for Y, in the world determined by setting $X = x$ (where each of X and Y could be a collection of variable), defines the potential response variable Y_x, taking value $Y_x(\mathbf{u})$ if $\mathbf{U} = \mathbf{u}$. The joint distribution of the (U_i) now determines a joint probabilistic model across all the actual variables and all their other-worldly counterparts, thus allowing us to evaluate (conditional on evidence about observed variables in this world) counterfactual queries about what would have happened in other worlds, defined by interventions different from those actually made. Functional DAG models thereby support inferences about causes of, or explanations for, observed effects. This is the emphasis of the later parts of Pearl (2000).

7 Causes of effects

The apparent advantage of the PR approach (and its variations as above) over the DT approach is that it alone has the expressive power to formulate counterfactual conditional queries, and thus CoE-type causal queries. However, this greater expressive power does not, of itself, mean that the formulation and analysis of such queries is straightforward. In Dawid (2000) I expressed serious disquiet over existing approaches to addressing CoE. My point was this. Any model (e.g., a PR model, or a Pearlian causal DAG) built for this purpose will of necessity have to say something about "other possible worlds" and their relationship to the one we inhabit. Philosophers have long mused and disputed over the difficulties that beset attempts to imbue such necessarily non-falsifiable counterfactual propositions with meaning (see e.g., Collins et al. (2004)), and I am not aware of any generally accepted way of adjudicating counterfactual disagreements. But in applications to statistical causal reasoning the counterfactual assumptions are often made unthinkingly—e.g., the typically unremarked but crucial assumption that one (usually of many possible) functional DAG representation of the purely associational relationships between a certain set of variables can be interpreted according to Pearl's causal semantics.

Once we have chosen such a model, applying it to extract inferences about CoE may be relatively simple, which may well induce the mistaken impression that our problem is easily solved. However, there will always be a choice of models, entirely indistinguishable in this world, but saying different things about other worlds—and so reaching different conclusions about CoE, even though we will never be able to tell them apart. This puts a heavy burden on the modeller, who in order to be able to sell his causal conclusions needs to be able to justify, in necessarily non-empirical terms, why he chose this model rather than another, observationally equivalent, one. While this may not be an utterly hopeless task—some relevant considerations are discussed in e.g., Lewis (1973); Pollock (1976); Halpern and Pearl (2005), as well as in Dawid (2000)—it does pose a tricky challenge, which should not be ignored (although it usually is).

8 Effects of causes

Since at this point I have no new proposals for addressing the above issues, I shall say no more here about CoE. But this does not mean I have nothing further to say about potential response modelling, since it is the fact that far and away its biggest sphere of application of PR has been in addressing EoC queries—even though these do not require counterfactual reasoning, and, as we shall see, there is no difficulty in applying DT to such cases.

For example, a major concern of EoC inference is to account for "confounding": the possibility that the way the world appears to behave when we merely observe it might be a poor guide to how it would behave under intervention. The following quotes typify the strength of adhesion to the PR approach for such purposes (where again the context will show that the authors are loosely using the term "counterfactual" when they in fact mean potential response):

> "How is it possible to draw a distinction between causal relations and non-causal associations? In order to meet this concern a further element must be added to the definition—a counterfactual" (Parascandola and Weed, 2001)

> "Probabilistic causal inference (of which Dawid is an advocate) in observational studies would inevitably require counterfactuals" (Höfler, 2005b)

There are however one or two dissenting voices, including this author:

> "For making inference about the likely effects of applied causes, counterfactual arguments are unnecessary and potentially misleading" (Dawid, 2000)

> "Counterfactuals are a hot topic in economics today... I shall argue that on the whole this is a mistake" (Cartwright, 2006)

I believe that—notwithstanding the fact that virtually all development of the field to date has been within the PR framework—DT is better suited than PR to address all meaningful problems of EoC. But before expanding on this point, let us see how PR goes about the problem.

9 EoC: Potential response approach

9.1 Causal effect

Within the PR framework, a seemingly natural approach to interpreting "the causal effect of X on Y" is by means of a direct comparison between the *values* of Y_0 and Y_1: for example, their difference $Y_1 - Y_0$, or their ratio Y_1/Y_0. This is also termed the *individual causal effect*, ICE, since it will have a value (generally unknown) for each unit in the population. When we assign a joint distribution (be it over the population, or epistemological) to the pair (Y_0, Y_1), this will induce a distribution for ICE.

However this definition lays us open to what has been termed the "fundamental problem of causal inference" (henceforth FPCI) (Holland, 1986): since we can never lift both curtains of the diptych at once, there is no possible world in which such a causal effect could be measured—its value could be available only to an omniscient god. For us mere mortals, at least one of the component terms Y_0 and Y_1 has the status of "missing data": information that we would like to have access to, but don't. It is perhaps no accident that the PR approach to causal inference was introduced and developed by Rubin shortly after he had developed important analyses of the general statistical problem of making inferences in the face of missing data (Rubin, 1976), and that the PR approach to causal inference treats it as a special application of this technology (van der Laan and Robins, 2003).

9.2 Experiment

Suppose we have conducted a randomised clinical trial, whereby the selection of the treatment X is effected by tossing a coin, after which the response is observed. In the PR interpretation, this means that we will observe Y_0 if in fact $X = 0$ and Y_1 if $X = 1$. We are interested in what can be said about the distribution of the ICE, defined here as $Y_1 - Y_0$.

The "randomness" of the coin toss might be regarded as justifying an assumption that it is entirely independent of what might be engraved on the diptych. Mathematically this can be stated (expressed using the conditional independence notation of Dawid (1979)) as:

(8) $X \perp\!\!\!\perp (Y_0, Y_1)$,

which reads: "Treatment X is independent of the pair (Y_0, Y_1) of potential responses." Now from the data where $X = 0$ we can estimate the conditional distribution of Y_0, given $X = 0$; and under condition (9) this will be the same as the marginal distribution of Y_0. Similarly we can estimate the marginal distribution of Y_1. We can therefore make any inference that depends only on these marginal distributions. In particular, we can estimate both expectations, $E(Y_0)$ and $E(Y_1)$, and hence their difference—which happens to be the same as $E(Y_1 - Y_0)$, i.e., E(ICE), the *average causal effect*, ACE. Hence under assumption (9) we can estimate ACE. This might be regarded as a successful application of PR.

However, making only small variations to the above we find our way barred by FPCI. Thus suppose (as seems entirely unexceptionable, and indeed desirable) we were interested in some other aspect of the distribution of ICE than just its mean: for example, its variance. This can not be determined from the marginal distributions of Y_0 and Y_1 alone, which, even assuming (8), is all we can estimate. Hence we could never obtain enough information from our experimental data—however extensive—to estimate such a parameter.

Or suppose we were to redefine ICE as Y_1/Y_0—seemingly just as meaningful a comparison as $Y_1 - Y_0$. Since the expectation of Y_1/Y_0 can not be expressed in terms of the marginal distributions for Y_0 and Y_1, we could not now estimate the associated ACE.

9.3 Observation

Often our data are collected under observational conditions, where we do not ourselves have the option of assigning or randomising the treatments to the units. For example, we might just get to see what treatments were given, and with what outcomes, to a collection of patients in some hospital; and for all we know the doctor may have chosen to give treatment $X = 1$ to the healthier looking, and $X = 0$ to the sicklier, patients. In this case we would expect to see better results under treatment 1 even if the treatment had absolutely no effect: this is the problem of *confounding*.

Again the PR approach involves the three variables (X, Y_0, Y_1), assumed to have a joint distribution. If we conceive of the pair (Y_0, Y_1) of potential responses for any unit as engraved on a diptych, just waiting for one of its curtains to be lifted, we might assume that the values, and hence the joint distribution, of this pair are the same in this observational regime as in the interventional one.

In certain special circumstances, we may still feel able to justify the assumption (8). In this case we can again estimate ACE from the observational data. In fact all that the argument for this in § 9.2 needs is the weaker assumption of independence of X and each potential response, separately:

(9) $\quad X \perp\!\!\!\perp Y_i \quad (i = 0, 1)$

(although it might be hard to justify (9) in the absence of property (8).) When (9) can be assumed, we may say that there is *no confounding* (of the effect of X on Y).

However, this case is very much the exception in observational studies: the possibility of confounding means we would generally have to allow for dependence between the values engraved on the diptych and the actually assigned treatment X. In that case we can no longer safely assume (9), and are thus not justified in interpreting $E(Y_1 \mid X = 1)$—which we can estimate from the patients receiving treatment 1 in the observational study—with the unconditional mean-value $E(Y_1)$, which is the quantity of direct inferential relevance. In particular, without further input or assumptions we can not even estimate ACE.

9.4 (Un)confounders

Suppose that, although we are not willing to assume the no-confounding property (9), there is an additional variable (or collection of variables) C in the problem, jointly distributed with (X, Y_0, Y_1), such that, for $i = 0, 1$, we can reasonably assume:

(10) $X \perp\!\!\!\perp Y_i \mid C$.

This says: "Treatment X is independent of each potential response, after conditioning on C". That is, (9) is valid, and thus there is no confounding, within strata defined by the value of C. For example, C might represent the complex of patient health indicators that the doctor takes into account when selecting which treatment to apply. A quantity C with the property (10) is usually called a (potential) *confounder*[12]—though I prefer to call it an *unconfounder*, since conditioning on it eliminates the problem of confounding. Note that there may be more than one choice of unconfounder in a problem.

If we can observe an unconfounder C then, arguing essentially as in § 9.2, we can estimate $E(\text{ICE} \mid C)$—and thus the expectation of this quantity, which is just ACE. However, in the absence of knowledge of C we may still be subject to confounding bias.

The identification of the appropriate condition (10) to handle the problem of confounding could be considered a success of the PR approach. It is doubtless consideration of such problems as this that prompted the quotations reproduced in § 8.

---

[12] When C is multivariate such a description is often applied to just one component of C, but this property is of little interest in itself.

10 EoC: Decision-theoretic approach

We now describe how the problem of confounding can be addressed within DT, without making any use of potential responses. This is possible (contrary, it would seem, to the beliefs of many working statisticians) because EoC reasoning only requires hypothetical conditioning, which can be effected within DT, rather than counterfactual reasoning, which can not.

10.1 Experiment

Consider first an experimental setting as in §9.2. Since we no longer have the two versions Y_0 and Y_1 of Y, we can not now even define the ICE. Consequently the focus of our interest must be different from that of PR.

What we do have is two conditional distributions for Y, given by their densities $p(y \mid X = 0)$ and $p(y \mid X = 1)$. We can in principle estimate these to any desired degree of accuracy from experimental data. And then we can make any desired comparison between these distributions. This difference in focus can be described as follows: In the PR approach we are concerned with *uncertainty about the comparison* of the outcome values associated with each of the treatments; while in DT we are concerned with *comparison of the uncertainty* about the outcome resulting from each of the treatments. And now there is no FPCI since nothing is missing.

Whichever aim you feel is more appropriate, it is worth emphasising again that all we can actually estimate from data is the pair of conditional distributions: $p(y \mid X = 0)$ (equivalent to $p(y_0 \mid X = 0)$ in the PR formulation) and $p(y \mid X = 1)$ (equivalent to $p(y_1 \mid X = 1)$). So on purely pragmatic grounds the comparison of uncertainties is privileged as the only kind of message we can directly extract from data.

This distributional level of information is in any case sufficient for the pragmatic purpose of deciding on what treatment to give a new patient (assumed similar to those in the experiment). Suppose we measure the value of obtaining outcome y by the utility $u(y)$. Standard statistical decision theory says that we should choose between options on the basis of their expected utilities. If we apply treatment $X = 0$ the expected utility is calculated under the distribution $p(y \mid X = 0)$; similarly for treatment $X = 0$ we perform the expectation calculation using $p(y \mid X = 1)$. The requisite comparison of these two expected utility values therefore comes under the heading of *comparison of the uncertainty*. For the purposes of treatment assignment, nothing is gained by thinking about e.g., ICE, the comparison of the *values* of the potential responses under the two (in fact mutually exclusive) choices.

10.2 Observation

From this viewpoint, the essential problem of confounding is that the conditional distributions of Y given X in some observational regime from which we have data need not be the same as the "interventional" distributions that are relevant for the above treatment choice problem. In order to express this clearly, we introduce a further variable, the *regime indicator* F. This has values $\emptyset$, 0 and 1 to indicate, respectively, that we are in the observational regime; or the interventional regime under treatment 0; or the interventional regime under treatment 1. Thus F has the logical status of a parameter rather than a random variable.

In general there is no reason to expect the desired interventional distribution of the response to treatment x, $p(y \mid X = x, F = x)$,[13] to be the same as the corresponding observational distribution $p(y \mid X = x, F = \emptyset)$: different circumstances and contexts naturally result in different distributions. The happy circumstance that these do indeed coincide can be formally expressed by the conditional independence property:

(11) $Y \perp\!\!\!\perp F \mid X$.

This then is the appropriate representation of "no confounding" in the DT framework: when valid it directly supports the use of data gathered under the observational regime for the purpose of interventional decision-making.

One pragmatic advantage of condition (11) is that it can be checked empirically by gathering data under both regimes. On the contrary (9) can never be checked empirically.

10.3 Unconfounders

In DT, the property that a variable C acts as an unconfounder is expressed by the following pair of conditions:

(12) $Y \perp\!\!\!\perp F \mid (X, C)$

(13) $C \perp\!\!\!\perp F$.

Condition (12) is just a conditional version of (11); this parallels the replacement of (9) by (10) in the PR formulation. However we have also introduced a further condition, (13), requiring that C have the same distribution in both regimes. No such explicit condition was needed for PR, because in that framework there was only one version of C and only one joint distribution for the variables; thus the analogue of (13) holds automatically in PR.

---

[13] The condition $X = x$ is in fact redundant because it is implied by $F = x$.

Under (12) and (13) we have (assuming discrete distributions for simplicity):

(14) $\quad p(y \mid F = x) = \sum_c p(y \mid X = x, C = c, F = x)\, p(C = c \mid F = x)$

(15) $\qquad\qquad = \sum_c p(y \mid X = x, C = c, F = \emptyset)\, p(C = c \mid F = \emptyset)$

where (14) uses the fact that $F = x$ implies $X = x$, and (15) uses (12) and (13) in turn. When C is observed in addition to X and Y, we can estimate the terms entering (15) from observational data. Then (14) allows us to estimate the desired *interventional* distribution of Y when treatment $X = x$ is assigned.

Again it is possible to check the conditions (12) and (13) empirically from data collected under the various regimes. In contrast (10) has no empirically testable content.

An extension

In some circumstances we might be willing to assume property (12), but not property (13): that is, we expect the variable C to have different distributions in the different regimes. Then we can not use (15) as it stands. But we might know, or be able to estimate from another source, the term $p(C = c \mid F = x)$[14] in (14), i.e., the density of C in the target interventional regime. Then we could apply a simple variant of (15), with the final term replaced by this density.

It is not clear (at any rate to this author) how one might best formulate and tackle such an extension using the machinery of PR: probably it would be necessary to introduce multiple "potential versions" (C_x) of C.

11 Other EoC problems

There is a very wide range of purely EoC-type problems that have been tackled using PR, and there is no doubt that this has led to important advances in our treatment of causal inference. However in every case that has so far been re-examined from the DT standpoint the latter has provided a more straightforward route to the answer, at the same time simplifying and rendering more intelligible the conditions needed to justify taking that route. Some examples include:

---

[14] Although the method could be applied even more generally, one would typically expect this term not to depend on the specific value x of the applied treatment—but nevertheless that common interventional density of C could differ from its density $p(C = c \mid F = \emptyset)$ in the observational regime.

- Estimating the effects of dynamic interventions (Dawid and Didelez, 2005)

- Setting bounds on causal effects when patients may not comply with the recommended treatment (Dawid, 2003)

- Definition of direct and indirect effects (Geneletti, 2005; Didelez et al., 2006)

- Defining and estimating "the effect of treatment on the treated" (Geneletti, 2005)

- Instrumental variable estimation (Didelez and Sheehan, 2005)

11.1 Dynamic treatment plans

We illustrate the DT approach for the first problem above: dynamic interventions. A generic patient presents with symptoms L_0, on the basis of which a treatment A_0 is chosen. Later a health outcome indicator L_1 is measured, after which a further treatment A_1 can be chosen and applied. Finally we observe the response variable Y. We have observational data, and wish to know when, and how, we can use this to make inferences about the effects of various (static or dynamic) treatment plans.

PR approac

Robins (Robins, 1986, 1987), working within the PR framework, identified a set of assumptions under which this can be done: the appropriate calculation is then effected by what he termed the *G-computation formula*.

Specifically, consider a *dynamic treatment plan g*, which specifies how to choose the first decision A_0 as a function, $g(L_0)$ say, of the presenting symptom L_0, and likewise the later decision A_1 as a function, $g(L_0, L_1)$ say, of the then available data (L_0, L_1). Corresponding to any such plan g we are required to conceive of potential versions of all the outcome (i.e., non-action) variables: (L_{0g}, L_{1g}, Y_g). As is typical of the PR approach, all such potential variables, for all plans g under consideration, are regarded as having simultaneous existence and a joint probability distribution. We need to relate the potential variables to the actual observed variables (L_0, A_0, L_1, A_1, Y), which is done as follows. Suppose that, in the observational regime, we are about to observe some outcome variable (i.e., L_0, L_1, or Y). Suppose further that, for any *earlier* action variable, its observed value happens to be the same as that prescribed by plan g, applied to the information available at the relevant decision point. Then we assume that that realised value of the upcoming outcome variable will be identical with its potential value under the operation of g.

Mathematically, this is stated as:

$$
(16) \quad \left\{ \begin{array}{l} A_0 = g(L_0) \\ A_0 = g(L_0),\ A_1 = g(L_0, L_1) \end{array} \right. \begin{array}{l} \Rightarrow \\ \Rightarrow \end{array} \begin{array}{l} L_0 = L_{0g} \\ L_1 = L_{1g} \\ Y = Y_g. \end{array}
$$

Having set up this framework, the additional condition allowing estimation of the "causal effect" of plan g by means of the G-computation formula is:

> In the observational regime, each action variable is independent of all future potential observables under plan g, given any history to date that is a possible history under g.

This is expressed mathematically as:

$$
(17) \quad \left\{ \begin{array}{llll} A_0 & \perp\!\!\!\perp & (L_{1g}, Y_g) & |\ L_0 = l_0 \\ A_1 & \perp\!\!\!\perp & Y_g & |\ L_0 = l_0, A_0 = g(l_0), L_1 = l_1, \end{array} \right.
$$

and can be regarded as a generalisation of the "no-confounding" condition (9) to this dynamic setting.

DT approach

The DT approach is quite different. We introduce a non-random regime indicator G, whose values include all the interventional plans under consideration as well as the observational regime $\emptyset$. We do not introduce different *versions* of the variables in the problem. Instead, we consider their varying *distributions* across the regimes. In this setting, an assumption sufficient for application of the G-computation formula is:

> The distribution of each non-action variable, given the past history, is the same in all regimes.

Mathematically:

$$
(18) \quad \left\{ \begin{array}{llll} L_0 & \perp\!\!\!\perp & G & \\ L_1 & \perp\!\!\!\perp & G & |\ L_0, A_0 \\ Y & \perp\!\!\!\perp & G & |\ L_0, A_0, L_1, A_1. \end{array} \right.
$$

Again, this can be regarded as generalising the relevant static "no-confounding" condition, viz (11). Although it is a matter of personal taste, I find the machinery required for PR analysis of this problem both over-complex and mystifying, and conditions such as (17) impossible to think about. By contrast, the DT machinery is simple and intelligible, as are the conditions (18). From a pragmatic point of view, these DT conditions are meaningful (as the PR conditions are not) because they could be tested by collecting suitable data. Another advantage of DT is that there is no difficulty whatsoever in allowing an element of randomisation into a treatment plan, whereas it is difficult to see how to extend the PR approach to allow for this.

12 Philosophical comparisons

We here conduct a general comparative analysis of the PR and DT formal frameworks in the context of inference about "effects of causes". It is important to try to keep separate the various dimensions of such a comparison (although there is inevitably some overlap). We shall conduct comparisons in the following philosophical arenas: ontology, expressiveness, epistemology and pragmatics. For purely interpretive comparisons we make some further distinctions between PR and its mathematically equivalent variations introduced in §6.

12.1 Ontology

Ontology is about what we consider is "really out there". Thus philosophers disagree as to whether *causes* and *probabilities* are properties of the real world, or of our beliefs about the world. Important though such broad questions certainly are[15], here our ontological concerns will be more focused.

PR approach

As we have seen in §5.1, the PR approach conceives of the simultaneous existence of all the potential responses to all the treatments being considered. In cases such as considered in §11.1, such multiplication of values affects certain other variables also.

In PR it is further assumed that the value Y_x of the potential response under treatment x will be the same, no matter which treatment is actually applied, nor which regime (e.g., observational or interventional) is operating.[16]

These assumptions can be regarded as expressing a kind of "fatalism" (Dawid, 2000, §7): whatever we may do, Y_x can not change (metaphorically, the values are engraved on a diptych once and for all, never to be erased or overwritten). A still stronger property that is sometimes implicitly assumed is "determinism", which treats the value of Y_x for any unit as given by a function of certain attributes of that unit. Determinism holds in any functional model of the form (5) whenever the variable U is regarded as representing some property of the real world, existing prior to treatment. In particular, in interpretations of (7) it is common to regard U as such a pre-existing "latent variable" that has real-world existence, and could in principle be measured. The response is then to be regarded as completely determined, without any residual uncertainty, by the value of U together

---

[15] Some of the author's views on these matters can be found in Dawid (2000) (Rejoinder) and Dawid (2004).

[16] Some relaxation of this assumption, allowing for an additional random component at the point of observation, has been considered by e.g., Neyman (1935); Greenland (1987); Robins and Greenland (1989).

with the applied treatment x. Of course typically the value of U is unknown. Indeed typically even the specification and nature of the variable U is left completely open—thus neatly evading the potentially fatal problem that, if we could observe all the variables, we might find that there was no functional relationship between them after all.

As noted in Dawid (2000), this fatalist conception also underlies the approach to the problem of *treatment non-compliance* of Imbens and Rubin (1997). For this we are supposed able to classify patients as "compliers", "defiers", "always-takers" and "never-takers"—these categories being defined in terms of the pair of potential drug-taking responses, on being told to take, or not to take, the drug. The same idea underlies the method of "principal stratification" (Frangakis and Rubin, 2002).

DT approach

By contrast, DT makes do with a single version of each variable in the problem. Where appropriate, it considers a multiplicity of joint distributions for all the variables, to account for different possible regimes in which they are generated or observed. When interpreted as referring to actual variation across a population of units, it is these probability distributions that are being regarded as "really out there". This fully stochastic (rather than deterministic or fatalistic) view of the world is not essentially different from standard interpretations and uses of probability in non-causal areas of statistical inference. In DT we do not need to conceive of the relationships between variables (in any regime) as being functional (deterministic): in particular we can conceive of random processes, affecting the outcome of treatment X, developing after treatment application—a perfectly reasonable conception, but one that does not sit easily in the PR ontology.

As we have seen, in a PR "causal model" there is an assumption of *invariance* of the *value* of the potential response Y_x (across the different treatments and regimes under consideration). Moreover this invariance assumption is built into the framework and can not be evaded. By constrast, for DT a "causal model" involves invariance of (conditional) *distributions* (across regimes). An example of such an invariance assumption is given in § 10.3, where (12) expresses the equality of the conditional distributions of Y given X and C, and (13) the equality of the marginal distributions of C, across the observational and interventional regimes.

But DT never *obliges* us to make any connexions between distributions across different regimes, unless we consider it appropriate. It merely provides a convenient calculus to develop the consequences of such assumptions. For example, in the extension of unconfounding treated in § 10.3 we were able, without any essential difficulty, to drop the invariance assumption (13) that C has the same distribution in both regimes. The PR framework can not readily respond so flexibly.

12.2 Expressiveness

We have seen that in certain important respects PR has more expressive power that DT. Specifically, it allows us to formulate counterfactual queries, and address problems of "causes of effects", that are simply not expressible within DT. It might therefore be considered a "richer" language than that of DT. However, there are other respects in which the languge of DT can be considered richer than PR. Thus, to echo points in § 12.1, DT allows us to make arbitrary distinctions between different regimes, or to model "process uncertainty".

In any case, for purely EoC purposes, for which hypothetical reasoning suffices, the additional counterfactual expressiveness of PR serves no obvious purpose. There is nevertheless a common view (Wasserman, 2000) that, while the ontology of PR need not be taken seriously, its expressive power means we can make valuable *purely instrumental* use of it to solve EoC problems that other methods (such as DT) can not reach. However, while it is undeniable that PR has often been put to such instrumental use to excellent effect, I am still waiting to see an application where it genuinely makes the solution easier. On the contary, in most cases the extra machinery it involves simply gets in the way of a clear route to a solution, while the conditions that have to be invoked are impossible to interpret without acceptance of the ontology, and impossible to verify even with such acceptance. It is a tribute to the insights and intuitions of many workers in PR that they have so often been able to identify useful solutions to EoC problems, notwithstanding the confusing extra clutter PR brings along.

It is also necessary to exercise extreme care when formulating EoC problems using PR. For example, in § 9.2 we saw that the problem of estimating ACE can be meaningfully tackled by PR, in the sense that it provides the solution to a genuine DT problem; but an estimate of var(ICE) (which could certainly be obtained by PR, given suitable—but essentially arbitrary— assumptions) would be meaningless within DT. Wasserman (2000) regards this as a problem of non-identifiability (which it is), and says: "So long as we remain vigilant about non-identifiability, counterfactuals are not dangerous". However, such vigilance can be very demanding.

While there are some formal tools (Galles and Pearl, 1998) that can be applied to reduce certain apparently counterfactual assertions (both deterministic and probabilistic) to assertions about observables, thereby showing them to be empirically meaningful, these have limited application (for example, they are not applicable to var(ICE)); furthermore, failure to perform such a reduction still leaves the question of observability open. Without a foolproof decision criterion, the PR framework gives inadequate guidance as to when the answers it gives are or are not meaningful in DT, and consequently stands in danger of reaching "conclusions" that have no fully empirical justification.

There are some PR analyses that wholeheartedly embrace its freedom from the tyranny of the empirical—for instance, the approach to "treatment non-compliance" mentioned in § 12.1 above. In general there is no purely empirical way of identifying the "principal strata".

The "causal inferences" constructed rely on additional untestable assumptions (for example, that there are no defiers), and, being expressed in terms of within-stratum comparisons of potential responses, are not themselves empirically testable. In the absence of empirical support, the metaphysical underpinnings of this approach would appear to need more attention and justification than they typically get.

On the other hand, some apparently non-identifiable PR concepts, such as "the effect of treatment on the treated" (Heckman and Robb, 1985) do, somewhat surprisingly, turn out to have a sensible DT interpretation after all (Geneletti, 2005).

12.3 Epistemology

Epistemology relates to the question "What can we know?". The ingredients in a DT model are probability distributions over observable quantities in well-specified regimes. So long as we can conduct appropriate experiments or observational studies, these distributions are knowable (as much as anything in statistics is). PR, however, has as an essential ingredient a joint distribution for alternative versions of the same outcome variable under a variety of mutually incompatible conditions. There is no conceivable means by which such a joint distribution could be learned. True, it may be possible to learn certain specific aspects of it, such as the ACE— these are the "identifiable" quantities discussed above. If we can indeed exercise the diligence recommended by Wasserman (2000), we can ignore this epistemological problem. But as indicated above this is a non-trivial programme.

When, as is often the case, non-identifiable aspects of the joint PR distribution enter into our inferential conclusions, particular caution is needed. For then two analysts having formally distinct but observationally entirely indistinguishable PR models can be led to different conclusions from the same data (no matter how extensive or how carefully controlled). It is difficult to know how such a disagreement should be resolved. Fortunately this problem simply can not arise within the DT formulation.

12.4 Pragmatics

Pragmatics is about "fitness for purpose". Of course, that depends on our purpose. As conceived by DT, the purpose of EoC inference is to use past data to assist the future choice of treatment for a new unit. This requires that we understand very clearly the real-world meaning of terms such "observational regime" or "interventional regime", since there are many possible varieties of such regimes. Only when we know exactly what real quantities, interventions and regimes the terms in our model refer to does it make sense to consider the validity of such conditions as (12) and (13). This approach is thus very flexible, allowing full account to be taken of specific pragmatic aspects of the decision-problem we actually face and the kind of data actually available. Likewise concepts such as "no confounding", as expressed by (11), can be applied to much more general types of intervention or regime change than those that simply involve "setting" the value of X.

A PR approach to EoC, on the other hand, can not respond in this flexible way. Moreover, even when a PR model can be considered appropriate, the assumptions required to justify inference are often far removed from any pragmatic considerations, and the technical machinery is over-ponderous even when it can be applied to obtain a meangful answer. This is fitness for purpose in the same way that a sledgehammer is fit for nutcracking.

Agency

We have distinguished between observational and interventional regimes: in Pearl's words, between "seeing" and "doing". This distinction is fundamental, and indeed the "agency" theory of causality is part of mainstream pragmatist philosophy (Price, 2001). However from some philosophical standpoints agency is problematic: do we indeed have freedom to choose our action, or are we too just part of a self-contained overall system, with merely the illusion of free will? Any fully satisfactory analysis of decision-making would have to come to terms with this philosophy of "agency"; but for our current purposes it is adequate to rely on common sense understandings.

Within a DT model agency is explicitly represented by means of an interventional regime. Within PR it is implicit, but present none the less: the potential response Y_x represents what you would get to see if you decided to apply treatment x (and, indeed, in any other circumstance that resulted in this application).

However, DT can also be used fruitfully in situations where agency is not an issue. For example, we might have a diagnostic kit that gives a reading, Y, that is positive (+) in 95% of cases when a patient has disease D, and in 10% of cases when the patient does not. This could be applied in different countries or contexts, with varying disease incidence; then the "diagnostic probability" $\Pr(D \mid +)$ would also vary. Within DT, we could express the stability of the behaviour of the kit by the the condition $Y \perp\!\!\!\perp F \mid D$, where the regime indicator F now labels the various different countries. Whether such stability is or is not regarded as a "causal" concept is not really an issue: we have the technology to tackle such pragmatic problems.

Within PR we could again try to represent this stability by splitting Y into its two potential versions, keeping just one version of D. But since the distribution of D is then fixed, this ploy fails.

13 Conclusion

I have attempted to bring a general philosophical attitude to bear on the important issue of how statisticians should go about conducting causal inference. This is not a purely theoretical issue but can truly be a matter of life and death, for example in the manifold medical and legal problems where the machinery of statistical causality is employed. The current "state of the art" technology of potential response modelling has been examined and found wanting on a number of philosophical and pragmatic dimensions. I have argued that, for the many applications that focus on "effects of causes", a more straightforward application of probability and decision theory is simultaneously more satisfactory with regard to these philosophical criteria, and more useful, interpretable, tractable and reliable for methodological and practical application.

Acknowledgments

I am indebted to Sander Greenland, Judea Pearl and Donald Rubin, as well as two anonymous referees, for helpful comments on an earlier draft of this work.

A. Philip Dawid
Department of Statistical Science, University College London, UK.
dawid@stats.ucl.ac.uk

BIBLIOGRAPHY

Anderson, A. R. (1951). A note on subjunctive and counterfactual conditionals. *Analysis*, 12.

Cartwright, N. (2006). Counterfactuals in economics: A commentary. In O'Rourke, M., Campbell, J. K., and Silverstein, H., editors, *Explanation and Causation: Topics in Contemporary Philosophy*, volume 4. MIT Press, Cambridge, Massachusetts. To appear.

Chisholm, R. M. (1955). Law statements and counterfactual inferences. *Analysis*, 15:97–105.

Collins, J., Hall, N., and Paul, L. A., editors (2004). *Causation and Counterfactuals*. MIT Press, Cambridge, Massachusetts.
Dawid, A. P. (1979). Conditional independence in statistical theory (with Discussion). *Journal of the Royal Statistical Society, Series B*, 41:1–31.
Dawid, A. P. (1980). Conditional independence for statistical operations. *Annals of Statistics*, 8:598–617.
Dawid, A. P. (2000). Causal inference without counterfactuals (with Discussion). *Journal of the American Statistical Association*, 95:407–448.
Dawid, A. P. (2002). Influence diagrams for causal modelling and inference. *International Statistical Review*, 70:161–189. Corrigenda, *ibid.*, 437.
Dawid, A. P. (2003). Causal inference using influence diagrams: The problem of partial compliance (with Discussion). In Green, P. J., Hjort, N. L., and Richardson, S., editors, *Highly Structured Stochastic Systems*, pages 45–81. Oxford University Press.
Dawid, A. P. (2004). Probability, causality and the empirical world: A Bayes–de Finetti–Popper–Borel synthesis. *Statistical Science*, 19:44–57.
Dawid, A. P. and Didelez, V. (2005). Identifying the consequences of dynamic treatment strategies. Research Report 262, Department of Statistical Science, University College London.
Didelez, V., Dawid, A. P., and Geneletti, S. G. (2006). Direct and indirect effects of sequential treatments. UAI (to appear).
Didelez, V. and Sheehan, N. A. (2005). Mendelian randomisation and instrumental variables: What can and what can't be done. Research Report 05-02, Department of Health Sciences, University of Leicester.
Frangakis, C. E. and Rubin, D. B. (2002). Principal stratification in causal inference. *Biometrics*, 58:21–29.
Galles, D. and Pearl, J. (1998). An axiomatic characterization of causal counterfactuals. *Foundation of Science*, 3:151–182.
Geneletti, S. G. (2005). *Aspects of Causal Inference in a Non-Counterfactual Framework*. PhD thesis, Department of Statistical Science, University College London.
Glymour, C. (1986). Statistics and causal inference: Comment: Statistics and metaphysics. *Journal of the American Statistical Association*, 81:964–966.
Greenland, S. (1987). Interpretation and choice of effect measures in epidemiologic analysis. *American Journal of Epidemiology*, 125:761–768.
Halpern, J. Y. and Pearl, J. (2005). Causes and explanations: A structural-model approach. Part I: Causes. *British Journal for the Philosophy of Science*, 56:833–887.
Heckerman, D. and Shachter, R. (1995). Decision-theoretic foundations for causal reasoning. *Journal of Artificial Intelligence Research*, 3:405–430.
Heckman, J. and Robb, R. (1985). Alternative methods for estimating the impact of interventions. In Heckman, J. and Singer, B., editors, *Longitudinal Analysis of Labor Market Data*, pages 156–245. New York: Cambridge University Press.
Höfler, M. (2005a). The Bradford Hill considerations on causality: A counterfactual perspective. *Emerging Themes in Epidemiology*, 2(11).
Höfler, M. (2005b). Causal inference based on counterfactuals. *BMC Medical Research Methodology*, 5(28).
Holland, P. W. (1986). Statistics and causal inference (with Discussion). *Journal of the American Statistical Association*, 81:945–970.
Imbens, G. W. and Rubin, D. B. (1997). Bayesian inference for causal effects in randomized experiments with noncompliance. *Annals of Statistics*, 25:305–327.
Korb, K. (2004). Private communication.
Lewis, D. K. (1973). *Counterfactuals*. Blackwell, Oxford.
Milne, P. (1997). Bruno de Finetti and the logic of conditional events. *British Journal for the Philosophy of Sciencer*, 48:195–232.
Neyman, J. (1935). Statistical problems in agricultural experimentation (with Discussion). *Journal of the Royal Statistical Society, Supplement*, 2:107–180.

Parascandola, M. and Weed, D. L. (2001). Causation in epidemiology. *J. Epidemiol. Community Health*, 55:905–912.

Pearl, J. (2000). *Causality*. Cambridge University Press.

Pollock, J. L. (1976). *Subjunctive Reasoning*. D. Reidel Publishing Company, Dordrecht.

Price, H. (2001). Causation in the special sciences: The case for pragmatism. In Costantini, D., Galavotti, M. C., and Suppes, P., editors, *Stochastic Causality*, pages 103–120. CSLI Publications.

Raiffa, H. (1968). *Decision Analysis*. Addison-Wesley, Reading, Massachusetts.

Robins, J. M. (1986). A new approach to causal inference in mortality studies with sustained exposure periods—application to control of the healthy worker survivor effect. *Mathematical Modelling*, 7:1393–1512.

Robins, J. M. (1987). Addendum to "A new approach to causal inference in mortality studies with sustained exposure periods—application to control of the healthy worker survivor effect". *Comp. Math. Appl.*, 14:923–945.

Robins, J. M. and Greenland, S. (1989). The probability of causation under a stochastic model for individual risk. *Biometrics*, 45:1125–1138.

Rubin, D. B. (1974). Estimating causal effects of treatments in randomized and nonrandomized studies. *Journal of Educational Psychology*, 66:688–701.

Rubin, D. B. (1976). Inference and missing data. *Biometrika*, 63:581–590.

Rubin, D. B. (1978). Bayesian inference for causal effects: the role of randomization. *Annals of Statistics*, 6:34–68.

Spirtes, P., Glymour, C., and Scheines, R. (1993). *Causation, Prediction and Search*. Springer-Verlag, New York.

Stalnaker, R. C. (1968). A theory of conditionals. In Rescher, N., editor, *Studies in Logical Theory*, volume 2 of *American Philosophical Quarterly Monograph Series*, pages 98–112. Blackwell, Oxford.

Suppes, P. (1970). *A Probabilistic Theory of Causality*, volume 24 of *Acta Philosophica Fennica*. North-Holland, Amsterdam.

van der Laan, M. J. and Robins, J. M. (2003). *Unified Methods for Censored Longitudinal Data and Causality*. Springer-Verlag, New York.

Wasserman, L. A. (2000). Comment. *Journal of the American Statistical Association*, 95:442–443. In discussion of "Causal Inference Without Counterfactuals (with Discussion)" by Alexander Philip Dawid, 2000.

Winship, C. and Morgan, S. L. (1999). The estimation of causal effects from observational data. *Annual Reviews of Sociology*, 25:659–706.

INDEX

Absoluteness assumption, 469, 477, 479–481, 485
Abstraction, 428
ACE, *see* causal effect
Acyclicity, 489
Adequate causation, 415, 428, 431
Adjusting for confounding, 271, 272
Agency, 529
Algorithm
    classification, 55
    evolutionary, 47
    greedy search, 56
    PC, 124
Ancestral graph, 22
Anderson, A. R., 505
Antecedent conditions, 421, 435
Artefact-type probability (ART probability), 335–360
Arts and Humanities Citation Index, 1
Assignment operator, 132
Astronomy, 248, 249
Atoms, 421, 422
Average causal effect, 267, 275
Average effect theory, 202
Avogadro number, 449

Bach, J. S., 250
Background condition, 380, 384, 409
Background context, 475, 480
Background knowledge, 260, 261, 480, 483
Bateson, W., 244–246
Bayesian networks, 487–501

algorithm, 489
all-embracive, 467, 477–481, 485
assumptions of, 489
Belis, M., 8
Bell's inequalities, 76, 79, 80, 91, 96
Benzi, M., 11
Bernoulli, J., 250
Bias, 243, 260, 261
Biological factors, 219
Biology, 243–245, 248, 249, 260
Biomedical sciences, 211, 212
Biometry, 244, 245, 249
Bohm, D., 66, 67, 70, 104
Bopp, F., 446, 456
Bulmer, M., 255, 256
Butterfield, J., 91

c-function, 226
Calculi, 229
Calibration, 116, 117
Cancer epidemiology, 7
Capacity, 8, 300
Cartwright, N., 89, 94, 95, 146, 155, 447, 471, 516
Cartwright, N. and Reiss, J., 155
Case control, 282, 283
Causal background context, 471–473
Causal conditions, 418
Causal dualism, 363, 366, 436
    interpretation of, 364, 366–368, 377
Causal effect, 505, 517, 524

average (ACE), 518, 519, 527, 528
    direct, 523
    indirect, 523
    individual (ICE), 517–520, 527
Causal explanation, 396–397, 406
Causal field, 422
Causal inference, 22, 269, 287, 487–501
Causal interaction, 205, 206
Causal Markov Condition, 2, 4–7, 11, 474, 475, 489, 495
Causal model, 467–486
    probabilistic, 514
Causal net, 2, 474, 475
Causal ordering, 136, 166–168
Causal parameter, 269, 282, 283, 288
Causal pluralism, 320, 363, 368, 377, 379–385, 405, 407
Causal process, 210, 300
Causal relevance, 470–473, 476
Causal structure, 243
Causal sufficiency, 474
Causality, 110, 243, 319
    agency theory of, 6, 87, 100–103
    agency-related models of, 418
    backwards in time, 85, 87, 100, 101, 103
    by omission, 328
    causal-mechanical model of, 6, 435
    chance-lowering, 331
    concepts of, 247
    conceptual analysis of, 326
    conditional model of, 415, 419, 423, 427, 430, 435
    conditional view of, 425
    conserved quantity theory of, 8, 319
    contrastive, 505
    counterfactual account of, 6, 364, 416
    counterfactual-interventionist approach of, 416, 435
    dependence view of, 363, 366, 369, 375
    deterministic, 419
    empirical analysis of, 326
    epistemic theory of, 6
    graphical analysis of, 132
    independence theory (C), 83–85, 97, 99
    interventionist account of, 364, 373, 416
    intuitive notion of, 323
    metaphysics of, 368
    negative, 365, 367, 372
    physical notion of, 324
    positive statistical relevance, 332
    probabilistic, 6, 200, 319, 364, 425, 467, 468, 471–472, 503
    process theory of, 364, 371
    production view of, 363, 366, 369, 371
    realist understanding of, 370, 371
    singular, 331
    statistical, 503, 509, 530
    stochastic, 419
    token, 503
    type, 503
    unified model of, 436
Causally interpreted Bayesian nets, 2
Causation*, 320
    counterfactual theory of, 328
Causes of effects (CoE), 503, 506, 509, 515, 527

Chance, 306, 335–360, 478
    causally grounded, 340–342, 355–359
Chisholm, R. M., 505
Classical physics, 435
Closure, 428, 431
Cluster, 420
Cluster of conditions, 420
CoE, *see* causes of effects
Coefficient, 135
Coherence, 227
Collins, J., 515
Common cause, 439
    token, 452, 456, 458
    type, 456
    type vs. token, 448–450
Comparative experiment, 506, 517
Comparative linguistics, 463
Complete partial ancestral graph, 24
Conceptual analysis, 367, 376
Condition
    ceteris paribus, 421, 434
    consequent, 435
    necessary, 420, 430
    necessary and sufficient, 427, 428
    sine qua non, 394, 395
    sufficient, 420, 430
Conditional, 503–506
    counterfactual, 505, 506, 515
    hypothetical, 505, 506, 509, 520
Conditional event, 507, 510
Conditional independence, 267, 281
Conditional probability, 323
Confirmation
    degree of, 226
Confounder, 49
    potential, 519
    unconfounder, 519, 521, 526

Confounding, 263, 264, 270, 516, 518, 520, 521
    no confounding, 519, 521, 524, 529
Congressional voting data set, 55
Consistency test, 475
Constant, 133
Constraint, 429
    cytological, 247
    in SEM-modelling, 261
    practical, 432, 434
    statistical, 244, 248, 249, 251
    structural, 432
Context of inquiry, 471, 476, 480, 485
Context of occurrence, 471
Context unanimity, 203, 472
Contract law, 387, 389, 401, 409
Correa, E., 3
Correlation, 179, 183, 186, 191–194, 220
    auto, 189, 191
    nonsense, 179, 180, 190, 191
    population, 182
    sample, 182, 184, 186, 192–194
Correns, C., 245
Counterfactual, 268, 271, 284, 287, 416–418, 434, 503, 505, 506, 509–511, 513, 514, 516, 527
    dependence, 382, 385, 392–396, 400
    situations, 429
Counterfatual, 279
Covariance structure models, 217
Cowles Commission, 5, 131
Cowles, A., 5
Credibility, 453
Criminal law, 381, 386, 387, 389, 390, 394, 396, 401, 403,

409

d-separation, 246–248, 253, 259, 276
DAG, *see* directed acyclic graph
Darwin, C., 253, 456, 459
Data mining, 43
Datalism, 525
Dawid, A. P., 12, 503, 504, 508, 511, 515–517, 523, 525, 526
de Beauregard, C., 85
de Vries, H., 245
Decay example, 332
Decision theory (DT), 12, 503, 508, 509, 517, 520, 521, 524, 526, 528–530
Decision variable, 267, 287
Deduction, 492
Dependence, 243, 260
    probabilistic, 183
    conditional, 243, 260
    probabilistic, 180–182, 184
Design argument for the existence of God, 443
Determinism, 525
Deterministic phenomena, 305
Dickson, M., 104
Didelez, V., 7, 523
Dilution theory, 255, 256
Diptych, 510, 511, 517, 518, 525
Directed acyclic graph, 2, 121, 125, 126, 275, 284, 288, 473, 514
    causal, 504, 515
Disposition, 337, 339, 341, 342
    probabilistic, 337, 342
    sure-fire, 337, 343, 349
Divine curtain, 511, 517
Divine intervention, 5
Double prevention, 365
Dowe, P., 8, 10, 319, 415

Dretske, F., 453
Drouet, I., 11
DT, *see* decision theory
Dullstein, M., 9
Duress, 395
Dutch Book, 227
Dynamic treatment, 509, 523, 524

Econometrics, 109, 111, 161–177
    calibration approach, 110
Eells, E., 471, 472
Effect of treatment on the treated (ETT), 523, 528
Effects of causes (EoC), 503, 506, 509, 516, 517, 522, 527, 529, 530
Elementary errors
    hypothesis of, 249, 253, 254
Empirical analysis, 371
Empirical laws, 433
    vs. trends, 433
Endogenous, 512
Engle, R. F., 135
Entropy, *see* Maximum entropy principle, 441, 442, 453
Environmental risk factors, 219
EoC, *see* effects of causes
Epistemic causality, 482–484
Epistemology, 229, 504, 528
EPR (Einstein-Podolsky-Rosen) experiment, 4
    Bohm's version of, 66, 67
    correlations in, 65, 67, 69, 70, 76, 77, 79, 81–85, 88–91, 93–97, 100, 101, 103
    zigzag model of, 86, 87
Equipossibility, 226
Equivocal degree of belief, 233
Error variable, 512, 514
ETT, *see* effect of treatment on the treated
Event

token, 445
type, 445
Evidence, 453, 457, 459, 461, 463
Evolutionary biology, 446, 458, 463
Exclusion condition, 137
Exogenous, 512
Experiment, 27
Explanandum, 244–248, 250, 251, 258
Explanans, 244, 245, 248, 253
Explanation, 243, 245, 248, 249, 251, 254, 256–258
  causal, 382, 385, 392, 404–406, 418
Expressiveness, 527
Extension, 406, 407
External set, 143

Factorisation, 275, 280, 281
Factorizability, 76, 77, 79, 80, 90, 91, 96
Faithfulness condition, 11, 121, 122, 124, 126, 474, 477, 489
Falsificationism, 117
Fatalism, 526
Fennell, D., 5, 138, 155
Fidelity, 453, 462
Fine, A., 66, 94
Fisher, F.M., 147
FPCI, see fundamental problem of causal inference
Franck-Hertz experiment, 416, 420, 422, 423, 425, 427, 433, 435
Frangakis, C. E., 526
Freitas, A., 3
Frequency, 335–360
Function, 335–360
Functional model, 504
Functional relationship, 512–514, 525, 526

Fundamental problem of causal inference (FPCI), 517, 518

G-computation, 523, 524
Galles, D., 527
Galton, F., 7
Galton, F., 243–251, 253–261
Geneletti, S. G., 523, 528
Generalities, 435
Generic, 220
Genericity, 136
Genetic factors, 219
Genetics, 244, 245, 256, 260
GHZ (Greenberger, Horne and Zeilinger) experiment, 84, 97
Glymour, C., 188, 189, 503
Granger, C.W.J., 152
Granger-causality, 110, 119, 120, 126, 152
Graphical causal models, 3
Graphical model, 110, 121–123, 126
Greenland, S., 525
Grice, P., 469

Höfler, M., 510, 516
Hacking, I., 260
Hall, N., 9, 364–366, 515
Halpern, J. Y., 516
Hart, H., 470
Hausman, D., 65, 71, 76, 82, 97, 165
Heckerman, D., 508
Heckman, J., 157–158, 528
Heisenberg indeterminacy relations, 425
Heredity, 243–245, 247, 249, 251, 253–256, 258, 260
Hidden autonomy, 76, 80, 81, 92, 96
Hidden locality, 76, 80, 81, 91, 92, 96
Historical linguistics, 446, 454, 461

Hitchcock, C., 448–450
Hitler, A., 417, 429, 430, 434
Holland, P. W., 517
Homology, 459, 464
Homoplasy, 445, 450, 459, 464
Hoover, K., 135, 146, 148, 182, 184–189, 193, 194
Human agency, 421, 432
Hume, D., 419
Hypothesis
  ad hoc, 255, 256
Hypothesis of elementary errors, 249, 254
Hypothetical, 503, 505, 506, 508, 512, 514, 527
Hypothetical experiment, 417
Hypothetical induction, 492
Hypothetico-deductive, 117, 492, 498

IC algorithm, 489
ICE, *see* causal effect
Ideal gas law, 4
Ideal type, 428, 429
Idealisation, 428
Identifiability, 147, 527, 528
  and causality, 148
IF-THEN rules, 48
Imbens, G. W., 526
Incidence, 237
Independence, 84, 85, 97, 148, 243, 260, 508, 512, 513, 517, 518
  conditional, 243, 260, 508, 517, 519, 521, 524, 530
Indicative, 504, 507–509, 511
Individual causal effect, 268
Individual identities, 140
Inducing path, 21
Inductive generalizations, 432
Inference
  causal, 179, 180, 185–188, 190, 193, 194
  statistical, 182, 186, 187, 194
Instrumental variable, 263, 264, 278–280, 283, 284, 288, 523
Instrumentalism, 527
Intention, 388, 389, 391
Intervention, 134, 143, 161–177, 266–268, 271, 279, 280
  hypothetical, 416, 418
Intervention indicator, 287
INUS condition, 10, 415, 419, 469, 470
Invariance, 526

Jarrett, J., 76

Knowledge discovery in databases, 43
Kolmogorov, A.N., 319
Korb, K., 472, 505
Koyck transformation, 151

L'Episcopo, A., 8
Laboratory experiment, 425
Laplace, P. S., 249
Latent variable, 3
Law of ancestral heredity, 253–255, 258
Law of error, 248, 249
Law of nature, 417
Lawlike regularities, 431
Laws, 417, 428, 431, 432
Laws of nature, 418
Laws of science, 432
Leamer, E.E., 134
Legal causation, 10
Legal liability, 385–392, 395, 403–407
  types of, 385, 388–390, 392
Legal positivism, 386
Leray, P., 3

LeRoy, S., 5, 164
Leuridan, B., 7
Levels of context-dependence, 480
Lewis, D. K., 469, 510, 516
Linear regression, 245, 250
Linear structural equations, 279, 287
Linkage disequilibrium, 284
Longy, F., 9
Lucas critique, 114, 116
Lucas, R.E., 145

Mach, E., 1
Mackie, J., 415, 419, 427, 469, 470
Maes, S., 3
Mark transmission, 210
Markov condition, 89, 121–124, 126, 447, 473
Markov properties, 275
Marshack, J., 146
Maudlin, T., 76, 89
Maximal ancestral graph, 22
Maximum entropy principle, 228
McGarry, K., 3
Mechanism, 161–177, 243, 245, 247–249, 253, 260, 364, 369, 372
    causal, 243, 245, 247, 248, 253
    of heredity, 243, 245, 247–249, 253, 260
    social, 205
Meganck, S., 3
Mendel, G., 7, 244–248, 251, 253, 256, 257, 260
Mendelian randomisation, 263, 265, 277, 278, 280, 287, 288
Menzies, P., 469
Methodology, 492
Micro innocence (μ-innocence) principle, 87
Mill, J. S., 419, 422
Milne, P., 507

Minimality, 474, 476
Misconnection, 320
Missing data, 517
Mixing (of probability distributions), 185, 188
Modal, 504, 505, 509
Model, 431
    bilinear, 139
    ideal-typical, 436
    vector autoregressive (VAR), 110, 118, 119, 123, 125
Model fit, 260
Modularity, 5, 155, 161–177
Moneta, A., 5
Moral graph, 276, 281
Moral responsibility, 386–388, 390, 391, 395, 404, 406
Moralisation, 276
Multilevel analysis, 217
Mutation
    constant rates of, 462, 464
    uniform rates of, 464

N-Connection (NC) principle, 82, 83
Natural kind, 346, 352
Neutrality of statistics, 244, 260, 261
Neyman, J., 509, 525
No counfounding, see confounding
Non-compliance, 523, 526, 528
Normal distribution, 248–251, 254, 260
Normative element, 380, 382, 410

Objective Bayesianism, 482–484, 496
Objective chance, 306
Objective homogeneity thesis, 472
Objectivity, 229
Observational equivalence, 169, 170

Odds ratio, 236, 283
Omission, 365
    qualitative and quantitative, 400
Ontology, 504, 525, 527
Operationally meaningful, 139
Overdetermination, 365, 367, 375

Papineau, D., 477
Parameter, 134, 275
    deep, 145
Parascandola, M., 516
Pareto-dominance condition, 472
Parsimony, 229
Path analysis, 490
Pattern
    interestingness of, 44
    surprising, 44
    understandable, 43
Paul, N. A., 515
PC algorithm, 489
Pearl, J., 154–157, 162, 243, 261, 418, 473, 476, 504, 514–516, 527, 529
Pearson, K., 1, 245, 247, 250, 253
Perfect intervention, 5
Pheneticism, 450
Physical connection, 392–395
Physical probability (PHYS probability), 335–360
Physical sciences, 431
Physical structure, 432
Physics, 417
Platonic heaven, 510, 511
Pleiotropy, 285
Pollock, J. L., 504, 505, 516
Popper, K., 5, 431
Population distribution, 507, 511, 526
Population stratification, 285
Possible worlds, 509, 515, 517

Potential response (PR), 12, 504, 509–511, 513–518, 522–530
Potentially directed graph, 28
PR, *see* potential response
Practical possibility, 322
Pragmatics, 520, 521, 524, 529
pragmatics, 530
Prestructuring, 261
Presuppositions, 261
Price, H., 65, 71, 76, 81, 85, 92, 100, 529
Principal strata, 526, 528
Principle of indifference, 226
Principle of the common cause (PCC), 6, 10, 179–182, 184–189, 191–194
Prob. Deviation, 248
Probability, 243, 244, 260, 296, 320, 506, 507, 510, 511, 526, 528
    axioms of, 320
    Bayesian, 7, 457
    bivariate distribution, 511
    classical interpretation of, 7, 224–233
    conditional, 507–509, 512, 514, 515, 520, 526
    conditional density function, 507
    diagnostic, 530
    DYSF-ART, 9
    DYSF-PHYS, 9
    empirically-based subjectivist interpretation of, 224–233
    frequency interpretation of, 224–233
    frequency theory of, 7
    joint distribution, 511, 515, 523
    logical interpretation of, 7, 224–

233
  meaning of, 224
  objective, 477
  objective Bayesian interpretation of, 7, 224–235
  physical interpretation of, 224–233
  propensity interpretation of, 7, 224–233
  subjective interpretation of, 224–233, 479
  unconditional, 507
Probability calculus, 319
Probability distribution, 180, 185, 186, 188, 194
Probability model, 182
Problem of identification, 110, 112
Process, 134
Profligate causation, 396, 408
Propensity, 8, 299, 335–360, 478
  instantaneous, 306
Proportion, 237
Public health intervention, 266, 277, 283, 287, 288
Pundik, A., 10

Quantitative causal analysis, 217
Quantum axiom of reduction, 68
Quantum mechanics, 65–106, 415, 419, 420, 425, 435
Quantum non-separability, 69
Quantum singlet state, 69
Quantum statistical algorithm, 68
Quantum systems, 432
Quasi-causation, 382, 392–394, 396–398, 400, 404–406
Quetelet, A., 245, 248, 249, 258
Quetelismus, 248

Raiffa, H., 508
Random phenomena, 303
  discontinous behaviour of, 303
  qualitative changes of, 303
Randomisation, 517, 524
Rask, R., 446, 456
Real kind, 344, 346–350, 352, 354, 355, 357, 359
Reduced form, 131
Reductionism, 477–480
Reference class problem, 227
Regime, 522, 524–529
  indicator, 521, 524, 530
  interventional, 516, 518, 521, 522, 524, 526, 529
  observational, 518, 519, 521–524, 526, 529
Regime changes, 146
Regression, 150–151, 258
  filial, 251, 252, 255, 256
  fraternal, 251, 253
  mid-parental, 251
  parental, 251, 253
  towards the mean, 250, 251, 255–258, 260
Regularity, 415, 417, 431, 434
  lawlike, 418, 419
  patterns of, 430
  vs trends, 430
  vs. trend, 417, 418
  vs. trends, 431
Reichenbach, H., 8, 70–77, 80, 148, 182, 424, 440–445, 447, 451, 453, 458, 461
Reiss, J., 6
Relativity theory, 77, 86, 89, 91, 93, 99, 102
Reliability, 453
Reverse causation, 264, 270, 281
Risk, 236
Robb, R., 528
Robins, J. M., 509, 517, 523, 525
Rubin, D. B., 509, 517, 526
Rule interestingness measure, 49

Russell, B., 1, 4
Russo, F., 7
Rutherford's decay law, 433
Rutherford, E., 421

Salmon, W.C., 415, 446, 448, 461
Scattering experiments, 421
Scheines, R., 188, 474, 503
Schrödinger, E., 69
Science Citation Index Expanded, 1
Screening conditions, 448, 458
Screening-off condition, 246–248, 253, 259, 447
Selection, 396, 409
Selection effect, 281
SEM, 244, 260, 261
Semi-Markovian causal model, 3, 20
Shachter, R., 508
Sheehan, N. A., 523
Sheehan, N.A., 7
Simon, H., 5, 136, 166–168
Simple cause, 143
Simpson's paradox, 4, 54
Sims, C.A., 152
Single-case, 220
Singular causes, 296–299
Sober, E., 95, 179, 181–185, 188–191, 443, 445, 446, 448, 450–452, 455, 456, 458
Social institutions, 431
Social sciences, 200, 211, 212, 218, 248, 260, 415, 420, 421, 430, 431, 434, 435, 488
    naturalistic view of, 436
Social Sciences Citation Index, 1
Social system, 435
Social trend, 433
Social world, 422, 423
Spirtes, P., 188, 503

Spohn, W., 478
Stability, 122, 530
Stalnaker, R. C., 509
Standard deviation σ, 249
Statistical modelling, 218
Statistics, 243–245, 247, 248, 250, 256, 260, 261
    neutrality of, 244, 260, 261
Steel, D., 188–190, 194
Stochastic process, 180, 182
Structural coefficient, 131
Structural equation, 504, 514
Structural equation modelling, 3, 244, 512
Structural form, 131
Structural information, 432, 433
Structural models, 217
Structural view, 432
Suárez, M., 4
Subjunctive, 504, 506–509, 511, 513
Subset condition, 143
Succession, 419
Sufficiency, 141–144
Suppes, P., 8, 467–503
Supply-demand system, 131
Symbolic dimension, 431
Symmetry, 417

Taxation, 255, 256
Teleology, 442
Test
    but-for, 394
    NESS, 394
TETRAD, 121, 124
Thermodynamics
    second law of, 441, 443, 444
Time series, 6, 179, 180, 183, 186, 188–192
    co-integrated, 184
    integrated, 183, 184, 191
    non-stationary, 183, 184, 188–190, 193, 194

stationary, 184, 186, 188–191
strictly stationary, 183
trend stationary, 183, 184
weakly stationary, 183, 184
Tort law, 381, 387, 390, 391, 394, 396, 401, 409
Treatment-unit additivity (TUA), 513
Tucker, A., 10
Twardy, C., 472

Uncertainty, 506–509, 511–513
    about comparisons, 520
    comparison of, 520
    counterfactual, 509
    epistemic, 507, 511
    joint, 510, 511, 514
    process, 526, 527
Uncounfounder, *see* confounder
Underlying DAG, 19
Uniform rates, 456
User-defined general impressions, 48
Utility, 520

van der Laan, M. J., 517
van Fraassen, B.C., 65, 70, 76, 77, 80–82, 88, 89, 95, 96, 100, 449
Variable, 133
    endogenous, 133
    exogenous, 133
    external, 133
    internal, 133
Variation-free, 144
Variation-free condition, 134
Variational group, 452, 454, 455, 457–460
Variety, 229
Variety of petty influence, 249, 254

Wasserman, L. A., 527, 528

Web of Science, 1
Weber, E., 6
Weber, M., 10, 415, 421, 427, 428, 430
Weed, D. L., 516
Weimar Republic, 434
Weinert, F., 10
Weismann, A., 253
Williamson, J., 7, 481, 482, 496
Winship, C., 510
Woodward, J., 89, 165, 416, 418, 435
World War Two, 434

Yule, G. U., 181, 191

www.ingramcontent.com/pod-product-compliance
Ingram Content Group UK Ltd.
Pitfield, Milton Keynes, MK11 3LW, UK
UKHW021315180426
11947UKWH00015B/1248